PRINCIPLES OF NUCLEIC ACID STRUCTURE

PRINCIPLES OF NUCLEIC ACID STRUCTURE

SECOND EDITION

STEPHEN NEIDLE
The School of Pharmacy,
University College London, United Kingdom

MARK SANDERSON
Department of Metabolism, Digestion and Reproduction,
Faculty of Medicine, Imperial College London,
United Kingdom

ELSEVIER

ACADEMIC PRESS
An imprint of Elsevier

Academic Press is an imprint of Elsevier
125 London Wall, London EC2Y 5AS, United Kingdom
525 B Street, Suite 1650, San Diego, CA 92101, United States
50 Hampshire Street, 5th Floor, Cambridge, MA 02139, United States
The Boulevard, Langford Lane, Kidlington, Oxford OX5 1GB, United Kingdom

Notices

Knowledge and best practice in this field are constantly changing. As new research and experience
broaden our understanding, changes in research methods, professional practices, or medical
treatment may become necessary.

Practitioners and researchers must always rely on their own experience and knowledge in evaluating
and using any information, methods, compounds, or experiments described herein. In using such
information or methods they should be mindful of their own safety and the safety of others, including
parties for whom they have a professional responsibility.

To the fullest extent of the law, neither the Publisher nor the authors, contributors, or editors, assume
any liability for any injury and/or damage to persons or property as a matter of products liability,
negligence or otherwise, or from any use or operation of any methods, products, instructions, or ideas
contained in the material herein.

Library of Congress Cataloging-in-Publication Data
A catalog record for this book is available from the Library of Congress

British Library Cataloguing-in-Publication Data
A catalogue record for this book is available from the British Library

ISBN: 978-0-12-819677-9

For information on all Academic Press publications visit our website at
https://www.elsevier.com/books-and-journals

Publisher: Andre Gerhard Wolff
Editorial Project Manager: Susan Ikeda
Production Project Manager: Punithavathy Govindaradjane
Cover Designer: Mark Rogers

Typeset by TNQ Technologies

CONTENTS

Preface to the second edition

The first edition of this book was published 13 years ago. The goal at that time was to produce a reasonably comprehensive account of DNA and RNA structures, together with an introduction to their protein complexes. As well as describing the basic principles of DNA/RNA organisation, stereochemistry and recognition by small and large molecules, a central aim was to provide a resource providing information on the more significant structures and to the literature in the field. These defining principles remain in this new edition, as does much of the original text, although all sections of the book have been updated, with the addition of new text and references to significant studies published since 2008. Most of the figures have been re-drawn. We have also taken the opportunity to correct some minor errors and clarify several topics, following useful feedback from readers.

The underlying principles of the subject have not changed since 2008, but there have been many major advances in methodology, especially for studying the structure of large protein–nucleic acid biomolecular complexes by cryo-EM. The number of native DNA and RNA structures has continued to increase as has their accuracy and reliability, with few major changes in conclusions about them. Some other individual topics, notably the study of quadruplex nucleic acids and RNA interactions with proteins and small molecules, have seen very considerable growth. These are reflected in this new edition. Above all the number, variety and biological importance of protein–nucleic acid structures has mushroomed so that any one volume cannot any longer do full justice to this area. Mark Sanderson, with his strong background in both research and teaching on nucleic acid–protein interactions, has taken on the daunting task of updating the protein–DNA section and writing a new chapter on protein–RNA interactions. These chapters are necessarily selective in view of the greatly increased growth in both fields, but we hope that they provide accounts that will be useful to a wide range of readers.

The publishers have agreed to the inclusion of a good number of colour illustrations in the paper version, replacing many of the black-and-white figures of the first edition. On-line readers will find that all the structure figures are now in colour. We hope that these will enhance the clarity of the book. We also hope that readers will be encouraged to examine individual structures themselves, so to facilitate this, the on-line version of the book includes hyperlinks to entries in the Protein Data Bank in the many tables of structures. As before, we have included a list of molecular graphics programs that we find to be of especial use when visualising nucleic acids, as well as links to other useful resources.

Stephen Neidle
June 2021

Preface from the first edition

The years that have elapsed, since the previous version of this book was published, in 2001, have been momentous ones for nucleic acid studies. In 2003, we celebrated both the 50th anniversary of the discovery of the structure of the DNA double helix and the announcement of the determination of the sequence of the human genome. It might therefore be thought that the study of nucleic acid structure is itself now part of history and that there is little more to be known. The reality is very different; we have seen a number of profound new discoveries relating to both RNA and DNA structure, just in the first 7 years of this millennium. These significant advances in the subject have required not just a new edition but an expansion of many sections and a re-write of others.

The aim of the book is to provide an introduction to the underlying fundamental features and principles governing nucleic acid structures, as well as many of the structures themselves. It is hoped that this provides a firm foundation for subsequent studies of the structural biology and chemistry of nucleic acids. Its intended audience is at graduate level, and it is hoped that it will be of use to active researchers, and even to the more inquisitive final-year undergraduate students. The book does not attempt to be a comprehensive survey of all nucleic acid-containing structures. Instead, it concentrates on more general themes and focusses on those structures that illustrate a particular feature of interest or generality, especially in the context of their relevance to chemical, biological or pharmacological issues. I apologise in advance to those whose favourite structure has been ignored in favour of my own more subjective judgments.

The book emphasises those structures determined by X-ray crystallography, since this methodology continues to dominate the field in terms of size of molecule whose structure can be determined, as well as still providing the majority of high-resolution structures. The introduction to crystallography and other techniques is designed to provide the nonspecialist with sufficient understanding to read the primary literature, and most importantly, to be able to begin to judge the scope and quality of both experimental and theoretical structural studies. I have also expanded the reference and reading lists to provide a reasonably comprehensive guide to both the past and recent literature and have included information on a number of relevant websites.

Any book on molecular structure suffers from the disadvantage of not being able to adequately convey the three dimensionality of structures. The previous edition was associated with a dedicated Internet site, which enabled the structures to be examined interactively, and in a variety of display modes. The excellence of the many graphics programs freely available on the web, together with the molecular display tools available from the

Protein Data Bank and other web sites, makes a dedicated site no longer necessary, or even desirable. I have included tables of the PDB and NDB (Nucleic Acid Database) codes for a large number of representative structures, to aid the reader in speedily viewing a particular feature, or downloading a structure file for subsequent display and analysis on one's own desktop or laptop. I have also included a list of my own favourite molecular graphics programs that have nucleic acid-friendly features.

I am grateful to my wife Andrea and children Dan, Ben and Hannah for their constant support and encouragement in this and many other ventures, and to my colleagues, collaborators and students for their contributions, insights and discussions. Thanks also to my editor at Elsevier, Kirsten Funk, for all her hard work, patience and support.

Stephen Neidle
London, June 2007

CHAPTER 1

Methods for studying nucleic acid structure

1.1 Introduction

Our knowledge of DNA and RNA three-dimensional structure has advanced immeasurably since the elucidation of the first such structure, that of the DNA double helix in 1953 by Watson and Crick in conjunction with the X-ray fibre diffraction data of Franklin, Wilkins and Gosling. Fibre diffraction methods subsequently enabled the morphologies of a wide range of polymeric nucleic acid double helices to become established. More recently, the relationships between DNA primary sequence and the fine details of its molecular structure have become well understood, in large part from single-crystal and nuclear magnetic resonance (NMR) structural studies on defined-sequence oligonucleotides. However, DNA structure continues to surprise with its ability to exist in a wide variety of forms, such as left-handed, multiple-stranded helices and even catalytic DNAs. The study of RNA structure has a more recent history, which has revealed that RNA can fold in a wide variety of complex ways as well as occurring in double-helical form. There is also now a very large amount of experimental information on the structures of protein–DNA, protein–RNA and drug–DNA complexes, as well as larger multi-component complexes such as the ribosome, initiation complexes and chromatin.

The discovery of the double helix, as Watson and Crick realised, immediately provided fundamental new insights into the nature of genetic events. We now have extensive knowledge of both the detail and the variety of DNA and RNA structures themselves, together with the manner in which they are recognised by regulatory, repair and other proteins, as well as by small molecules. All this is giving us altogether more profound levels of understanding of the processes of gene regulation, transcription and translation, mutation/carcinogenesis and drug action at the atomic and molecular levels. We can now piece together how all this works in the context of eukaryotic chromatin, so some of the challenges over the past few years have been to study the structural biology of large-scale DNA–protein structural assemblies just as has been done for the ribosome. The advent of DNA CRISPR editing technology is having a profound effect on the practice of cell and molecular biology, and we are just now starting to see its impact on human genomic editing and disease modulation. Much is now known about the structural basis of CRISPR editing (see Chapter 7).

These advances in nucleic acid structural studies have been largely due to the increased power and sophistication of the experimental approach of X-ray crystallography, which have provided much of the highly detailed structural information to date. Single-crystal

Principles of Nucleic Acid Structure
ISBN 978-0-12-819677-9, https://doi.org/10.1016/B978-0-12-819677-9.00008-1

methods continue to play a major role and are reflected in the emphasis of this book. NMR spectroscopy, molecular modelling/simulation and chemical/biochemical probe techniques also play important parts in providing information on structure, dynamics and flexibility that can approach near-atomic resolution in at least some of its detail. The study of large macromolecular complexes such as chromatin in various re-modelling states has been revolutionised by the increasing availability of both hardware and software for cryogenic electron microscopy (cryo-EM). This has enabled near-atomic resolution studies to be undertaken on many such complexes. Traditional spectroscopic-based biophysical methods can provide important complementary information, mostly at the macroscopic level. More recently developed techniques, such as surface plasmon resonance (SPR) spectroscopy and single-molecule methods, are extending their power so that the gap is now diminishing between macroscopic data on nucleic acids which the more traditional methods provide and that at the atomic level. We provide a few key references to these techniques for the interested reader, but detailed descriptions of them and their applications are beyond the scope of this book.

Underpinning all of this progress in the nucleic acid structure field have been a number of significant technical advances. These have notably been in (i) the development of routine chemical methods for oligonucleotide synthesis and purification at the milligram and multi-gram level for both DNA and more demanding RNA sequences, (ii) the advent of efficient and routine cloning and expression systems for RNA- and DNA-binding proteins and for native RNA molecules longer than ca 50 nucleotides, at which point in vitro transcription is more efficient than chemical synthesis, (iii) continuing advances in detector technology, enabling rapid and accurate detection of low-energy diffraction data and (iv) the advent of increasingly powerful X-ray and laser synchrotron sources, enabling, for example, the analysis of very small crystals (as small as a few thousand Å) by the recently developed technique of serial crystallography (Pearson & Mehrabi, 2020; Thompson et al., 2020). This uses femtosecond pulses of extremely high X-ray intensity.

This chapter provides a brief introduction to these major structural methods, emphasising their scope as well as their limitations for nucleic acid structural studies.

1.2 X-ray diffraction methods for structural analysis
1.2.1 Overview

X-rays typically have a wavelength of the same dimensions as inter-atomic bonds in molecules (about 1.5 Å). Scattering (or diffraction) of X-rays by molecules in ordered matter is the result of interactions between the radiation and the electron distribution of each component atom. Typical **diffraction patterns** from DNA, in the form of fibres or single crystals, are shown in Figs. 1.1 and 1.2. Reconstruction of the internal molecular arrangement by analysis of the scattered X-rays, analogous to a lens focussing scattered

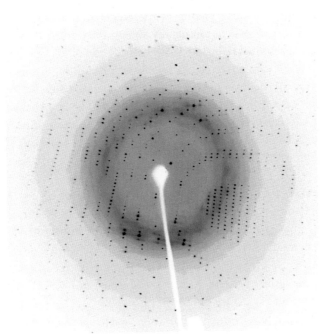

Figure 1.1 X-ray diffraction pattern from a single crystal of a drug—oligonucleotide complex, taken with an image plate and a conventional laboratory X-ray source. The resolution limit for this pattern is 1.6 Å.

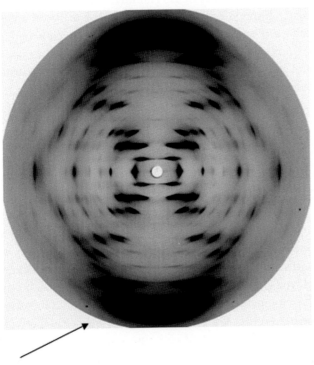

Figure 1.2 X-ray fibre diffraction pattern from a sample of calf thymus DNA, showing a characteristic B-form pattern. Provided by Professor Watson Fuller. The arrow indicates the 3.4 Å layer line.

light from a microscope sample, thus provides a picture of the **electron density** distribution in the molecule. This reconstruction is not generally straightforward, due to the loss of phase information from the individual reflected X-rays during the diffraction process. The **phase problem** needs to be solved (see below for a brief description of various methods of doing this) in order for the electron density to be calculated in three dimensions (as a Fourier series), which is commonly termed a **Fourier map.** The approximate equivalence of the wavelength of X-rays and bond distance, of ~1–1.5 Å, means that, in principle, the electron density of individual atoms in a molecule can be resolved, provided that the pattern of diffracted X-rays can be reconstituted into a real-space image.

The degree of electron density detail that can be visualised is dependent on the **resolution** of the recorded diffraction pattern. Resolution may be defined in terms of the shortest separation between objects (i.e. atoms or groups of atoms in a molecule) that can be observed in the electron density reconstituted from the diffraction pattern. The resolution limit (r) is governed by the maximum diffraction angle (θ) recorded for the diffraction data and the wavelength (λ) of the X-rays: r is defined as $\lambda/2\sin\theta$. At a resolution of 2.5 Å, individual atoms in a structure cannot be resolved in an electron density map although the shape and orientation of ring systems (e.g. base pairs) can be readily distinguished. These appear as elongated regions of electron density, with substituents being apparent as 'outgrowths' from the main density. At 1.5 Å, individual atoms are generally just about observable in a map, although only at about 1.0–1.2 Å or better are all atoms fully resolved and separated from each other (Fig. 1.3). There has been a marked increase in ultra-high-resolution studies ($>>$ 1.0 Å) in recent years due to the increasing use and world-wide availability of high-flux **synchrotron** sources of X-rays for structural biology studies. There are currently over 50 synchrotrons with beam lines dedicated to crystallography (see https://lightsources.org and https://en.wikipedia.org/wiki/List_of_synchrotron_radiation_facilities for further details). In 2005, of the 4515 macromolecular crystal structures submitted to the Protein Data Bank (PDB) (http://www.rcsb.org), data for 3398 (75.3%) were collected on synchrotron beam lines. The proportion in 2020 is 93.6% for approximately four times the number of deposited structures. Synchrotron beam lines have intensities of X-ray beams greater by many orders of magnitude than conventional laboratory X-ray sources. Synchrotron facilities have also enabled much smaller crystals than hitherto to be successfully analysed, with the new technique of serial crystallography being able to cope with micro-crystals. Most importantly, the ability to tune X-rays to differing wavelengths has provided the means whereby powerful methods of structure analysis can be employed (see below). Although not comparable with synchrotron beam intensities, the development of highly effective mirror-optics focussing systems for laboratory X-ray sources in recent years can enable good-quality preliminary diffraction data for screening purposes to be collected in-house, especially when larger crystals can be obtained. It is now almost universal practice to collect diffraction data

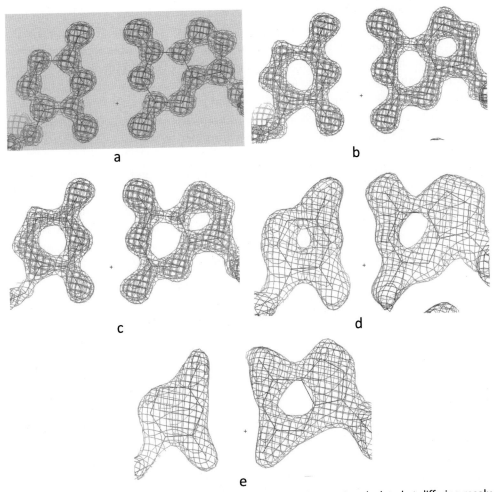

Figure 1.3 Calculated electron density in the plane of a C•G base pair, calculated at differing resolutions, showing the amount of atomic detail visible at particular resolutions: (a) 0.9 Å, (b) 1.25 Å, (c) 1.5 Å, (d) 2.0 Å and (e) 2.5 Å.

from macromolecules at liquid nitrogen temperatures using, for example, the technique of "**flash-freezing**", which tends to minimise crystal decay in the X-ray beam, and can improve the diffracting power of a given crystal.

The number of individual diffraction maxima observed from a crystalline or semi-crystalline sample depends directly on two factors: the resolution of the pattern and the size of the crystallographic **unit cell**. The number of unique reflections derived from these measurements also depends on the **symmetry** of the crystal. An ultra-high-resolution (0.70 Å) structure of a typical DNA 10-mer oligonucleotide crystal would give approximately 29,000 individual unique maxima (reflections) in a monoclinic space group. By contrast, a 12-mer oligonucleotide crystal in the well-studied orthorhombic space group $P2_12_12_1$ would give only some 3000 unique reflections at 2.2 Å resolution, which has historically been a common limit for oligonucleotide crystals with diffraction data collected using laboratory X-ray sources. The measured intensity of an individual reflection is proportional to its **structure amplitude**, or observed **structure factor**, which when combined with phase information results in the calculated structure factor. X-ray structures are always optimised (**refined**) against this observed structure amplitude data, for example, by a least-squares method (see below).

The accuracy and reliability of the resulting structure depends in part on the quantity and resolution of the diffraction data, as well as the quality of its measurement. Of key importance is the actual correctness of the structural model itself, both in gross outline (which is defined by the low to medium-resolution data) and in the detailed aspects of the structure (defined by the high-resolution data). The standard ways of assessing these factors for a given structural model are as follows:

- to calculate the crystallographic **R factor**, defined as $R = \Sigma ||F_o| - |F_c|| / \Sigma |F_o|$, summed over all observed reflections, where F_o and F_c are the observed and calculated structure factors. R, which is also termed the reliability index, is expressed as a percentage, or sometimes as a decimal.

- to calculate the so-called **free R factor** (R_{free}) for a small (typically 5%) sub-set of reflections, often chosen randomly. This set is not used in the refinement, and so the value of R_{free} is unbiased by the course of the refinement and any errors introduced during it.

The R value for a correctly refined structural model can range from <10% to >20%; in general, the lower the value, the more reliable is the model. Values for R_{free} are usually a few percent higher but are very sensitive to even small changes and errors in the model. R_{free} is often used to judge the completion of a structure analysis, especially in terms of the behaviour of water molecules located in electron density maps and added to the model during successive rounds of refinement. At a certain point, adding more water molecules may well reduce the value of R simply because the number of variables is increased in the least-squares refinement. However, if the value of R_{free} increases, then these 'water molecules' are not physically real. It is also common practice to calculate Fourier maps with parts of a structure omitted to verify that these parts re-appear in the map at the stereochemically correct positions.

1.2.2 Fibre diffraction methods

Historically, helical DNA and RNA structures were first analyzed by fibre diffraction techniques (Marvin, 2017; Franklin & Gosling, 1953). Polymeric nucleic acids, directly extracted from cell nuclei, have not been crystallised as single crystals capable of three-dimensional structure analysis. Initial diffraction patterns of poor quality were obtained from DNA by Astbury in 1937/38, but significant progress was not made until the early 1950s, by Wilkins, Franklin and their associates. They made DNA into oriented fibres, when the act of 'pulling' such a fibre orients the nucleic acid helix along the direction of the fibre. These fibres can have exactly repetitious helical dimensions even though the underlying naturally occurring (often genomic) nucleic acid sequences in them are not simple repeats, and thus the sequence information in these nucleic acid molecules is lost. An important exception occurs with the use of synthetic polynucleotides of known, simply repeating sequence such as poly(dA-dT)•poly(dA-dT).

Natural and synthetic polynucleotides can form fibres with varying degrees of internal order, having one- or two-dimensional para-crystalline arrays in the fibre, with the latter usually having the greater order because of their non-random sequences. These differing degrees of order are reflected in their X-ray diffraction patterns, with natural double-helical DNA and RNA molecules usually having a degree of order along the **helix axis** but being randomly oriented with respect to each other. This gives rise to an X-ray diffraction pattern with characteristic spots and streaks of intensity — for example, the 'helical cross' diffraction pattern, which is characteristic of B-type DNA double helices. Such patterns can be analysed to give the **helical dimensions** of pitch, rise and number of residues per helical turn, as well as defining the overall helical type (A, B, etc.). Even the best ordered of para-crystalline polynucleotide fibres give at most only a few hundred individual diffraction maxima, corresponding to a typical maximum resolution of about 2.5 Å.

It has not been in general possible to analyse these patterns of fibre diffraction intensities, determine phases and derive a molecular structure ab initio, since the patterns are averages from all the nucleotide units in a helical repeat. However, the recently developed methodology of Coherent Diffraction Imaging has enabled, in a test case, the DNA double helix to be reconstructed directly from the well-known fibre diffraction pattern of B-DNA (Franklin & Gosling, 1953), albeit at a low resolution of 3.4 Å (Latychevskaia & Fink, 2018). In traditional fibre diffraction analyses the pattern is fitted to a model using a least-squares procedure (Arnott, 1970a). This enables conformational details of the averaged (mono- or di-nucleotide) repeat to be varied and optimised. The correctness and quality of the model may be assessed using the standard crystallographic R factor. Values for R of 0.15—0.25 indicate that the calculated diffraction pattern agrees well with the observed one, and that the model is physically reasonable in terms of its stereochemistry.

An important question is whether the attainment of good agreement for these criteria necessarily means that the phase problem for these fibre structures has been uniquely solved. The process of analysis assumes a particular starting model and other models might in principle fit a data set at least moderately well, especially since atomic levels of resolution are not available from fibre diffraction. This question led some years ago to several suggestions of alternative structures to the Watson—Crick anti-parallel double helix for DNA. Only on detailed examination was it found that none of the alternative base-pairing arrangements could be fitted in an acceptable manner to the observed diffraction data, as defined by the R factor and other tests (Arnott, 1970b). This, together with their numerous close intra-molecular contacts, enabled these alternative structures to be conclusively rejected and the anti-parallel Watson—Crick double helix accepted as the sole model that can acceptably fit the B-DNA observed diffraction data.

It is striking, in spite of the limitations of fibre diffraction methods, that their charac-terisations of idealised DNA and RNA double helical geometry in terms of helical type (A, B, etc.), have been found to be closely in accord with the large number of subsequent single-crystal analyses of short helical DNA and RNA segments, even at much higher resolutions, as well as of DNA/RNA helices in multi-subunit protein complexes such as the ribosome and nucleosomes. Fibre diffraction analysis also has the considerable advantage of being able to readily study conformational transitions under a range of envi-ronmental conditions. The A ↔ B transition of duplex DNA observed with variations in relative humidity is the classic example of this technique.

1.2.3 Single-crystal methods

By contrast with fibre analyses, single-crystal X-ray crystallographic methods can deter-mine the complete three-dimensional molecular structures of biological macromolecules without necessarily recourse to any preconceived model, provided the molecules are discrete and not the effectively infinite disordered polymers of nucleic acid fibres. Single crystals can be thus defined as ordered arrays of discrete and identical molecules in three dimensions.

Crystallisation of many DNA and RNA oligonucleotides has historically been chal-lenging, being in the past sometimes more dependent on chance than systematic scientific study. This has changed radically with the advent of manual (and increasingly) automated methods to rapidly and systematically screen a wide set of crystallising conditions (Ducruix & Giegé, 2004; Holbrook & Holbrook, 2010). Commercial crystallisation kits focussing on oligonucleotides are now widely available with pre-prepared solutions of a wide range of concentrations and types of counter-ion, buffer and precipitating agents, so that many crystallising trials can be set up with minimal effort. This approach is also useful for finding alternative crystal forms if initial trials produce crystals with poor diffraction or exceptionally large unit cells. The use of robotic crystallisation methods is now routine. These enable rapid, large-scale screening of crystallisation conditions to be undertaken, which is especially useful when dealing with 'difficult' molecules such as large DNAs, RNAs or protein—nucleic acid multi-subunit complexes. Many nucleic acid crystallographers have developed specialised sets of conditions for their own speciality, notable examples being in the RNA and ribozyme field (for example,

Ke & Doudna, 2004; Reyes et al., 2009). The HELIX crystallisation screen (Viladoms & Parkinson, 2014) can be used for a wide range of DNAs and RNAs and incorporates 96 conditions adopted from successful crystallisation conditions deposited in the PDB, ordered into five sub-classes dependent on oligonucleotide type and size. This screen avoids the use of sodium cacodylate, an organoarsenic-containing buffer, in view of the known interference of arsenic signals in anomalous dispersion data-collection experiments.

The range of resolution reported for single-crystal studies of oligonucleotides spans from 0.55 to 3.0 Å (Fig. 1.3). Thus, the highest-resolution oligonucleotide structures have true atomic resolution and accordingly are of corresponding accuracy (≤ 0.02 Å for distances and $\leq 0.2°$ for angles) in respect of derived geometric parameters. A typical 2.5 Å resolution structure analysis, by contrast, would have distances reliable to about ± 0.3 Å and angles to about $\pm 5°$. It is normally necessary to use constraints to standard bond geometries during the crystallographic refinement process of non-atomic resolution crystal structures. This means that it is not only non-bonded and intermolecular distances but also conformational and base morphological features that have to be interpreted with care, and likely errors and uncertainties taken into account. Hydrogen atoms are only directly observed in electron density maps from high-resolution oligonucleotide analyses (better than 1.0 Å resolution), and so hydrogen-bonding schemes (especially those involving water molecules) normally have to be inferred.

X-ray diffraction patterns from oligonucleotide crystals can be analysed, and their underlying molecular structures solved ab initio by the standard heavy-atom multiple and single **isomorphous replacement** (MIR and SIR) phasing methods of macromolecular crystallography. These do not presume any particular structural model and hence do not bias the resulting structure, for example, to have all Watson–Crick base pairing in a double-stranded oligonucleotide. However, a number of such heavy-atom derivatives are required for satisfactory MIR phasing, which are not always readily obtained, especially for helical nucleic acids. It is now routine to solve a structure with a single derivative by means of a combination of phasing from isomorphous replacement and anomalous scattering at a single wavelength.

The availability of tunable wavelength X-ray facilities at many high-flux synchrotron facilities has enabled the technique of phasing by **Multi-wavelength Anomalous Diffraction** (MAD) to be used. This uses a single appropriate heavy atom, which can absorb X-rays to differing extents at different wavelengths; phases and hence electron density maps can be directly calculated from such data. These maps, when obtained at high resolution, are sometimes of remarkably high quality, revealing complete structures at the outset. It has been fortunate for the development of nucleic acid crystallography that bromine and iodine atoms, which can be readily chemically attached to uracil bases, provide excellent anomalous diffraction signals. This powerful method has been the method of choice, limited only by the availability of sufficient tunable synchrotron beam time. Several alternatives to the use of bromine or iodine have been found, which have been needed when it has been found that the halogen–uracil bond is rapidly cleaved in the X-ray beam but are increasingly the method of choice. Examples include using a single nucleotide with a thio-containing backbone (to bind a mercury heavy-atom derivative) or a phosphoroselenoate to replace a phosphate (Wilds et al., 2002). It is possible to replace oxygen by selenium at the 2′ position of a thymine or uridine nucleotide (Jiang et al., 2007; Pallan & Egli, 2007a), in the non-bridging backbone (Pallan & Egli, 2007b), or at the thymidine 4-position (Salon et al., 2007). Oligonucleotides incorporating this selenium modification produce crystals that grow much more rapidly and having higher diffraction quality than do bromine-derivatised oligonucleotides (Lin et al., 2011). It is possible to utilise the anomalous signal of phosphorus atoms to phase nucleic acid structures when ultra–high-resolution (>1.0Å) diffraction data of high redundancy are available (Dauter & Adamiak, 2001). Phasing using a single-wavelength anomalous scatter (SAD phasing) is also now common and has been used with, for example, zinc ions (Hou & Tsodikov, 2019), phosphate groups (Luo et al., 2014; Zhang, El Omari, et al., 2020, Zhang, Zheludev, et al., 2020) and manganese (Pfoh et al., 2009).

Alternatively, it is possible to take account of the fact that many nucleic acid structures crystallise in an arrangement isomorphous to structures previously determined (for example, by heavy-atom phasing) or are presumed to contain a well-defined structural motif such as a double helix. These structures can often be solved by molecular replacement or 'search' methods, which assume at least part of the structure and attempt to locate it in the crystallographic unit cell. Problems have occasionally arisen with this approach, when, for example, a helix has been correctly oriented within the unit cell but its position is incorrectly indicated, being systematically related to the correct one, for example, by a simple translation of a base pair. Search methods become increasingly challenging with a decreasing fraction of known geometry in a structure, and heavy-atom methods then become advisable. They are also difficult when the correct geometry of the search fragment is not precisely known, and then the correct rotational and translational solution becomes unclear. New protein–nucleic acid crystal structures are usually solved by heavy-atom MIR, MAD, and SAD methods, as are an increasing number of new types of oligonucleotide crystal structure. It is fortunate that the key crystal structures of a B and a Z-DNA oligonucleotide have been solved ab initio by heavy-atom methods (see Chapter 3), thereby ensuring a firm and unambiguous basis for subsequent molecular replacement analyses of a large number of DNA oligonucleotide structures.

A more recently developed approach to DNA and RNA crystallisation and structure analysis has been the use of racemic oligonucleotides (Mandal et al., 2014). The availability of unnatural L-nucleotides and the consequent ability to synthesise all-L-oligonucleotides have been driven by their resistance to nuclease degradation in cells and thus their potential use as antisense reagents. For crystallographic purposes, simply mixing stoichiometric equivalents of D and L enantiomers has been found to yield useable crystals when using a standard crystal screening approach. These normally crystallise in a low symmetry centrosymmetric space group, which facilitates crystal packing and thus crystallisation. Structure determination and refinement is made more straightforward since the crystallographic phases are necessarily one of two values and thus phasing errors are less than with the natural sequences, which always crystallise in non-centrosymmetric space groups with general phases. This has been demonstrated for the non-self-complementary Pribnow box prokaryotic promoter sequence, which has hitherto been challenging to crystallise and analyse. High-quality crystals were readily obtained, in striking contrast to the poor quality low-resolution ones with all-D-natural stereochemistry (Mandal et al., 2016).

It is increasingly possible to obtain true atomic resolution ultra-high-quality synchrotron diffraction data on oligonucleotides whose crystals are exceptionally well ordered and diffract to significantly better than 1.0 Å resolution, providing the opportunity for phasing methods that do not rely on any heavy atoms being required. Several pioneering studies have shown that 'direct' methods, which employ mathematical relationships between phases, may be used in favourable cases to compute phases (Han, 2001), and then electron density maps from native structures. However, these methods are challenging for molecules of greater than 1000 kDa in size.

Macromolecular crystal structures are normally optimised with respect to the observed diffraction data by non-linear least-squares fitting procedures, which formally minimise the differences for the structure factors between observed and calculated models. This is the process of crystallographic refinement. When the diffraction data do not extend to atomic resolution, it is necessary to incorporate restraining information from established stereochemical and structural features (such as bond lengths and angles, planar geometry of the DNA bases and preferred torsion angles). The standard compilation of these data (Parkinson et al., 1996) has used data from accurate small-molecule crystal structures in the Cambridge Crystallographic Data Base. Those for the sugar moiety have been updated (Kowiel et al., 2020), using the more accurate small-molecule data that have become available in the intervening 24 years. The use of constraints and restraints is necessary since the total number of experimental structure amplitude observations may not sufficiently exceed the total number of variables (positional parameters and a temperature factor for each atom) for the least-squares fitting to work in a robust manner. Typically, a medium-resolution structure when refined without suitable constraints and restraints will increasingly diverge from a starting point of a physically meaningful approximate structure to one with significant unacceptable stereochemistry. The known geometric features are used to set up intra-molecular constraints and restraints between them and so improve the initial models by 1. reducing the number of variables so that individual atoms are no longer able to freely move during a refinement and 2. enabling acceptable stereochemistry to be retained. One of the early and widely used programs for macromolecular refinement, X-PLOR/CNS, used empirical energy terms as part of the minimised function to ensure optimal intra- and inter-molecular geometry. The technique of simulated annealing has been adopted from molecular dynamics (MD) as an effective way of refining structures when large-scale (>1 Å) atomic movements are required, since conventional least-squares methods are inherently incapable of effecting such large changes. The program REFMAC5 (Murshudov et al., 2011) is now extensively employed for DNA and RNA refinement. It uses a maximum likelihood technique rather than least-squares or simulated annealing. This systematically adjusts the positional and thermal parameters that describe a model structure, with respect to the experimental data, using R and R_{free} to monitor the process.

Oligonucleotide and oligonucleotide—protein crystals are heavily hydrated, with often over 50% solvent content. It is typical in medium-resolution structures for only a small fraction of these water molecules to be located in electron density maps, largely because their high mobility smears their electron density to below the signal-to-noise level of these maps. The majority of water molecules reported in these structures are unsurprisingly the least mobile ones, which are directly hydrogen bonded to the structure — these are the 'first-shell' water molecules (see Chapters 3 and 5 for detailed discussions of water arrangements in nucleic acid structures). The ways in which molecules pack in the crystal are sometimes of importance when examining structural features, since considerations of efficient packing can readily force parts of molecules to interact one with another by hydrogen bonding and van der Waals interactions and consequently possibly modify some features of otherwise flexible conformation.

The quality and reliability of an oligonucleotide crystal structure are not straightforward to assess, especially for a non-crystallographer. Yet, judgements on these factors are critical when undertaking and using structural comparisons and analyses. The important crystallographic parameters of quality (R, R_{free}) have been outlined above. Of at least equal significance are the derived stereochemical features — examination of these is a reliable guide to quality (Das et al., 2001; Wlodawer et al., 2008).

Particular features to examine in a structure include the following:

- close non-bonded intra- and inter-molecular contacts that are less than the sum of the van der Waals radii of the atoms involved. A small number of close contacts may indicate minor errors in a structure; larger numbers are suggestive of significant errors or even an incorrect structure.
- the distribution of values for torsion angles around single bonds. Eclipsed ($\sim 0°$) values are indicative of problems in refinement
- hydrogen bonds with distances appreciably outside the accepted ranges of $\sim 2.7-3.2$ Å
- estimates of error in atomic positions
- the quality of the electron density for individual groups and atoms
- values of atomic temperature factors, especially for water molecules

Checks on the purely structural features are equally applicable when examining structural models derived from NMR analyses. In practice, all new crystal and NMR structures are rigorously checked for internal consistency and potential errors when they are being deposited in the PDB, and any problems are drawn to the attention of the investigators. However, it remains the case that a significant number of older structures retain some problematic features that have not been corrected (Westhof, 2016). All checked data are openly available. The conformational flexibility of phosphate groups has been observed in some high-resolution structures and needs to be taken into account should it be observed in difference Fourier electron density maps (Sunami et al., 2017).

1.3 NMR methods for studying nucleic acid structure and dynamics

The underlying principle of NMR is the detection in a magnetic field of those atomic nuclei in a molecule, which have nuclear spin. Protons are abundant in nucleic acids and oligonucleotides and fortunately have readily detectable spin signals. These signals, termed chemical shifts, are dependent on the shielding effect of neighbouring protons, and thus can be used to determine the chemical environment of a proton once they can be unequivocally assigned as arising from particular atoms. Examples of highly characteristic, conformation-dependent chemical shifts are those arising from the protons on a deoxyribose sugar, which vary according to the pucker of the sugar (see Chapter 3). Other nuclei are less sensitive than protons and have low natural abundance, but the high field strengths of magnets now widely available for NMR studies are making ^{13}C- and ^{15}N-enriched oligonucleotides amenable to detailed studies. NMR studies of oligonucleotides have been extensively used to examine interactions with proteins and small molecules, by monitoring characteristic changes in certain chemical shifts.

The magnetic interactions between a pair of protons give rise to an NMR spin—spin coupling constant which is directly related to the dihedral angle between them, by the Karplus relationship. Hence, measurement of coupling constants provides direct and reliable information on sugar puckers and on part of the backbone conformation (notably the glycosidic angle between sugar and base) in a nucleotide or oligonucleotide. Use of ^{13}C- and ^{13}N-labelled oligonucleotides enables coupling constants to be determined for otherwise inaccessible backbone torsion angles.

NMR methods enable structures to be determined in solution, largely by means of measurements of proton—proton coupling constants and through-space nuclear Overhauser effect (NOE)-derived distances using 2D NMR methods. Solution-phase studies have the obvious advantage that molecules do not have to be crystallised, which is often the major (and highly frustrating!) limitation to the analysis of a macromolecule by X-ray crystallography. There is also the apparent advantage that a structure determined in solution may be more relevant to physiological processes than an X-ray crystallographic study in the solid state. However, the two techniques should not be considered as alternatives. Rather they are complementary, providing distinct information. For example, NMR results emphasise the flexible nature of DNA and RNA molecules and the fact that individual groups such as sugars are dynamically in motion. It is notable that parallel observations of sequence-dependent effects in several oligonucleotide sequences have been reported from both crystallographic and NMR studies, although differences in detail are sometimes apparent.

There are several limitations to the accuracy and reliability of NMR methods as applied to nucleic acids. By contrast with crystallography, there is a limitation on the size of problem that can be analysed in detail, due to the increase in the number of signals with molecular weight and consequent overlap of chemical shifts. The NOE is only significant for protons, and so phosphate geometry is not directly defined by it. Hence, nucleotide backbone conformations are in principle incompletely defined by standard ^{1}H NMR methods. Since an NOE intensity is proportional to the inverse sixth power of a proton–proton distance, it is a short-range effect, which is only significant at distances less than 5–6 Å. Longer distances important in DNA structure, such as groove width, may not be derived directly. Most reliability from NMR experiments can be placed on those features of DNA structure undergoing the least motion, especially base pairing, sugar pucker, as well as the location of ligand binding sites. Detailed aspects of especially sequence-dependent structure have been a matter of considerable controversy. This is in large part because of the relatively small number of NOE data available compared to the several thousands of X-ray intensities from a typical medium resolution crystallographic analysis. The consequent relative under-determination of NMR-derived DNA structures means that their effective 'resolution' is probably ~3–4 Å, with those parts of a structure providing the most NOE data being the most reliable. This situation is distinct from that for small (<25 kD) globular proteins, where the richness of NOE data arising from compact, closely packed amino acid residues, provides for highly reliable and detailed NMR structure determination more equivalent to that from high-resolution crystallographic analyses.

A standard approach in NMR structure determination is to use restrained MD methods and an assumed rough starting model. Commonly used programs have historically included X-PLOR and CNS, thus sharing parameters and algorithms with earlier crystallographic refinements. The distances established from individual NOE assignments are taken as constraints together with the NMR-derived conformational angles, to arrive at a plausible model or set of models. It is common practice to employ the NOE data in a semi-quantitative manner, with the NOE signals being assigned to three groups. These correspond to long (e.g. 4–6 Å), medium (3–5 Å) and short (2–3.5 Å) interatomic distances. The accuracy of an NMR structure can be assessed by back-calculating the NOE intensities and comparing them with the observed, in a manner analogous to that used in crystallography. However, the NMR 'R factor' can be a less reliable guide since the number of observations is so much less than in crystallography, and flexibility in part

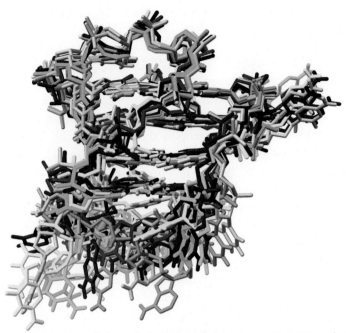

Figure 1.4 An ensemble of 10 structures from an NMR determination, of a DNA quadruplex formed by a sequence found in the human BCL2 promoter (PDB entry no. 2F8U) (Dai, J., Chen, D., Jones, R. A., Hurley, L. H., & Yang, D. (2006). NMR solution structure of the major G-quadruplex structure formed in the human BCL2 promoter region. *Nucleic Acids Research, 34*, 5133–5144). Note the close correspondence of the bases in the various structures, contrasting with the diversity of conformation and flexibility shown by the extra-helical groups.

of a structure may bias the R factor to give an apparently poor value. NMR studies produce an ensemble of related structures, rather than the single one from crystallography (Fig. 1.4). The distribution of some elements of a structure is a consequence of the small number of experimental observations of relative to the situation in a typical crystal structure. It also indicates the relative flexibility of different regions in a structure — typically in a nucleic acid, the base pairs are the least flexible component compared to extra-helical loop regions.

1.4 Cryo-EM methods for macromolecular complexes

Single-particle Cryo-EM has emerged as an increasingly powerful technique for determining the structures of large macromolecules and complexes at cryogenic temperatures using transmission electron microscopy. It does not require crystalline samples and uses image reconstruction methodology together with atomistic modelling when the data are not at near-atomic resolution. Cryo-EM was originally developed to study viruses and membrane proteins and was limited in scope because of radiation damage to samples. Improvements in detector technology (notably the CMOS detector technology (McMullan et al., 2016), automated data collection and image processing software have greatly extended its scope (Bhella, 2019; Lyumkis, 2019; Wu & Lander, 2020) so that it is now the major methodology for structural biological studies of complex systems, and near-atomic resolution structures are increasingly common. It is the method of choice since the challenges of crystallisation are no longer an impediment to successful structure determination, although its lengthy data collection time compared to conventional X-ray crystallography has to date limited its applicability in routine structure-based drug design. However, several recent cryo-EM structure determinations of drug complexes with ion channels have demonstrated the power of the approach and its potential for understanding and exploiting drug interactions with large biological assemblies (van Drie & Tong, 2020). True high-resolution enabling individual atoms to be resolved have been recently achieved for the protein apoferritin, at 1.25 Å, which has also enabled hydrogen atoms to be unambiguously visualised (Yip et al., 2020). Resolution beyond this point would require a very large number of particle images to be collected, well beyond what is currently practically feasible.

Cryo-EM is having a major impact in the nucleic acid structure field. A group of RNA structures (chiefly ribozymes and riboswitches) have been determined by this approach (Kappel et al., 2020), even though many RNAs were previously believed to be too small to be successfully imaged, assuming a lower size limit of 50 kDa for successful imaging. The group is comprised of 18 diverse RNAs with between 65 and 388 nucleotides (21−126 kDa). Structures were obtained for all those with ordered tertiary folds, with resolutions between 4.7 and 11 Å such that all the folds were unambiguously visible in the electron density, as were RNA grooves. Subsequent modelling into the density using the Ribosolve pipeline (Kappel et al., 2020) has enabled all-atom models to be unambiguously defined, so that conclusions could be made about issues such as riboswitch folds from different species. The determination of the structure of the 28 kDa frameshift stimulation element of the SARS-CoV-2 genome, at a resolution of 5.9 Å, has enabled a model for frameshifting in the viral RNA genome, as well as possible small molecule inhibitor binding sites (Zhang, El Omari, et al., 2020, Zhang, Zheludev, et al., 2020). The 2.9 Å structure of the SARS-CoV-2 polymerase is a landmark study of an active form of the enzyme complexed with two turns of double-stranded RNA. It shows the RNA in the centre of the complex and bound to a hitherto unseen long α-helical region covering much of the RNA (Hillel et al., 2020).

1.5 Molecular modelling and simulation of nucleic acids

Crystallographic analyses provide a quasi-static view of molecular structure. The process of X-ray data acquisition from a single crystal, even at a high-flux synchrotron source, can take typically several minutes (although Laue and other newer methods can enable time-resolved diffraction data to be collected in the time scales of chemical and biochemical events). The vast majority of crystal structures provide a time-averaged picture of molecular motions about the low-energy structure in the crystal (which is typically > 50% solvent). By contrast, molecular modelling techniques enable dynamic changes in structure and conformation to be calculated and visualised in terms of their effects on molecular energetics. These theoretical methods thus provide information complementary to the experimental techniques.

It is not feasible at present to compute conformational or energetic properties for significant lengths of nucleic acid sequence by ab initio quantum mechanics. Instead, empirical force-field methods are widely used. These have been derived from experimental data that describe the energetics of a DNA or RNA molecule in terms of the sum of a number of factors:

$$V(r^N) = \sum_{bonds} \frac{1}{2}k_b(l - l_0)^2 + \sum_{angles} \frac{1}{2}k_a(\theta - \theta_0)^2$$

$$+ \sum_{torsions} \frac{1}{2}V_n[1 + \cos(n\omega - \gamma)]$$

$$+ \sum_{j=1}^{N-1} \sum_{i=j+1}^{N} \left\{ 4\varepsilon_{i,j} \left[\left(\frac{\sigma_{ij}}{r_{ij}} \right)^{12} - \left(\frac{\sigma_{ij}}{r_{ij}} \right)^6 \right] + \frac{q_i q_j}{4\pi\varepsilon_0 r_{ij}} \right\}$$

- van der Waals non-bonded interactions, the 6–12 potential in the above formalism
- bond length and angle distortions
- barriers to rotation about single bonds
- Coulombic electrostatic contributions from full and partial electrostatic-potential derived atomic charges
- hydrogen bonding (often incorporated as an implicit part of the electrostatic component)

The more recently developed nucleic acid force fields can also include the contributions of polarisation effects, which can be especially important when examining the interactions of nucleic acids with drug or protein molecules.

The major empirical nucleic acid force fields have been incorporated into algorithms that can be used to minimise the conformation of a molecule with respect to its internal energy. This is the method of molecular mechanics, which in effect optimises local low-energy minima. Much more extensive explorations of conformations can be made by MD, which applies Newton's equations of motion to an empirical force field, for all atoms in a molecule. This technique is computer intensive; although with current desk-top workstations and high-end multi-processor Linux-based PCs, it is now routine to undertake realistic simulations of molecular motions, with the inclusion of large numbers of calculated solvent molecules. The widespread availability of supercomputers and large multi-processor clusters has enabled an increasing number of simulations to be performed over thousands of picoseconds of molecular movements, with many studies of >10 ns (1 ns $=$ 1000 ps) duration. An increasing number of simulations reached and exceeded the 1 μs limit, which is still faster than most biochemical events, but the rate of increase in computer power makes true biological simulations an attainable goal. The report (Jung et al., 2019) of a simulation of the GATA4 gene is the first atomistic simulation of an entire gene locus (with a billion atoms) and used an array of 130,000 processor cores. This ambitious study included 83 kilobases of DNA complexed in 427 nucleosomes, in an arrangement constructed from experimental ultra-structural data followed by coarse-grain modelling.

MD can enable barriers between local energy minima to be traversed, unlike molecular mechanics. It is widely used in conjunction with distance geometry data (from NOEs) to derive plausible DNA/RNA and nucleic acid—protein structures from these NMR measurements. The technique of simulated annealing is often employed, in which a structure is first simulated by MD at a high temperature, when it can overcome high energy barriers between conformations. The system is then gradually 'cooled', when the most likely energy states become populated.

Solvent and counter-ions are normally routinely incorporated into dynamics simulations, their positions having been generated by Monte Carlo algorithms. The earlier use of a distance-dependent dielectric model to compensate for the lack of solvent often led to unstable structures during MD simulation. Even with the inclusion of a simple solvent model, care is needed to check that nucleic acid simulations remain stable due to the inability of the force field to adequately account for long-range electrostatic forces and solvation effects. This is typically seen in the breaking apart of structures after a period of simulation, which can be several hundred picoseconds. The use of Ewald and particle-mesh Ewald summation methods has led to a marked improvement in the stability of nucleic acid systems even in long time-scale simulations. Validation against both NMR and X-ray crystallographic structures is an important activity, especially with the advent of numerous accurate and reliable experimental structures. Simulations with explicit solvent are computationally expensive, but the use of the reliable generalised Born solvation model usually enables DNA to be modelled without the need for explicit water and counter-ions molecules to be present. This treatment can adequately account for solvent continuum effects and solute—solvent electrostatic polarisation.

The widespread application of MD methods has greatly increased in recent years, due not only to the ready availability of high-performance computing facilities, even on one's desk-top, but also to the widespread availability of well-validated academic and commercially derived computer programs, most with graphical frontends for the display of results. Several modelling and simulation programs are in common use which have their force fields specifically parameterised for nucleic acids and their components. An early comparison of simulations with different force fields applied to a B-DNA decamer system (Reddy et al., 2003) indicated that disconcertingly not all of the simulations resulted in the same overall B-type helix, although many parameters in these force fields have been updated since this study was undertaken (see, for example, Soares et al., 2005).

The most widely used and validated nucleic acid-focussed molecular simulation programs are as follows:

- AMBER ('Assisted Model building and Energy Refinement'), originally from the laboratory of the late P. A. Kollman, University of California, San Francisco. The force fields incorporated in recent releases of this package have been especially well validated for nucleic acids (Case et al., 2005; Cheatham & Case, 2013; Cornell et al., 1995; Wang et al., 2004) and have been generally taken to be the gold standard. The current version (2020) of the package is AMBER20. Several updated AMBER force fields are now available, that has been validated for improving the representation of particular features of a given nucleic acid, such as backbone geometry (Pérez et al., 2007), with the Parmbsc1 force field being the most recent (Galindo-Murillo et al., 2016; Ivani et al., 2016). AMBER is available from https://www.ambermd.org. The AMBER force field has also been incorporated into the PHENIX crystallographic refinement (Moriarty et al., 2020), leading to improved crystal structures, especially in terms of non-bonded contacts and hydrogen bonds, probably on account of the superior treatment of electrostatic interactions with this force field.

- CHARMM ('Chemistry at Harvard Macromolecular Mechanics'), originally from the laboratory of M. Karplus, Harvard University, obtainable from www.charmm.org. CHARMM is also the forerunner of the X-PLOR and CNS programs, which were amongst the first programs for crystallographic and NMR structure refinement that incorporated energy terms. Further details are available from https://en.wikipedia.org/wiki/CHARMM. The most recent release was in 2015, v. 41. The AMBER force fields for nucleic acids (Bsc1 for DNA and OL3 for RNA) are now available for use in CHARMM (Lee et al., 2020). Versions 27 and 36 have updated nucleic acid parameterisations (Hart et al., 2012).

- GROMOS ('Groningen Molecular Simulation'), (Soares et al., 2005) developed in the laboratory of W. F. van Gunsteren and H. J. C. Berendsen, ETH Zurich. The most recent version of the package is GROMOS 11 version 1.4.0, released in February 2018. Obtainable from http://www.gromos.net.

Several studies, for example Dans et al. (2017) have examined behaviour of nucleic acid structures during and at the end of long time-scale simulations in terms of accuracy (compared to X-ray structures) and stability, for several well-studied oligonucleotides such as the Dickerson—Drew dodecamer (Chapter 3). For example, Galindo-Murillo et al. (2016)

compared results on nucleic acids from the AMBER and CHARMM force fields, using the recent bsc1 and OL15 modifications to AMBER. They found that the former was better able to reproduce the averaged experimental structure, over an aggregate of 14 ms of simulation time, with a rms of <1 Å, compared to the 1.3 Å rms with the CHARMM force field. A longer 40 base pair DNA sequence has been used in a study of force field effects on DNA flexibility and bending (Minhas et al., 2020), using two AMBER DNA force fields (parmbsc0 and parmbsc1) and two from relatively recent CHARMM updates, CHARMM-26 and -27.

Other simulation programs are listed in
https://en.wikipedia.org/wiki/Comparison_of_nucleic_acid_simulation_software.

1.6 Chemical, enzymatic and biophysical probes of structure and dynamics

Enzymes such as DNase I cleave the phosphodiester bond in a DNA duplex at every nucleotide position, although the cutting efficiency is markedly dependent on sequence, and by implication, on sequence-related structural features. Cleavage may be blocked by protein or drug binding. Hence, DNase I can be used to determine sites of binding along a DNA sequence as well as to assess possible effects of individual sequences on DNA structure. Chemical cleaving agents such as hydroxyl radicals can give similar information. Since these are much smaller molecules than cleavage enzymes, their effects on DNA structure are less perturbing and sequence dependent. Other types of chemical probe can attack specific base sites. These can be useful in defining the precise sites of protection resulting from drug or protein binding to a DNA sequence (see, for example, Azad et al., 2018; Lazarovici et al., 2013).

These foot-printing methods have the important advantage over the fine-structure techniques of crystallography and NMR, of being applicable to long (up to several thousand base pair) DNA sequences and thus of being more directly relevant to DNA in the cell. Hence, the use of chemical and enzymatic probes for DNA provides a way of obtaining at least some molecular-level data on otherwise inaccessible structural problems in DNA—protein and drug recognition. Foot-printing can also be quantitated to provide kinetic and thermodynamic information at specific sites, although other techniques, notably SPR and isothermal titration calorimetry, generally provide more accurate data.

Other structural methods that are beyond the scope of this book include atomic force microscopy, which is a powerful method of imaging large nucleic acids and protein—nucleic acid complexes. Single-molecule force measurement methods can monitor events such as structural transitions, folding and nucleic acid enzymology, providing information that complements and extends that from the fine-structure approaches described in this book.

1.7 Sources of structural data

The results of a crystal structure, fibre diffraction analysis or NMR study are most useful as a set of atomic coordinates. Those from fibre diffraction are available either in the primary literature or in various review chapters and compilations (see the 3DNA database and also Chapter 3). Crystallographic coordinates may be obtained from a database, provided that they have been deposited in the first instance. This is no longer a problem since almost all journals now insist on deposition by publication, although authors still retain the right to up to a year's delay before public release of a data set. All available oligonucleotide crystal and NMR structures are in the successor to the Brookhaven Protein Data Bank, the RCSB (Research Collaboration for Structural Biology) PDB (universally known as the PDB) (http://www.rcsb.org) and in the PDB in Europe portal at the European Bioinformatics Institute (EBI) (http://www.ebi.ac.uk/pdbe). Depositions are accompanied by structure factor data, although this may not be available for the more historic structures. The PDB also contains a number of modelled structures. The Cambridge Crystallographic Database (which is primarily for small molecules) also contains coordinate data on a small number of historic oligonucleotide structures (http://www.ccdc.cam.ac.uk). The Nucleic Acid Database (NDB) is a comprehensive relational database for nucleic acid crystallographic data, at Rutgers University, USA (Berman et al., 2002; Narayanan et al., 2014) (http://ndbserver.rutgers.edu), which is intimately connected to the PDB. The NDB provides a set of powerful tools for the comparative study of nucleic acid structural features involving conformation, base and base-pair morphology, enabling detailed analyses of trends and features to be undertaken. The PDB (and the NDB) contain all deposited protein–nucleic acid structures, as well as many NMR-derived ones. Many NMR structures are also available in the BioMagResBank Repository at http://www.bmrb.wisc.edu, which is now affiliated with the PDB.

As of January 2021, the number of deposited structures in the PDB and NDB was

	Nucleic acids	Nucleic acid–protein complexes
X-ray	2128	7151
NMR	1310	269
Electron microscopy	63	1589

These databases are accessible via the internet from their primary sites or from mirror sites worldwide.

1.8 Visualisation of nucleic acid molecular structures

Molecular structures are best viewed interactively on a tablet, computer or phone screen, rather than as flat representations on a page. This book provides PDB identification codes for many of the structures discussed in this chapter. The reader can then easily access and display individual structures, either by direct access to the PDB, NDB or the EBI and the extensive visualisation tools available in them or by downloading coordinates from the PDB/NDB/EBI and inputting them into a molecular visualisation or modelling program, such as one of those listed here.

Many such programs are now available, either commercially or as freeware. Many have been implemented on multiple platforms. Listed here are some recommended freeware programs that can be used on a PC (Windows, Android or a Unix/Linux dialect) or an Apple desktop computer and which are nucleic acid structure friendly in that they can cope with nucleic acid residues and can also display nucleic acids in cartoon form. Increasingly, structures are visualised on mobile devices: laptop, tablet or phone. Most of the available programs can directly download files from the PDB.

1. **RasMol** was the first freely available PC molecular graphics program and is still used. The latest version, RasMol 2.7.5, was released in 2009. It is available for WINDOWS systems only, from http://www.bernstein-plus-sons.com/software/rasmol/.

2. **UCSF Chimera**, at http://www.cgl.ucsf.edu/chimera/. Developed by the Computer Graphics Laboratory, University of California, San Francisco, USA (Couch et al., 2006). The latest version, 1.15, was released in November 2020. It runs on WINDOWS, LINUX, and MacOS operating systems. The next-generation program **ChimeraX**, which will eventually supersede Chimera, had a 1.1 production release in September 2020. It is available at https://www.cgl.ucsf.edu/chimerax/ and also runs on WINDOWS, LINUX and MacOS operating systems (Pettersen et al., 2020).

3. **PyMOL**, at https://pymol. org/. Developed by DeLano Scientific, California, USA. The current commercial version is 2.3.4 (2019). Earlier versions such as 2.1.0 may also be available as free open source. **PyMOL** runs on WINDOWS, LINUX and MacOS operating systems. An extension of **PyMOL** to show schematics of nucleic acid structures is available (Lu, 2020).

4. **VMD** (Visual Molecular Dynamics), at http://www.ks.uiuc.edu/Research/vmd/. Developed by the Theoretical and Molecular Biophysics Group, University of Illinois at Urbana, Champaign, USA (Humphrey et al., 1996). The most recent release is version 1.9.3. It runs on WINDOWS, LINUX/UNIX and MacOS operating systems. A more recent pre-release version, 1.9.4, was made available in 2020.

5. **Biovia Discovery Studio Visualizer** has well-developed nucleic acid cartoon features and runs on WINDOWS and LINUX operating systems. It is available from https://discover.3ds.com/discovery-studio-visualizer-download.

6. A selection of molecular visualisation apps is available for portable devices:
   (i) for Apple iPhones and iPad tablets, from the Applestore
   (ii) for Android phones and tablets, from Google Play Store

1.8.1 The structural visualisations in this book

Structures have been drawn using a range of different representations — sometimes more than one has been used in a single figure. Chimera, ChimeraX and Discovery Visualizer have been our programs of choice, to illustrate

- Cartoons in which a phosphodiester backbone is represented by a ribbon, with the 5' → 3' indicated by an arrow Bases and sugars are shown as filled slabs
- Diagrams with bonds shown as lines or solid sticks
- Figures with bonds shown as solid sticks and atoms as small spheres
- van der Waals representations, having atoms drawn as spheres with radii set at their van der Waals values
- Surface representations, showing the solvent-excluded surface of a molecule

References

Arnott, S. (1970a). The geometry of nucleic acids. *Progress in Biophysics and Molecular Biology, 21*, 265—319.

Arnott, S. (1970b). Crystallography of DNA: Difference synthesis supports Watson-Crick base pairing. *Science, 167*, 1694—1700.

Azad, R. N., Zafiropoulos, D., Ober, D., Jiang, Y., Chiu, T. P., Sagendorf, J. M., Rohs, R., & Tullius, T. D. (2018). Experimental maps of DNA structure at nucleotide resolution distinguish intrinsic from protein-induced DNA deformations. *Nucleic Acids Research, 46*, 2636—2647.

Berman, H. M., Westbrook, J., Feng, Z., Iype, L., Schneider, B., & Zardecki, C. (2002). The nucleic acid database. *Acta Crystallographica Section D, 58*, 889—898.

Bhella, D. (2019). Cryo-electron microscopy: An introduction to the technique, and considerations when working to establish a national facility. *Biophysical Reviews, 11*, 515—519.

Case, D. A., Cheatham, T. E., III, Darden, T., Gohlke, H., Luo, R., Merz, K. M., Jr., Onufriev, A., Simmerling, C., Wang, B., & Woods, R. J. (2005). The AMBER biomolecular simulation programs. *Journal of Computational Chemistry, 26*, 1668—1688.

Cheatham, T. E., III, & Case, D. A. (2013). Twenty-five years of nucleic acid simulations. *Biopolymers, 99*, 969—977.

Cornell, W. D., Cieplak, C. I., Bayly, I. R., Gould, I. R., Merz, K. M., Ferguson, D. M., Spellmeyer, D. C., Fox, T., Caldwell, J. W., & Kollman, P. A. (1995). A second generation force field for the simulation of proteins, nucleic acids and organic molecules. *Journal of the American Chemical Society, 117*, 5179—5197.

Couch, G. S., Hendrix, D. K., & Ferrin, T. E. (2006). Nucleic acid visualization with UCSF Chimera. *Nucleic Acids Research, 34*, e29.

Dans, P. D., Ivani, I., Hospital, A., Portella, G., González, C., & Orozco, M. (2017). How accurate are accurate force-fields for B-DNA? *Nucleic Acids Research, 45*, 4217—4230.

Das, U., Chen, S., Fuxreiter, M., Vaguine, A. A., Richelle, J., Berman, H. M., & Wodak, S. J. (2001). Checking nucleic acid crystal structures. *Acta Crystallographica, D57*, 813—828.

Dauter, Z., & Adamiak, D. A. (2001). Anomalous signal of phosphorus used for phasing DNA oligomer: Importance of data redundancy. *Acta Crystallographica, D57*, 990—995.

van Drie, J. H., & Tong, L. (2020). Cryo-EM as a powerful tool for drug discovery. *Bioorganic & Medicinal Chemistry Letters, 30*, 127524.

Ducruix, A., & Giegé, R. (Eds.). (2004). *Crystallisation of nucleic acids and proteins* (2nd ed.). Oxford: Oxford University Press.

Franklin, R. E., & Gosling, R. G. (1953). Molecular configuration in sodium thymonucleate. *Nature, 171*, 740—741.

Galindo-Murillo, R., Robertson, J. C., Zgarbová, M., Šponer, J., Otyepka, M., Jurečka, P., & Cheatham, T. E., III. (2016). Assessing the current state of amber force field modifications for DNA. *Journal of Chemical Theory and Computation, 12*, 4114—4127.

Han, G. W. (2001). Direct-methods determination of an RNA/DNA hybrid decamer at 1.15 Å resolution. *Acta Crystallographica, D57*, 213–218.

Hart, K., Foloppe, N., Baker, C. M., Denning, E. J., Nilsson, L., & MacKerell, A. D., Jr. (2012). Optimization of the CHARMM additive force field for DNA: Improved treatment of the BI/BII conformational equilibrium. *Journal of Chemical Theory and Computation, 8*, 348–362.

Hillen, H. S., Kokic, G., Farnung, L., Dienemann, C., Tegunov, D., & Cramer, P. (2020). Structure of replicating SARS-CoV-2 polymerase. *Nature, 584*, 154–156.

Holbrook, E. L., & Holbrook, S. R. (2010). In *Encyclopedia of life sciences*. Chichester, UK: John Wiley and Sons Ltd.

Hou, C., & Tsodikov, O. V. (2019). Utilizing guanine-coordinated $Zn^{(2+)}$ ions to determine DNA crystal structures by single-wavelength anomalous diffraction. *Acta Crystallographica, D75*, 32–40.

Humphrey, W., Dalke, A., & Schulten, K. (1996). VMD - visual molecular dynamics. *Journal of Molecular Graphics, 14*, 33–38.

Ivani, I., Dans, P. D., Noy, A., Pérez, A., Faustino, I., Hospital, A., Walther, J., Andrio, P., Goñi, R., Balaceanu, A., Portella, G., Battistini, F., Gelpí, J. L., González, C., Vendruscolo, M., Laughton, C. A., Harris, S. A., Case, D. A., & Orozco, M. (2016). Parmbsc1: A refined force field for DNA simulations. *Nature Methods, 13*, 55–58.

Jiang, J., Sheng, J., Carrasco, N., & Huang, Z. (2007). Selenium derivatization of nucleic acids for crystallography. *Nucleic Acids Research, 35*, 477–485.

Jung, J., Nishima, W., Daniels, M., Bascom, G., Kobayashi, C., Adedoyin, A., Wall, M., Lappala, A., Phillips, D., Fischer, W., Tung, C. S., Schlick, T., Sugita, Y., & Sanbonmatsu, K. Y. (2019). Scaling molecular dynamics beyond 100,000 processor cores for large-scale biophysical simulations. *Journal of Computational Chemistry, 40*, 1919–1930.

Kappel, K., Zhang, K., Su, Z., Watkins, A. M., Kladwang, W., Li, S., Pintilie, G., Topkar, V. V., Rangan, R., Zheludev, I. N., Yesselman, J. D., Chiu, W., & Das, R. (2020). Accelerated cryo-EM-guided determination of three-dimensional RNA-only structures. *Nature Methods, 17*, 699–707.

Ke, A., & Doudna, J. A. (2004). Crystallization of RNA and RNA-protein complexes. *Methods, 34*, 408–414.

Kowiel, M., Brzezinski, D., Gilski, M., & Jaskolski, M. (2020). Conformation-dependent restraints for polynucleotides: The sugar moiety. *Nucleic Acids Research, 48*, 962–973.

Latychevskaia, T., & Fink, H. W. (2018). Three-dimensional double helical DNA structure directly revealed from its X-ray fiber diffraction pattern by iterative phase retrieval. *Optics Express, 26*, 30991–31017.

Lazarovici, A., Zhou, T., Shafer, A., Dantas Machado, A. C., Riley, T. R., Sandstrom, R., Sabo, P. J., Lu, Y., Rohs, R., Stamatoyannopoulos, J. A., & Bussemaker, H. J. (2013). Probing DNA shape and methylation state on a genomic scale with DNase I. *Proceedings of the National Academy of Sciences of the United States of America, 110*, 6376–6381.

Lee, J., Hitzenberger, M., Rieger, M., Kern, N. R., Zacharias, M., & Im, W. (2020). CHARMM-GUI supports the Amber force fields. *The Journal of Chemical Physics, 153*, 035103.

Lin, L., Sheng, J., & Huang, Z. (2011). Nucleic acid X-ray crystallography via direct selenium derivatization. *Chemical Society Reviews, 40*, 4591–4602.

Lu, X.-J. (2020). DSSR-enabled innovative schematics of 3D nucleic acid structures with PyMOL. *Nucleic Acids Research, 48*, e74.

Luo, Z., Dauter, M., & Dauter, Z. (2014). Phosphates in the Z-DNA dodecamer are flexible, but their P-SAD signal is sufficient for structure solution. *Acta Crystallographica, D70*, 1790–1800.

Lyumkis, D. (2019). Challenges and opportunities in Cryo-EM single-particle analysis. *Journal of Biological Chemistry, 294*, 5181–5197.

Mandal, P. K., Collie, G. W., Kauffmann, B., & Huc, I. (2014). Racemic DNA crystallography. *Angewandte Chemie International Edition in English, 53*, 14424–14427.

Mandal, P. K., Collie, G. W., Srivastava, S. C., Kauffmann, B., & Huc, I. (2016). Structure elucidation of the Pribnow box consensus promoter sequence by racemic DNA crystallography. *Nucleic Acids Research, 44*, 5936–5943.

Marvin, D. A. (2017). Fibre diffraction studies of biological macromolecules. *Progress in Biophysics and Molecular Biology, 127*, 43–87.

McMullan, G., Faruqi, A. R., & Henderson, R. (2016). Direct electron detectors. *Methods in Enzymology, 579*, 1–17.

Minhas, V., Sun, T., Mirzoev, A., Korolev, N., Lyubartsev, A. P., & Nordenskiöld, L. (2020). Modeling DNA flexibility: Comparison of force fields from atomistic to multiscale levels. *Journal of Physical Chemistry B, 124,* 38–49.

Moriarty, N. W., Janowski, P. A., Swails, J. M., Nguyen, H., Richardson, J. S., Case, D. A., & Adams, P. D. (2020). Improved chemistry restraints for crystallographic refinement by integrating the Amber force field into Phenix. *Acta Crystallographica, D76,* 51–62.

Murshudov, G. N., Skubak, P., Lebedev, A. A., Pannu, N. S., Steiner, R. A., Nicholls, R. A., Winn, M. D., Long, F., & Vagin, A. A. (2011). REFMAC5 for the refinement of macromolecular crystal structures. *Acta Crystallographica, D67,* 355–367.

Narayanan, B. C., Westbrook, J., Ghosh, S., Petrov, A. I., Sweeney, B., Zirbel, C. L., Leontis, N. B., & Berman, H. M. (2014). The nucleic acid database: New features and capabilities. *Nucleic Acids Research, 42,* D114–D122.

Pallan, P. S., & Egli, M. (2007a). Selenium modification of nucleic acids: Preparation of oligonucleotides with incorporated 2'-SeMe-uridine for crystallographic phasing of nucleic acid structures. *Nature Protocols, 2,* 647–651.

Pallan, P. S., & Egli, M. (2007b). Selenium modification of nucleic acids: Preparation of phosphoroselenoate derivatives for crystallographic phasing of nucleic acid structures. *Nature Protocols, 2,* 640–646.

Parkinson, G., Vojtechovsky, J., Clowney, L., Brünger, A. T., & Berman, H. M. (1996). New parameters for the refinement of nucleic acid-containing structures. *Acta Crystallographica, D52,* 57–64.

Pearson, A. R., & Mehrabi, P. (2020). Serial synchrotron crystallography for time-resolved structural biology. *Current Opinion in Structural Biology, 65,* 168–174.

Pérez, A., Marchán, I., Svozil, D., Sponer, J., Cheatham, T. E., III., Laughton, C. A., & Orozco, M. (2007). Refinement of the AMBER force field for nucleic acids: Improving the description of α/γ conformers. *Biophysical Journal, 92,* 3817–3829.

Pettersen, E. F., Goddard, T. D., Huang, C. C., Meng, E. C., Couch, G. S., Croll, T. I., Morris, J. H., & Ferrin, T. E. (2020). UCSF ChimeraX: Structure visualization for researchers, educators, and developers. *Protein Science.* https://doi.org/10.1002/pro.3943

Pfoh, R., Cuesta-Seijo, J. A., & Sheldrick, G. M. (2009). Interaction of an echinomycin-DNA complex with manganese ions. *Acta Crystallographica, F65,* 660–664.

Reddy, S. Y., Leclerc, F., & Karplus, M. (2003). DNA polymorphism: A comparison of force fields for nucleic acids. *Biophysical Journal, 84,* 1421–1449.

Reyes, F. E., Garst, A. D., & Batey, R. T. (2009). Strategies in RNA crystallography. *Methods in Enzymology, 469,* 119–139.

Salon, J., Sheng, J., Jiang, J., Chen, G., Caton-Williams, J., & Huang, Z. (2007). Oxygen replacement with selenium at the thymidine 4-position for the Se base pairing and crystal structure studies. *Journal of the American Chemical Society, 129,* 4862–4863.

Soares, T. A., Hunenberger, P. H., Kastenholz, M. A., Krautler, V., Lenz, T., Lins, R. D., Oostenbrink, C., & van Gunsteren, W. F. (2005). An improved parameter set for the GROMOS force field. *Journal of Computational Chemistry, 26,* 725–737.

Sunami, T., Chatake, T., & Kono, H. (2017). DNA conformational transitions inferred from re-evaluation of $m|Fo| - D|Fc|$ electron-density maps. *Acta Crystallographica, D73,* 600–608.

Thompson, M. C., Yeates, T. O., & Rodriguez, J. A. (2020). Advances in methods for atomic resolution macromolecular structure determination. *F1000Res, 9,* F1000. Faculty Rev-667.

Viladoms, J., & Parkinson, G. N. (2014). HELIX: A new modular nucleic acid crystallization screen. *Journal of Applied Crystallography, 47,* 948–955.

Wang, J., Wolf, R. M., Caldwell, J. W., Kollman, P. A., & Case, D. A. (2004). Development and testing of a general AMBER force field. *Journal of Computational Chemistry, 25,* 1157–1174.

Westhof, E. (2016). Perspectives and pitfalls in nucleic acids crystallography. *Methods in Molecular Biology, 1320,* 3–8.

Wilds, C. J., Pattanayek, R., Pan, C., Wawrzak, Z., & Egli, M. (2002). Selenium-assisted nucleic acid crystallography: Use of phosphoroselenoates for MAD phasing of a DNA structure. *Journal of the American Chemical Society, 124,* 14910–14915.

Wlodawer, A., Minor, W., Dauter, Z., & Jaskolski, M. (2008). Protein crystallography for non-crystallographers, or how to get the best (but not more) from published macromolecular structures. *FEBS Journal, 275,* 1–21.

Wu, M., & Lander, G. C. (2020). Present and emerging methodologies in cryo-EM single-particle analysis. *Biophysical Journal, 119*, 1281–1289.

Yip, K. M., Fischer, N., Paknia, E., Chari, A., & Stark, H. (2020). Atomic-resolution protein structure determination by cryo-EM. *Nature, 587*, 157–161.

Zhang, Y., El Omari, K., Duman, R., Liu, S., Haider, S., Wagner, A., Parkinson, G. N., & Wei, D. (2020). Native de novo structural determinations of non-canonical nucleic acid motifs by X-ray crystallography at long wavelengths. *Nucleic Acids Research, 48*, 9886–9898.

Zhang, K., Zheludev, I. N., Hagey, R. J., Wu, M. T., Haslecker, R., Hou, Y. J., Kretsch, R., Pintilie, G. D., Rangan, R., Kladwang, W., Li, S., Pham, E. A., Bernardin-Souibgui, C., Baric, R. S., Sheahan, T. P., Souza, V. D., Glenn, J. S., Chiu, W., & Das, R. (2020). Cryo-electron microscopy and exploratory antisense targeting of the 28-kDa frameshift stimulation element from the SARS-CoV-2 RNA genome. *bioRxiv*. https://doi.org/10.1101/2020.07.18.209270

Further reading

General

Blackburn, G. M., Gait, M. J., Loakes, D., & Williams, D. M. (Eds.). (2006). *Nucleic acids in chemistry and biology* (3rd ed.). Cambridge, UK: The Royal Society of Chemistry (An excellent survey of nucleic acid function, with some emphasis on structural aspects).

Bloomfield, V. A., Crothers, D. M., & Tinoco, I., Jr. (2000). *Nucleic acids. Structures, properties and functions*. Sausalito, California: University Science Books (An indispensable book for much background material, emphasising the biophysical chemistry of nucleic acids, but covering many structural aspects as well).

Calladine, C. R., Drew, H. R., Luisi, B., & Travers, A. A. (2004). *Understanding DNA* (3rd ed.). London: Academic Press (An excellent account of selected areas of DNA structure, written with real understanding of the complexity of DNA structure and its biological consequences).

Egli, M. (2004). Nucleic acid crystallography: Current progress. *Current Opinion in Structural Biology, 8*, 580–591 (A good summary of many of the topics covered in the first part of this book, covering the literature up to the beginning of the 21st century, but still very useful).

Lilley, D. M. J., & Dahlberg, J. E. (Eds.). (1992). *Methods in enzymology: DNA structures* (Vol. 211, p. 212). San Diego: Academic Press (A comprehensive compilation of reviews on many aspects of DNA structure, emphasising the methodologies of the early 1990s).

Neidle, S. (Ed.). (1999). *Oxford handbook of nucleic acid structure*. Oxford: Oxford University Press (A detailed reference book on atomic-level DNA and RNA structures in the crystalline and solution states as studied by X-ray and NMR methods, providing much early data, especially on fibre diffraction models of DNA and RNA).

Saenger, W. (1984). *Principles of nucleic acid structure*. Berlin: Springer-Verlag (Dated in many areas, but still an important reference book, providing a wealth of background information and detail, and historic perspective).

Sinden, R. R. (1994). *DNA structure and function*. San Diego: Academic Press.

The history of DNA structure

The story of the discovery of the double-helical structure of DNA is in itself a minor industry, not least of the continuing controversies surrounding the discovery and its protagonists. This short selection of articles and books, taken together, provides most of the story. Brief reviews can also be found in the 50th anniversary Double Helix supplement published by Nature in 2003. The first two articles in this list have been written by experts in nucleic acid fibre diffraction who were involved in the important series of studies that rigorously defined the details of the fibre-diffraction structures in the decade after 1953.

Arnott, S. (2006). Historical article: DNA polymorphism and the early history of the double helix. *Trends in Biochemical Sciences, 31*, 349—354.

Fuller, W. (2003). Who said helix? *Nature, 424*, 876—878 (A short but illuminating account of the discovery of the double-helix structure).

Judson, H. F. (2013). *The eight day of creation: Makers of the revolution in biology* (enlarged ed.). Cold Spring Harbor Press (An excellent account of the origins and history of molecular biology).

Klug, A. (2004). The discovery of the double helix. *Journal of Molecular Biology, 335*, 3—26 (A more detailed article, which i(ncludes material not previously generally available).

Maddox, B. (2003). *Rosalind Franklin: The dark lady of DNA*. London: Harper Collins (An excellent biography that combines scientific accuracy with real insight).

Olby, R. (1974). *The path to the double helix*. London: Macmillan Press (An authoritative and detailed account of the history of the discovery of the double helical structure and the background of earlier structural studies).

Watson, J. D. (1968). *The double helix*. London: Weidenfeld and Nicolson (A highly personalized and entertaining account of the discovery of DNA).

Wilkins, M. H. F. (2005). *The third man of the double helix*. Oxford: Oxford University Press (An autobiographical account from one of the principal protagonists in the double helix story, which provides some unexpected insights into the events of 1953).

Williams, G. (2019). *Unravelling the double helix: The lost heroes of DNA*. London: Weidenfeld and Nicolson (A useful account of the contributions of the largely forgotten scientists whose contributions helped to lay the foundations for the double helix).

More detailed accounts of particular topics can be found in the following

X-ray crystallography

Blow, D. M. (2002). *Outline of crystallography for biologists*. Oxford: Oxford University Press.

Dauter, Z. (2005). Efficient use of synchrotron radiation for macromolecular diffraction data collection. *Progress in Biophysics and Molecular Biology, 89*, 153—172.

Egli, M. (2016). Diffraction techniques in structural biology. *Current Protocols in Nucleic Acid Chemistry, 65*, 7.13.1—7.13.41.

Ennifar, E. (Ed.). (2016). *Nucleic acid crystallography: Methods and protocols (methods in molecular biology)*. New Jersey: Humana Press.

McPherson, A. (2009). *Introduction to macromolecular crystallography* (2nd ed.). New Jersey: Wiley-Blackwell.

Mooers, B. H. M. (2009). Crystallographic studies of DNA and RNA. *Methods, 47*, 168—176.

Rhodes, G. (2006). *Crystallography made crystal clear* (3rd ed.). San Diego: Academic Press.

Rossmann, M. G., & Arnold, E. (Eds.). (2001). *International tables for xray crystallography, volume F, crystallography of biological macromolecules*. Dordrecht: Kluwer.

Rupp, B. (2009). *Biomolecular crystallography: Principles, practice, and application to structural biology*. New York: Garland Science.

Sherwood, D., & Cooper, J. (2015). Crystals, X-rays and proteins. In *Comprehensive protein crystallography*. Oxford: Oxford University Press.

NMR methods

Barnwell, R. P., Yang, F., & Varani, G. (2017). Applications of NMR to structure determination of RNAs large and small. *Archives of Biochemistry and Biophysics, 628*, 42—56.

Ganser, L. R., Kelly, M. L., Herschlag, D., & Al-Hashimi, H. M. (2019). The roles of structural dynamics in the cellular functions of RNAs. *Nature Reviews Molecular Cell Biology, 20*, 474—489.

Lane, A. N. (1995). Determination of fast dynamics of nucleic acids by NMR. *Methods in Enzymology, 261*, 413—435.

Latham, M. P., Brown, D. J., McCallum, & Pardi, A. (2005). NMR methods for studying the structure and dynamics of RNA. *ChemBioChem, 6*, 1492—1505.

Nuclear magnetic resonance of biological macromoleculesJames, T. L. (Ed.). *Methods in Enzymology*, , (2005) 394.

Oh, K. I., Kim, J., Park, C. J., & Lee, J. H. (2020). Dynamics studies of DNA with non-canonical structure using NMR spectroscopy. *International Journal of Molecular Sciences, 21*, 2673.

Patel, D. J., Shapiro, L., & Hare, D. R. (1987). DNA and RNA: NMR studies of conformations and dynamics in solution. *Quarterly Reviews of Biophysics, 20*, 35–112.

Salmon, L., Yang, S., & Al-Hashimi, H. M. (2014). Advances in the determination of nucleic acid conformational ensembles. *Annual Review of Physical Chemistry, 65*, 293–316.

Wemmer, D. E. (1991). The applicability of NMR methods to solution structure of nucleic acids. *Current Opinion in Structural Biology, 1*, 452–458.

Wüthrich, K. (1986). *NMR of proteins and nucleic acids*. New York: John Wiley.

Molecular modelling and simulation methods

Beveridge, D. L., & McConnell, K. J. (2000). Nucleic acids: Theory and computer simulation. *Current Opinion in Structural Biology, 10*, 182–196.

Cheatham, T. E., III (2004). Simulation and modeling of nucleic acid structure, dynamics and interactions. *Current Opinion in Structural Biology, 14*, 360–367.

MacKerell, A. D., Jr. (2004). Empirical force fields for biological macromolecules: Overview and issues. *Journal of Computational Chemistry, 25*, 158–1604.

McCammon, J. A., & Harvey, S. C. (1987). *Dynamics of proteins and nucleic acids*. Cambridge, UK: Cambridge University Press.

Schlick, T. (2010). *Molecular modeling and simulation: An interdisciplinary guide* (2nd ed.). New York: Springer.

Schlick, T. (2012). In *Innovations in biomolecular modeling and simulation* (Vol. 1, p. 2). Cambridge, UK: RSC Publishing.

Šponer, J., Bussi, G., Krepl, M., Banáš, P., Bottaro, S., Cunha, R. A., Gil-Ley, A., Pinamonti, G., Poblete, S., Jurečka, P., Walter, N. G., & Otyepka, M. (2018). RNA structural dynamics as captured by molecular simulations: A comprehensive overview. *Chemical Reviews, 118*, 4177–4338.

Šponer, J., Shukla, M. K., Wang, J., & Leszczynski, J. (2017). Computational modeling of DNA and RNA fragments. In *Handbook of computational chemistry* (pp. 1803–1826). Springer.

Šponer, J., Šponer, J. E., Mládek, A., Banáš, P., Jurečka, P., & Otyepka, M. (2013). How to understand quantum chemical computations on DNA and RNA systems? A practical guide for non-specialists. *Methods, 64*, 3–11.

van Gunsteren, W. F., & Berendsen, H. J. C. (1990). Computer simulation of molecular dynamics: Methodology, applications, and perspectives in chemistry. *Angewandte Chemie International Edition, 29*, 992–1023.

Chemical, enzymatic, biophysical and single-molecule probes of nucleic acid structure

Bockelman, U. (2004). Single-molecule probes of nucleic acid structure. *Current Opinion in Structural Biology, 14*, 368–373.

Fox, K. R. (Ed.). (2012). *Drug-DNA interaction protocols* (2nd ed.). New Jersey: Humana Press.

Nguyen, B., Tanious, F. A., & Wilson, W. D. (2007). Biosensor-surface plasmon resonance: Quantitative analysis of small molecule-nucleic acid interactions. *Methods, 42*, 150–161.

Parker, S., & Tullius, T. D. (2011). DNA shape, genetic codes, and evolution. *Current Opinion in Structural Biology, 21*, 342–347.

Tullius, T. D., & Greenbaum, J. A. (2005). Mapping nucleic acid structure by hydroxyl radical cleavage. *Current Opinion in Chemical Biology, 9*, 127–134.

Wilson, W. D. (2002). Analyzing biomolecular interactions. *Science, 295*, 2103–2104.

The nucleic acid and protein databases

Berman, H. M., Gelbin, A., & Westbrook, J. (1996). Nucleic acid crystallography: A view from the nucleic acid database. *Progress in Biophysics and Molecular Biology, 66*, 255–288.

WWPDB Consortium. (2019). Protein Data Bank: The single global archive for 3D macromolecular structure data. *Nucleic Acids Research, 47*, D520–D528.

CHAPTER 2

The building blocks of DNA and RNA

2.1 Introduction

Nucleic acids were discovered by Friedrich Miescher in 1871. Chemical degradation studies on material extracted from cell nuclei established that it was acidic and contained phosphorus (Chargaff & Davidson, 1955). It was subsequently shown that the high molecular weight 'nucleic acid' was composed of individual acid units, termed nucleotides. Four distinct types were isolated — guanylic, adenylic, cytidylic and thymidylic acids. These could be further cleaved to phosphate groups and four distinct nucleosides. The latter were subsequently identified as consisting of a deoxypentose sugar and one of four nitrogen-containing heterocyclic bases. Thus, each repeating unit in a nucleic acid polymer comprises these three units linked together — a phosphate group, a sugar and one of the four bases.

The bases are planar aromatic heterocyclic molecules and are divided into two groups — the pyrimidine bases thymine and cytosine and the purine bases adenine and guanine. Their major tautomeric forms are shown in Fig. 2.1. Thymine is replaced by uracil in ribonucleic acids, which also have an extra hydroxyl group at the $2'$ position of their (ribose) sugar groups. The standard nomenclature for the atoms in nucleic acids, as approved by the International Union of Biochemistry, is shown in Figs. 2.1 and 2.2. Accurate bond length and angle geometries for all bases, nucleosides and nucleotides have been well established by X-ray crystallographic analyses. Structural surveys (Clowney et al., 1966; Gelbin et al., 1966) have calculated mean values for these parameters (which define their equilibrium values) from the most reliable structures at the time in the Cambridge, Protein Data Bank (PDB) and Nucleic Acid Databases. These have been incorporated in several implementations of the AMBER and CHARMM force fields widely used in molecular mechanics and dynamics modelling (see Chapter 1) and in computer packages for both crystallographic and NMR structural analyses (Parkinson et al., 1996). This widely cited analysis has been more recently updated with detailed geometric data on bases from small-molecule crystal structures, ultra-high-resolution oligonucleotide crystal structures and accurate quantum mechanics calculations (Gilski et al., 2019). Accurate crystallographic analyses, at very high resolution, can also directly yield quantitative information on the electron density distribution in a molecule, and hence on individual partial atomic charges. These charges and geometries for nucleosides are best obtained by high-quality ab initio quantum mechanical calculations (see, for example, Kruse et al., 2019; Kruse & Šponer, 2015). The interested reader is also referred to the detailed manual of the AMBER simulation

Principles of Nucleic Acid Structure
ISBN 978-0-12-819677-9, https://doi.org/10.1016/B978-0-12-819677-9.00004-4

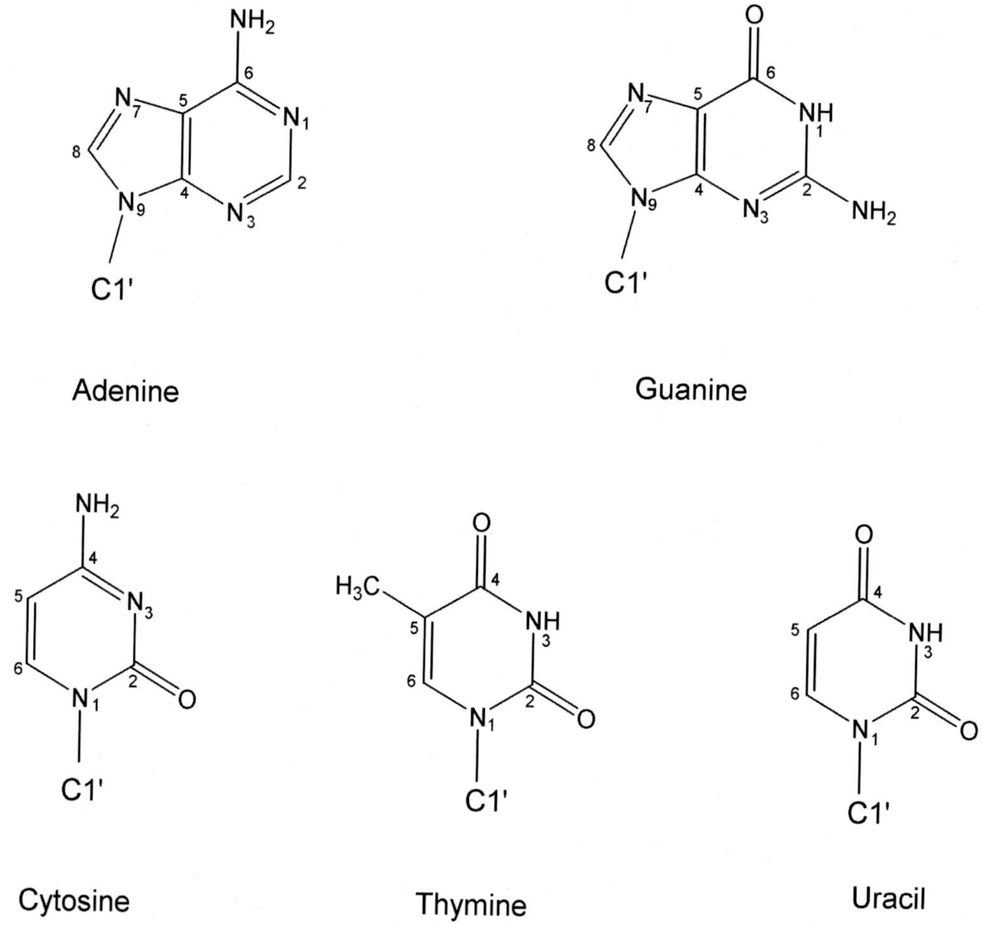

Figure 2.1 The five bases of DNA and RNA.

program (Chapter 1) for information on the various parameters including the atomic charges used in its DNA and RNA force fields (https://ambermd.org/doc12/Amber20.pdf).

Individual nucleoside units are linked together in a nucleic acid in a linear manner, through phosphate groups attached to the 3′ and 5′ positions of the sugars (Fig. 2.2). Hence, the full repeating unit in a nucleic acid is a 3′,5′-nucleotide.

Nucleic acid and oligonucleotide sequences use single-letter codes for the five unit nucleotides − A, T, G, C and U. The two classes of bases can be abbreviated as Y (pyrimidine) and R (purine). Phosphate groups are usually designated as p. A single oligonucleotide chain is conventionally numbered from the 5′ end, for example, ApGpCpTpTpG has the 5′ terminal adenosine nucleoside, with a free hydroxyl at its 5′ position, and thus the 3′ end guanosine has a free 3′ terminal hydroxyl group. The

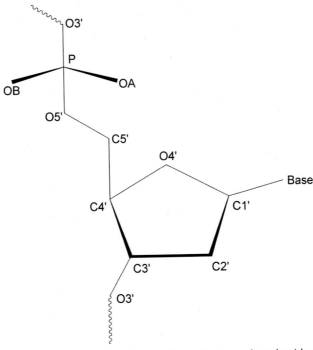

Figure 2.2 The organisation of repeating units in a polynucleotide chain.

letter p, to denote intervening phosphate groups, is frequently omitted when a sequence is written down. Chain direction is sometimes emphasised with 5′ and 3′ labels. Thus, an anti-parallel double-helical sequence can be written as

$^{5'}$CpGpCpGpApApTpTpCpGpCpG
$^{3'}$GpCpGpCpTpTpApApGpCpGpC

or simply as

CGCGAATTCGCG.

Structural publications on DNA usually prefix a sequence with 'd', as in d(CGAT), to emphasise that the oligonucleotide is a deoxyribose one rather than being an oligoribonucleotide. The prefix 'r' to denote ribonucleosides, ribonucleotides and their oligomers is often used.

The bond between sugar and base is known as the glycosidic bond. Its stereochemistry is important. In natural nucleic acids, the glycosidic bond is always β, that is, the base is above the plane of the sugar when viewed onto the plane and therefore on the same face of the plane as the 5′ hydroxyl substituent (Fig. 2.3a). The absolute stereochemistry of other substituent groups on the deoxyribose sugar ring of DNA is defined such that when viewed end-on with the sugar ring oxygen atom O4′ at the rear (Fig. 2.4a), the hydroxyl group at the 3′ position is below the ring and the hydroxymethyl group

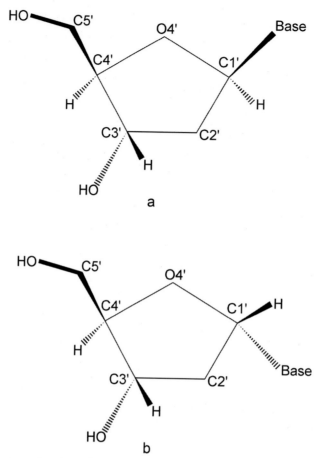

Figure 2.3 (a) The stereochemistry of a natural β-nucleoside. Solid bonds are coming out of the plane of the page, towards the reader. Dashed bonds are going away from the reader. (b) The stereochemistry of an α-nucleoside.

at the 4′ position is above it. A unit nucleotide can have its phosphate group attached either at the 3′ or 5′ ends and is thus termed either a 3′ or a 5′ nucleotide. It is chemically possible to construct α-nucleosides and from them α-oligonucleosides, which have their bases in the 'below' configuration relative to the sugar rings and their other substituents (Fig. 2.3b). These are much more resistant to nuclease attack than standard natural β-oligomers and have been used as antisense oligomers to mRNAs on account of their superior intracellular stability.

2.2 Base pairing

The realisation that the planar bases can associate in particular ways by means of hydrogen bonding was a crucial step in the elucidation of the structure of DNA. The important

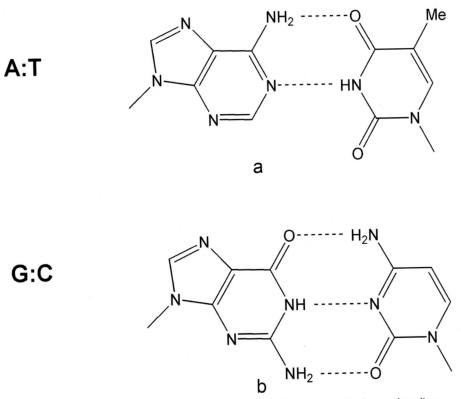

A:T

G:C

Figure 2.4 (a) A•T and (b) G•C base pairs, showing Watson—Crick hydrogen bonding.

early experimental data of Chargaff showed that the molar ratios of adenine:thymine (A•T) and cytosine:guanine in natural DNA were both unity. This led to the proposal by Crick and Watson that in each of these pairs the purine and pyrimidine bases are held together by specific hydrogen bonds, to form planar base pairs. In native double-helical DNA, the two bases in a base pair necessarily arise from two separate strands of DNA (with inter-molecular hydrogen bonds) and so hold the DNA double helix together (Watson & Crick, 1953).

The A•T base pair has two hydrogen bonds compared to the three in a guanine:cytosine (G•C) one (Fig. 2.4). (The initial Watson—Crick model assigned just two hydrogen bonds to the G•C pair and the third hydrogen bond emerged some 3 years later (Wain-Hobson, 2006)). Fundamental to the Watson—Crick arrangement is that the sugar groups are both attached to the bases on the same side of the base pair. As will be seen in Chapter 3, this defines the mutual positions of the two sugar—phosphate strands in DNA itself. The two base pairs are required to be almost identical in dimensions by the Watson—Crick model. High-resolution (0.8—0.9 Å) X-ray crystallographic analyses of the ribodi-nucleoside monophosphate duplexes r(GpC) and r(ApU) by A. Rich and colleagues in

Table 2.1 Hydrogen bond distances (in Å) in Watson–Crick base pairs (Rosenberg et al., 1976; Seeman et al., 1976), in the crystalline state. Estimated standard deviations are in parentheses.

U•A:	N3—H … N1	2.835(8)
	O4 … H—N6	2.940(8)
C•G:	O2 … H—N2	2.86(1)
	N3 … H—N1	2.95(1)
	N4—H … O6	2.91(1)

the early 1970's (Rosenberg et al., 1976; Seeman et al., 1976) established accurate geometries for these A•T and G•C base pairs (Table 2.1). The structure determinations showed that there are only small differences in size between the two types of base pairings, as indicated by the distance between glycosidic carbon atoms in a base pair. The $C1'$ … $C1'$ distance in the G•C base pair structure is 10.67Å and 10.48Å in the A•U-containing dinucleoside.

Theoretical studies on base pairing have enabled estimates to be made of the extra stability conferred on a G•C base pair by the extra hydrogen bond. Molecular dynamics simulations of both base pairs in an aqueous environment have given the free energies of A•T and G•C base pairs as −4.3 and −5.8 kcal/mol (Stofer et al., 1999). A more recent ab initio quantum mechanical study (Szatyłowicz & Sadlej-Sosnowska, 2010) has calculated the energies of the individual hydrogen bonds in these two base pairs, and has found, unsurprisingly, that non-bonded and interaction energies are sensitive to the details of the methodology used, although all such methods, which include geometry optimisation, are sensitive to even small differences in geometry. An alternative, reversed Watson–Crick A•T base pairs involving a 180° rotation of one base, has similar stability to the standard Watson–Crick A•T. An estimate of base pair energies, using ab initio quantum mechanical methods (Sponer et al., 2004), has shown broad agreement with values calculated from the AMBER force field in use at that time. This reinforces the reliability of the force-field approaches that are widely used and which emphasise van der Waals and electrostatic point-centred charge contributions, even though they do not take account of polarisation and charge transfer effects. The interaction enthalpies of hydrated A•T and G•C base pairs have been calculated as 14.0 kcal/mol and 27.0 kcal/mol, respectively, (i.e. 7–9 kcal/mol per hydrogen bond), using AMBER together with density functional theory calculations (Liu et al., 2006). Other A•T base pair arrangements have been predicted by theory (Gould & Kollman, 1994) to be more stable than Watson–Crick pairing. However, it has long been believed that these alternatives would require significant changes in backbone conformation in order to be accommodated within a B-DNA duplex. It is striking that these alternatives have not until recently been observed in normal duplex DNA, suggesting that the requirements for optimal base stacking and minimal backbone distortions greatly favour the standard Watson–Crick arrangement. Transient occurrence of Hoogsteen A•T base pairing in normal duplex DNA has been observed using NMR relaxation methods (Alvey et al., 2014) and the energy landscapes for the necessary base flipping have been simulated using molecular dynamics approaches (Chakraborty & Wales, 2018; Ray & Andricioaei, 2020). This is a consequence of thermal motions in DNA and has been estimated to have a 0.1% occurrence and a lifetime of ca 5 ms. It is striking though that Hoogsteen base pairing has been captured in several DNA complexes of human DNA polymerase (see, for example, Nair et al., 2004).

2.3 Base and base pair flexibility

The individual bases in a nucleic acid are flat, but base pairs (and consecutive bases on an individual strand) can show considerable flexibility. This flexibility is to some extent dependent on the nature of the bases and base pairs themselves but is more related to their base-stacking environments. Thus, descriptions of base morphology have become important in describing and understanding many sequence-dependent features and deformations of nucleic acids. The former features are often considered primarily at the dinucleoside local level, whereas longer-range effects, such as helix bending, can also be analysed at a more global level.

A number of rotational and translational parameters have been devised to describe these geometric relations between bases and base pairs, which were originally defined in 1989 (the 'Cambridge Accord'). These definitions, together with the Cambridge Accord sign conventions, are given below. Confusingly, two distinct types of approaches have been reported in the earlier literature to calculate these parameters (Lu et al., 1999) — the Cambridge Accord did not define a single unambiguous convention for their calculation. In one approach, the parameters are defined with respect to a global helical axis, which need not be linear. Another uses a set of local axes, one per dinucleotide step. Also, a variety of definitions of local and global axes have been used. The overall effect for most undistorted structures is fortunately that only a minority of parameters appear to have very different values depending on the method of calculation, using several of the widely available programs (see below). Before detailing these, we introduce the parameters themselves. The major and minor grooves are the indentations in nucleic acid double helices formed as a consequence of the asymmetry of base pairs. The C1'-N9 (purine) and C1'-N1 (pyrimidine) base-sugar bonds are by convention on the minor groove side, so that the C6/N7 (purine) and C4 (pyrimidine) base atoms and their substituents are on the major groove side. The grooves and their characteristics are described further in Chapter 3.

   (i) For individual base pairs.
      (a) **Propeller twist** (ω) between bases is the dihedral angle between normals to the bases, when viewed along the long axis of the base pair (Fig. 2.5). The angle has a negative sign under normal circumstances, with a clockwise

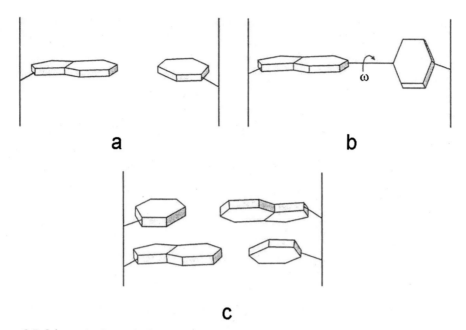

a   **b**

c

Figure 2.5 Schematic views of a base pair with (a) zero and (b) high propeller twist. Panel (c) shows the effects of propeller twist on two successive base pairs.

rotation of the nearer base when viewed down the long axis. The long axis for a purine—pyrimidine base pair is defined as the vector between the C8 atom of the purine and the C6 of a pyrimidine in a Watson—Crick base pair. Analogous definitions can be applied to other non-standard base pairings in a duplex including purine—purine and pyrimidine—pyrimidine ones.

(b) Buckle (κ) is the dihedral angle between bases, along their short axis, after propeller twist has been set to $0°$ (Fig. 2.6). The sign of buckle is defined as positive

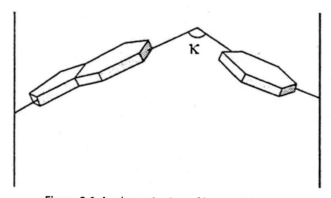

Figure 2.6 A schematic view of base-pair buckle.

if the distortion is convex in the direction $5' \rightarrow 3'$ of strand 1. The change in buckle for succeeding steps, termed **cup**, has been found to be a useful measure of changes along a sequence. Cup is defined as the difference between the buckle at a given step, and that of the preceding one.

(c) **Inclination** (η) is the angle between the long axis of a base pair and a plane perpendicular to the helix axis. This angle is defined as positive for right-handed rotation about a vector from the helix axis towards the major groove.

(d) **X and Y displacements** define translations of a base pair within its mean plane in terms of the distance of the mid-point of the base pair long axis from the helix axis. X displacement is towards the major groove direction, when it has a positive value. Y displacement is orthogonal to this and is positive if towards the first nucleic acid strand of the duplex.

(ii) For base pair steps.

(e) **Helical twist** (Ω) is the angle between successive base pairs, measured as the change in orientation of the $C1'-C1'$ vectors on going from one base pair to the next, projected down the helix axis (Fig. 2.7). For an exactly repetitious double helix, helical twist is $360°/n$, where n is the unit repeat defined above.

(f) **Roll** (ρ) is the dihedral angle for rotation of one base pair with respect to its neighbour, about the long axis of the base pair. A positive roll angle opens up a base pair step towards the minor groove (Fig. 2.8). **Tilt** (τ) is the corresponding dihedral angle along the short (i.e. x-axis) of the base pair.

(g) **Slide** is the relative displacement of one base pair compared to another, in the direction of nucleic acid strand one (i.e. the Y displacement), measured between the mid-points of each C6—C8 base pair long axis.

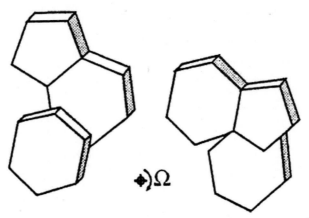

Figure 2.7 View down two successive base pairs, showing the helical twist angle between them.

Figure 2.8 Views of two successive base pairs, (a) with 0° roll angle between them and (b) with a positive roll angle. Panel (c) shows a view of positive roll along the long axis of the base pairs.

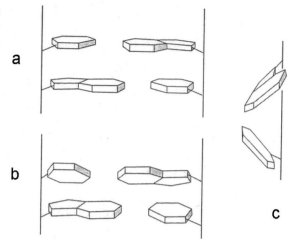

A standardised coordinate reference frame for the calculation of these parameters has been proposed (Lu & Olson, 1999) and has been endorsed by the successor to the Cambridge Accord, the 1999 Tsukuba Accord (Olson et al., 2001). Details of the reference frame are also available on the Nucleic Acid Database web site at http://ndbserver/ rutgers/edu/ndbmodule/standards/index.html.

This reference frame is unambiguous and has the advantage of being able to produce values for most local base-pair and base-step parameters which are almost independent of the algorithm used. A notable exception is rise, which is especially sensitive to the definition of origin and to small changes in buckle and roll. The right-handed reference frame used is shown in Fig. 2.9. It has the x-axis directed towards the major groove along the pseudo twofold axis of an idealised Watson–Crick base pair (shown as •). The y-axis is along the long axis of the base pair, parallel to the C1' ... C1' vector. The position of

Figure 2.9 Reference frame for idealised helical DNA, showing the axis origin at (•), along the pseudo twofold axis between two successive base pairs.

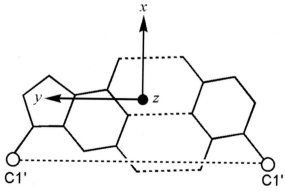

the origin is dependent on the geometry of the bases and the base pair, which have been taken from the published compilations (Clowney et al., 1996; Gelbin et al., 1996).

Early programs for calculating these parameters include:

- *FREEHELIX*, the successor to the original helical parameter program NEWHELIX, calculates an optimum linear helical axis, so that the derived parameters are basically global ones (Dickerson, 1998). Local bending can also be calculated (Goodsell & Dickerson, 1994).

- *CEHS* (Lu et al., 1997) also focusses on local parameters, and like FREEHELIX, uses the C6 … C8 base-pair vector as the *y*-axis for the reference frame. Roll and tilt are commuted into a single variable RollTilt (Γ).

The two currently most widely used computer programs for base and base-pair morphology calculations, which use the parameter definitions adopted by the Tsukuba Accord, are as follows:

- *CURVES+*, the successor to *CURVES* (Blanchet et al., 2011; Lavery & Sklenar, 1988, 1989; Lavery et al., 2014), calculates an optimised global helix axis which can be (and invariably is), curved. This is especially useful for irregular structures. The reference frame is defined at the base rather than the base-pair level. Local parameters can also be obtained with this program, and irregular non-standard base-pairing arrangements and non-duplex structures can be accommodated. Groove dimensions are also calculated with CURVES+. CURVES+ is available for downloading at https://bisi.ibcp.fr/tools/curves_plus/and a web-based service is at http://curvesplus.bsc.es/analyse.

- *3DNA* (Li et al., 2019), the successor to RNA (Babcock et al., 1994), uses the reference frame convention specified in the Tsukuba agreement and calculates local helical parameters. It can also be used for non-standard bases and non-duplex structures and can calculate groove widths. *3DNA* is available for downloading at https://x3dna.org/and a web-based service is at http://web.x3dna.org/, which also incorporates a visualiser program and a useful generator of exact nucleic acid helices derived from fibre diffraction studies (see Chapter 3).

Readers of crystallographic or NMR structure analysis literature should be aware of the axis definitions that have been used in parameter calculations. Ambiguities can be clarified by use of the NDB, which can produce tables of base helical parameters calculated with *3DNA* or either of the two recent programs outlined above.

2.4 Sugar puckers

The five-membered deoxyribose sugar ring in DNA is inherently non-planar. This non-planarity is termed puckering. The precise conformation of a deoxyribose ring can be completely specified by the five endocyclic torsion angles within it (Fig. 2.10). The ring puckering arises from the effect of non-bonded interactions between

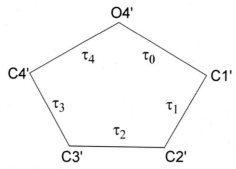

Figure 2.10 The five internal torsion angles in a ribose ring.

substituents at the four-ring carbon atoms — the energetically most stable conformation for the ring has all substituents as far apart as possible. Thus, different substituent atoms would be expected to produce differing types of puckering. The puckering can be described by either

(i) a simple qualitative description of the conformation in terms of atoms deviating from ring co-planarity or

(ii) precise descriptions in terms of the ring internal torsion angles.

In principle, there is a continuum of inter-convertible puckers, separated by energy barriers. These various puckers are produced by systematic changes in the ring torsion angles. The puckers can be succinctly defined by the parameters P and τ_m (Altona & Sundaralingam, 1972). The value of P, the phase angle of pseudorotation, indicates the type of pucker since P is defined in terms of the five torsion angles $\tau_0-\tau_4$

$$\tan P = \frac{(\tau_4 + \tau_1) - (\tau_3 + \tau_0)}{2*\tau_2*(\sin 36° + \sin 72°)}$$

and the maximum degree of pucker, τ_m, by

$$\tau_m = \tau_2/\cos P$$

The pseudorotation phase angle can take any value between $0°$ and $360°$. If τ_2 has a negative value, then $180°$ is added to the value of P. The pseudorotation phase angle is commonly represented by the pseudorotation wheel, which indicates the continuum of ring puckers (Fig. 2.11). Values of τ_m indicate the degree of puckering of the ring; typical experimental values from crystallographic studies on mononucleosides are in the range $25-45°$. The five internal torsion angles are not independent of each other, and so to a good approximation, any one angle τ_j can be represented in terms of just two variables:

$$\tau_j = \tau_m \cos[P + 0.8\pi(j - 2)]$$

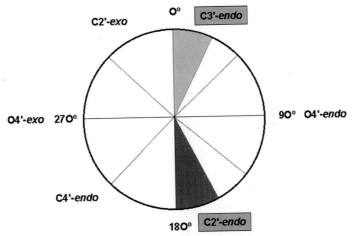

Figure 2.11 The pseudorotation wheel for a deoxyribose sugar. The shaded areas indicate the preferred ranges of the pseudorotation angle for the two principal sugar conformations.

An on-line interactive facility (PROSIT) is available to calculate pseudorotation parameters (Sun et al., 2004), at https://www.cactus.nci.nih.gov/prosit/.

Many distinct deoxyribose ring pucker geometries have been observed experimentally, by X-ray crystallography and NMR techniques. When one ring atom is out of the plane of the other four, the pucker type is an envelope one. More commonly, two atoms deviate from the plane of the other three, with these two either side of the plane. It is usual for one of the two atoms to have a larger deviation from the plane than the other, resulting in a twist conformation. The direction of atomic displacement from the plane is important. If the major deviation is on the same side as the base and C4′–C5′ bond, then the atom involved is termed *endo*. If it is on the opposite side, it is called *exo*. The most commonly observed puckers in crystal structures of isolated nucleosides and nucleotides are either close to C2′-*endo* or C3′-*endo* types (Fig. 2.12a and b). In practice, these pure envelope forms are rarely observed, largely because of the differing substituents on the ring. Consequently, the puckers are then best described in terms of twist conformations. When the major out-of-plane deviation is on the *endo* side, there is a minor deviation on the opposite, *exo* side. The convention used for describing a twist deoxyribose conformation is that the major out-of-plane deviation is followed by the minor one, for example, C2′-*endo*, C3′-*exo*. The C2′-*endo* family of puckers have P values in the range 140–185°; in view of their position on the pseudorotation wheel, they are sometimes termed S (south) conformations. The C3′-*endo* domain has P values in the range −10 to +40°, and its conformation is termed N (north). These more geographical terms are mostly used by NMR spectroscopists.

The pseudorotation wheel implies that deoxyribose puckers are free to interconvert. In practice, there are energy barriers between major forms. The exact size of these barriers

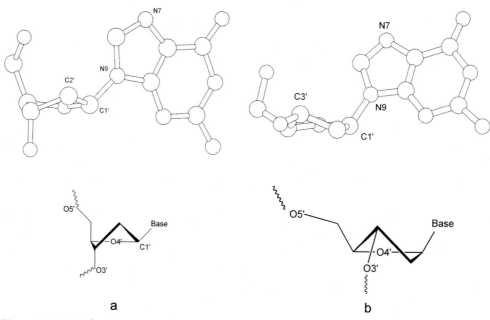

Figure 2.12 (a) C2′-*endo* sugar puckering for a guanosine nucleoside, viewed along the plane of the sugar ring. (b) C3′-*endo* sugar puckering for a guanosine nucleoside, with the sugar viewed in the same direction as in panel (a).

has been the subject of considerable study (see, for example, Harvey & Prabhakaran, 1986; Jana & MacKerell, 2015; Nester & Plazinski, 2020; Olson, 1981; Olson & Sussman, 1982). The consensus is that the barrier height is dependent on the route around the pseudorotation wheel. For inter-conversion of C2′-*endo* to C3′-*endo*, the preferred pathway is via the O4′-*endo* state, with a barrier of 2–5 kcal/mol found from an analysis of a large body of experimental data (Olson & Sussman, 1982) and a somewhat smaller (potential energy) value of 1.5 kcal/mol, from a molecular dynamics study (Harvey & Prabhakaran, 1986). The former value, being an experimental one, represents the total free energy for inter-conversion. These values also broadly concur with those found by application of various ab initio quantum mechanical methods, which have given values of 2.5–4.0 kcal/mol (Foloppe et al., 2001). A detailed molecular dynamics study with a wide range of substituents on the furanose ring (Nester & Plazinski, 2020) has revealed the sensitivity of the (small) energy differences between the barrier heights to the position and nature of substituents on the furanose ring. Barrier heights are generally in accord with results from an earlier molecular dynamics study (Harvey & Prabhakaran, 1986). A more recent molecular dynamics analysis (Li & Szostak, 2014) suggests that the free energy barrier between the C2′-*endo* and C3′-*endo* pseudorotation is 3 kcal/mol, in accord with the mid-point of the earlier estimate based on experimental data (Olson & Sussman, 1982).

Relative populations of puckers can be monitored directly by NMR measurements of the ratio of coupling constants between H1′-H2′ and H3′-H4′ protons. These show that in contrast to the 'frozen-out' puckers found in the solid-state structures of nucleosides and nucleotides, there is rapid inter-conversion in solution. Nonetheless, the relative populations of the major puckers are dependent on the type of base attached. Purines show a preference for the C2′-endo pucker conformational type, whereas pyrimidines favour C3′-endo. Deoxyribose nucleosides are primarily (>60%) in the C2′-endo form and ribonucleosides favour C3′-endo. The latter are significantly more restricted in their mobility; this has significance for the structures of oligoribonucleotides (see Chapter 6). These differences in puckering equilibria and hence in their relative populations in solution and in molecular dynamics simulations are reflected in the patterns of puckers found in surveys of crystal structures (Murray-Rust & Motherwell, 1978). Again, this is a demonstration of the complementarity of information provided by the different structural techniques. Sugar pucker preferences have their origin in the non-bonded interactions between substituents on the sugar ring and to some extent on their electronic characteristics. For example, the C3′-endo pucker (Fig. 2.7b) would have hydroxyl substituents at the 2′ and 3′ positions further apart than with C2′-endo pucker; hence, the preference of the former by ribonucleosides. A recent survey of sugar geometry from available high-resolution crystal structures (Kowiel et al., 2020) has found some variability in furanose ring bond lengths and angles, depending on pucker and the nature of the attached nucleobase with spreads of ca 0.015 Å and 1.7° from mean values. These findings will undoubtedly be used in due course for updates to crystallographic and NMR restraint libraries as well as force field parameterisations.

Numerous crystallographic and NMR studies have found correlations between sugar pucker and several backbone conformational variables, both in isolated nucleosides/nucleotides and in oligonucleotide structures. These are discussed later in this chapter. Changes in sugar pucker are important determinants of oligo- and polynucleotide structure because they can alter the orientation of C1′, C3′ and C4′ substituents, resulting in major changes in backbone conformation and overall structure, as indeed is found (Chapter 3).

2.5 Conformations about the glycosidic bond

The glycosidic bond links a deoxyribose sugar and a base, being the C1′-N9 bond for purines and the C1′-N1 bond for pyrimidines. The torsion angle χ around this single bond can in principle adopt a wide range of values, although as will be seen, structural constraints result in marked preferences being observed. Glycosidic torsion angles are defined in terms of the four atoms:

    O4′-C1′-N9-C4 for purines

    O4′-C1′-N1-C2 for pyrimidines

Theory has predicted two principal low-energy domains for the glycosidic angle, in accord with experimental findings for many nucleosides and nucleotides. The *anti* conformation has the N1, C2 face of purines and the C2, N3 face of pyrimidines directed away from the sugar ring (Fig. 2.13a) so that the hydrogen atoms attached to C8 of purines and C6 of pyrimidines are lying over the sugar ring. Thus, the Watson–Crick hydrogen-bonding groups of the bases are directed away from the sugar ring. These orientations are reversed for the *syn* conformation, with these hydrogen-bonding groups now oriented towards the sugar and especially its O5′ atom (Fig. 2.13b). A number of

Figure 2.13 (a) A guanosine nucleoside with the glycosidic angle χ set in an *anti* conformation. (b) A guanosine nucleoside in the *syn* conformation.

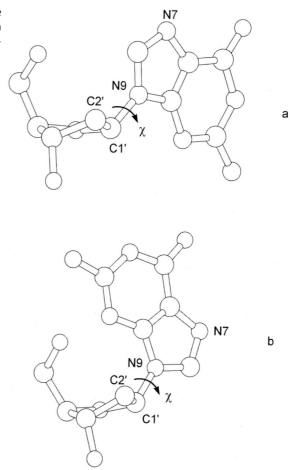

crystal structures of *syn* purine nucleosides have found hydrogen bonding between the O5$'$ atom and the N3 base atom, which would stabilise this conformation. Otherwise, for purines, the *syn* conformation is slightly less preferred than the *anti*, since there are fewer non-bonded steric clashes in the latter case. The principal exceptions to this rule are guanosine-containing nucleotides, which have a small preference for the *syn* form because of favourable electrostatic interactions between the exocyclic N2 amino group of the guanine base and the 5$'$ phosphate atom. For pyrimidine nucleotides, the *anti* conformation is preferred over the *syn*, because of unfavourable contacts between the O2 oxygen atom of the base and the 5$'$-phosphate group. The results of quantum mechanics energy calculations on all four DNA (and RNA) nucleotides in both *syn* and *anti* forms are qualitatively in accord with these observations (Foloppe et al., 2002), with the syn conformers of each nucleotide being of consistently higher energy than anti ones. A dependence on the nature of the nucleobase and on the γ backbone angle was also noted. A more recent quantum mechanical study (Zgarbová et al., 2011) using several different high-level methodologies, whilst agreeing with the *syn/anti* order, found that the glycosidic angle χ profile is sensitive to various factors, notably the level of theory used, and importantly, the solvation around the nucleotide.

The sterically preferred ranges for the two domains of glycosidic angles are as follows:
Anti: $-120 > \chi > 180°$
Syn: $0 < \chi < 90°$

Values of χ in the region of about $-90°$ are often described as 'high *anti*'. There are pronounced correlations between sugar pucker and glycosidic angle, which reflect the changes in non-bonded clashes produced by C2$'$-*endo* versus C3$'$-*endo* puckers. Thus, *syn* glycosidic angles are not found with C3$'$-*endo* puckers due to steric clashes between the base and the H3$'$ atom, which points towards the base in this pucker mode.

2.6 The backbone torsion angles and correlated flexibility

The phosphodiester backbone of an oligonucleotide has six variable torsion angles (Fig. 2.14), designated α .... ζ, in addition to the five internal sugar torsions τ_0 .... τ_4 and the glycosidic angle χ. As will be seen, a number of these have highly correlated values (and therefore correlated motions in a solution environment). Steric considerations alone dictate that the backbone angles are restricted to discrete ranges (Olson, 1982a,b; Sundaralingam, 1969) (Fig. 2.15) and are accordingly not free to adopt any value between 0 and 360°. Fig. 2.15 uses a conformational wheel to show these preferred values, which are directly readable from their positions around the wheel. The fact that angles α, β, γ and ζ each have three allowed ranges, together with the broad range for angle ε that includes two staggered regions, leads to a large number of possible low-energy conformations for the unit nucleotide, especially when glycosidic angle and sugar pucker flexibility is taken into account. In reality, only a small number of DNA oligonucleotide and

Figure 2.14 The backbone torsion angles in a unit nucleotide. Each rotatable bond is indicated by a curved *arrow*.

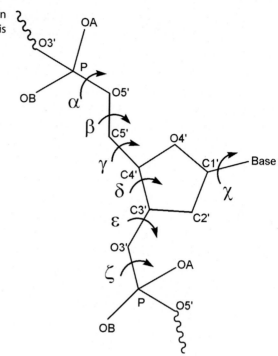

Figure 2.15 Conformational wheel showing the allowed ranges of backbone torsion angles (shaded) in nucleosides, nucleotides, deoxy-oligonucleotides and deoxypoly-nucleotides.

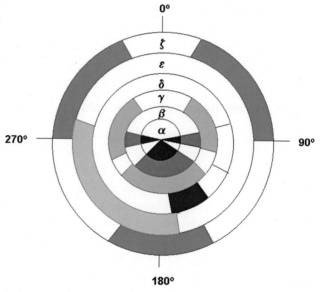

polynucleotide structural classes have actually been observed out of this large range of possibilities; this is doubtless in large part due to the restraints imposed by Watson–Crick base pairing on the backbone conformations when two DNA strands are intertwined together. In contrast, crystallographic and NMR studies on many standard and modified mononucleosides and nucleotides have shown their considerably greater conformational diversity, in accord with the possibilities indicated in Fig. 2.15. For these, backbone conformations in the solid state and in solution are not always in agreement; the requirements for efficient packing in the crystal can often overcome the modest energy barriers between different values for a given torsion angle. Many large RNA molecules are characterised by a wide range of base–base interactions, which can therefore adopt a wide variety of backbone conformations and conformational classes (Schneider et al., 2004) (see Chapter 6 for a detailed discussion of RNA conformational variability).

A common convention for describing these backbone torsion angles is to term values of ~60° as *gauche* + (g$^+$), −60° as *gauche* $^-$ (g$^-$) and ~180° as *trans* (t). Thus, for example, angles α (about the P–O5′ bond) and γ (the exocyclic angle about the C4′–C5′ bond) can be in the g$^+$, g$^-$ or t conformations. The two torsion angles around the phosphate group itself, α and ζ, have been found to show a high degree of flexibility in various dinucleoside crystal structures, with the tg$^-$, g$^-$g$^-$ and g$^+$g$^+$ conformations all having been observed (Kim et al., 1973). As will be described in Chapter 3, only the g$^-$g$^-$ conformation can place successive nucleotide units in arrangements that have their bases in potential hydrogen-bonding positions with respect to a second nucleotide strand. Thus, this is the phosphate conformation for DNA and RNA double helices. The torsion angle β, about the O5′–C5′ bond, is almost always found to be *trans*. All three possibilities for the γ angle have been observed in nucleoside crystal structures, although the g$^+$ conformation predominates in right-handed oligo- and polynucleotide double helices. The *trans* conformation for γ places the 5′-phosphate group in quite a distinct position with respect to the deoxyribose ring. The torsion angle δ around the C4′–C3′ bond adopts values that relate to the pucker of the sugar ring, since the internal ring torsion angle τ_3 (also around this bond) has a value of about 35° for C2′-*endo* and about 40° for C3′-*endo* puckers. δ is about 75° for C3′-*endo* and about 150° for C2′-*endo* puckers.

An alternative nomenclature for torsion angle ranges used by some workers in the nucleic acid field is that common in organic chemistry (the Klyne–Prelog system). In this, the *syn* (s) designation is given to angles clustered around 0° and *anti* (a) for those around 180°. Intermediate angles are defined as ± *synclinal* (±sc) for around ± 60° and ± *anticlinal* (±ac) for around ± 120°.

There are many well-established correlations involving pairs of these backbone torsion angles, as well as sugar pucker and glycosidic angle (Svozil et al., 2008). These have been found (Altona & Sundaralingam, 1972) in mononucleosides and nucleotides (which are inherently more flexible in solution as well as being more subject to packing forces in the crystal) and more recently, in oligonucleotides (Packer & Hunter, 1998;

Schneider et al., 1997; Svozil et al., 2008; Varnai et al., 2002). A comprehensive analysis of the available structural data has found 44 distinct DNA step conformational classes (Schneider et al., 2018) — see Chapter 3 for a further discussion of this topic. The existence of such correlations is important: it means that the atomic motions in oligo- and polynucleotides follow concerted patterns of inter-dependence. In general, these correlations are due to the diminution of non-bonded contacts that occur with some conformations. Some of the more important correlations that have been observed in mononucleosides and nucleotides are as follows:

- Between sugar pucker and glycosidic angle χ, especially for pyrimidine nucleosides. C3′-endo pucker is usually associated with median-value anti glycosidic angles, whereas C2′-endo puckers are commonly found with high anti χ angles. Syn glycosidic angle conformations show a marked preference for C2′-endo sugar puckers.
- The C4′–C5′ torsion angle γ is correlated with the glycosidic angle and to some extent with sugar pucker and backbone angle α (Schneider et al., 1997; Varnai et al., 2002). Anti glycosidic angles tend to correlate with g^+ conformations for γ.

Backbone torsion angles for all structures in the PDB are available from the NDB and from the CURVES + or 3DNA programs. The DNATCO web site https://dnatco. datmos.org/ also provides these data for any structure and in addition provides up-to-date information on the (currently) 96 DNA and RNA conformational classes that it has identified from analyses of almost 120,000 dinucleotide steps in over 2000 structures in the PDB (Černý, Božíková, Malý, et al., 2020, Černý, Božíková, Svoboda, et al. 2020), readily enabling assignments to be made by the user for any given structure.

References

(1989). Definitions and nomenclature of nucleic acid structure parameters. *The EMBO Journal, 8*, 1—4.

Altona, C., & Sundaralingam, M. (1972). Conformational analysis of the sugar ring in nucleosides and nucleotides. A new description using the concept of pseudorotation. *Journal of the American Chemical Society, 94*, 8205—8212.

Alvey, H. S., Gottardo, F. L., Nikolova, E. N., & Al-Hashimi, H. M. (2014). Widespread transient Hoogsteen base pairs in canonical duplex DNA with variable energetics. *Nature Communications, 5*, 4786.

Babcock, M. S., Pednault, E. P. D., & Olson, W. K. (1994). Nucleic acid structure analysis. Mathematics for local Cartesian and helical structure parameters that are truly comparable between structures. *Journal of Molecular Biology, 237*, 125—156.

Blanchet, C., Pasi, M., Zakrzewska, K., & Lavery, R. (2011). CURVES+ web server for analyzing and visualizing the helical, backbone and groove parameters of nucleic acid structures. *Nucleic Acids Research, 39*, W68—W73.

Černý, J., Božíková, P., Malý, M., Tykač, M., Biedermannová, L., & Schneider, B. (2020). Structural alphabets for conformational analysis of nucleic acids available at dnatco.datmos.org. *Acta Crystallographica, D76*, 805—813.

Černý, J., Božíková, P., Svoboda, J., & Schneider, B. (2020). A unified dinucleotide alphabet describing both RNA and DNA structures. *Nucleic Acids Research, 48*, 6367—6381.

Chakraborty, D., & Wales, D. J. (2018). Energy landscape and pathways for transitions between Watson-Crick and Hoogsteen base pairing in DNA. *Journal of Physical Chemistry Letters, 9*, 229—241.

Chargaff, E., & Davidson, J. N. (1955). *The nucleic acids. Chemistry and biology* (Vol. 1). New York: Academic Press.

Clowney, L., Jain, S. C., Srinivasan, A. R., Westbrook, J., Olson, W. K., & Berman, H. M. (1996). Geometric parameters in nucleic acids: Nitrogenous bases. *Journal of the American Chemical Society,* 118, 509–518.

Dickerson, R. E. (1998). DNA bending: The prevalence of kinkiness and the virtues of normality. *Nucleic Acids Research,* 26, 1906–1926.

Foloppe, N., Hartmann, B., Nilsson, L., & MacKerell, A. D., Jr. (2002). Intrinsic conformational energetics associated with the glycosyl torsion in DNA: A quantum mechanical study. *Biophysical Journal,* 82, 1554–1569.

Foloppe, N., Nilsson, L., & MacKerell, A. B., Jr. (2001). Ab initio conformational analysis of nucleic acid components: Intrinsic energetic contributions to nucleic acid structure and dynamics. *Biopolymers,* 61, 61–76.

Gelbin, A., Schneider, B., Clowney, L., Hsieh, S.-H., Olson, W. K., & Berman, H. M. (1996). Geometric parameters in nucleic acids: Sugar and phosphate constituents. *Journal of the American Chemical Society,* 118, 519–529.

Gilski, M., Zhao, J., Kowiel, M., Brzezinski, D., Turner, D. H., & Jaskolski, M. (2019). Accurate geometrical restraints for Watson-Crick base pairs. *Acta Crystallographica, B75,* 235–245.

Goodsell, D. S., & Dickerson, R. E. (1994). Bending and curvature calculations in B-DNA. *Nucleic Acids Research,* 22, 5497–5503.

Gould, I. R., & Kollman, P. A. (1994). Theoretical investigation of the hydrogen bond strengths in guanine-cytosine and adenine-thymine base pairs. *Journal of the American Chemical Society,* 116, 2493–2499.

Harvey, S. C., & Prabhakaran, M. (1986). Ribose puckering: Structure, dynamics, energetics, and the pseudorotation cycle. *Journal of the American Chemical Society,* 108, 6128–6136.

Jana, M., & MacKerell, A. D. (2015). CHARMM drude polarizable force field for aldopentofuranoses and methyl-aldopentofuranosides. *Journal of Physical Chemistry B,* 119, 7846–7859.

Kim, S.-H., Berman, H. M., Seeman, N. C., & Newton, M. D. (1973). Seven basic conformations of nucleic acid structural units. *Acta Crystallographica, B29,* 703–710.

Kowiel, M., Brzezinski, D., Gilski, M., & Jaskolski, M. (2020). Conformation-dependent restraints for polynucleotides: The sugar moiety. *Nucleic Acids Research,* 48, 962–973.

Kruse, H., Banáš, P., & Šponer, J. (2019). Investigations of stacked DNA base-pair steps: Highly accurate stacking interaction energies, energy decomposition, and many-body stacking effects. *Journal of Chemical Theory and Computation,* 15, 95–115.

Kruse, H., & Šponer, J. (2015). Towards biochemically relevant QM computations on nucleic acids: Controlled electronic structure geometry optimization of nucleic acid structural motifs using penalty restraint functions. *Physical Chemistry Chemical Physics,* 17, 1399–1410.

Lavery, R., Moakher, M., Maddocks, J. H., Petkeviciute, D., & Zakrzewska, K. (2014). Conformational analysis of nucleic acids revisited: Curves+. *Nucleic Acids Research,* 37, 5917–5929.

Lavery, R., & Sklenar, H. (1988). The definition of generalized helicoidal parameters and of axis curvature for irregular nucleic acids. *Journal of Biomolecular Structure & Dynamics,* 6, 63–91.

Lavery, R., & Sklenar, H. (1989). Defining the structure of irregular nucleic acids: Conventions and principles. *Journal of Biomolecular Structure & Dynamics,* 6, 655–667.

Li, S., Olson, W. K., & Lu, X. J. (2019). Web 3DNA 2.0 for the analysis, visualization, and modeling of 3D nucleic acid structures. *Nucleic Acids Research,* 47, W26–W34.

Li, L., & Szostak, J. W. (2014). The free energy landscape of pseudorotation in 3'-5' and 2'-5' linked nucleic acids. *Journal of the American Chemical Society,* 136, 2858–2865.

Liu, D., Wyttenbach, T., & Bowers, M. T. (2006). Hydration of mononucleotides. *Journal of the American Chemical Society,* 128, 15155–15163.

Lu, X.-J., & Olson, W. K. (1999). Resolving the discrepancies among nucleic acid conformational analyses. *Journal of Molecular Biology,* 285, 1563–1575.

Lu, X.-J., Babcock, M. S., & Olson, W. K. (1999). Overview of nucleic acid analysis programs. *Journal of Biomolecular Structure & Dynamics,* 16, 833–843.

Lu, X.-J., El Hassan, M. A., & Hunter, C. A. (1997). Structure and conformation of helical nucleic acids: Analysis program (SCHNAaP). *Journal of Molecular Biology, 273*, 668–680.

Murray-Rust, P., & Motherwell, S. (1978). Computer retrieval and analysis of molecular geometry. III. Geometry of the β-1′-aminofuranoside fragment. *Acta Crystallographica, B34*, 2534–2546.

Nair, D. T., Johnson, R. E., Prakash, S., Prakash, L., & Aggarwal, A. K. (2004). Replication by human DNA polymerase-iota occurs by Hoogsteen base-pairing. *Nature, 430*, 377–380.

Nester, K., & Plazinski, W. (2020). Deciphering the conformational preferences of furanosides. A molecular dynamics study. *Journal of Biomolecular Structure and Dynamics, 38*, 3359–3370.

Olson, W. A. (1981). Three-state models of furanose pseudorotation. *Nucleic Acids Research, 9*, 1251–1262.

Olson, W. K. (1982a). How flexible is the furanose ring? 2. An updated potential energy estimate. *Journal of the American Chemical Society, 104*, 278–286.

Olson, W. K. (1982b). In S. Neidle (Ed.), *Topics in nucleic acid structure, Part 2* (pp. 1–79). London: Macmillan Press.

Olson, W. K., Bansal, M., Burley, S. K., Dickerson, R. E., Gerstein, M., Harvey, S. C., Heinemann, U., Lu, X.-J., Neidle, S., Shakked, Z., Sklenar, H., Suzuki, M., Tung, C.-S., Westhof, E., Wolberger, C., & Berman, H. M. (2001). A standard reference frame for the description of nucleic acid base-pair geometry. *Journal of Molecular Biology, 313*, 229–237.

Olson, W. K., & Sussman, J. L. (1982). How flexible is the furanose ring? 1. A comparison of experimental and theoretical studies. *Journal of the American Chemical Society, 104*, 270–278.

Packer, M. J., & Hunter, C. A. (1998). Sequence-dependent DNA structure: The role of the sugar-phosphate backbone. *Journal of Molecular Biology, 280*, 407–420.

Parkinson, G., Vojtechovsky, J., Clowney, L., Brünger, A. T., & Berman, H. M. (1996). New parameters for the refinement of nucleic acid-containing structures. *Acta Crystallographica, D52*, 57–64.

Ray, D., & Andricioaei, I. (2020). Free energy landscape and conformational kinetics of Hoogsteen base pairing in DNA vs. RNA. *Biophysical Journal, 119*, 1568–1579.

Rosenberg, J. M., Seeman, N. C., Day, R. O., & Rich, A. (1976). RNA double-helical fragments at atomic resolution. II. The crystal structure of sodium guanylyl-3′,5′-cytidine nonahydrate. *Journal of Molecular Biology, 104*, 145–167.

Szatyłowicz, H., & Sadlej-Sosnowska, N. (2010). Characterizing the strength of individual hydrogen bonds in DNA base pairs. *Journal of Chemical Information and Modeling, 50*, 2151–2161.

Schneider, B., Božíková, P., Nečasová, I., Čech, P., Svozil, D., & Černý, J. (2018). A DNA structural alphabet provides new insight into DNA flexibility. *Acta Crystallographica, D74*, 52–64.

Schneider, B., Moravek, Z., & Berman, H. M. (2004). RNA conformational classes. *Nucleic Acids Research, 32*, 1666–1677.

Schneider, B., Neidle, S., & Berman, H. M. (1997). Conformations of the sugar-phosphate backbone in helical DNA crystal structures. *Biopolymers, 42*, 113–124.

Seeman, N. C., Rosenberg, J. M., Suddath, F. L., Kim, J. P., & Rich, A. (1976). RNA double-helical fragments at atomic resolution. I. The crystal and molecular structure of sodium adenylyl-3′,5′-uridine hexahydrate. *Journal of Molecular Biology, 104*, 109–144.

Sponer, J., Jurecka, P., & Hobza, P. (2004). Accurate interaction energies of hydrogen-bonded nucleic acid base pairs. *Journal of the American Chemical Society, 126*, 10142–10151.

Stofer, E., Chipot, C., & Lavery, R. (1999). Free energy calculations of watson-crick base pairing in aqueous solution. *Journal of the American Chemical Society, 121*, 9503–9508.

Sun, G., Voigt, J. H., Filippov, I. V., Marquez, V. E., & Nicklaus, M. C. (2004). PROSIT: Pseudo-rotational online service and interactive tool, applied to a conformational survey of nucleosides and nucleotides. *Journal of Chemical Information and Computer Sciences, 44*, 1752–1762.

Sundaralingam, M. (1969). Stereochemistry of nucleic acids and their constituents. IV. Allowed and preferred conformations of nucleosides, nucleoside mono-, di-, tri-, tetraphosphates, nucleic acids and polynucleotides. *Biopolymers, 7*, 821–860.

Svozil, D., Sponer, J. E., Marchan, I., Pérez, A., Cheatham, T. E., 3rd, Forti, F., Luque, F. J., Orozco, M., & Sponer, J. (2008). Geometrical and electronic structure variability of the sugar-phosphate backbone in nucleic acids. *Journal of Physical Chemistry B, 112*, 8188–8197.

Varnai, P., Djuranovic, D., Lavery, R., & Hartmann, B. (2002). Alpha/gamma transitions in the B-DNA backbone. *Nucleic Acids Research, 30*, 5398–5406.

Wain-Hobson, S. (2006). The third bond. *Nature, 439*, 539.

Watson, J. D., & Crick, F. H. C. (1953). Molecular structure of nucleic acids; a structure for deoxyribose nucleic acid. *Nature, 171*, 737–738.

Zgarbová, M., Otyepka, M., Sponer, J., Mládek, A., Banáš, P., Cheatham, T. E., 3rd, & Jurečka, P. (2011). Refinement of the Cornell et al. nucleic acids force field based on reference quantum chemical calculations of glycosidic torsion profiles. *Journal of Chemical Theory and Computation, 7*, 2886–2902.

Further reading

Jeffrey, G. A., & Saenger, W. (1991). *Hydrogen bonding in biological structures. Chapters 15, 16.* Berlin: Springer-Verlag.

Šponer, J., Leszczynski, J., & Hobza, P. (2001). Electronic properties, hydrogen bonding, stacking, and cation binding of DNA and RNA bases. *Biopolymers, 61*, 3–31.

Šponer, J., Mládek, A., Šponer, J. E., Svozil, D., Zgarbová, M., Banáš, P., Jurečka, P., & Otyepka, M. (2012). The DNA and RNA sugar-phosphate backbone emerges as the key player. An overview of quantum-chemical, structural biology and simulation studies. *Physical Chemistry Chemical Physics, 14*, 15257–15277.

Voet, D., & Rich, A. (1970). The crystal structures of purines, pyrimidines and their intermolecular complexes. *Progress in Nucleic Acid Research & Molecular Biology, 10*, 183–265.

CHAPTER 3

DNA structure as observed in fibres and crystals

3.1 Structural fundamentals

3.1.1 Helical parameters

The fibre diffraction method for determining the structures of natural and synthetic polymeric DNA (and RNA) molecules has been outlined in Chapter 1. Examination of X-ray diffraction photographs of oriented fibres obtained from them enables the basic parameters defining the dimensions of the repeating helix, to be calculated. Measurement of the spacing between the layer lines directly gives the reciprocal of the helical pitch P, defined as the distance, parallel to the helix axis, between successive nucleotide units per complete turn of the helix (Fig. 3.1). The presence of a meridional reflection for a fibre that is oriented perpendicular to the direction of the X-ray beam indicates the presence of a structural periodicity in that direction. Analysis of the meridional reflection gives the regular repeat distance d (the helical rise if the nucleotide units are perpendicular to the helix axis). The unit repeat n, the number of nucleotide units in a single full turn of helix, that is per complete pitch P, is thus P/d. Helical rise is the distance between successive nucleotide units, projected to be parallel to the helix axis if the units are not strictly perpendicular to this axis. The diameter of the helix can be derived from the dimensions of the closely packed unit cell projection down the fibre axis.

3.1.2 Base-pair morphological features

Base pairs are not necessarily exactly planar, and indeed are rarely so. This is in part because the geometric requirements of hydrogen bonding are not especially stringent, and the angle subtended at the hydrogen atom (donor .... H — acceptor) can deviate by up to about 35° from linearity without appreciable loss of hydrogen bond energy. Other factors such as the need to avoid steric clash in some base pair ... base pair non-bonded interactions within a helix can result in distortions of base pairs from planarity, since these do not distort hydrogen bonds beyond these energetically favourable limits. These deviations from planarity can take place in several ways that can be categorised into two groups of local base-pair morphological features (Dickerson et al., 1989; Olson et al., 2001), as described in Chapter 2. The first category involves individual base pairs. It specifies both the relationship of one base relative to another within a base pair (for example, propeller twist and buckle) and the relationship of the base pair to global axes in the helix. The second category is concerned with the relative relationship of base pairs in successive base pairs

Principles of Nucleic Acid Structure
ISBN 978-0-12-819677-9, https://doi.org/10.1016/B978-0-12-819677-9.00007-X

53

Figure 3.1 Schematic DNA double helix, with the backbones shown in ribbon representation.

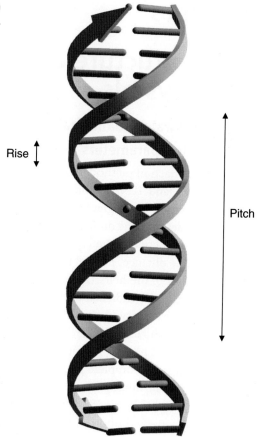

Rise

Pitch

(base pair steps) — for example, helical twist and roll. It thus defines the relative relationships of successive base pairs in a structure. Inevitably, the second set of features incorporate the first, so correlations between them can be expected.

3.2 Polynucleotide structures from fibre diffraction studies

3.2.1 Classic DNA structures

Information on the dimensions of polynucleotide double helices was initially derived from the diffraction patterns of DNA fibres at high (92%) relative humidity, which showed a readily interpretable and characteristic Maltese cross-pattern of strong diffraction intensities (Fig. 1.2). Examination of this pattern, which arises from the stacked base-pair repeats of the B-DNA double helix, led directly to the original Watson–Crick model in 1953 (Watson & Crick, 1953). Subjecting DNA fibres to other conditions produces rather different diffraction patterns. Notably, lower (65%–75%) relative humidity conditions result in the A-DNA pattern, which typically has many more diffraction maxima, indicating greater crystallinity and hence order in the fibres. Several other forms have subsequently been found (see Section 2.2), most of which are sub-classes of the A and B double helices, the two most important ones for random sequence natural DNAs. Their exactly repetitious structures are often referred to as 'canonical DNA' ones. Double-helical DNA is thus highly polymorphic, the differing forms corresponding to distinct yet inter-convertible molecular structures. Some DNA structural types have only been found with defined sequence polynucleotides rather than with random-sequence DNA (the standard A and B forms can occur with most sequences). Under appropriate experimental conditions, the diffraction pattern of such a polynucleotide is best fitted to a repeating unit, that rather than being a simple mononucleotide, can be a dinucleotide or even an oligonucleotide. Some of these repetitious fibres produce semi-crystalline diffraction patterns that give intensity data to significantly higher resolution than classic calf thymus DNA fibres. In these instances, the relatively high number of observed diffraction intensities can result in structures that are as reliable as those of standard A and B polynucleotide helices.

B-DNA, the classic structure first described by the Watson and Crick model, has subsequently been refined (Arnott, 1999) using the linked atom least-squares procedure developed by Arnott and his colleagues. The helical repeat is a single nucleotide unit, with necessarily all the nucleotides in the structure having the conformation of this unit. Helical parameters for B- and other DNA polynucleotides are given in Table 3.1. The backbone conformation (Table 3.2) has high *anti* glycosidic angles and C2'-*endo* sugar puckers (Fig. 3.2).

The right-handed double helix has 10 base pairs per complete turn, with the two polynucleotide chains wound anti-parallel to each other (Fig. 3.3a and c) and linked

Table 3.1 Selected helical parameters for various polymorphs of DNA polynucleotides, derived from fibre diffraction studies.

	Unit repeat	Rise (Å)	Helical twist (°)	Base-pair propeller twist (°)	Base step roll (°)	Base-pair inclination (°)
A	11	2.54	32.7	−10.5	0.0	22.6
B	10	3.38	36.0	−15.1	0.0	2.8
C	28/3	3.31	38.6	−1.8	0.0	−8.2
D	8	3.01	45.0	−21.0	0.0	−13.0
Z (C)	6	7.25	−49.3	8.3	5.6	0.1
Z (G)	6	7.25	−10.3	8.3	−5.6	0.1

Values are mostly from Arnott, S. (1999). In S. Neidle (Ed.), *Oxford Handbook of nucleic acid structure* (p. 1). Oxford University Press.

Table 3.2 Backbone conformational angles in (°) for various polymorphs of DNA polynucleotides.

	α	β	γ	δ	ε	ζ	χ
A	−52	175	42	79	−148	−75	−157
B	−30	136	31	143	−141	−161	−98
C	−37	−160	37	157	161	−106	−97
D	−59	156	64	145	−163	−131	−102
Z (C)	−140	−137	51	138	−97	82	−154
Z (G)	52	179	−174	95	−104	−65	59

Taken from Arnott, S. (1999). In S. Neidle (Ed.), *Oxford Handbook of nucleic acid structure* (p. 1). Oxford University Press.

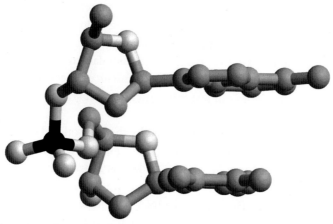

Figure 3.2 Conformation of two successive nucleotides in one strand of canonical B-DNA, from fibre diffraction analysis.

by Watson—Crick A•T and G•C base pairs. The paired bases are almost exactly perpendicular to the helix axis, and they are stacked over the axis itself (Fig. 3.3b). Consequently, the base-pair separation is the same as the helical rise — 3.4 Å. An important consequence of the Watson—Crick base pairing arrangement is that the two deoxyribose sugars linked to an individual base pair are on the same side of it. So, when successive base pairs are stacked on each other in the helix, the gap between these sugars forms continuous indentations in the surface that wind along, parallel to the sugar—phosphodiester chains. These indentations are termed **grooves**. Groove width can be defined in various ways. A commonly used convention defines it as the perpendicular distance between phosphate groups on opposite strands, minus the van der Waals diameter of a phosphate group (5.8 Å), although it is equally valid in some circumstances to use pairs of other atoms, such as C1′ or O4′. Groove depths are normally defined in terms of the differences in cylindrical polar radii between phosphorus and N2 guanine or N6 adenine atoms, for minor and major grooves, respectively.

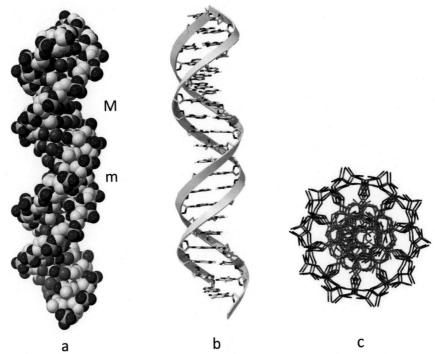

Figure 3.3 The structure of canonical B-DNA, from fibre diffraction analysis. (a) Space-filling representation, showing major and minor grooves. (b) Cartoon representation, with the backbones shown in ribbon form. (c) View down the helix axis, showing the 10-fold helicity and the disposition of base pairs straddling the axis.

The asymmetry in the base pairs results in two parallel types of groove whose dimensions (especially, their depths) are related to the distances of base pairs from the axis of the helix and their orientation with respect to the axis. The B-DNA wide major groove (Fig. 3.3a and Table 3.3) is almost identical in depth to the much narrower minor groove, which has the hydrophobic hydrogen atoms of the sugar groups forming its walls. In general, the major groove is richer in base substituents — O6, N6 of purines and N4, O4 of pyrimidines compared to the minor one. This, together with the steric differences between the two, has important consequences for interaction with other molecules (see Chapters 4—7). Base-pair and base step morphologies for several polynucleotide helices are given in Table 3.1.

The A-DNA duplex also has a single nucleotide unit as the helical repeat (Fig. 3.4). This has C3′-*endo* sugar puckers, which brings consecutive phosphate groups on the nucleotide chains closer together — 5.9 Å compared to the 7.0 Å in B-form DNA and alters the glycosidic angle from high *anti* to *anti*. Consequently, the base pairs are twisted and tilted with respect to the helix axis (Fig. 3.5a and c) and are displaced nearly 5 Å

Table 3.3 Groove dimensions (in Å) in polynucleotide DNA double helices, from fibre diffraction analyses (taken from Arnott, 1999 for the A—D helices) and from single-crystal data (Wang et al., 1979, 1981) for the Z helix.

Polymorph	Major groove		Minor groove	
	Width	Depth	Width	Depth
A	2.2	13.0	11.1	2.6
B	11.6	8.5	6.0	8.2
C	10.5	7.6	4.8	7.9
D	9.6	6.2	0.8	7.4
Z	8.8	3.7	2.0	13.8

Figure 3.4 Conformation of two successive nucleotides in one strand of canonical A-DNA, from fibre diffraction analysis.

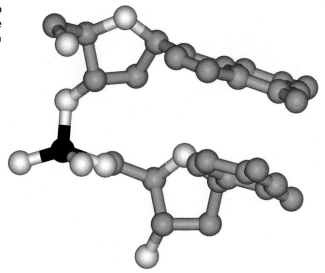

from it, in striking contrast to the B helix. The helical rise is much reduced to 2.54 Å, compared to 3.4 Å for canonical B-DNA. The A-helix is wider than the B one and has an 11 base-pair helical repeat. The combination of base pair tilt with respect to the helix axis and base-pair displacement from the axis results in very different groove characteristics for the A double helix compared to the B form (Fig. 3.5b). This also results in the centre of the A double helix being a hollow cylinder (Fig. 3.5b). The major groove is now deep and narrow, and the minor one is wide and very shallow.

3.2.2 DNA polymorphism in fibres

The structural transition from A to B forms in 'mixed-sequence' DNA is induced by changes in the relative humidity surrounding the DNA fibres. Other forms can be

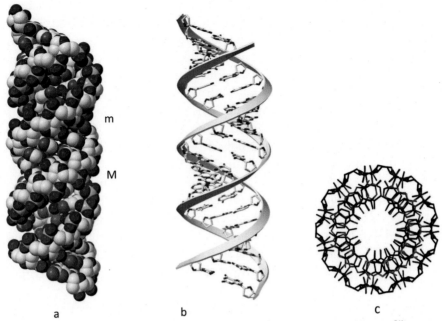

a   b   c

Figure 3.5 The structure of canonical A-DNA, from fibre diffraction analysis. (a) Space-filling representation, showing major and minor grooves. (b) Cartoon representation, with the backbones shown in
ribbon form. (c) View down the helix axis, showing the 11-fold helicity and the disposition of base
pairs around the axis and the void running through the centre of the helix.

produced by fine control of this condition (the C form) or from defined sequence polynucleotides (the D and Z forms). For example, there are a number of members of both
the A and B families that differ from the 'canonical' forms outlined above, in large part
because of the degree of over-winding of the helices (for the B type family), and have
considerable differences in pitch, base–pair tilt with respect to the helix axis and hence
groove characteristics. There are variants on the A-DNA duplex that have a wide major
groove (Arnott et al., 1982), reminiscent of that in a B helix, rather than that of the narrow standard A groove. Changes in groove width can thus be achieved at little energy
cost.

The C form of DNA results from relatively low humidity conditions for a DNA fibre
and is over-wound relative to B-DNA, with 9.3 residues per turn. The overall appearance of this helix and the dimensions of the grooves resembles B-type DNA rather than
the A form morphologies. The D form cannot be adopted by random sequence native
DNAs and has been observed in crystalline fibres of the alternating purine/pyrimidine
polynucleotides poly(dA-dT)•poly(dA-dT) and poly(dI-dT)•poly(dI-dT), where I is
inosine (the rare 8-oxo-purine). The structure of this polymorph has not been unambiguously defined, with models proposed ranging from left-handed seven- and eightfold

helices to a more conventional eightfold right-handed one. Structural parameters for this right-handed model are given in Tables 3.1—3.3. Observations of the reversible structural transition from B to D forms in crystalline fibres have been made using time-resolved X-ray diffraction by means of a high-intensity synchrotron source of X-rays (Mahendrasingam et al., 1986). The observation of a gradual rather than an abrupt increase in helical pitch, from 24 to 34 Å, is consistent only with a transition in which there is no change in helix handedness, and so a left-handed D-DNA model can be rejected.

The Z form is an authenticated left-handed structure (Arnott et al., 1980). It exists in the alternating sequence poly(dC-dG)•poly(dC-dG) and is presumed to be the structure formed by this sequence in solution in high salt (>2.5M NaCl) conditions. This left-handed DNA has a dinucleotide repeat (Fig. 3.6) with quite distinct nucleoside conformations for the guanosine compared to the cytosine residues. Each individual repeat has a helical rise of 7.25 Å, so that the rise between successive base pairs is half of this — 3.63 Å (Fig. 3.7). Z-DNA is discussed further in Section 5 of this chapter; it was discovered in a defined sequence oligonucleotide crystal structure as well as by the analysis of the fibre diffraction pattern produced from the synthetic polynucleotide poly(dC-dG)•poly(dC-dG).

The polymorphism of DNA structure that is apparent in polymeric fibres is suggestive of an underlying flexibility in DNA structure. This is a natural consequence of the large number of backbone and sugar conformational variables, which, together with base-pair flexibility, can result in double helices that are structurally distinct yet are equivalent in energetic terms. Fibre diffraction studies can only hint at the fine detail of this flexibility. So, further significant advances in DNA structure studies have taken place with the advent of single-crystal and nuclear magnetic resonance (NMR) analyses of a large number of defined oligonucleotide sequences, which, as described in Chapter 1, provide structural data that are not averaged over a sequence. These studies have provided

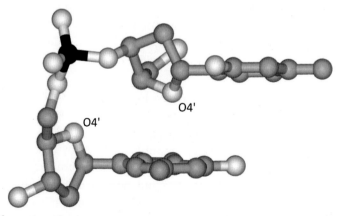

Figure 3.6 Conformation of two successive nucleotides in one strand of canonical Z-DNA, from fibre diffraction analysis.

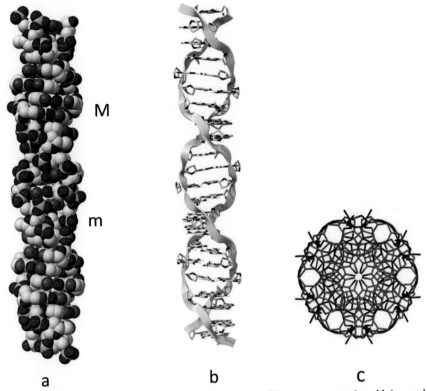

Figure 3.7 The Z-DNA left-handed double helix. (a) Space-filling representation. Major and minor grooves are indicated. (b) In the same orientation, but in ribbon representation to emphasise the discontinuities in the backbone. (c) View down the helix axis, showing the disposition of base pairs around the axis, and the reduced diameter of the helix compared to the A and B forms.

much detail not only of the structures themselves, but also have enabled the principles to be determined that relate DNA structural features to sequence and environment, and ultimately to biological function in association with DNA-binding proteins (Chapter 7). The symmetry and pseudo-symmetry inherent in fibre diffraction helices and in many other nucleic acid structures has been highlighted by a discussion of the local and global symmetry twofold axes in base pairs and strand orientations (Heinemann & Roske, 2020).

3.3 B-DNA oligonucleotide structure as seen in crystallographic analyses

3.3.1 The Dickerson–Drew dodecamer

The single-crystal structure of the self-complementary dodecanucleotide d(CGCGAATTCGCG)$_2$ was determined by multiple isomorphous replacement

Figure 3.8 (a, b) Two ribbon views of the structure of the Dickerson-Drew dodecamer (PDB id 1BNA), from the first crystallographic analyses of this sequence (Drew & Dickerson, 1981; Wing et al., 1980). The two views are rotated by ~ 45° and show the narrowing of the minor groove in the central region of the sequence as well as slight irregularities in the backbone conformation.

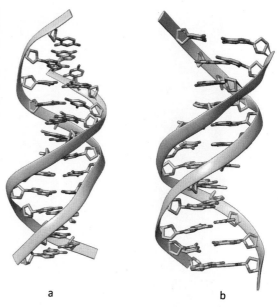

a   b

methods (Wing et al., 1980). This structure analysis is of historic importance in that without recourse to any preconceived model, it revealed a complete turn of an anti-parallel right-handed B-DNA double helix (Fig. 3.8a), and thus unequivocally demonstrated the correctness of the Watson—Crick model for DNA structure. This, the 'Dickerson—Drew' sequence has subsequently been widely studied by a variety of other experimental and theoretical techniques that have complemented, generally confirmed and sometimes extended the original structural results. The original crystallographic analysis, at 1.9 Å resolution, also revealed sequence-dependent structural features that could not be observed in fibre diffraction studies of averaged sequence B-DNA, which has all the nucleotides in an identical conformation. These features are as follows:

- a narrow minor groove in the 5'-AATT region (Fig. 3.8a and b), which at its extreme point is only 3.2 Å wide compared to 6.0 Å in averaged canonical B-DNA from fibre diffraction studies. The major groove at this point in the dodecamer structure is exceptionally wide (12.7 Å).

- a well-ordered and regular network of water molecules in this A/T region of the minor groove (Drew & Dickerson, 1981). This network has been termed the 'spine of hydration'. It involves hydrogen bonding of first-shell water molecules to the O2 atom of thymine and the N3 atom of adenine such that each water molecule spans the two dodecamer strands. These water molecules are linked together by further, second-shell waters. Evidence for the persistence of this structured water arrangement in solution has come from NMR studies (Kubinec & Wemmer, 1992), from several subsequent high-resolution crystallographic studies (Lercher et al., 2014; Vlieghe et al., 1999) and from a number of molecular dynamics simulation studies (see Rueda et al., 2004 for a comprehensive survey of earlier work). The spine has been confirmed to exist at room temperature in solution using chiral non-linear vibrational spectroscopy (McDermott et al., 2017) — the previous crystallographic data were obtained at cryogenic temperatures. Minor-groove hydration and the structure of the water spine are discussed in detail in Chapter 5.

- differences in the values for local base pair and base-step parameters at different points along the sequence (Dickerson & Drew, 1981). Most notable are the high propeller twist values for the four central A•T base pairs (Fig. 3.9).

- a wide distribution of values for sugar puckers, glycosidic angles and backbone torsion angles. The value of a given torsion angle is to a considerable degree dependent on the sequence context of the nucleotide involved. For example, values for torsion angles ε and ζ are distributed in two sub-groups within their broad ranges. Both angles are normally in the *trans, gauche⁻* domain, but are *gauche⁻, trans* when the nucleotide is followed by a purine. This alternative phosphate conformation has been designated B_{II}, with the standard *trans, gauche⁻* phosphate conformation being termed B_I. The B_{II} conformation is associated with increased minor groove width, both in the original dodecamer and in several oligonucleotide crystal structures that have been

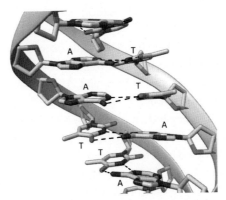

Figure 3.9 View of the central region in the Dickerson—Drew dodecamer, showing the high propeller twist of the A•T base pairs.

subsequently determined. Its occurrence is mostly at the dinucleotide steps GpC, CpG, CpA, TpG and TpA (Madhumalar & Bansal, 2005). Several other correlations between backbone torsion angles were found in the original Dickerson—Drew structure, and subsequently confirmed in other oligonucleotide structures, such as those between angles α and γ, and between angle δ and glycosidic angle χ. The range of values for the δ angle is itself sugar-pucker dependent; this important correlation confirms and extends earlier findings from surveys of mononucleoside structures (Chapter 2, Section 2.5), to the oligonucleotide level

- the average features of the Dickerson—Drew structure are similar to those for canonical B-DNA from fibre diffraction. For example, the average helical twist is 35.9° and there are 10.1 base pairs per helical turn.

- the helix is not straight but is bent by about 19° in the major groove direction. This bending is apparent in Fig. 3.8a.

3.3.2 Other studies of the Dickerson—Drew dodecamer

The role played by lattice interactions in the dodecanucleotide crystal structure has been examined in detail by several studies. For example, comparison of nine different dodecamers in three space groups has shown that the key parameters of minor-groove width and propeller twist for the central 5′-AATT region are largely unaffected by crystal packing factors (Dickerson et al., 1994). This is to be expected since in the $P2_12_12_1$ original space group for the Dickerson—Drew (and many other) dodecamers only the terminal two nucleotides at each end are involved in inter-molecular interactions with other duplexes. Thus, the central helical region of six to eight base pairs is unconstrained, and its features are intrinsic to this sequence. In a more demanding test of the generality of the features, it has been found that d(CGCGAATTCGCG) can crystallise in alternative space groups such as the trigonal $R3$ or $P3_212$ (Table 3.4), with distinct packing arrangements, especially when under the influence of particular metal ions such as calcium (Acosta-Reyes et al., 2020; Johansson et al., 2000; Liu & Subirana, 1999; Minasov et al., 1999).

In all these structures, the key features of base and base-pair morphology are still apparent, although sequence-dependent features tend to differ less from canonical values. These together with the characteristic of the narrow minor groove being filled by a spine of hydration provide good evidence for the general correctness of the original structure. All these structures contain both dodecanucleotide self-complementary strands in the crystallographic asymmetric unit and the resulting duplex is not perfectly twofold symmetric, by contrast with its structure in solution. A crystal form (Johansson et al., 2000)

Table 3.4 Selected B-DNA crystal structures. The second strand is listed for sequences that are non-self-complementary.

Sequence	PDB id	Resolution, Å	Space group
d(CGCGAATTCGCG)	1BNA	1.90	$P2_12_12_1$
d(CGCGAATTCGCG)	436D	1.10	$P2_12_12_1$
d(CGCGAATTCGCG)	4C64	1.32	$P2_12_12_1$
d(CGCGAATTCGCG)	1EHV	1.80	$P3_212$
d(CGCAAAAAAGCG) +	1D98	2.50	$P2_12_12_1$
d(CGCTTTTTTGCG)			
d(CGCAAAAATGCG) +	1BDN	2.60	$P2_12_12_1$
d(CGCATTTTTGCG)			
d(CCGCTAGCGG)	1DCV	2.50	$C2$
d(CCAACGTTGG)	5DNB	1.40	$C2$
d(CCAGTACTGG)	1D8G	0.74	$C2$
d(ACCGCCGGCGCC) +	330D	2.70	$H3$
d(GGCGCCGGCGGT)			
d(CATGGGCCCATG)	1DC0	1.30	$P4_12_12$
d(CGCAAATTTGCG)	1S2R	1.53	$P2_12_12_1$
d(CGCTGGAAATTTCCAGC)	1SGS	1.60	$H3_2$
d(CTTTTTAAAAGAAAAG)	1N4L	2.00	$P2_12_12$
d(ACACTACAATGTTGCAAT) +	3BSE	1.60	$H3$
d(GTATTGCAACATTGTAGT)			
d(ATATAT)	1GQU	2.50	$P2_1$
d(CGAATTAATTCG)	5M68	2.64	$P3_2$
Racemic d(CGCTATAATGCG) +	5EWB	1.69	$P2_1/c$
Racemic d(CGCATTATAGCG)			

in the trigonal space group $P3_212$ has the duplex sitting on a crystallographic twofold axis, so that exact twofold symmetry is imposed on the structure (Fig. 3.10). Features such as the narrow minor groove and hydration spine are still apparent, but now they are twofold symmetric, reflecting the exact symmetry implied by the self-complementary Dickerson–Drew sequence.

The status of the Dickerson–Drew structure, as the paradigm for many of the fine details of B-DNA structure generally, implies that its features are considered as being highly reliable, especially by non-experts in crystallography. It is important to bear in mind that the original structure was determined to a resolution that is only moderate by current standards and refined by methods that are now considered to be sub-optimal. Thus, the accuracy and precision of this analysis and the derived geometric parameters are lower than more recent studies on this and other sequences. These have used (i) high-quality, highly redundant diffraction data from synchrotron sources, (ii) libraries

Figure 3.10 The structure of the dodecamer d(CGCGAATTCGCG) in space group $P3_212$, (PDB id 1EHV) where the duplex utilises a twofold crystallographic symmetry axis (Johansson et al., 2000). The terminal nucleotides are not base-paired within the duplex; instead they are hydrogen bonded to other duplex molecules in the crystal lattice.

of high-precision nucleotide geometry and (iii) refinement methods that are able to overcome false local minima. The reassuring picture that has emerged is one in which the central concept of sequence-dependent structure is retained, although the fine detail may differ from conclusions based on the older medium-resolution structures.

In more recent studies of the sodium and magnesium forms of the Dickerson—Drew sequence, crystal structures have been determined to 1.1 and 1.4 Å, respectively (Lercher et al., 2014; Minasov et al., 1999; Shui et al., 1998; Tereshko et al., 1999). These structures confirm many of the original features observed by Dickerson et al., apart from some details in the assignment of solvent molecules in the minor groove, with a larger number of solvent molecules being well defined. One analysis is consistent with primary solvent sites in the spine of hydration sharing between sodium cations and water molecules (the arrangement is described in detail in Chapter 5). Analysis of minor groove width in the Dickerson—Drew structure using molecular dynamics simulations has led to the suggestion (Hamelberg et al., 2000; Hamelberg et al., 2001) that it changes as a result of cation interactions and is thus dependent on the nature of the cation, as well as on sequence in the minor groove. The structure of the Dickerson—Drew sequence in solution has been examined in detail in solution by NMR techniques (see, for example, Wu et al., 2003), which have tended to suggest that there is some conformational heterogeneity, consistent with the picture from several of the more recent high-resolution X-ray structures. This conclusion has been supported and extended by a combined NMR and low-angle X-ray analysis (Schwieters & Clore, 2007), which has provided reliable data on the amplitudes of motion for B_I/B_{II} phosphate motions, sugar pucker and base morphology.

3.3.3 Other B-DNA oligonucleotide structures

The crystal structures of standard B-type oligonucleotides mostly have averaged features that closely approximate those of averaged sequence fibre diffraction B-DNA. Few are more than 12 base pairs in length. The majority are of self-complementary palindromic sequences (Table 3.4). They fall into two principal categories. The majority are either dodecamers, many of which are isomorphous to the Dickerson—Drew structure and therefore pack identically in the crystal in the space group $P2_12_12_1$, or they are decamers, which crystallise in a variety of space groups, some of which give very high-resolution (0.74—1.4 Å) structures. The isomorphism of the dodecamers provides a convenient framework to analyse systematic changes in the central base pairs, whilst keeping invariant the 3′ and 5′ terminal sequences, which are involved in the crystal packing arrangements. Several decamers crystallise in forms with the 10-mer duplexes packed end-to-end in the crystal, in a quasi-helical manner. The Dickerson—Drew-related sequence d(CGCAATTGCG) has been crystallised in no less than five different space groups, showing some differences in crystal packing (Valls et al., 2004). These result in differences in the terminal base pairs, whilst the central CAATTG sequence has an almost invariant structure in all five structures, providing good evidence that its features are intrinsic to the sequence.

Many crystal structures have also been examined that have changes to the central GAATTC sequence of the Dickerson—Drew dodecamer. When the sequence is a purely alternating one, as in d(CGCATATATGCG) (Yoon et al., 1988), the minor groove width is equally narrow as in the Dickerson—Drew structure, although the narrowest points do not coincide. This 5′-ATATAT sequence shows a pronounced alternation of several of its base pair and step parameters, especially roll; propeller twist values for the A•T base pairs are consistently lower than those in the Dickerson—Drew sequence.

The structure of the duplex formed by d(CGCAAAAAAGCG) and its complementary sequence (Nelson et al., 1987) is remarkable for several reasons. It is a model sequence for an A-tract, which in natural DNA has anomalous structural and dynamic properties, it cannot be readily be re-constituted into nucleosomes and is intrinsically bent when in phase with a helical repeat (see Section 3.6 in this chapter). The structure has high propeller twists for the A•T base pairs, of up to 26°, with a network of bifurcated three-centre hydrogen bonds connecting them (Fig. 3.11). These bifurcated hydrogen bonds are in the major groove and involve atoms N6 of adenine and O4 of thymine bases. They have been suggested to be major factors in stabilising the high propeller twists for the A•T base pairs, and thus in stiffening the A-tract, at least in this structure. The roll, tilt and slide values for the ApT steps in the structure are all close to 0°, with bending (towards the major groove) occurring at the ends of the A-tract rather than within it.

Bending in the major groove direction has also been observed in the structure of the duplex formed by d(CGCAAAAATGCG) and its complementary strand (DiGabriele & Steitz, 1993), although the situation here is complicated by the presence of two distinct orientations of the duplex in the crystal lattice. This study has concluded that the bending found in the isomorphous dodecamer crystal structures is more a consequence of crystal packing forces than of intrinsic properties of the A-tract sequences. The crystal structure of d(CGCAAATTTGCG) has been determined at medium (2.2 Å) (Edwards et al., 1992) and high (1.4 Å) resolution (Woods et al., 2004). Both structures show structured water molecules in the minor groove, although there are differences in groove width and water organisation. These may reflect not only the superior resolution of the latter structure but also its improved data quality and the many improvements in refinement over a 12-year period. Even so, both structures concur in not finding significant three-centre hydrogen bonding in the A/T region, suggesting that these effects may not be inherently

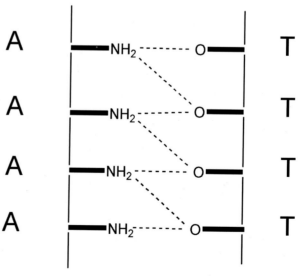

Figure 3.11 Schematic view of three-centre major groove bifurcated hydrogen bonding involving A•T base pairs.

required to stabilise propeller twist and a narrow minor groove. Rather, propeller twist is more a consequence of base stacking preferences.

The weight of evidence supports the concept of a narrow minor groove being preferred in A/T regions. However, this is only a preference, and not a definitive rule obeyed under all circumstances. On the one hand, the duplex formed by d(CGCAAGCTGGCG) has an average minor groove width of about 7 Å in the 5′-AGCT region (Webster et al., 1990). This is close to that of averaged fibre diffraction B-DNA, as is the G/C-rich minor groove in the d(ACCGGCGCCACA) duplex (Timsit et al., 1989). On the other hand, the high-resolution (1 Å) structures of the decamers d(CCAACGTTGG) and d(CCAGCGCTGG) both show a pronounced narrowing in the central region of the minor groove (Chiu & Dickerson, 2000), even though individual base-pair propeller twists are generally no greater than about 12°. The flexibility of such structures is further shown in the crystal structure (Kielkopf et al., 2000) of the decamer d(CCAGTACTGG). This is notable in being one of the very few DNA duplex crystal structures to be determined at true atomic resolution (0.74 Å), so that the accuracy and precision of atomic coordinates and derived parameters is fully established (The average esds of bond lengths and angles in this structure are 0.028 Å and 2.2°, respectively). The structure shows that several regions of the backbone adopt more than one conformation, an effect that is undetected at lower resolution. These alternative conformations do not affect the general pattern of minor groove narrowing that is centred around the short two base-pair A-tract. These A•T base pairs have high propeller twists of −20.3° and −18.8°, whereas the two adjacent C•G ones have values of −8.4° and −9.7°, respectively.

The non-palindromic sequence d(ACCGGCGCCACA) (Timsit et al., 1989) is a hot spot for frameshift mutagenesis by several chemical carcinogens. The structure is non-isomorphous with the other dodecamers and is remarkable for several features. Three G•C base pairs in the centre of the helix are partially opened, with exceptionally high propeller twists and some loss of G•C hydrogen bonding. It has been suggested that this represents the intermediate stage prior to full base-pair opening. Analysis of a low temperature form of the crystals has shown (Timsit et al., 1991) that cooling produces a one base-pair shift of the major groove base-pairing that is favoured by CA tracts. The crystal packing involves helices interacting together by means of groove–backbone interactions (Timsit & Moras, 1991), which are analogous to true DNA–DNA junctions such as the Holliday junction involved in recombination processes (see Chapter 4). This pattern contrasts with the packing observed in crystal structures of all-A/T oligonucleotides, most of which adopt a closely similar pattern of stacked parallel duplexes, often with staggered ends so that quasi-infinite helices are produced (Campos et al., 2006). Such an arrangement also resembles the packing in polynucleotide fibres, and analogous orientational disorder may be observed, so that detailed single-crystal analyses are not always possible, as in the structure of d(ATATATATATAT), whose fibrous diffraction pattern has been interpreted in terms of a hexanucleotide-repeat coiled coil structure with a pitch of 218.6 Å (Campos et al., 2005). The crystal structure of the pure A/T sequence d(ATATAT) is remarkable in showing an anti-parallel duplex with Hoogsteen rather than Watson–Crick base pairs (Abrescia et al., 2002). The right-handed double helix has overall similarity to the standard B-form in that the sugar puckers are C2'-*endo*, there are 10.6 base pairs per turn and the base pairs are perpendicular to the helix axis (Fig. 3.12). The glycosidic angles for all the adenosine nucleotides are required to be in the *gauche*$^+$ range for the *syn* conformation, as indeed is observed.

Crystallographic analyses of a number of the diverse sequences that are able to crystallise as decamer duplex helices (for example, Goodsell et al., 1995; Heinemann et al., 1992; Lipanov et al., 1993; Quintana et al., 1992) have provided further evidence for variations in sequence-dependent structural features being real effects rather than artefacts. Comparisons of the structures show that the parameters helical twist, roll and rise are not independent of each other, but that there are several correlations between them. For example, rise is linearly related to helical twist. The TpA step, even though it is consistently found to be partially unstacked in all the B-DNA oligonucleotide structures where it occurs, has highly variable sequence and crystal environment-dependent behaviour. An extreme and surprising example of this is in the sequence d(CGATATATCG),

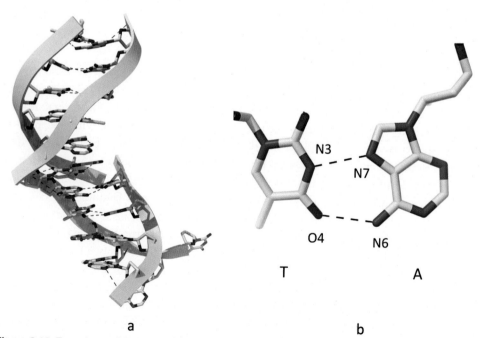

Figure 3.12 Two views of the d(ATATAT) structure, PDB id 1RSB (Abrescia et al., 2002). (a) showing the two anti-parallel duplexes, (b) one of the A•T Hoogsteen base pairs in the structure.

which has a wide minor groove in the alternating AT region (Yuan et al., 1992), albeit with a characteristic alternation of helical twist angles that is a consequence of the alternation of good base stacking at ApT steps combined with the poor stacking at the TpA ones. Thus, although alternating A/T sequences generally have a marked tendency to produce narrow minor grooves in B-DNA, their inherent structural flexibility means that this tendency can be readily overcome under the influence of TpA unstacking. The existence of the B_{II} phosphate conformation, which often occurs at dinucleotide steps in which the second base is a purine, is itself a factor in producing a widened minor groove (Hartmann et al., 1993). However, the B_I/B_{II} backbone conformational flexibility observed in several high-resolution crystal structures (Kielkopf et al., 2000; Schuerman & Van Meervelt, 2000) suggests that this is less a factor than the inherent properties of the bases and base steps themselves.

Oligonucleotide crystallographic studies using X-rays have until recently generally not able to define the positions of hydrogen atoms in relatively mobile water molecules, even in the higher resolution crystal structures. The methodology of neutron diffraction can do so since neutrons are strongly scattered by hydrogen (and deuterium) atoms. This technology has the disadvantages that high-flux neutron sources with diffraction facilities are only available in a small number of centres worldwide, and that large crystals are required, with a volume >1 mm^3, which tends to limit the observable resolution. Crystals of up to 2.8 mm^3 volume have been grown from D_2O solution (Arai, Chatake, Minezaki, et al., 2002) for the sequence d(CCATTAATGG), which has been previously studied by X-ray methods (Goodsell et al., 1994). The neutron diffraction analysis to 3.0 Å (Arai et al., 2005) has revealed the presence and orientation of water molecules bridging between bases, which contribute to the somewhat complex spine of hydration found in the minor groove of this structure.

One of the goals of studies on the relationship between DNA sequence and structure has been to relate to DNA sequences in their biological context. This requires knowledge of both native structures and as bound protein complexes. Only a small number of such native B-DNA crystal structures have been determined that have direct biological relevance, and for them, interesting patterns of sequence-dependent flexibility have emerged. A prominent example is that of DNA bending shown by the DNA sequences binding to the human papillomavirus E2 protein (see Section 3.6.3). The DNA sequence that binds the transcription factor NF-κB is a 10 base-pair palindromic site, which has been incorporated into the 17-mer sequence, and its crystal structure has been determined (Huang et al., 2005). This shows significant differences compared to the identical DNA sequence when bound to the NF-κB protein (Chen et al., 1998). The flanking G:C regions, which have the most contacts with the protein, buckled and opened in the complex but not in the native structure. The large roll angle at the centre of the A/T tract in the complex produces significant curvature (see Section 3.6.3), but not in the native structure. The narrow minor groove in this structure has a spine of hydration.

It is remarkable though how few native DNA crystal structures are of the most abundant biological DNA sequences (Subirana & Messeguer, 2010). An exception is the 16-mer non-self-complementary crystal structure of an enhancer element sequence from *Drosophila melanogaster* which is involved in gene expression defining sex determination in this species (Narayana & Weiss, 2009). The central region of this sequence, d(AATGTT), is also represented in many eukaryotic transcription factor binding sites. The crystal structure (Fig. 3.13) reveals a continuous 16 base-pair B-DNA 10.7 residues per turn helix, with the two nucleotides at each end being disordered and not visible in the electron density. The central 8 base-pair region has a narrow and deep minor groove, which is enlarged towards both ends of the helix, the width changing from a minimum of 4.1—7.3 Å. This central minor groove has a characteristic one-dimensional spine of hydration, which is unaffected by the presence of a C.G base pair. The central 12 base-pair segment is almost straight, with bending towards the major groove direction at each end such that the overall structure is bent by ca 15°. The majority of A.T base pairs exhibit high propeller twist, in contrast to the majority of C.G base pairs. Overall, the sequence dependency of the structure appears to be related to the thermal motion, and ultimately to its recognition by regulatory proteins.

A novel host—guest approach has been developed for studying DNA sequences that are unable to crystallise on their own (Coté et al., 2000). This utilises the N-terminal fragment of the Moloney murine leukaemia virus reverse transcriptase, as a sequence-neutral 'host' to bind to a wide range of 16-mer DNA sequences, including ones with mismatches and bound drug molecules (see Section 5.6.2). The resulting structures have a

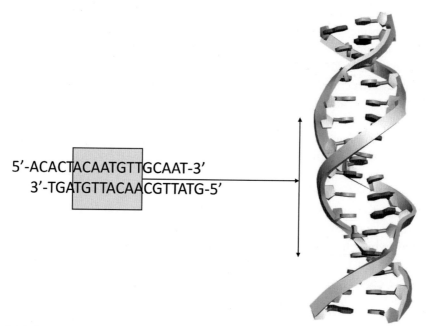

5'-ACACTACAATGTTGCAAT-3'
3'-TGATGTTACAACGTTATG-5'

Figure 3.13 Sequence and crystal structure of the 16 base-pair sex-specific enhancer element (Narayana & Weiss, 2009), PDB id 3BSE.

dumbbell-like appearance with a protein molecule at each end connected by the oligonucleotide duplex. All crystallise in the same space group, $P2_12_12$, but their sheer size appears to allow for flexibility in the DNA such that sequence-dependent effects are apparent. A study of a DNA analogue of the HIV-1 polypurine A-tract, the asymmetric sequence d(CTTTTTAAAAGAAAAG) with its complementary sequence, has shown (Coté et al., 2003) that in spite of the presence of the three A-tracts, the structure is essentially straight. Its features are closely similar to several other related sequences in distinct environments, again strongly supportive of the notion that DNA local structure is intrinsic to its sequence rather than to its environment, both as a native structure and often when bound to a sequence-neutral protein. The host—guest approach has been used in a comparative analysis of four self-complementary 16-mer sequences, which has enabled properties of particular dinucleotide steps highlighted, such as the positive roll angle of the CpA step (Montaňo et al., 2006).

As outlined in Chapter 1, a novel approach to determining oligonucleotide structure has been the use of racemic mixtures of sequence, to give centrosymmetric crystal structures such as that of the sequence d(CCGGTACCGG) (Mandal et al., 2014). Four crystal structures of the non-self-complementary duplex formed from d(CGCTATAATGCG) and its complement d(CGCATTATAGCG) (Mandal et al., 2016) have been determined and illustrate the power of the technique to crystallise otherwise often inaccessible duplex structures. The central d(TATAAT) region is the consensus sequence of the prokaryotic promoter so-called 'Pribnow box' and is thus of considerable biological relevance. Remarkably, in all four crystal forms, this central region has few crystal-packing contacts, supporting the concept that its features are intrinsic to the sequence. The base pairs in the central region have low propeller twists, which increase towards the 5' and 3' ends, as do helical twist values. The narrow minor grooves (minimum width of 3.3—3.5 Å) have short conserved water spines, analogous to that in the original Dickerson—Drew structure.

3.3.4 Sequence-dependent features of B-DNA; their occurrence and their prediction

The central question that comparative studies on the large number of DNA crystal structures seek to answer is whether and to what extent the observed sequence-dependent structural effects are real or are artefacts of (a) crystal packing effects, with adjacent helices generally having close contacts, or (b) the random and systematic errors that accompany X-ray (and NMR) structures, where the likely errors of medium-resolution structures are in the same range as the differences in some parameters from canonical mean.

Variations in the local base-pair and step parameters propeller twist, roll and helical twist for two dodecamer structures are illustrated in Fig. 3.14 (see Chapter 2 for definitions). Sequence-dependent effects are most reproducible for the central six to eight residues of these sequences since the 3' and 5' termini of the dodecamer structures are the

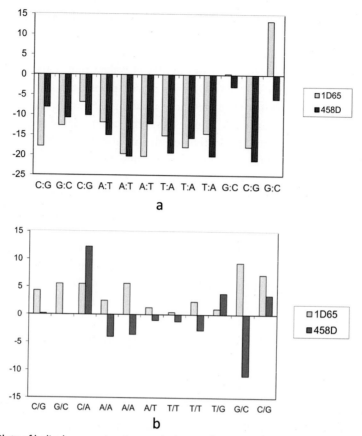

Figure 3.14 Plots of helical parameters in two dodecanucleotide crystal structures (PDB ids 1D65 and 458D), (a) showing propeller twist and (b) showing base-pair roll. Note that the values in the central region are closely similar, within experimental error.

regions most involved in packing interactions with other molecules in the crystal lattice. The pattern of base and base-pair morphological parameters is also consistent with those derived from NMR data in solution using the method of residual dipolar couplings (Trantirek et al., 2000). The most significant variations in values for these parameters from canonical values are as follows:

- the local helical twist varies by up to 15°, with pyrimidine-3′,5′-purine steps having lower than average values and purine-3′,5′-pyrimidine steps having higher than average ones. These differences correlate with differences in the susceptibility of the sugar–phosphate backbone bonds in this sequence to be cleaved by the enzyme DNAase I (Lomonossoff et al., 1981), with the TpC step, which has a high twist, being the most readily cleaved. The mean twist angle in the crystal structure is 36°, in accord with the fibre diffraction value

- propeller twists are significantly greater for A•T base pairs than for G•C ones, by an average of 5—7°
- roll angles for pyrimidine-3′,5′-purine steps have positive values, since they open up towards the minor groove, whereas purine-3′,5′-pyrimidine steps have negative roll angles with major groove opening

There is good evidence that at least some sequence-dependent features seen in the dodecamer structures have direct relevance to 'real' DNA sequences (see also the preceding section) and are not restricted to short-length crystal structures, especially the larger-scale features such as groove width variation. For example, it has been found (Drew & Travers, 1984) that there are variations in the ability of the DNase I enzyme to cleave the phosphodiester backbone of the 160 base pair *Escherichia coli tyr*T promotor within runs of A•T base pairs. These variations correlate with changes in minor groove width that occur in them when TpA steps are present or absent. A detailed study (Azad et al., 2018) using hydroxyl radical foot-printing to probe minor groove width in a 399 base-pair sequence has found that a region of (computationally predicted) narrow minor groove width that is present in naked DNA is in the majority of the 11 protein-binding sites probed, likely to be recognised by a DNA-binding protein and maintained in its protein—DNA complex.

The sequence-dependent changes in helical and propeller twist, roll and slide were initially rationalised in terms of steric clashes between substituent atoms on individual bases — the Calladine Rules (Calladine, 1982; Calladine & Drew, 1984). These rules are based on the hypothesis that the observed structural effects derive from opposite strand purine—purine steric hindrance, and thus give an essentially mechanical view of DNA structural features as arising from the interplay of rigid-body Newtonian forces. For purine-3′,5′-pyrimidine steps, these clashes are between major-groove substituent atoms O6 of guanine and N6 of adenine, and between minor-groove N2 of guanine and N3 of adenine for pyrimidine-3′,5′-purine steps. On this basis, little steric clash would be predicted for purine-3′,5′-purine and pyrimidine-3′,5′-pyrimidine steps. The rules suggest that clashes are avoided by combinations of changes in twist, roll and slide. Their relative proportions depend on the nature of the bases involved:

- a decrease in local helical twist decreases minor-groove clashes,
- an increase in roll angle in the groove with clashes,
- separation apart of successive purines by means of an increase in slide between them
- a decrease in propeller twist. The changes in slide result in alterations in the backbone angle δ, which thus has a higher value for purine compared to pyrimidine nucleosides

The Calladine Rules, which have reasonable predictive capability for some (though by no means all) B-DNA structures in addition to d(CGCGAATTCGCG), only have limited applicability to other helical types, especially A forms where there is inherently high propeller twist. The inability of the rules to take factors into account, such as hydration, electrostatics and base stacking, does mean that they only provide a first-order approximation to the understanding of sequence-dependent effects, even in B-DNA structures.

The rules provide a means of understanding the structural behaviour of dinucleotide base steps, of which there are just 10 distinct types. The rules also imply that each step is largely independent of its neighboring base step, in reality a considerable over-simplification. Extension to four-base sequences would require structural information for 136 distinct possible combinations. The establishment of the first B-DNA oligonu-cleotide structure provided information on only a very small number of these possible nearest-neighbour combinations of DNA sequence, and so there have been concerted attempts to experimentally determine structures for other sequences, both relating to and distinct from the Dickerson–Drew one. Even though many nucleic acid crystal structures containing B-DNA have subsequently been determined, a high proportion of the 136 tetranucleotide combinations have not been sampled experimentally. Accord-ingly, several approaches have been used in attempts to circumvent this apparent barrier, including a focus on dinucleotide rather than tetranucleotide steps (see below). An early theoretical study of sequence context effects at the tetranucleotide level (Packer et al., 2000) used an experimental database with 66 out of the 136 included. This study suggests that A/T-containing dinucleotide steps tend to be significantly more context indepen-dent than C/G-containing ones.

Many studies have examined relationships between parameters and their underlying structural basis, which have been able to use the increasing number of B-DNA structures available in the Protein Data Bank/Nucleic Acid Database (PDB/NDB) by searching for correlations between parameters (Subirana & Faria, 1997; Suzuki et al., 1997). Although the available structures do not enable all possible flanking sequences to be examined (see Section 3.4), it does appear that they only influence some individual base steps, notably pyrimidine-$3'$,$5'$-purine ones (Subirana & Faria, 1997). The sequence dependence of he-lical twist angles has been examined in a set of 38 structures (Gorin et al., 1995), which has demonstrated a significant correlation between twist and roll, as a consequence of ste-ric clashes, some of which are in addition to the purine–purine clashes defined by the Calladine Rules (Calladine, 1982). These authors have introduced an empirical clash strength function, which whilst able to reliably predict the twist of most steps, fails with some CpA ones. An ambitious attempt to crystallographically map out sequence de-pendency has been made with the inverted repeat sequence motif d(CCnnnN$_6$N$_7$N$_8$GG), where N$_6$, N$_7$ and N$_8$ are A, T, C and G (Hays et al., 2005). 63 out of the 64 permutations were crystallised, resulting in 37 crystal structures from 29 different sequences. This set contained A- and B-DNA structural types, as well as several involving four-stranded Holliday junctions. The approach has enabled sequence-dependent effects to be identified and separated from the consequences of crystal-packing forces.

Considerations of base—base interactions, using empirical energy functions (Elsawy et al., 2005; Hunter, 1993; Hunter & Lu, 1997), have been a fruitful approach to understanding DNA sequence-dependent structure. These energy functions include terms for van der Waals and electrostatic interactions, with charge distributions in bases being calculated (Farwar et al., 2006; Hunter, 1993), on the basis of a π-electron model rather than with the more commonly used point-charge approach. Experimental values for rise, roll and tilt, which are especially dependent on base stacking, are well reproduced by this approach. An extension of this approach has used a genetic algorithm to find global minima (Packer & Hunter, 2001) for a set of 30 diverse oligonucleotide structures, all of which also have known crystal structures. Most of them are well predicted, with the sequence-dependent trends in base step and pair morphologies mostly being well reproduced. When backbone and base variability is taken into account (Farwar et al., 2006), the average root-mean-square deviation (RMSD) comparing prediction to experiment is 2.24 Å. Since the accuracy of the structures themselves is unknown, correlations with computational predictions of the details of sequence dependency do need large samples of structures if they are to have statistical reliability. A comprehensive study of stacking energies in 10 B-DNA base pair steps has used accurate quantum chemistry methodology (Kruse et al., 2019). This is valuable in providing a benchmark for other lower-cost computations and empirical force-field (for example, AMBER based) methods.

Empirical energy functions have been used to theoretically examine all 32,896 unique octamer DNA sequences (Gardiner et al., 2003, 2004) and suggest (i) that the overwhelming majority have B-like conformations; A-type duplexes are likely to have consecutive guanosine residues in the sequence and (ii) that structural diversity is much less than sequence diversity. Importantly, though, it does appear that there are differences, at least in terms of flexibility, between promoter and non-promoter sequences. Long timescale molecular dynamics simulations (up to $10\,\mu s$) of the high-quality Dickerson—Drew dodecamer crystal and NMR structures (Dans et al., 2019) using the PARMBSC1 force field have found that helical parameters, groove dimensions and torsion angles sampled in the simulations in all cases matched the values from the experimental structures. Coarse-grained methods can simulate many more and much longer sequences than atomistic molecular dynamics. This approach has been used to study all 136 unique tetranucleotide DNA sequences, embedded in 56-mer long duplexes (Walther et al., 2020). The timescale difference between the two methodologies is around 10^5. Remarkably, the results closely reproduced those structures for which experimental data are available, with an average RMSD of ca 0.3 Å per base pair, thus giving confidence in the predictive power of this approach.

A multi-laboratory approach (Dans et al., 2019) has combined molecular dynamics data on 13 oligomers, each containing 18 base pairs that covering all tetranucleotide combinations. It used a standardised protocol (Pasi et al., 2014) and force field for the simulations and found that many parameters tend to be correlated, in accordance with earlier findings. For example, shift slide and roll twist are negatively correlated, whereas shift tilt and slide twist are positively correlated. Ionic environment was also found to be a factor. The analysis was able to develop an updated set of Calladine Rules that can predict base morphologies for any desired sequence: predictions are available on the web server https://mmb.irbbarcelona.org/miniABC/.

Backbone conformational angles show considerable variability in B-DNA structures with respect to sequence (Table 3.5). Even within the original Dickerson—Drew structure, individual angles typically show a >45° spread of values along the sequence. The central question, relevant both to variations in backbone angles and to base morphological parameters, is whether these differences reflect (i) true sequence-dependent variability, (ii) the influence of crystal packing interactions or (iii) are a reflection of a lack of precision in the crystal structure determinations. Surveys of conformational angles (Schneider et al., 1997) using the crystal structures available at the time showed that each angle occupies a discrete range (Table 3.4), with some such as ζ occupying two regions. The angles have Gaussian distributions, suggesting that the variations reflect both experimental inaccuracies and real fluctuations. A subsequent survey (Sims & Kim, 2003) has taken conformational information on 6253 individual dinucleoside phosphates, also from oligonucleotide structural entries in the NDB. These data have been comprehensively mapped using a multi-dimensional scaling method, which shows that the data separate into nine major conformational types, corresponding to A-, B- and Z-type clusters. The distributions of torsion angles correspond well with those found six years previously (Schneider et al., 1997), using a smaller data set. Interestingly, analysis of the more recent data set also revealed several minor conformational clusters, corresponding to non-canonical features. More recently an extensive study using 7739 nucleotides in 447 high-quality structures (Svozil et al., 2008) has continued this type of cluster analysis with the finding of an increased number of conformational clusters, with the B_I conformation being overwhelmingly the most common. This study has also obtained average torsion angle values (Table 3.5) from a sub-set of 118 native DNA structures, which are within the same broad regions as the canonical fibre diffraction values, but unsurprisingly are non-identical. The large amount of structural data have also enabled the conformational preferences for many, though not all, to be defined with high reliability. Again, the B_I class was found to be the dominant type.

It has been suggested, with some justification, that the simple classifications of nucleic acid double helical types into mostly A or B forms is insufficient for adequate descriptions of the many (mostly sequence dependent) variations observed in crystal and NMR structures. The experimental data on DNA oligonucleotide structures do not apparently point to clear sequence dependency relationships involving backbone conformation and helical parameters, in broad agreement with an early theoretical analysis (Packer & Hunter, 1998), which concludes that roll, tilt and rise are inherently backbone independent.

Table 3.5 Ranges and average values of backbone torsion angles (°) observed in high-resolution B-DNA crystal structures.

	α	β	γ	δ	ε	ζ	χ
(a) Range	270–330	130–200	20–80	70–180	160–270	150–210 230–300	200–300
Mean values	298	176; B_I 146; B_{II}	48	128; B_I 144; B_{II}	184; B_I 246; B_{II}	265; B_I 174; B_{II}	258; B_I, Pu 241; B_I, Py 271; B_{II}, Py
(b) Average B_I	299.0 ± 0.9	179.3 ± 1.0	48.4 ± 0.6	132.8 ± 1.0	181.7 ± 1.0	263.2 ± 0.8	250.3 ± 1.1
Average B_{II}	296.2 ± 1.3	143.1 ± 1.3	46.0 ± 0.9	143.0 ± 0.9	251.1 ± 2.1	168.0 ± 1.4	277.8 ± 1.4
(c) Canonical values	330	136	31	143	219	199	262

(a) From Schneider et al. (1997). The ranges corresponding to B_I and B_{II} phosphate conformations are indicated. (b) Average values from the more extensive survey by Svozil et al. (2008). (c) Canonical B-DNA fibre diffraction values, from Table 3.2

Helical twist is dependent on backbone conformation, though not in a straightforward way. A structural alphabet has been proposed (Černý, Božíková, Malý, et al., 2020; Černý, Božíková, Svoboda, et al., 2020; Schneider et al., 2018) that provides a framework for an altogether more detailed set of descriptors. Assignments for structures in the PDB and users' own structures can also be examined online at https://dnatco.datmos.org/. This approach is based on the definition of a dinucleotide step by 12 parameters — (i) seven backbone torsions, (ii) two torsion angles around the glycosidic bonds, (iii) one through-space pseudo-torsion angle between the two glycosidic bonds and (iv) two through-space inter-atomic distances, involving C1′—C1′ and N1/N9—N1/N9 atoms. As a result of analysing ca 120,000 dinucleotide steps in ca 2000 structures in the PDB, a total of 96 distinct conformational classes (plus a 97th for unassigned classes) have been identified, across all DNA and RNA structures in the PDB. These classes represent distinct dinucleotide conformers and can be grouped into 15 categories (Černý, Božíková, Malý, et al., 2020), providing a detailed sequence-dependent description of local structural variability. The dinucleotide alphabet approach can be applied to all categories of nucleic acid structure, and not solely duplexes. However, it does not address issues of sequence dependency of base and base-pair morphology.

3.4 A-DNA oligonucleotide crystal structures

DNA double helices of the A type are produced in fibres of random-sequence DNA under conditions of low humidity (see Section 3.2 above), and in solution when the water activity is reduced by the addition of various alcohols. The fibre diffraction studies suggest that certain runs of sequence, such as alternating G•C base pairs, have a tendency to be A-form. Single-crystal analyses of several lengths of oligonucleotide sequences have determined many to be of an A-type helix, especially when they have high G/C content. However, for most such structures, it is now apparent that this does not reflect so much any inherent structural preference, as the crystal packing requirements for these particular lengths of oligonucleotide duplexes. The high-alcohol crystallisation conditions commonly used in oligonucleotide crystallography may also be a contributory factor. Nonetheless, these structures have provided invaluable insights into the flexibility and conformational preferences of A-DNA helices in general, even if their significance in biological terms is not always apparent. There is at least one instance where A-type structures are of undoubted importance: this is in the case of RNA—DNA hybrids, formed, for example, during transcription and replication. Since RNA helices are always constrained to be in an A form, then DNA—RNA hybrids would be expected to be similarly constrained. This has been found to be the case by both fibre diffraction studies of RNA—DNA hybrid polynucleotides and by single-crystal analyses. For example, the structure of the duplex formed by r(GCG)d(TATACGC) shows that the DNA strand has adopted a conformation close to that of duplex RNA, with an 11-fold helix (average helical twist of 33°), C3'-*endo* sugar puckers and an A-type helical backbone conformation (Wang et al., 1982).

3.4.1 A-form octanucleotides

Many self-complementary octanucleotides, of widely varying sequence type, have been crystallised (Table 3.6) and found to have A-DNA family structures. As with B-type oligomers, structural features averaged over a number of such sequences are close to those of fibre diffraction A-DNA. By contrast with B-DNA oligomer crystal structures, backbone torsion angles tend to have narrow ranges of values, reflecting the greater conformational rigidity of the A-form. Some of the A-form octamer sequences crystallise in the tetragonal space group $P4_32_12$; most of the others are in the hexagonal space group $P6_1$. It has been possible to vary conditions such that some sequences such as d(GGGCGCCC) (Shakked et al., 1989) and d(GTGTACAC) (Jain et al., 1991) have been crystallised in both forms. The hexagonal form has also been found for sequences such as d(GGGGCCCC) (McCall et al., 1985), d(GGGTACCC) (Eisenstein et al., 1990) and d(GGGATCCC) (Lauble et al., 1988) as well as for a number of mismatch variants. All have helical and conformational features within the general A class, but with some marked local variations, in particular in the overall helical repeat, in minor groove widths and in base step parameters. The wide minor groove of ~15 Å in d(GGGGCCCC) contrasts with the more typical value of 9.7—9.8 Å in d(GGGTACCC), which is close to that for canonical fibre diffraction A-DNA, of 11.1 Å. Major groove width cannot be fully assessed in an octanucleotide duplex, since eight rather than seven phosphate groups are required in order to calculate the shortest phosphate—phosphate inter-strand distance. Approximated major groove widths show very wide variation, ranging from 5 Å in the

Table 3.6 Selected A-DNA crystal structures.

Sequence	PDB id	Resolution, Å	Space group
d(GGGCGCCC)	2D94	1.70	$P4_32_12$
d(GGGCGCCC)	1VT8	1.90	$P6_1$
d(CCCCGGGG)	187D	2.25	$P4_32_12$
d(ACGTACGT)	243D	1.90	$P4_32_12$
d(CCGGGCCGG)	321D	2.15	$P2_12_12_1$
d(GCGTACGTACGC)	117D	2.55	$P6_122$
d(CGCCCGCGGGCG)	399D	1.90	$P2_12_12_1$
d(CATGGGCCCATG)	1DC0	1.30	$P4_12_12$
d(AGGGGCGGGGCT)	1ZJE	2.10	$P2_12_12_1$
d(CCGGGGTACCCCGG)	5WV7	1.41	$P4_12_12$
d(CTACGCGCGTAG)	5MVK	1.53	$P3_221$
d(CCCGGGTACCCGGG)	5IYG	1.70	$P4_12_12$
d(CCCCGGTACCGGGG)	4OKL	1.65	$P4_12_12$

low-temperature form of d(GGGCGCCC) (Clark et al., 1990) to over 12 Å in d(GGGGCCCC) (McCall et al., 1986), compared to the very narrow value of 2.2 Å from fibre diffraction. The tetragonal form, with sequences having a pyrimidine-3',5'-purine step at positions 4 and 5, that is at the centre of the helix, shows significant deviations from standard A-form backbone geometry at this point, with torsion angles α and γ having *transoid* values rather than the normal *gauche*$^-$ and *gauche*$^+$ ones, respectively. This discontinuity has been observed in several structures, for example, that of d(ATGCGCAT) (Clark et al., 1990). It is likely that this effect is a consequence of the requirements imposed by hydration and crystal packing in the tetragonal unit cell, rather than being an intrinsic property of these sequences.

The 9-base pair binding site of the Sp1 transcription factor is GC rich, and three crystal structures of this sequence, d(GGGGCGGGG) embedded within 12 or 13 base-pair sequences have been reported, with all having the ends participating in overhanging base pairs in the crystal lattice (Dohm et al., 2005), and all forming A-type duplexes. There is some variation in detailed conformation between them, even though all have such standard A-type features as 11 base-pairs per helical turn, again indicative of a degree of local flexibility that is markedly less than for B-type structures.

3.4.2 Do A-form oligonucleotides occur in solution? Crystal packing effects

Are crystal packing factors the driving force behind all octanucleotide structures being in the A form? Solution NMR studies on sequences such as d(ATGCGCAT) (Clark et al., 1990) have demonstrated that they show B-family behaviour in solution even though the crystal structures are unequivocally of the A type. This dichotomy between crystal and solution environmental constraints is also shown by the sequence d(GGATGGGAG), which forms part of the binding site for the transcription factor TFIIIA that forms part of the initiation complex for transcription of the 5S RNA gene in *Xenopus*. The (low-resolution) 3 Å crystal structure of the duplex formed by this sequence, and its complementary strand, shows an A-family helix (McCall et al., 1986), with average major and minor-groove widths of 14.8 and 15 Å, respectively, and a helical repeat of 11.5 base pairs. These values closely correspond to those for A'-RNA from fibre diffraction studies. Nuclease digestion studies, showing a repeat of 11.4 base pairs per turn, together with circular dichroism measurements (Fairall et al., 1989), are consistent with the assignment of a structure for the d(GGATGGGAG) sequence as well as for the complete 54 base pair TFIIIFA binding site, that is not classical B-DNA. On the other hand, NMR and other biophysical methods all indicate that the nonamer in solution has normal C2'-*endo* sugar puckers and B-DNA range glycosidic angles (Aboul-ela et al., 1988).

Short A-DNA double helices are insufficient in length for the major groove to be fully formed, and so estimates of its dimensions are necessarily approximate. X-ray analyses of two dodecamers (in two distinct space groups) have each revealed the structure of a complete turn of A-DNA double helix, as well as conformational features that are not subject to the crystal packing factors of the octanucleotides. The structure of the sequence d(CCCCCGCGGGGG), comprising both alternating and non-alternating nucleotides (Verdaguer et al., 1991), has average backbone conformational angles and helical parameters that are remarkably close to those in A-DNA fibres, with some local backbone angle variations, for example, *trans* α and γ angles at one of the two CpG steps. This structure, together with others such as that of the A-DNA decamer d(ACCGGCCGGT) (Frederick et al., 1989), is consistent with the notion that very GC-rich sequences have a higher tendency to form A-type duplexes. The sequence d(CCGTACGTACGG), of which two-thirds consists of G•C base pairs, likewise has an A-DNA structure (Bingman et al., 1992), this time without any *transoid* α/γ angles. The average major groove narrow width in this structure, of 3.5 Å, is only slightly wider than the 2.2 Å value from fibre diffraction. A high-resolution crystallographic study of the sequence d(AGGGGCCCCT) in two distinct space groups concludes (Gao et al., 1999) that conformation is less conserved in this circumstance than when different A-type sequences crystallise in the same space group. The crystal structure of the sequence d(CTACGCGCGTAG), with three consecutive d(CpG) steps (Fig. 3.15a), shows that it forms an A-DNA duplex, whereas in solution NMR data unequivocally demonstrate that this sequence forms a B-type helix (Hardwick et al., 2017), in spite of the absence of A/T base pairs in the central region. The 14-mer sequence d(CCGGGGTACCCCGG) also forms an A-type helix in the crystalline state (Karthik et al., 2017). The structure (Fig. 3.15b and c) has an overall helical twist of 29.7°

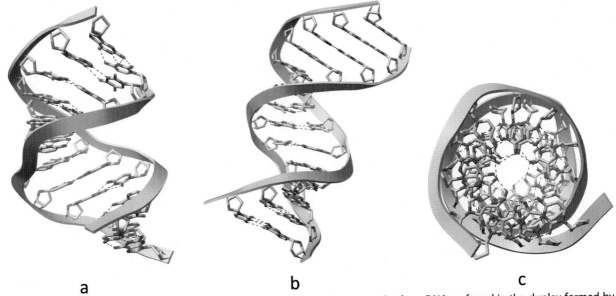

a   b   c

Figure 3.15 (a) Cartoon representation of the structure of a complete turn of A-form DNA, as found in the duplex formed by d(CTACGCGCGTAG), PDB id 5MVK (Hardwick et al., 2017). (b) Cartoon representation of the crystal structure of the 14-mer sequence d(CCGGGGTACCCCGG) (Karthik et al., 2017), PDB id 5WV7. (c) A view of 5WV7, looking down the helix axis.

(implying ca 12 base pairs per turn) and an average rise of 3.34 Å, compared to fibre diffraction values of 32.7° and 2.54 Å (Table 3.1). As Fig. 3.15b shows, these values result in this helix having some B-DNA-like morphological features, although it does have the characteristic A-like appearance when viewed down the helix axis (Fig. 3.15c). This has been explained (Karthik et al., 2017) by the division of the structure into three short helices, with the two end helices having non-standard A-like features, a possible consequence of crystal packing effects.

Changes in crystallisation conditions do not necessarily result in changes to the resulting helical structure. This has been demonstrated in a study (Finley & Luo, 1998) of the structure of the sequence d(GACCGCGGTC), crystallised under two contrasting conditions of ionic strength. This sequence, which is half of the human papilloma virus E2 binding site, forms an A-type helix with closely similar helical parameters under both conditions. However, there is a marked difference in the degree of helical bending, with that grown under higher salt conditions being bent by 31° in the major groove direction compared to 15° for the low-salt form. Such bending may be relevant to the need for flexibility when some sequences bind to their cognate proteins, as we shall see in Chapter 7. Hydration in A-DNA oligonucleotides is generally extensive, with observations of phosphates and bases hydrogen bonding to water molecules (Tippin & Sundaralingam, 1997). Most solvent ordering occurs in the major groove, with polygons of water molecules being observed in some crystal structures (Eisenstein & Shakked, 1995). These polygon arrangements are especially apparent at high resolution (Egli et al., 1998).

3.4.3 The A ↔ B transition in crystals

There are nearly 150 A-DNA crystal structures in the NDB, which provide a very detailed view of this DNA form and the myriad of deviations from canonical ideality (notably in major groove width) that are possible (see, for example, Wahl & Sundaralingam, 1997). It may be thought that there is then little more to learn about A-DNA. However, the capacity of DNA structure to surprise is still apparent. The crystal structure of the dodecamer d(CGCCCGCGGGGCG) is remarkable in that parts of the structure demonstrate true hybrid character (Malinina et al., 1999). The central octamer region has an A-DNA conformation, but with a high degree of bending around the central purine-3′,5′-pyrimidine base step (Fig. 3.16). This 65° bending results in some phosphate groups becoming very close to each other. By contrast, the terminal nucleotides have many B-DNA characteristics, such as B-like sugar puckers, backbone conformational angles and base slide. This structure thus has features akin to an A ↔ B transition, albeit with distinct conformations at different points along the sequence. Whether this structure is a consequence of the influence of crystal packing forces or innate tendencies of this sequence is not clear.

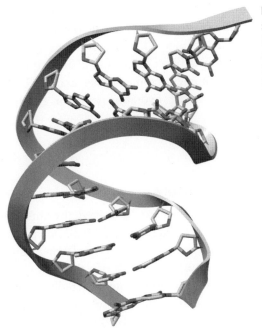

Figure 3.16 Cartoon representation of the structure of the duplex d(CGCCCGCGGGCG), PDB id 399D, showing an A-DNA helix having some B-DNA features, with the large bend at the centre of the structure being apparent (Malinina et al., 1999).

The high-resolution structure of d(CATGGGCCCATG) has features that are throughout its length intermediate between A and B (Ng et al., 2000), and thus resembles a trapped intermediate in the A $\leftrightarrow$ B transition. There is a significant base shift, producing an A-like channel in the centre of the helix, and groove widths are intermediate between the A and B ideal values. A complementary view is provided by the analysis of six d(GGCGCC) duplex hexanucleotide crystal structures with varying degrees of cytosine bromination or methylation (Ng & Dickerson, 2001; Vargason et al., 2001). This results in differing crystal forms and structures that progressively show features that span A, B and plausible intermediate states. For example, the major groove becomes progressively deeper and values for the slide parameter become increasingly negative on proceeding from the B-type to A-type structures. It has been suggested (Ng & Dickerson, 2002) that a factor in the A/B continuum seen in the d(CATGGGCCCATG) crystal structure is the tendency of GpG steps in this sequence to prefer A-like features.

Several instances have been described above when a sequence crystallises in an A-form, yet as shown by NMR methods, the B-form consistently dominates in solution. Thus, the presence of the A-form in the crystal is often a consequence of crystal packing forces (see, for example, Hardwick et al., 2017; Karthik et al., 2017). However, the

A-form is not an artefact. The historic importance of the recognition that A-DNA poly-nucleotide fibres are produced in lower humidity environments than B-DNA in the determination of the double-helix has been well documented. It is increasingly apparent that the A-form as observed both in single crystals and fibres can be found in appropriate hydrophobic environments, for example, in the genomes of some double-stranded DNA viruses that have evolved to be protected from excess hydration (see, for example, DiMaio et al., 2015).

3.5 Z-DNA — left-handed DNA

3.5.1 The Z-DNA hexanucleotide crystal structure

The first oligonucleotide duplex single-crystal structure to be solved (by multiple isomorphous replacement methods) was that of the alternating pyrimidine—purine sequence d(CGCGCG) in 1979, at the very high resolution of 0.9 Å (Wang et al., 1979). This structure is remarkable in that even though it consists of a duplex formed by the two anti-parallel hexamer strands, the helix is (quite unexpectedly) a left-handed one. The hexamer Z-DNA duplex crystal structure has subsequently been re-determined with a range of cations, and sometimes at exceptionally high resolution (for example, at 0.6 Å) (Tereshko et al., 2001: Fig. 3.17). The backbone of all Z-DNAs is characteristically irregular compared to A- or B-DNA since the dC and dG residues

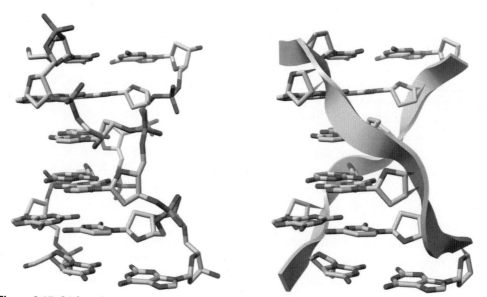

Figure 3.17 Stick and cartoon views of the structure (PDB id 1I0T) of the ultra-high-resolution structure (0.6 Å) of the Z-DNA sequence d(CGCGCG) (Tereshko et al., 2001).

have very distinct conformations (Fig. 3.6), resulting in a 'zig-zag' arrangement of phosphate groups — hence, the helix has been termed Z-DNA. The same left-handed arrangement has also been found (Arnott et al., 1980) in fibres of the alternating polynucleotide poly(dC-dG)•poly(dC-dG) — see Section 2.2 above and Fig. 3.7a and b. Z-DNA was in effect discovered some years earlier during the course of circular dichroism studies on poly (dG-dC).poly(dG-dC), when it was observed (Pohl & Jovin, 1972) that on increasing the ionic (as salt) concentration beyond ~4M NaCl, the CD spectrum became inverted from its standard B-DNA-associated shape. This major spectral change indicates a transition to a quite different conformational form, which is now accepted to be Z-DNA. Several other physico-chemical techniques have subsequently been used to study this left-handed structure in addition to X-ray techniques (Rich et al., 1984).

3.5.2 Overall structural features

Z-DNA oligonucleotides (and the poly(dG-dC)•poly(dG-dC) polynucleotide) have the following conformational characteristics:

- the purine (deoxyguanosine) nucleosides have *syn* glycosidic angles with a χ range of 55—80° (mean 60°), together with C3′-*endo* sugar puckers. The conformation about the α/γ backbone torsion angles is *gauche*$^+$, *trans* and *gauche*$^-$ about angle ζ.
- the pyrimidine (deoxycytidine) nucleosides have *anti* glycosidic angles with a χ range of −145 to −160° (mean −152°), and C2′-*endo* sugar puckers. The α/γ conformation is *trans*, *gauche*$^+$ and that about angle ζ is *gauche*$^+$.
- the G•C base pairs are of standard Watson—Crick type.

A consequence of these differences between purine and pyrimidine nucleosides is that the helical repeating unit is forced to be the CpG dinucleoside, rather than the mononucleoside one in standard right-handed canonical A- and B-DNA. The distinct α/ζ phosphate conformations of the two nucleosides in Z-DNA result in the characteristic zig-zag appearance of the backbone. Detailed examination of d(CGCGCG) in various crystal forms (Wang et al., 1981) has shown that there is a secondary backbone conformational family with a distinct set of phosphate orientations. This secondary conformation is termed Z_{II}, with the standard type as described above being termed Z_I. The Z_{II} conformation results in purines having a *gauche*$^+$ rather than a *gauche*$^-$ value for the ζ backbone torsion angle and a *gauche*$^-$ rather than a *trans* value for angle ε. Z_{II} pyrimidines have (rather smaller) changes in angles α and β compared to Z_I ones. Occurrence of the Z_{II} conformation at a particular residue is probably related to the coordination of the phosphate group at this point to a hydrated magnesium ion (Egli et al., 1991; Wang et al., 1981). Z-DNA oligonucleotides are usually crystallised in the presence of magnesium and spermine ions. Thus, the Z_{II} conformation does not occur at this point in the sequence in the pure-spermine, magnesium-free structure of d(CGCGCG). Neutron diffraction studies on d(CGCGCG) have shown that the water molecules in the

Z-DNA minor groove are well ordered, whereas those in the major groove are much less so (Chatake et al., 2005).

3.5.3 The Z-DNA helix

The Z-DNA double helix is to a large degree represented by the structure of the hexa-nucleotide duplex d(CGCGCG); however, there are slight differences between individual CpG units, quite apart from the occurrence of Z_I and Z_{II} forms. Idealised helices for both of these forms have been generated (Wang et al., 1981). The fibre diffraction model for Z-DNA is closest to the Z_I helix. This Z helix has a 44.6 Å pitch with 12 base pairs per helical turn and a diameter of ~18 Å (Fig. 3.7), making it slightly slimmer than a B-DNA helix (with a diameter of ~20 Å). The helical twists for two base-pair steps CpG and GpC are quite different as a consequence of the asymmetry in guanosine and cytidine conformations, with the CpG step having an exceptionally small twist and almost no base stacking between the two C•G base pairs in the step. A further consequence of the *syn* guanosine conformation is that the base pairs are not positioned astride the helix axis, as in B-DNA. Rather, their edges are at the surface of the helix. The N7 and C8 atoms of the imidazole 5-membered ring of guanine and (to a lesser extent) the C5 atom of cytosine actually protrude onto the helical surface of what would be the major groove, making the surface slightly convex at this point (Fig. 3.17). In other words, Z-DNA does not have a major groove at all. Its minor groove is very narrow and deep, and lined with phosphate groups (Table 3.3). The differences in phosphate orientation between Z_I and Z_{II} helices result in the latter having a ~1 Å wider minor groove.

3.5.4 Other Z-DNA structures

Z-DNA oligonucleotides have been crystallised (Table 3.7) with a range of related sequences, including pure analogues such as d(CGCGCGCG) (Fujii et al., 1985) and d(CGCGCGCGCG) (Ban et al., 1996), as well as variants where changes have been made to base and sequence type. All show the resilience of the Z-DNA structural entity. The central CpG can be replaced by TpA whilst maintaining the left-handed structure (Wang et al., 1984). This shows that Z-DNA can tolerate some A•T base pairs, although they do tend to destabilise the structure. The cytosines are required to be 5-methylated in order for stabilisation of the d(CGTACG) sequence, and in general modified cytosines are necessary if a Z-DNA structure contains A•T base pairs. It has been suggested (Wang et al., 1984) that A•T base pairs are not able to take part in the ordered Z-DNA groove hydration, which plays an important role in maintaining the integrity of the Z-DNA structure (Egli et al., 1991), by contrast with G•C base pairs. Hence, A•T base pairs cannot by themselves form a Z-DNA structure. The slight preference of guanosine nucleosides (unlike adenosine) to adopt the *syn* rather than the *anti* glycosidic conformation (see Chapter 2) is also a factor in the GC preference of Z-DNA. It is

Table 3.7 Selected Z-DNA-containing crystal structures. All the sequences are self-complementary. Structures 2ACJ and 5ZUO contain B-Z junctions.

Sequence	PDB id	Resolution, Å	Space group
d(CGCGCG)	1I0T	0.60	$P2_12_12_1$
d(CGCGCG) + Mg^{2+}, spermine	2DCG	0.90	$P2_12_12_1$
d(GCGCGCGCGC)	279D	1.90	$P6_522$
d(CACGTG) + propanediamine	2F8W	1.20	$P2_12_12_1$
d(GTCGCGCGCCATAAACC)	2ACJ	2.60	$P6_1$
d(CGCGCGCGCGCG)	4OCB	0.75	C2
d(GTCGCGCGCCATAAACC) + d(ACGGTTTATGGCGCGCG)	2ACJ	2.60	$P6_1$
d(GTCGCGCGCGATAAACC) + d(ACGGTTTATCGCGCGCG)	5ZUO	2.90	$P6_1$
Racemic d(CGCGCG)	5JZQ	0.78	C2/c

then surprising that reversal of the central pyrimidine-3′,5′-purine sequence (i.e. so that it is no longer purely alternating) in the structure of d(CGATCG), still retains a Z-DNA type structure (Wang et al., 1985), albeit with the thymidines and adenines adopting *syn* and *anti* glycosidic conformations, respectively. This energetically unfavourable Z structure was only stabilised by having the cytosine bases methylated or brominated at the five position of the base. Nonetheless, its existence suggest that the requirement for the formation of a Z-DNA structure is not so much that the sequence should be an alternating C/G-rich one, but that it retains the key feature of alternating *syn* and *anti* nucleosides. It is likely that replacement of thymine by uracil would increase Z-form stability, since the methyl group of the former presents a significance hydrophobic hindrance to solvent ordering around the Z helix. This has been borne out by subsequent structural studies on uracil-containing Z sequences (Schneider et al., 1992). In general, non-alternating sequences can form Z-DNA structures in the crystalline state when the non-alternating pyrimidines are able to adopt a *syn* conformation (Basham et al., 1999). Many of the more recent high-resolution Z-DNA crystal structures have revealed the presence of statistical disorder along the helix axis (Karthik et al., 2019), or sometimes disorder in the packing of helices in the crystal, akin to the disorder observed in most nucleic acid fibres (Luo et al., 2017). A 1 Å cryo-neutron diffraction study of d(CGCGCG), which has identified a significant number of associated water molecules, also found small, sub-Ångstrom positional disorder in the positions of several bases (Harp et al., 2018). The crystal structure of the decamer d(CGCGCGCGCG) has been determined at very high resolution (0.75 Å) using the anomalous scattering of backbone phosphorus atoms to determine the structure (Fig. 3.18). Surprisingly, this was possible even though 6 or 11 phosphate groups are disordered and were modelled in double positions.

Figure 3.18 Cartoon representation of the crystal structure of the Z-DNA sequence d(CGCGCGCGCGCG), PDB id 4OCB (Luo et al., 2014).

The structure has both Z_I and Z_{II} phosphate orientations, with the bases and sugars in well-defined singular positions.

The structure of the B–Z junction was initially revealed in a crystal structure (Ha et al., 2005) of a 15 base-pair sequence containing within it a CG hexanucleotide Z-DNA motif. This was tethered in the Z-form by being bound to a domain of the Z-DNA binding protein human $hZ\alpha_{ADAR1}$, with the effect that eight base pairs are in the Z-form and six are in a B-DNA conformation (Fig. 3.19). Strikingly, one A•T base pair at the junction is not formed and the adenine and thymine bases are displaced so that they are flipped out into the exterior of the helix, presumably in order to relieve the steric strain at the junction. This has enabled the sharp turn of the B–Z backbone at the junction to be formed, and continuous base stacking to be preserved throughout the length of the duplex. It is notable that the right-handed helical segment of the structure has a standard regular B-DNA conformation. These overall features are conserved in three more recent co-crystal structures of B–Z junctions, also with the human $hZ\alpha_{ADAR1}$ protein, but using a range of sequences (Kim et al., 2018). As before, continuous base stacking

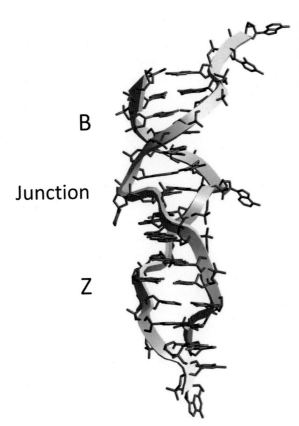

Figure 3.19 Structure (PDB id 2ACJ) of an oligonucleotide having a B-Z junction (Ha et al., 2005). The B (right-handed) and Z (left-handed) parts of the helix are indicated, together with the B-Z junction region.

B

Junction

Z

in the B and Z sections was observed together with extrusion of the A•T base pair. The RMSD values between the earlier B-Z structure and these three are 0.70, 0.36 and 0.45 Å.

3.5.5 Biological aspects of Z-DNA

The accessibility of the C8 (and to a lesser extent the N7) position of guanine in Z-DNA renders it susceptible to attack by several electrophilic compounds at these positions. Examples include the N-aryl carcinogens acetylaminofluorene and aflatoxin, as well as C8-bromination. All stabilise Z-DNA and promote the B to Z structural transition. Thus, Z-DNA tracts within genomic sequences could act as mutational hot-spots for these agents. The fact that Z-DNA formation is also facilitated by methylation at the C5 position of cytosine, a possible mechanism in some eukaryotic organisms for regulating gene transcription suggests that Z-DNA could play a role in gene regulation.

Z-DNA is a higher-energy polymorph of DNA and in its linear form is less stable than either A- or B-forms, requiring high salt or alcohol concentrations for maintenance of its structure in solution. These reduce electrostatic interactions between inter-strand phosphate groups, which in Z-DNA are much closer together than in B-DNA (7.7 Å in the Z_I form compared to 11.7 Å across the minor groove in B-DNA). Z-DNA sequences can also be stabilised at physiological salt levels when incorporated in under-wound, negatively supercoiled covalently closed-circular DNA, for example, in plasmid DNA. Even low (~1–2%) levels of Z-forming sequence can significantly change supercoiling properties. It has been proposed (Rahmouni & Wells, 1992) that a role for Z-DNA in vivo is to act as a signal for the induction of transcription via supercoiling.

The classic Z-forming sequence $d(CG)_n$ is much rarer in eukaryotic than in prokaryotic organisms. However, the sequence $d(CA/GT)_n$, which can also form left-handed DNA in negatively supercoiled plasmids, is much more prevalent in eukaryotic genomes. The overriding questions of whether Z-DNA actually exists naturally, and if so what its function may be, have not as yet been fully answered (Rich & Zhang, 2003). It is on balance, likely though that Z-DNA can be formed in cells under appropriate circumstances. A number of proteins that bind to Z-DNA have been isolated, and several co-crystal structures have been determined. Following the crystal structure analysis of a Z-DNA complex with the Zα high-affinity binding domain of an RNA editing enzyme (Schwartz et al., 1999), a subsequent database search for other proteins with the same Z-binding motif found a closely similar domain in a tumour-associated protein (DLM-1). The crystal structure (Schwartz et al., 2001) of a Z-DNA complex with this shows closed structural analogy with the Zα complex. In both, the critical protein to Z-DNA contacts are made by several conserved amino acid residues. The crystal structure of a Z-DNA complex with a poxvirus protein that also contains the Z-binding motif shows nine amino acid residues in contact with the DNA (Ha et al., 2004). In all these instances, the nature of the protein is consistent with the notion that Z-DNA can play a role in the regulation of transcription/translation. Several other Z-binding protein crystal structures have been subsequently reported, such as the Z-DNA-binding domain of the *Danio rerio* protein kinase PKZ (Ravichandran et al., 2019; Subramani et al., 2016).

However, there is still a lack of definitive proof that Z-DNA has a biological function, although several well-established lines of evidence point to its involvement in numerous biological processes. For example, it has been proposed that it fulfills a function to maintain negative DNA winding. Several Z-DNA binding proteins have been identified that are involved in immune responses (Kesavardhana et al., 2020; Nichols et al., 2021), although the precise role of left-handed Z-DNA in these remains to be determined. Z-forming sequences are widely distributed in eukaryotic organisms and they can induce genetic instability in both yeast and mammalian cells, which are resolved by the nucleotide excision repair pathway (McKinney et al., 2020). This study developed plausible molecular models for B-Z junction interactions with the ERCC1-XPF and MSH2-MSH3 repair protein assemblies, which were supportive of the cleavage patterns and other experimental data in this study.

3.6 Bent DNA

3.6.1 DNA periodicity in solution

DNA under physiological solution conditions is generally assumed to be in a B-type conformation, even when combined with chromosomal proteins. There are at first sight some important differences between the B structures as found in the crystal and fibre, and that in solution, even as naked DNA. The structural studies all point to a helical repeat of 10.0—10.1 residues per turn for averaged sequence DNA. However, plasmid mobility, nuclease digestion and hydroxyl radical cleavage experiments (Tullius & Dombroski, 1985) have given a rather different value, of close to 10.5 base pairs per turn, again averaged over many base pairs. This apparent discrepancy is only partly resolved by the observation that helical periodicity in solution is markedly sequence dependent (Rhodes & Klug, 1981). Cleavage at a particular point along a DNA sequence depends on the local helical twist, and hence groove width, in a manner that correlates with the features that have been found in the Dickerson—Drew dodecamer, both in the crystal and in solution by NMR.

3.6.2 A-tracts and bending

The DNA double helix is usually thought of as a straight molecule. The original Watson—Crick model is of a straight helix, since it was based on data from 'straight' DNA fibres. In fact, it must be bent or curved in many of its functional roles, especially when interacting with particular proteins. DNA is even deformed from linearity in many of the dodecamer crystal structures, being typically bent by about 19°. There has been much speculation on the nature of DNA bending in such situations as in the nucleosome component of chromatin (which has 145 base pairs wound around a histone protein core) and closed-circular DNA, prior to determinations of nucleosome crystal structures. Some sequences have an inherent tendency to bend independently of there being any protein present. The best studied such sequences are those comprising runs of adenines on one strand and thymines on the other (variously termed A-tracts or A/T tracts).

Intrinsic DNA bending and curvature was discovered (Crothers & Shakked, 1999; Marini et al., 1982) in restriction fragments of kinetoplast DNA, which have extensive repeats of A-tracts within them, each having about five to seven base pairs. These fragments showed highly anomalous gel electrophoresis behaviour, with very slow migration through the gel, as if the DNA had a much higher molecular weight than was the case. Bending occurs when sequences or motifs such as 5'-AAAAAA are repeated in phase with the DNA helical repeat itself (Crothers et al., 1990; Hagerman, 1990), that is, every 10 or 11 base pairs. The phasing is critical since it forces the tract to be always on the same side of the helix, thereby enabling the individual small bends associated with each tract to add up to a significant amount of total bending. It has been estimated (Koo et al., 1990) that an individual $(A)_6$ tract bends a helix by about 17–21°. A typical bending sequence, from kinetoplast DNA, is as follows:

5'-...CC<u>AAAAA</u>TGTC<u>AAAAAA</u>TAGGC<u>AAAAAA</u>TGCC<u>AAAAAT</u>...

Two distinct structural models have been suggested to account for these observations of bending. In one, a combination of gel retardation and NMR studies have been used (Nadeau & Crothers, 1989), resulting in a structural model that has bending towards the minor groove direction of the helix. The compression and closure of the minor groove is in part a result of negative A•T base pair tilt within the A-tract itself. In this, the 'junction' model, bending is caused by the abrupt change in structure when the A-tract meets the more normal (straight) B-DNA intervening sequence, resulting in the helix axes of the two segments being at an angle to each other (Fig. 3.20a). Abrupt changes in either tilt or roll can produce this situation. Thus, in the junction model, the A-tracts are themselves not bent. By contrast, the 'wedge' model (Ulanovsky & Trifonov, 1987) has continuous bending along the A-tract (Fig. 3.20b) as a result of changes in roll at each

Figure 3.20 Schematic representations of bending models for A-tract DNA. (a) junction model. (b) wedge model.

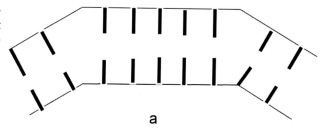

a

b

dinucleotide step — this model has also been suggested for the continuous (induced) bending of general sequence DNA when wrapped around histones in nucleosomes. The structural basis for bending has been controversial, with their being continuing debate on the roles played by base-pair propeller twist, roll, buckle and slide both within and flanking A-tracts themselves.

The advent of relevant crystal and NMR structures has provided structural data for distinguishing between these models, with increasing reliability for the various base and base pair morphological parameters, although no bent structure is currently available at atomic resolution.

3.6.3 Structures showing bending

Several oligonucleotide crystal structures contain short dA•dT sequences that might be thought of as models for A-tracts, as discussed in Sections 3.1 and 3.2 in this chapter. These have narrow minor grooves with ordered water molecules, high propeller twists for the A•T base pairs, together with, in two cases (Nelson et al., 1987; DiGabriele & Steitz, 1993), bifurcated three-centre hydrogen bonds involving adjacent A•T base pairs (Fig. 3.11). Such hydrogen bonding was, however, not observed (Edwards et al., 1992; Woods et al., 2004) in the higher resolution structures determined for d(CGCAAATTTGCG). The $(A)_6•(T)_6$ tract in one crystal structure (Nelson et al., 1987) is itself little bent, lending support to the junction model. All three crystal structures have an overall bending towards the major groove direction, in contrast to the inference from gel studies, that A-tracts are bent towards the minor groove. Crystal packing forces have been invoked to account for this discrepancy (DiGabriele et al., 1989). More recently determined structures (see below) are probably better models for A-tracts.

There is some evidence that A-tracts do not inherently require three-centre A•T hydrogen bonds at all positions for curvature to be produced, in addition to the structures mentioned above. For example, sequences having inosine•cytosine (I•C) base pairs (these cannot form major-groove three-centre hydrogen bonds since inosine lacks an N6 group) still result in bent structures (Diekmann et al., 1992). A combined crystallographic and gel electrophoresis study has been undertaken on a number of A-tract model oligonucleotides with varying numbers of I•C base pairs in the tract (Shatzky-Schwatz et al., 1997). This showed that high propeller twist and bending is maintained when there is isolated I•C substitution for A•T but abolished for pure I-tracts. Interestingly, this analysis also suggested the existence of attractive three-centre NH...H interactions (Luisi et al., 1998).

The degree of bending in a DNA structure can be calculated with several of the helix morphology programs outlined in Chapter 2. The use of normal vector plots is especially useful for visualising bending (Dickerson, 1998; Goodsell & Dickerson, 1994). This procedure uses the unit vectors perpendicular to the least-squares plane from a base pair and enables both local and global bending to be analysed. The method was used to show that the Dickerson–Drew dodecamer (Dickerson & Drew, 1981; Drew & Dickerson, 1981; Wing et al., 1980) is bent by 19°, albeit in the major groove direction. The crystal structure of the decamer d(CATGGCCATG) shows distinct curvature features. The helix is not straight but has a smooth 23° bend over the four central base pairs (Goodsell et al., 1993). This intrinsic bend is consistent with a straight A-tract model, although the roll compression of the major groove suggests a third type of model distinct from the wedge or junction models. The crystal structure (Hizver et al., 2001) of the dodecamer d(ACCGAATTCGGT) is non-isomorphous with the Dickerson–Drew dodecamer, since it crystallises in the triclinic space group P1 rather than the normal $P2_12_12_1$ one for dodecamers. This sequence is a biologically relevant one — it is the high-affinity A-tract binding site of the E2 papillomavirus protein. There are three independent molecules in the crystallographic asymmetric unit, providing three independent views of the preferred structure of this sequence. All three are closely similar, with RMSDs ranging from 0.5 to 0.7 Å, strongly suggesting that their structure represents a global minimum. They have an average curvature of 9°, with very narrow minor grooves in the 5'-AATT regions and highly twisted A•T base pairs. The magnitude and direction of curvature is in accord with solution data on this sequence. The bending is produced by a combination of minor-groove compression in the A-tract and major groove compression of the flanking C•G base pairs. These are a consequence of the inherent features of the central octamer duplex sequence — high propeller twists for the A•T base pairs and buckling at the A-tract junctions.

The structure of the duplex formed by the non-self-complementary dodecamer d(GGCAAAAAACGG), bound to its complementary strand d(CCGTTTTTTGCC), has been analysed by NMR methods (MacDonald et al., 2001). This A-tract structure is bent by 19°, in accord with gel retardation measurements. The A•T base pairs in the A-tract have high propeller twist, with most of the bending occurring in the short flanking C/G sequences. The A-tract itself in this structure has a small amount (5°) of intrinsic bend. The junction bends have been ascribed to combinations of tilt and roll. NMR methods have also shown (Hud et al., 1999) that ammonium ions are bound in the A-tract minor groove and may play a role in the bending. These ions cannot be distinguished from water molecules in oligonucleotide structures by current X-ray methods. The NMR structure of the same decamer duplex as that reported by MacDonald et al. (2001) has also been analysed by NMR methods (Barbič et al., 2003). This study has found an average global bend of ~9° into the minor groove for the inner eight base pairs (Fig. 3.21), with roll and tilt from several base pairs contributing to this value. An NMR-based comparison has been made between A-tract sequences with a central ApT and TpA step (Stefl et al., 2004). This has concluded that whereas the large negative roll at the ApT step produces a bend towards the minor groove that are in phase with the bends at the A-tract junctions, the TpA step has a positive roll with bending towards the major groove, cancelling out the junction bends so that there is no overall bend for this sequence (Table 3.8).

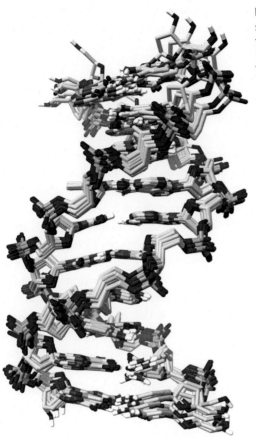

Figure 3.21 The ensemble of NMR-derived structures (Barbič et al., 2003), PDB id 1NEV, for the duplex formed by d(GGCAAAAAACGG) and its complementary strand d(CCGTTTTTTGCC). The structure has an average global bend of ~9° into the minor groove for the inner eight base pairs.

Table 3.8 Selected bent DNA crystal and NMR structures. The second strand sequences are listed for non-self-complementary structures.

Sequence	PDB id	Resolution Å	Space group
d(CGCAAAAAAGCG) + d(CGCTTTTTTGCG)	1D98	2.50	$P2_12_12_1$
d(CGCAAAAATGCG) + d(CGCATTTTTGCG)	1BDN	2.60	$P2_12_12_1$
d(GCAAAATTTTGC)	1RVH	NMR	—
d(CGTTTTAAAACG)	1RVI	NMR	—
d(GGCAAAACGG) + d(CCGTTTTGCC)	1NEV	NMR	—

It is striking that the macroscopic degree of bending observed in these structures accords reasonably well with solution measurements and with the overall features of the junction model. Some features are common to all the structures such as a very narrow minor groove in the A-tract coupled with high A•T base pair propeller twists. Yet, the fine detail of how bending is achieved at and close to the junctions, as well as the precise degree of bending, differ from one structure to another (Dickerson, 1999). This suggests that there is inherent flexibility of the junctions in A-tract structures, and that there are several equi-energetic bent arrangements that can be readily accessed, albeit in differing ways. This is supported by a survey of bending both in native oligomers and in DNA—protein complexes (Young et al., 1995), which also finds evidence for the straight A-tract model. There has been some debate as to whether, and to what extent, the bending in the crystallographically derived structures is affected by the near-universal use of 2-methyl-2,4-pentanediol (MPD) in crystallisation experiments, where its role is to act as a precipitating agent by decreasing the aqueous solubility of oligonucleotides. Electrophoresis mobility (Sprous et al., 1995) and free-radical cleavage (Ganunis et al., 1996) experiments in the presence of MPD suggest that high concentrations do produce a change in A-tract structure and decreased bending, although these 'low resolution' methods clearly cannot specify the molecular nature of the changes. On the other hand, most of the A-tract model crystal structures do show solution-relevant bending, even though MPD may in some circumstances qualitatively affect its degree (Dickerson et al., 1996). A simulation study (Rohs et al., 2005) using a Monte Carlo simulation approach with oligonucleotide sequences that bend when bind to the E2 protein from papillomaviruses has resulted in excellent correspondence with the earlier crystal structures (Hizver et al., 2001).

3.6.4 The structure of poly dA•dT

The structure of poly dA•dT itself has been the subject of several studies. Fibre diffraction methods have defined a structure, the 'heteronomous' model (Arnott et al., 1983) having distinct backbone conformations and sugar puckers for the two dA and dT strands. The results of subsequent analyses (Aymami et al., 1989; Chandrasekaran & Radha, 1992), although differing in detail, are consistent with this structure, with both strands having B-like conformations, albeit with a narrow minor groove. Propeller twists for the A•T base pairs are likely to be significant (~15°), but not as high as in the earlier crystal structures, thus reducing the possibility of three-centre hydrogen bonding. Molecular dynamics studies (Fritsch & Westhof, 1991; Sherer et al., 1999; Sprous et al., 1999; Strahs & Schlick, 2000) similarly indicate that this type of hydrogen bonding is not of primary importance in stabilising oligo or poly dA•dT, but merely that it can occur when the geometric circumstances allow.

3.7 Concluding remarks

Fibre diffraction studies on DNA polynucleotides have long established the inherent flexibility of the double helix. These have also defined the major polymorphs in terms of A, B and Z duplexes. The fact that the majority of polynucleotide fibre structures can be readily inter-convertible under appropriate environmental conditions has also shown us that they actually represent low-energy points along a continuum of structures. The very large number of oligonucleotide crystal structures now known is mostly representative of this continuum, with the important caveat that it is sometimes not possible to distinguish between true fluctuations from the canonical mean, and experimental uncertainties, especially when these are not fully defined. The aim of establishing a straightforward and universal code for correspondence between sequence and structure remains elusive, and indeed may not exist. Instead, these structures are telling us much about DNA flexibility, its low-energy states and the pathways between them. Succeeding chapters describe how this flexibility can be exploited by interactions with other molecules, small and large.

References

Aboul-ela, F., Varani, G., Walker, G. T., & Tinoco, I., Jr. (1988). The TFIIIA recognition fragment d(GGATGGGAG).d(CTCCCATCC) is B-form in solution. *Nucleic Acids Research, 16*, 3559—3572.

Abrescia, N. G. A., Thompson, A., Huynh-Dinh, T., & Subirana, J. A. (2002). Crystal structure of an antiparallel DNA fragment with Hoogsteen base pairing. *Proceedings of the National Academy of Sciences of the United States of America, 99*, 2806—2811.

Acosta-Reyes, F. J., Pagan, M., Fonfría-Subirós, E., Saperas, N., Subirana, J. A., & Campos, J. L. (2020). The influence of Ni^{2+} and other ions on the trigonal structure of DNA. *Biopolymers, 8*, e23397.

Arai, S., Chatake, T., Minezaki, Y., & Niimura, N. (2002). Crystallisation of a large single crystal of a B-DNA decamer for a neutron diffraction experiment by the phase-diagram method. *Acta Crystallographica, D58*, 151—153.

Arai, S., Chatake, T., Ohhara, T., Kurihara, K., Tanaka, I., Suzuki, N., Fujimoto, Z., Mizuno, H., & Niimura, N. (2005). Complicated water orientations in the minor groove of the B-DNA decamer d(CCATTAATGG)$_2$ observed by neutron diffraction experiments. *Nucleic Acids Research, 33*, 3017—3024.

Arnott, S. (1999). In S. Neidle (Ed.), *Oxford handbook of nucleic acid structure* (p. 1). Oxford: Oxford University Press.

Arnott, S., Chandrasekaran, R., Birdsall, D. L., Leslie, A. G. W., & Ratliffe, R. L. (1980). Left-handed DNA helices. *Nature, 283*, 743—745.

Arnott, S., Chandrasekaran, R., Hall, I. H., & Puigjaner, L. C. (1983). Heteronomous DNA. *Nucleic Acids Research, 11*, 4141—4155.

Arnott, S., Chandrasekaran, R., Hall, I. H., Puigjaner, L. C., Walker, J. K., & Wang, M. (1982). DNA secondary structures: Helices, wrinkles, and junctions. *Cold Spring Harbor Symposia on Quantitative Biology, 47*, 53—65.

Aymami, J., Coll, M., Frederick, C. A., Wang, A. H.-J., & Rich, A. (1989). The propeller DNA conformation of poly(dA).poly(dT). *Nucleic Acids Research, 17*, 3229—3245.

Azad, R. N., Zafiropoulos, D., Ober, D., Jiang, Y., Chiu, T. P., Sagendorf, J. M., Rohs, R., & Tullius, T. D. (2018). Experimental maps of DNA structure at nucleotide resolution distinguish intrinsic from protein-induced DNA deformations. *Nucleic Acids Research, 46*, 2636—2647.

Ban, C., Ramakrishnan, B., & Sundaralingam, M. (1996). Crystal structure of the self-complementary 5'-purine start decamer d(GCGCGCGCGC) in the Z-DNA conformation. I. *Biophysical Journal, 7,* 1215–1221.

Barbič, A., Zimmer, D. P., & Crothers, D. M. (2003). Structural origins of adenine-tract bending. *Proceedings of the National Academy of Sciences of the United States of America, 100,* 2369–2373.

Basham, B., Eichman, B. F., & Ho, P. S. (1999). In S. Neidle (Ed.), *Oxford handbook of nucleic acid structure* (p. 199). Oxford: Oxford University Press.

Bingman, C. A., Zon, G., & Sundaralingam, M. (1992). Crystal and molecular structure of the A-DNA dodecamer d(CCGTACGTACGG). Choice of fragment helical axis. *Journal of Molecular Biology, 227,* 738–756.

Calladine, C. R. (1982). Mechanics of sequence-dependent stacking of bases in B-DNA. *Journal of Molecular Biology, 161,* 343–352.

Calladine, C. R., & Drew, H. R. (1984). A base-centred explanation of the B-to-A transition in DNA. *Journal of Molecular Biology, 178,* 773–782.

Campos, J. L., Urpi, L., Sanmartin, T., Gouyette, C., & Subirana, J. A. (2005). DNA coiled coils. *Proceedings of the National Academy of Sciences of the United States of America, 102,* 3663–3666.

Campos, J. L., Valls, N., Urpi, L., Gouyette, C., Sanmartin, T., Richter, M., Alechaga, E., Santaolalla, A., Baldini, R., Creixell, M., Ciurans, R., Skokan, R., Pous, J., & Subirana, J. A. (2006). Overview of the structure of all-AT oligonucleotides: Organization in helices and packing interactions. *Biophysical Journal, 91,* 892–903.

Černý, J., Božíková, P., Malý, M., Tykač, M., Biedermannová, L., & Schneider, B. (2020a). Structural alphabets for conformational analysis of nucleic acids available at dnatco.datmos.org. *Acta Crystallographica Section D, 76,* 805–813.

Černý, J., Božíková, P., Svoboda, J., & Schneider, B. (2020b). A unified dinucleotide alphabet describing both RNA and DNA structures. *Nucleic Acids Research, 48,* 6367–6381.

Chandrasekaran, R., & Radha, A. (1992). Structure of poly d(A).poly d(T). *Journal of Biomolecular Structure & Dynamics, 10,* 153–168.

Chatake, T., Tanaka, I., Umino, H., Arai, S., & Niimura, N. (2005). The hydration structure of a Z-DNA hexameric duplex determined by a neutron diffraction technique. *Acta Crystallographica, D61,* 1088–1098.

Chen, Y. Q., Ghosh, S., & Ghosh, G. (1998). A novel DNA recognition mode by the NF-kappa B p65 homodimer. *Nature Structural Biology, 5,* 67–73.

Chiu, T. K., & Dickerson, R. E. (2000). A crystal structures of B-DNA reveal sequence-specific binding and groove-specific bending of DNA by magnesium and calcium. *Journal of Molecular Biology, 301,* 915–945.

Clark, G. R., Brown, D. G., Sanderson, M. R., Chwalinski, T., Neidle, S., Veal, J. M., Jones, R. L., Wilson, W. D., Zon, G., Garman, E., & Stuart, D. I. (1990). Crystal and solution structures of the oligonucleotide d(ATGCGCAT)2: A combined X-ray and NMR study. *Nucleic Acids Research, 18,* 5521–5528.

Coté, M. L., Pflomm, M., & Georgiadas, M. M. (2003). Staying straight with A-tracts: A DNA analog of the HIV-1 polypurine tract. *Journal of Molecular Biology, 330,* 57–74.

Coté, M. L., Yohannan, S. J., & Georgiadis, M. M. (2000). Use of an N-terminal fragment from moloney murine leukemia virus reverse transcriptase to facilitate crystallization and analysis of a pseudo-16-mer DNA molecule containing G-A mispairs. *Acta Crystallographica, D56,* 1120–1131.

Crothers, D. M., & Shakked, Z. (1999). In S. Neidle (Ed.), *Oxford handbook of nucleic acid structure* (p. 455). Oxford: Oxford University Press.

Crothers, D. M., Haran, T. E., & Nadeau, J. G. (1990). Intrinsically bent DNA. *Journal of Biological Chemistry, 265,* 7093–7096.

Dans, P. D., Balaceanu, A., Pasi, M., Patelli, A. S., Petkevičiūtė, D., Walther, J., Hospital, A., Bayarri, G., Lavery, R., Maddocks, J. H., & Orozco, M. (2019). The static and dynamic structural heterogeneities of B-DNA: Extending Calladine-Dickerson rules. *Nucleic Acids Research, 47,* 11090–11102.

Dickerson, R. E., et al. (1989). Definitions and nomenclature of nucleic acid structure parameters. *The EMBO Journal, 8,* 1–4.

Dickerson, R. E. (1998). DNA bending: The prevalence of kinkiness and the virtues of normality. *Nucleic Acids Research, 26,* 1906–1926.

Dickerson, R. E. (1999). In S. Neidle (Ed.), *Oxford handbook of nucleic acid structure.* Oxford: Oxford University Press.

Dickerson, R. E., & Drew, H. R. (1981). Structure of a B-DNA dodecamer: II. Influence of base sequence on helix structure. *Journal of Molecular Biology, 149,* 761–786.

Dickerson, R. E., Goodsell, D., & Kopka, M. L. (1996). MPD and DNA bending in crystals and in solution. *Journal of Molecular Biology, 256,* 108–125.

Dickerson, R. E., Goodsell, D. S., & Neidle, S. (1994). …the tyranny of the lattice…. *Proceedings of the National Academy of Sciences of the United States of America, 91,* 3579–3583.

Diekmann, S., Mazzaralli, J. M., McLaughlin, L. W., von Kitzing, E., & Travers, A. A. (1992). DNA curvature does not require bifurcated hydrogen bonds or pyrimidine methyl groups. *Journal of Molecular Biology, 225,* 729–738.

DiGabriele, A. D., Sanderson, M. R., & Steitz, T. A. (1989). Crystal lattice packing is important in determining the bend of a DNA dodecamer containing an adenine tract. *Proceedings of the National Academy of Sciences of the United States of America, 86,* 1816–1820.

DiGabriele, A. D., & Steitz, T. A. (1993). A DNA dodecamer containing an adenine tract crystallizes in a unique lattice and exhibits a new bend. *Journal of Molecular Biology, 231,* 1024–1039.

DiMaio, F., Yu, X., Rensen, E., Krupovic, M., Prangishvili, D., & Egelman, E. H. (2015). A virus that infects a hyperthermophile encapsidates A-form DNA. *Science, 348,* 914–917.

Dohm, J. A., Hsu, M.-H., Hwu, J.-R., Huang, R. C. C., Moudrianakis, E. N., Lattman, E. E., & Gittis, A. G. (2005). Influence of ions, hydration, and the transcriptional inhibitor P4N on the conformations of the Sp1 binding site. *Journal of Molecular Biology, 349,* 731–744.

Drew, H. R., & Dickerson, R. E. (1981). Structure of a B-DNA dodecamer: III. Geometry of hydration. *Journal of Molecular Biology, 151,* 535–556.

Drew, H. R., & Travers, A. A. (1984). DNA structural variations in the *E. coli* tyrT promoter. *Cell, 37,* 491–502.

Edwards, K. J., Brown, D. G., Spink, N., Skelly, J. V., & Neidle, S. (1992). Molecular structure of the B-DNA dodecamer d(CGCAAATTTGCG)$_2$ an examination of propeller twist and minor-groove water structure at 2·2Å resolution. *Journal of Molecular Biology, 226,* 1161–1173.

Egli, M., Tereshko, V., Teplova, M., Minasov, G., Joachimiak, A., Sanishvili, R., Weeks, C. M., Miller, R., Maier, M. A., An, H., Cook, P. D., & Manoharan, M. (1998). X-ray crystallographic analysis of the hydration of A- and B-form DNA at atomic resolution. *Biopolymers, 48,* 234–252.

Egli, M., Williams, L. D., Gao, Q., & Rich, A. (1991). Structure of the pure-spermine form of Z-DNA (magnesium free) at 1-Å resolution. *Biochemistry, 30,* 11388–11402.

Eisenstein, M., & Shakked, Z. (1995). Hydration patterns and intermolecular interactions in A-DNA crystal structures. Implications for DNA recognition. *Journal of Molecular Biology, 248,* 662–678.

Eisenstein, M., Frolow, F., Shakked, Z., & Rabinovich, D. (1990). The structure and hydration of the A-DNA fragment d(GGGTACCC) at room temperature and low temperature. *Nucleic Acids Research, 18,* 3185–3194.

Elsawy, K. M., Hodgson, M. K., & Caves, L. S. D. (2005). The physical determinants of the DNA conformational landscape: An analysis of the potential energy surface of single-strand dinucleotides in the conformational space of duplex DNA. *Nucleic Acids Research, 33,* 5749–5762.

Fairall, L., Martin, S., & Rhodes, D. (1989). The DNA binding site of the Xenopus transcription factor IIIA has a non- B-form structure. *The EMBO Journal, 8,* 1809–1817.

Farwar, J., Packer, M. J., & Hunter, C. A. (2006). Prediction of atomic structure from sequence for double helical DNA oligomers. *Biopolymers, 81,* 51–61.

Finley, J. B., & Luo, M. (1998). X-ray crystal structures of half the human papilloma virus E2 binding site: d(GACCGCGGTC). *Nucleic Acids Research, 26,* 5719–5727.

Frederick, C. A., Quigley, G. J., Teng, M.-K., Coll, M., van der Marel, G., van Boom, J. H., Rich, A., & Wang, A. H.-J. (1989). Molecular structure of an A-DNA decamer d(ACCGGCCGGT). *European Journal of Biochemistry, 181,* 295–307.

Fritsch, V., & Westhof, E. (1991). Three-center hydrogen bonds in DNA: Molecular dynamics of poly(dA).poly(dT). *Journal of the American Chemical Society, 113,* 8271–8277.

Fujii, S., Wang, A. H.-J., Quigley, G. J., Westerink, H., van der Marel, G., van Boom, J. H., & Rich, A. (1985). The octamers d(CGCGCGCG) and d(CGCATGCG) both crystallize as Z-DNA in the same hexagonal lattice. *Biopolymers, 24*, 243–250.

Ganunis, R., Guo, H., & Tullius, T. D. (1996). Effect of the crystallizing agent 2-methyl-2,4-pentanediol on the structure of adenine tract DNA in solution. *Biochemistry, 35*, 13729–13732.

Gao, Y.-G., Robinson, H., & Wang, A. H.-J. (1999). High-resolution A-DNA crystal structures of d(AGGGGCCCCT). An A-DNA model of poly(dG) x poly(dC). *European Journal of Biochemistry, 261*, 413–420.

Gardiner, E. J., Hunter, C. A., Lu, X. J., & Willett, P. (2004). A structural similarity analysis of double-helical DNA. *Journal of Molecular Biology, 343*, 879–889.

Gardiner, E. J., Hunter, C. A., Packer, M. J., Palmer, D. S., & Willett, P. (2003). Sequence-dependent DNA structure: A database of octamer structural parameters. *Journal of Molecular Biology, 332*, 1025–1035.

Goodsell, D. S., & Dickerson, R. E. (1994). Bending and curvature calculations in B-DNA. *Nucleic Acids Research, 22*, 5497–5503.

Goodsell, D. S., Grzeskowiak, K., & Dickerson, R. E. (1995). Crystal structure of C-T-C-T-C-G-A-G-A-G. Implications for the structure of the Holliday junction. *Biochemistry, 34*, 1022–1029.

Goodsell, D. S., Kaczor-Grzeskowiak, M., & Dickerson, R. E. (1994). Crystal structure of C-C-A-T-T-A-A-T-G-G. Implications for bending of B-DNA at A-T steps. *Journal of Molecular Biology, 239*, 79–86.

Goodsell, D. S., Kopka, M. L., Cascio, D., & Dickerson, R. E. (1993). Crystal structure of CATGGCCATG and its implications for A-tract bending models. *Proceedings of the National Academy of Sciences of the United States of America, 90*, 2930–2934.

Gorin, A. A., Zhurkin, V. B., & Olson, W. K. (1995). B-DNA twisting correlates with base-pair morphology. *Journal of Molecular Biology, 247*, 34–48.

Ha, S. C., Lokanath, N. K., Van Quyen, D., Wu, C. A., Lowenhaupt, K., Rich, A., Kim, Y. G., & Kim, K. K. (2004). A poxvirus protein forms a complex with left-handed Z-DNA: Crystal structure of a Yatapoxvirus Zalpha bound to DNA. *Proceedings of the National Academy of Sciences of the United States of America, 101*, 14367–14372.

Ha, S. C., Lowenhaupt, K., Rich, A., Kim, Y. G., & Kim, K. K. (2005). Crystal structure of a junction between B-DNA and Z-DNA reveals two extruded bases. *Nature, 437*, 1183–1186.

Hagerman, P. J. (1990). Sequence-directed curvature of DNA. *Annual Review of Biochemistry, 59*, 755–781.

Hamelberg, D., McFail-Isom, L., Williams, L. D., & Wilson, W. D. (2000). Flexible Structure of DNA: Ion dependence of minor-groove structure and dynamics. *Journal of the American Chemical Society, 122*, 10513–10520.

Hamelberg, D., Williams, L. D., & Wilson, W. D. (2001). Influence of the dynamic positions of cations on the structure of the DNA minor groove: Sequence-dependent effects. *Journal of the American Chemical Society, 123*, 7745–7755.

Hardwick, J. S., Ptchelkine, D., El-Sagheer, A. H., Tear, I., Singleton, D., Phillips, S. E. V., Lane, A. N., & Brown, T. (2017). 5-Formylcytosine does not change the global structure of DNA. *Nature Structural & Molecular Biology, 24*, 544–552.

Harp, J. M., Coates, L., Sullivan, B., & Egli, M. (2018). Cryo-neutron crystallographic data collection and preliminary refinement of left-handed Z-DNA d(CGCGCG). *Acta Crystallographica, F74*, 603–609.

Hartmann, B., Piazzola, D., & Lavery, R. (1993). BI-BII transitions in B-DNA. *Nucleic Acids Research, 21*, 561–568.

Hays, F. A., Teegarden, A., Jones, Z. J. R., Harms, M., Raup, D., Watson, J., Cavaliere, E., & Ho, P. S. (2005). How sequence defines structure: A crystallographic map of DNA structure and conformation. *Proceedings of the National Academy of Sciences of the United States of America, 102*, 7157–7162.

Heinemann, U., Alings, C., & Bansal, M. (1992). Double helix conformation, groove dimensions and ligand binding potential of a G/C stretch in B-DNA. *The EMBO Journal, 11*, 1931–1939.

Heinemann, U., & Roske, Y. (2020). Symmetry in nucleic-acid double helices. *Symmetry, 12*, 737.

Hizver, J., Rozenberg, H., Frolow, F., Rabinovich, D., & Shakked, Z. (2001). DNA bending by an adenine-thymine tract and its role in gene regulation. *Proceedings of the National Academy of Sciences of the United States of America, 98*, 8490–8495.

Huang, D.-B., Phelps, C. B., Fusco, A. J., & Ghosh, G. (2005). Crystal structure of a free kappaB DNA: Insights into DNA recognition by transcription factor NF-kappaB. *Journal of Molecular Biology, 346,* 147−160.

Hud, N. V., Sklenář, & Feigon, J. (1999). Localization of ammonium ions in the minor groove of DNA duplexes in solution and the origin of DNA A-tract bending. *Journal of Molecular Biology, 286,* 651−660.

Hunter, C. A. (1993). Sequence-dependent DNA structure: The role of base stacking interactions. *Journal of Molecular Biology, 230,* 1025−1054.

Hunter, C. A., & Lu, X.-J. (1997). DNA base-stacking interactions: A comparison of theoretical calculations with oligonucleotide X-ray crystal structures. *Journal of Molecular Biology, 265,* 603−619.

Jain, S., Zon, G., & Sundaralingam, M. (1991). Hexagonal crystal structure of the A-DNA octamer d(GTGTACAC) and its comparison with the tetragonal structure: Correlated variations in helical parameters. *Biochemistry, 30,* 3567−3576.

Johansson, E., Parkinson, G., & Neidle, S. (2000). A new crystal form for the dodecamer C-G-C-G-A-A-T-T-C-G-C-G: Symmetry effects on sequence-dependent DNA structure. *Journal of Molecular Biology, 300,* 551−561.

Karthik, S., Mandal, P. K., Thirugnanasambandam, A., & Gautham, N. (2019). Crystal structures of disordered Z-type helices. *Nucleosides, Nucleotides & Nucleic Acids, 38,* 279−293.

Karthik, S., Thirugnanasambandam, A., Mandal, P. K., & Gautham, N. (2017). Crystal structure of d(CCGGGGTACCCCGG)$_2$ at 1.4 Å resolution. *Acta Crystallographica, F73,* 259−265.

Kesavardhana, S., Malireddi, R. K. S., Burton, A. R., Porter, S. N., Vogel, P., Pruett-Miller, S. M., & Kanneganti, T. D. (2020). The Zα2 domain of ZBP1 is a molecular switch regulating influenza-induced PANoptosis and perinatal lethality during development. *Journal of Biological Chemistry, 295,* 8325−8330.

Kim, D., Hur, J., Han, J. H., Ha, S. C., Shin, D., Lee, S., Park, S., Sugiyama, H., & Kim, K. K. (2018). Sequence preference and structural heterogeneity of BZ junctions. *Nucleic Acids Research, 46,* 10504−10513.

Kielkopf, C. L., Ding, S., Kuhn, P., & Rees, D. C. (2000). Conformational flexibility of B-DNA at 0.74 Å resolution: d(CCAGTACTGG)$_2$. *Journal of Molecular Biology, 296,* 787−801.

Koo, H.-S., Drak, J., Rice, J. A., & Crothers, D. M. (1990). Determination of the extent of DNA bending by an adenine-thymine tract. *Biochemistry, 29,* 4227−4234.

Kruse, H., Banáš, P., & Šponer, J. (2019). Investigations of stacked DNA base-pair steps: Highly accurate stacking interaction energies, energy decomposition, and many-body stacking effects. *Journal of Chemical Theory and Computation, 15,* 95−115.

Kubinec, M. G., & Wemmer, D. E. (1992). NMR evidence for DNA bound water in solution. *Journal of the American Chemical Society, 114,* 8739−8740.

Lauble, H., Frank, R., Blöcker, H., & Heinemann, U. (1988). Three-dimensional structure of d(GGGATCCC) in the crystalline state. *Nucleic Acids Research, 16,* 7799−7816.

Lercher, L., McDonough, M. A., El-Sagheer, A. H., Thalhammer, A., Kriaucionis, S., Brown, T., & Schofield, C. J. (2014). Structural insights into how 5-hydroxymethylation influences transcription factor binding. *Chemical Communications, 50,* 1794−1796.

Lipanov, A., Kopka, M. L., Kaczor-Grzeskowiak, Quintana, J., & Dickerson, R. E. (1993). Structure of the B-DNA decamer C-C-A-A-C-I-T-T-G-G in two different space groups: Conformational flexibility of B-DNA. *Biochemistry, 32,* 1373−1389.

Liu, J., & Subirana, J. A. (1999). Structure of d(CGCGAATTCGCG) in the presence of Ca^{2+}) ions. *Journal of Biological Chemistry, 274,* 24749−24752.

Lomonossoff, G. P., Butler, P. J. G., & Klug, A. (1981). Sequence-dependent variation in the conformation of DNA. *Journal of Molecular Biology, 149,* 745−760.

Luisi, B., Orozco, M., Sponer, J., Luque, F. J., & Shakked, Z. (1998). On the potential role of the amino nitrogen atom as a hydrogen bond acceptor in macromolecules. *Journal of Molecular Biology, 279,* 1123−1136.

Luo, Z., Dauter, M., & Dauter, Z. (2014). Phosphates in the Z-DNA dodecamer are flexible, but their P-SAD signal is sufficient for structure solution. *Acta Crystallographica, D70,* 1790−1800.

Luo, Z., Dauter, Z., & Gilski, M. (2017). Four highly pseudosymmetric and/or twinned structures of d(CGCGCG)$_2$ extend the repertoire of crystal structures of Z-DNA. *Acta Crystallographica, D73,* 940–951.

MacDonald, D., Herbert, K., Zhang, X., Polgruto, T., & Lu, P. (2001). Solution structure of an A-tract DNA bend. *Journal of Molecular Biology, 301,* 1081–1098.

Madhumalar, A., & Bansal, M. (2005). Sequence preference for BI/BII conformations in DNA: MD and crystal structure data analysis. *Journal of Biomolecular Structure & Dynamics, 23,* 13–27.

Mahendrasingam, A., Forsyth, V. T., Hussain, R., Greenall, R. J., Pigram, W. J., & Fuller, W. (1986). Time-resolved X-ray diffraction studies of the B in equilibrium D structural transition in the DNA double helix. *Science, 233,* 195–197.

Malinina, L., Fernandez, L. G., Huynh-Dinh, T., & Subirana, J. A. (1999). Structure of the d(CGCCCGCGGGCG) dodecamer: A kinked A-DNA molecule showing some B-DNA features. *Journal of Molecular Biology, 285,* 1679–1690.

Mandal, P. K., Collie, G. W., Kauffmann, B., & Huc, I. (2014). Racemic DNA crystallography. *Angewandte Chemie International Edition in English, 53,* 14424–14427.

Mandal, P. K., Collie, G. W., Srivastava, S. C., Kauffmann, B., & Huc, I. (2016). Structure elucidation of the Pribnow box consensus promoter sequence by racemic DNA crystallography. *Nucleic Acids Research, 44,* 5936–5943.

Marini, J. C., Levene, S. D., Crothers, D. M., & Englund, P. T. (1982). Bent helical structure in kinetoplast DNA. *Proceedings of the National Academy of Sciences of the United States of America, 79,* 7664–7668.

McCall, M., Brown, T., & Kennard, O. (1985). The crystal structure of d(G-G-G-G-C-C-C-C) a model for poly(dG) · poly(dC). *Journal of Molecular Biology, 183,* 385–396.

McCall, M., Brown, T., Hunter, W. N., & Kennard, O. (1986). The crystal structure of d(G-G-A-T-G-G-G-A-G) forms an essential part of the binding site for transcription factor IIIA. *Nature, 322,* 661–664.

McDermott, M. L., Vanselous, H., Corcelli, S. A., & Petersen, P. B. (2017). DNA's chiral spine of hydration. *ACS Central Science, 3,* 708–714.

McKinney, J. A., Wang, G., Mukherjee, A., Christensen, L., Sankara Subramanian, S. H., Zhao, J., & Vasquez, K. M. (2020). Distinct DNA repair pathways cause genomic instability at alternative DNA structures. *Nature Communications, 11,* 236.

Minasov, G., Tereshko, V., & Egli, M. (1999). Atomic-resolution crystal structures of B-DNA reveal specific influences of divalent metal ions on conformation and packing. *Journal of Molecular Biology, 291,* 83–99.

Montaño, S. P., Coté, M. L., Roth, M. J., & Georgiadas, M. M. (2006). Crystal structures of oligonucleotides including the integrase processing site of the Moloney murine leukemia virus. *Nucleic Acids Research, 34,* 5353–5360.

Nadeau, J. G., & Crothers, D. M. (1989). Structural basis for DNA bending. *Proceedings of the National Academy of Sciences of the United States of America, 86,* 2622–2626.

Narayana, N., & Weiss, M. A. (2009). Crystallographic analysis of a sex-specific enhancer element: Sequence-dependent DNA structure, hydration, and dynamics. *Journal of Molecular Biology, 385,* 469–490.

Nelson, H. C. M., Finch, J. T., Luisi, B. F., & Klug, A. (1987). The structure of an oligo(dA)-oligo(dT) tract and its biological implications. *Nature, 330,* 221–226.

Ng, H.-L., & Dickerson, R. E. (2001). DNA structure from A to B. *Proceedings of the National Academy of Sciences of the United States of America, 98,* 6986–6989.

Ng, H.-L., & Dickerson, R. E. (2002). Mediation of the A/B-DNA helix transition by G-tracts in the crystal structure of duplex CATGGGCCCATG. *Nucleic Acids Research, 30,* 4061–4067.

Ng, H.-L., Kopka, M. L., & Dickerson, R. E. (2000). The structure of a stable intermediate in the A ← → B DNA helix transition. *Proceedings of the National Academy of Sciences of the United States of America, 97,* 2035–2039.

Nichols, P. J., Bevers, S., Henen, M., Kieft, J. S., Vicens, Q., & Vögeli, B. (2021). Recognition of non-CpG repeats in Alu and ribosomal RNAs by the Z-RNA binding domain of ADAR1 induces A-Z junctions. *Nature Communications, 12,* 793.

Olson, W. K., Bansal, M., Burley, S. K., Dickerson, R. E., Gerstein, M., Harvey, S. C., Heinemann, U., Lu, X.-J., Neidle, S., Shakked, Z., Sklenar, H., Suzuki, M., Tung, C.-S., Westhof, E.,

Wolberger, C., & Berman, H. M. (2001). A standard reference frame for the description of nucleic acid base-pair geometry. *Journal of Molecular Biology, 313*, 229–237.

Packer, M. J., Dauncey, M. P., & Hunter, C. A. (2000). Sequence-dependent DNA structure: Tetranucleotide conformational maps. *Journal of Molecular Biology, 295*, 85–103.

Packer, M. J., & Hunter, C. A. (1998). Sequence-dependent DNA structure: The role of the sugar-phosphate backbone. *Journal of Molecular Biology, 280*, 407–420.

Packer, M. J., & Hunter, C. A. (2001). Sequence-structure relationships in DNA oligomers: A computational approach. *Journal of the American Chemical Society, 123*, 7399–7406.

Pasi, M., Maddocks, J. H., Beveridge, D., Bishop, T. C., Case, D. A., Cheatham, T., 3rd, Dans, P. D., Jayaram, B., Lankas, F., Laughton, C., Mitchell, J., Osman, R., Orozco, M., Pérez, A., Petkevičiūtė, D., Spackova, N., Sponer, J., Zakrzewska, K., & Lavery, R. (2014). μABC: A systematic microsecond molecular dynamics study of tetranucleotide sequence effects in B-DNA. *Nucleic Acids Research, 42*, 12272–12283.

Pohl, F. M., & Jovin, T. M. (1972). Salt-induced co-operative conformational change of a synthetic DNA: Equilibrium and kinetic studies with poly (dG-dC). *Journal of Molecular Biology, 67*, 375–396.

Quintana, J., Grzeskowiak, K., Yanagi, K., & Dickerson, R. E. (1992). Structure of a B-DNA decamer with a central T-A step: C-G-A-T-T-A-A-T-C-G. *Journal of Molecular Biology, 225*, 379–395.

Rahmouni, A. R., & Wells, R. D. (1992). Direct evidence for the effect of transcription on local DNA supercoiling in vivo. *Journal of Molecular Biology, 223*, 131–144.

Ravichandran, S., Subramani, V. K., & Kim, K. K. (2019). Z-DNA in the genome: From structure to disease. *Biophysical Reviews, 11*, 383–387.

Rhodes, D., & Klug, A. (1981). Sequence-dependent helical periodicity of DNA. *Nature, 292*, 378–380.

Rich, A., Nordheim, A., & Wang, A. H.-J. (1984). The chemistry and biology of left-handed Z-DNA. *Annual Review of Biochemistry, 53*, 791–846.

Rich, A., & Zhang, S. (2003). Timeline: Z-DNA: The long road to biological function. *Nature Reviews Genetics, 4*, 566–572.

Rohs, R., Sklenar, H., & Shakked, Z. (2005). Structural and energetic origins of sequence-specific DNA bending: Monte Carlo simulations of papillomavirus E2-DNA binding sites. *Structure, 13*, 1499–1509.

Rueda, M., Cubero, E., Laughton, C. A., & Orozco, M. (2004). Exploring the counterion atmosphere around DNA: What can be learned from molecular dynamics simulations? *Biophysical Journal, 87*, 800–811.

Schneider, B., Božíková, P., Nečasová, I., Čech, P., Svozil, D., & Černý, J. (2018). A DNA structural alphabet provides new insight into DNA flexibility. *Acta Crystallographica Section D, 74*, 52–64.

Schneider, B., Ginell, S. L., Jones, R., Gaffney, B., & Berman, H. M. (1992). Crystal and molecular structure of a DNA fragment containing a 2-aminoadenine modification: The relationship between conformation, packing, and hydration in Z-DNA hexamers. *Biochemistry, 31*, 9622–9628.

Schneider, B., Neidle, S., & Berman, H. M. (1997). Conformations of the sugar-phosphate backbone in helical DNA crystal structures. *Biopolymers, 42*, 113–124.

Schuerman, G. S., & Van Meervelt, L. (2000). Conformational flexibility of the DNA backbone. *Journal of the American Chemical Society, 122*, 232–240.

Schwartz, T., Behike, J., Lowenhaupt, K., Heinemann, U., & Rich, A. (2001). Structure of the DLM-1-Z-DNA complex reveals a conserved family of Z-DNA-binding proteins. *Nature Structural Biology, 8*, 761–765.

Schwartz, T., Rould, M. A., Lowenhaupt, K., Herbert, A., & Rich, A. (1999). Crystal structure of the Zα domain of the human editing enzyme ADAR1 bound to left-handed Z-DNA. *Science, 284*, 1841–1845.

Schwieters, C. D., & Clore, G. M. (2007). A physical picture of atomic motions within the Dickerson DNA dodecamer in solution derived from joint ensemble refinement against NMR and large-angle X-ray scattering data. *Biochemistry, 46*, 1152–1166.

Shakked, Z., Guerstein-Guzikevich, G., Eisenstein, M., Frolow, F., & Rabinovich, D. (1989). The conformation of the DNA double helix in the crystal is dependent on its environment. *Nature, 342*, 456–460.

Shatzky-Schwatrz, M., Arbuckle, N. D., Eistenstein, M., Rabinovich, D., Bareket-Samish, A., Haran, T. E., Luisi, B. F., & Shakked, Z. (1997). X-ray and solution studies of DNA oligomers and implications for the structural basis of A-tract-dependent curvature. *Journal of Molecular Biology, 267*, 595—623.

Sherer, E. C., Harris, S. A., Soliva, R., Orozco, M., & Laughton, C. A. (1999). Molecular dynamics studies of DNA A-Tract structure and flexibility. *Journal of the American Chemical Society, 121*, 5981—5991.

Shui, X., McFail-Isom, L., Hu, G. G., & Williams, L. D. (1998). The B-DNA dodecamer at high resolution reveals a spine of water on sodium. *Biochemistry, 37*, 8341—8355.

Sims, G. E., & Kim, S.-H. (2003). Global mapping of nucleic acid conformational space: Dinucleoside monophosphate conformations and transition pathways among conformational classes. *Nucleic Acids Research, 31*, 5607—5616.

Svozil, D., Kalina, J., Omelka, M., & Schneider, B. (2008). DNA conformations and their sequence preferences. *Nucleic Acids Research, 36*, 3690—3706.

Sprous, D., Young, M. A., & Beveridge, D. L. (1999). Molecular dynamics studies of axis bending in d(G$_5$-(GA$_4$T$_4$C)$_2$-C$_5$) and d(G$_5$-(GT$_4$A$_4$C)$_2$-C$_5$): Effects of sequence polarity on DNA curvature. *Journal of Molecular Biology, 285*, 1623—1632.

Sprous, D., Zacharias, W., Wood, Z. A., & Harvey, S. C. (1995). Dehydrating agents sharply reduce curvature in DNAs containing A tracts. *Nucleic Acids Research, 23*, 1816—1821.

Stefl, R., Wu, H., Ravindranathan, S., Sklenář, V., & Feigon, J. (2004). DNA A-tract bending in three dimensions: Solving the dA$_4$T$_4$ vs. dT$_4$A$_4$ conundrum. *Proceedings of the National Academy of Sciences of the United States of America, 101*, 1177—1182.

Strahs, D., & Schlick, T. (2000). A-tract bending: Insights into experimental structures by computational models. *Journal of Molecular Biology, 301*, 643—663.

Subirana, J. A., & Messeguer, X. (2010). The most frequent short sequences in non-coding DNA. *Nucleic Acids Research, 34*, 1172—1181.

Subirana, J. A., & Faria, T. (1997). Influence of sequence on the conformation of the B-DNA helix. *Biophysical Journal, 73*, 333—338.

Subramani, V. K., Kim, D., Yun, K., & Kim, K. K. (2016). Structural and functional studies of a large winged Z-DNA-binding domain of *Danio rerio* protein kinase PKZ. *FEBS Letters, 590*, 2275—2285.

Suzuki, M., Amano, N., Kakinuma, J., & Tateno, M. (1997). Use of a 3D structure data base for understanding sequence-dependent conformational aspects of DNA. *Journal of Molecular Biology, 274*, 421—435.

Tereshko, V., Minasov, G., & Egli, M. (1999). The Dickerson-Drew B-DNA dodecamer revisited at atomic resolution. *Journal of the American Chemical Society, 121*, 470—471.

Tereshko, V., Wilds, C. J., Minasov, G., Prakash, T. P., Maier, M. A., Howard, A., Wawrzak, Z., Manoharan, M., & Egli, M. (2001). Detection of alkali metal ions in DNA crystals using state-of-the-art X-ray diffraction experiments. *Nucleic Acids Research, 29*, 1208—1215.

Timsit, Y., & Moras, D. (1991). Groove-backbone interaction in B-DNA. Implication for DNA condensation and recombination. *Journal of Molecular Biology, 221*, 919—940.

Timsit, Y., Vilbois, E., & Moras, D. (1991). Base-pairing shift in the major groove of (CA)n tracts by B-DNA crystal structures. *Nature, 354*, 167—170.

Timsit, Y., Westhof, E., Fuchs, R. P. P., & Moras, D. (1989). Unusual helical packing in crystals of DNA bearing a mutation hot spot. *Nature, 341*, 459—462.

Tippin, D. B., & Sundaralingam, M. (1997). Comparison of major groove hydration in isomorphous A-DNA octamers and dependence on base sequence and local helix geometry. *Biochemistry, 36*, 536—543.

Trantírek, L., Urbášek, M., Štefl, R., Feigon, J., & Sklenár, V. (2000). A method for direct determination of helical parameters in nucleic acids using residual dipolar couplings. *Journal of the American Chemical Society, 122*, 10454—10455.

Tullius, T. D., & Dombroski, B. A. (1985). Iron (II) EDTA used to measure the helical twist along any DNA molecule. *Science, 230*, 679—681.

Ulanovsky, L., & Trifonov, E. N. (1987). Estimation of wedge components in curved DNA. *Nature, 326*, 720—722.

Valls, N., Wright, G., Steiner, R. A., Murshudov, G. N., & Subirana, J. A. (2004). DNA variability in five crystal structures of d(CGCAATTGCG). *Acta Crystallographica, D60*, 680—685.

Vargason, J. M., Henderson, K., & Ho, P. S. (2001). A crystallographic map of the transition from B-DNA to A-DNA. *Proceedings of the National Academy of Sciences of the United States of America, 98*, 7265—7270.

Verdaguer, N., Aymami, J., Fernández-Forner, D., Fita, I., Coll, M., Huynh-Dinh, T., Igolen, J., & Subirana, J. A. (1991). Molecular structure of a complete turn of A-DNA. *Journal of Molecular Biology, 221*, 623—635.

Vlieghe, D., Turkenberg, J. P., & Van Meervelt, L. (1999). B-DNA at atomic resolution reveals extended hydration patterns. *Acta Crystallographica, D55*, 1495—1502.

Wahl, M. C., & Sundaralingam, M. (1997). Crystal structures of A-DNA duplexes. *Biopolymers, 44*, 45—63.

Walther, J., Dans, P. D., Balaceanu, A., Hospital, A., Bayarri, G., & Orozco, M. (2020). A multi-modal coarse grained model of DNA flexibility mappable to the atomistic level. *Nucleic Acids Research, 48*, e29.

Wang, A. H.-J., Fujii, A., van Boom, J. H., van der Marel, G. A., van Boeckel, S. A. A., & Rich, A. (1982). Molecular structure of r(GCG)d(TATACGC): A DNA–RNA hybrid helix joined to double helical DNA. *Nature, 299*, 601—604.

Wang, A. H.-J., Gessner, R. V., van der Marel, G. A., van Boom, J. H., & Rich, A. (1985). Crystal structure of Z-DNA without an alternating purine-pyrimidine sequence. *Proceedings of the National Academy of Sciences of the United States of America, 82*, 3611—3615.

Wang, A. H.-J., Hakoshima, T., van der Marel, G., van Boom, J. H., & Rich, A. (1984). AT base pairs are less stable than GC base pairs in Z-DNA: The crystal structure of d(m^{5}CGTAm^{5}CG). *Cell, 37*, 321—331.

Wang, A. H.-J., Quigley, G. J., Kolpak, F. J., Crawford, J. L., van Boom, J. H., van der Marel, G., & Rich, A. (1979). Molecular structure of a left-handed double helical DNA fragment at atomic resolution. *Nature, 282*, 680—686.

Wang, A. H.-J., Quigley, G. J., Kolpak, F. J., van der Marel, G., van Boom, J. H., & Rich, A. (1981). Left-handed double helical DNA: Variations in the backbone conformation. *Science, 211*, 171—176.

Watson, J. D., & Crick, F. H. C. (1953). Molecular structure of nucleic acids; a structure for deoxyribose nucleic acid. *Nature, 171*, 737—738.

Webster, G. D., Sanderson, M. R., Skelly, J. V., Neidle, S., Swann, P. F., Li, B. F., & Tickle, I. J. (1990). Crystal structure and sequence-dependent conformation of the A.G mispaired oligonucleotide d(CGCAAGCTGGCG). *Proceedings of the National Academy of Sciences of the United States of America, 87*, 6693—6697.

Wing, R. M., Drew, H. R., Takano, T., Broka, C., Tanaka, S., Itakura, K., & Dickerson, R. E. (1980). Crystal structure analysis of a complete turn of B-DNA. *Nature, 287*, 755—758.

Woods, K. K., Maehigashi, T., Howerton, S. B., Sines, C. C., Tannenbaum, S., & Williams, L. D. (2004). High-resolution structure of an extended A-tract: [d(CGCAAATTTGCG)]$_2$. *Journal of the American Chemical Society, 126*, 15330—15331.

Wu, Z., Delagio, F., Tjandra, N., Zhurkin, V. B., & Bax, A. (2003). Overall structure and sugar dynamics of a DNA dodecamer from homo- and heteronuclear dipolar couplings and 31P chemical shift anisotropy. *Journal of Biomolecular NMR, 26*, 297—315.

Yoon, C., Privé, G. G., Goodsell, D. S., & Dickerson, R. E. (1988). Structure of an alternating-B DNA helix and its relationship to A-tract DNA. *Proceedings of the National Academy of Sciences of the United States of America, 85*, 6332—6336.

Young, M. A., Ravishanker, G., Beveridge, D. L., & Berman, H. M. (1995). Analysis of local helix bending in crystal structures of DNA oligonucleotides and DNA-protein complexes. *Biophysical Journal, 68*, 2454—2468.

Yuan, H., Quintana, J., & Dickerson, R. E. (1992). Alternative structures for alternating poly(dA-dT) tracts: The structure of the B-DNA decamer C-G-A-T-A-T-A-T-C-G. *Biochemistry, 31*, 8009—8021.

Further reading

Arnott, S. (1970). The geometry of nucleic acids. *Progress in Biophysics and Molecular Biology, 6*, 265—319.

Arnott, S. (2006). DNA polymorphism and the early history of the double helix. *Trends in Biochemical Sciences, 31*, 349—354.

Berman, H. M. (1997). Crystal studies of B-DNA: The answers and the questions. *Biopolymers, 44*, 23—44.

Berman, H. M., Gelbin, A., & Westbrook, J. (1996). Nucleic acid crystallography: A view from the nucleic acid database. *Progress in Biophysics and Molecular Biology, 66*, 255—288.

Dickerson, R. E. (1992). DNA structure from A to Z. *Methods in Enzymology, 211*, 67—111.

Egli, M. (2004). Nucleic acid crystallography: Current progress. *Current Opinion in Chemical Biology, 8*, 580—591.

Egli, M., & Pallan, P. S. (2010). The many twists and turns of DNA: Template, telomere, tool, and target. *Current Opinion in Structural Biology, 20*, 1—14.

Grzeskowiak, K. (1996). Sequence-dependent structural variation in B-DNA. *Chemical Biology, 3*, 785—790.

Haran, T. E., & Mohanty, U. (2009). The unique structure of A-tracts and intrinsic DNA bending. *Quarterly Reviews of Biophysics, 42*, 41—81.

Mooers, B. H. M. (2009). Crystallographic studies of DNA and RNA. *Methods, 47*, 168—176.

Shakked, Z., & Rabinovich, D. (1986). The effect of the base sequence on the fine structure of the DNA double helix. *Progress in Biophysics and Molecular Biology, 47*, 159—195.

CHAPTER 4

Non-standard and higher-order DNA structures: DNA—DNA recognition

4.1 Mismatches in DNA

4.1.1 General features

Base pairing in DNA is often considered solely in terms of Watson—Crick hydrogen bonding, but numerous other arrangements are possible, and many have been observed experimentally. These are described in this section.

Sixteen distinct arrangements for A·T and G·C base pair hydrogen bonding are in principle possible, though not all have been observed. An important non-Watson—Crick base-pairing pattern was first described by Karst Hoogsteen in 1963 and bears his name. Hoogsteen and reverse Hoogsteen pairs (Fig. 4.1a and b) have been found in several crystal structures of adenine·thymine and adenine·uracil complexes, as well as guanine·guanine pairs in higher-order quadruplex structures (see Section 4.3). These arrangements involve atoms N6 and N7 of adenine rather than the N1 and N6 atoms of Watson—Crick hydrogen-bonding — so Hoogsteen hydrogen bonding with thymine is with the major groove edge of the adenine base, at right angles from the base edge that forms Watson—Crick pairs in classic A- and B-DNA double helices. Thus, Hoogsteen pairing implies that the adenine base is in a *syn* glycosidic angle conformation if the resulting base pair is to be present in an anti-parallel DNA duplex.

Mismatched base-pairing in DNA can arise naturally during replication. The possible arrangements are: G·A, G·T, A·A, G·G, T·T, C·C, T·C or A·C. If left unchecked, these mismatches will lead to mutations and nonsense or incorrect gene products following cellular replication. Cells, however, have evolved numerous enzymatic repair mechanisms to recognise mis-pairs, which then attempt to repair them. In general, recognition of a mismatch is followed by duplex unwinding and enzymatic excision of the mis-pair. The efficiency of these processes depends on the nature of the mispairing, and often, on sequence context. The details of these mechanisms have been extensively studied in *Escherichia coli* and are increasingly well understood in higher organisms. Many structural studies have been reported on proteins involved in DNA repair, together with complexes including various damaged DNA sequences. DNA repair proteins are discussed in Chapter 7; here we focus on the nature of the initially modified or damaged DNA itself. It has been suggested that many such lesions result in a decrease in base stacking (Yang, 2006), which itself may be the first recognition signal for initiating repair. It is likely that recognition of local backbone deformities in B-form DNA is important, perhaps as a

Principles of Nucleic Acid Structure
ISBN 978-0-12-819677-9, https://doi.org/10.1016/B978-0-12-819677-9.00003-2

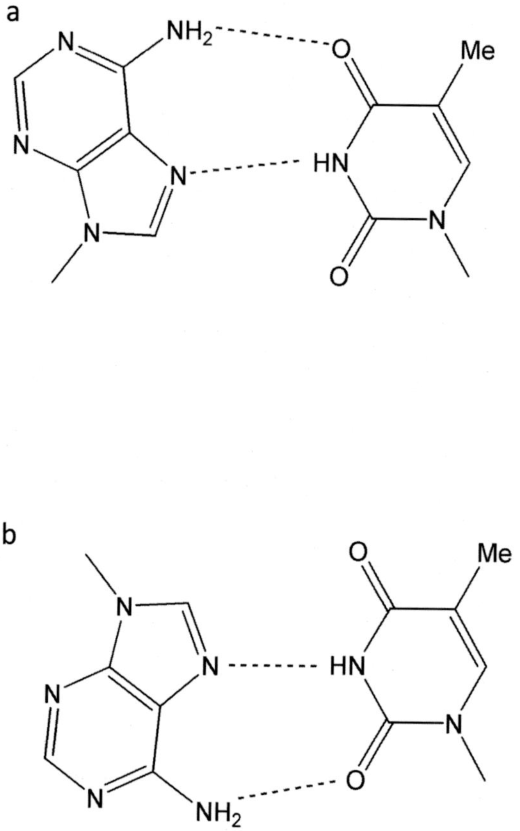

Figure 4.1 Structures of Hoogsteen and reverse Hoogsteen A·T mismatched base pairs.

discrete bulge, a looped-out base or extensive deformation over several base pairs. Sheared base pairs have one base displaced in plane relative to the other, so that mispairing then occurs (Chou et al., 2003). These can be formed during replication or mutation, and may also form in expanded tandem-repetitive DNA sequences in a variety of inherited genetic disorders.

Structural studies on DNA mis-pairs themselves have addressed the questions of:

- the nature of the hydrogen bonding involved in the pairings
- how these affect the local and global structure of duplex DNA
- the possible relationships between structure, sequence context of the mis-pair and efficiency of mis-match repair

Mismatches generally destabilise duplex DNA. This can be monitored by the extent to which the temperature is decreased at which the transition from a double helical to an unstructured arrangement occurs. (The transition temperature is termed the melting temperature T_m and the change is designated as the ΔT_m). Of the eight mis-pairs listed above, the A·G one has been the best-studied and is the one that is focused upon here. Structural studies have also been reported on oligonucleotides in A, B and Z forms with G·T, I(inosine)·T, U·G, I·C and I·A base pairs. The mode of action of the anticancer drug 6-thioguanine (S6G) involves its incorporation into DNA following metabolic conversion to the corresponding nucleotide triphosphate. S6G·C or S6G·T base pairs are closely structurally similar to their unmodified G·C or G·T counterparts (Bohon & de los Santos, 2003; Somerville et al., 2003), albeit significantly less stable (Bohon & de los Santos, 2005).

Mutagenesis (which can eventually lead to carcinogenesis) can occur when external agents such as methylating compounds produce covalent adducts with DNA bases. Structural work on covalently modified oligonucleotides requires milligram quantities of adducts, both for X-ray and NMR studies. This can now be routinely achieved, and numerous structural studies (especially using NMR techniques) have been performed on oligonucleotides incorporating, for example, methylated bases (Patel, 1992) and metabolites of carcinogens such as benzo[a]pyrene. Of particular interest has been the mutagenic methylation of the O6 atom in guanine, which is produced by a variety of environmental and chemotherapeutic agents.

4.1.2 Purine:purine mis-matches

Crystal structures have been determined for several self-complementary oligonucleotide sequences with A·G base pairs (Table 4.1). These show a variety of base pairing arrangements (Fig. 4.2). That with both nucleosides in an *anti* conformation necessarily forces the inter-strand C1′ ... C1′ distance to be further apart than in a standard B-DNA Watson—

Table 4.1 Some purine:purine mismatched oligonucleotide crystal structures, with the mismatched base pairs shown in bold. Distances are in Å.

Sequence	Glycosidic angles	C1′...C1′ distances	PDB id
5′-CGC**G**AATTA**G**CG GCG**A**TTAA**G**CGC	G(*anti*)·A(*syn*)	10.6, 10.8	112D
5′-CGC**A**AATT**G**GCG GCG**G**TTAA**A**CGC	A(*anti*)·G(*syn*)	10.8 (mean)	111D
5′-CGC**A**AGCT**G**GCG GCG**G**TCGA**A**CGC	A(*syn*)·G(*anti*)	10.7 (mean)	1DNM
5′-CCAA**G**ATT**G**G GG**T**TAGAACC	A(*anti*)·G(*anti*)	12.5	3DNB
5′-CGC**G**AATT**G**GCG GCG**G**TTAA**G**CGC	G(*anti*)·G(*syn*)	10.7, 11.2	1D80
5′-CC**G**AAT**G**AGG GG**A**GT**A**AGCC	A(*anti*)·G(*anti*)	8.7	1D9R

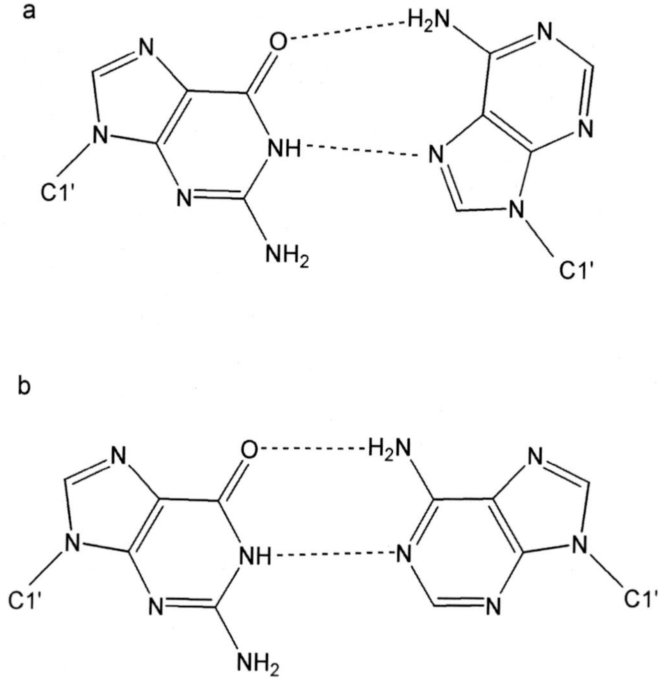

Figure 4.2 Structures of G(*anti*)·A(*anti*) and G(*anti*)·A(*syn*) mismatches.

Crick base pair (10.4 Å). This *anti, anti* form (Fig. 4.2a) has only been observed in one crystal structure (Leonard, Booth et al., 1990), where there are two consecutive A·G base pairs. Presumably the bulge in the helix produced by their excessive C1' ... C1' distances is relieved by the two A·G base pairs being well stacked on each other. For isolated purine:purine base pairs, it is easier to achieve normal C1' ... C1' distances, and hence greater helix stability, by having one base *anti* and the other in a *syn* arrangement (Fig. 4.2b). The base pairing is of the Hoogsteen type. Energetic considerations suggest that a *syn* conformation would be preferred for the guanine rather than the adenine base, but Table 4.1 shows that this is only sometimes the case.

All A·G-containing DNA oligonucleotides have B-form structures, with generally standard geometries and few distortions from ideality, except that the mis-pairs tend to be poorly stacked with their neighbouring base pairs (Table 4.1: Brown et al., 1986; Gao et al., 1999; Leonard, et al., 1990; Privé et al., 1987; Webster et al., 1990). Detailed examination of the stacking patterns has shown that the glycosidic conformation adopted by a mismatched base tends to maximise the stacking, and so overrides the slight energy penalty of the adenosine nucleoside being in a *syn* conformation. The nature of the sequence surrounding the mismatch is important, with most of the established structures following the rule that when the mis-paired base on the first strand is at the centre of the three-residue sequence 5′-Py-**Pu**-Pu, the central purine is then in an *anti* conformation. An oligonucleotide with G·G base pairs has also been found (Skelly et al., 1993) to have a B-like structure and an *anti, syn* arrangement for the mismatched bases (Fig. 4.3). This unusual mismatch is associated with B_{II} backbone phosphate conformations (see Chapter 3, Section 3.1), which have also been documented for some adjacent G·A mismatches (Chou et al., 1992). The B_{II} conformation results in altered phosphate group intrastrand distances; these changes from standard helical geometry may act as recognition signals for repair nucleases to identify and excise the mis-pairs at these points.

NMR studies of G·A-containing oligonucleotides have shown that the pattern of glycosidic angles seen in the crystal structures is not always maintained in solution, and some arrangements can be readily converted into others with changes in pH (Gao & Patel, 1988). Thus, the duplex of d(CGCAAATTGGCG)$_2$ adopts an A(*anti*)·G(*anti*) arrangement at high pH (Lane et al., 1991), yet an A$^+$(*anti*)·G(*syn*) one at low pH. It should be borne in mind that the typical NMR experiment uses different oligonucleotide solution conditions compared to those in crystallisation trials, with higher salt concentrations in the former and high hydrophobic-type solvent concentrations in the latter. These differences are likely to affect the conformational equilibria of equi-energetic mis-pairs in different ways, so care is needed in claiming that a particular structure (NMR or crystallographic) represents the situation in cellular conditions.

The variability of G·A base pairing arrangements (and hence their stability) is largely dependent on the optimisation of base-stacking interactions, and hence on the nature of adjacent bases in a sequence. This is graphically illustrated in the sequence d(ATGAGCGAATA)$_2$, with four G·A base pairs (Li et al., 1991a). NMR and molecular modelling show that these have N7…N2 and N3… N2 hydrogen bonds, with only this arrangement being capable of stabilisation by stacking interactions with adjacent bases.

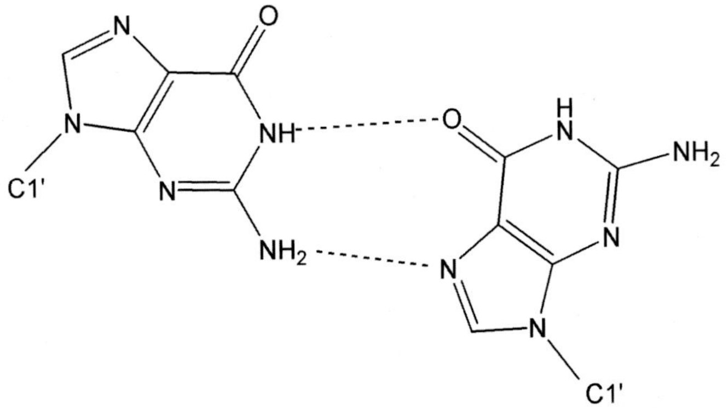

Figure 4.3 A G(*anti*)·G(*syn*) base mismatch, as observed in a mismatch dodecamer crystal structure (Skelly et al., 1993).

Thus, in general, the sequence 5′-Py-**GA**-Pu forms a particularly stable mis-paired unit (Li et al., 1991b), regardless of the details of the hydrogen bonding involved. This stability is seen to be retained in the high-resolution crystal structure (Gao et al., 1999) of the duplex formed by the sequence d(CC**GA**AT**GA**GG), which is a model for centromeric DNA, and has the repeated sequence d(GAATG)$_n$. These tandem G(*anti*)·A(*anti*) mismatches (Fig. 4.4) in centromeric DNA have been termed "sheared" on account of their side-by-side nature. The notable feature of this crystal structure is the extensive interstrand G/G and A/A base stacking, which helps to stabilise the unusual G·A base pairing. These features have also been observed in NMR structures of analogous sequences (Chou et al., 1997; Green et al., 1994). The stability of a particular A·G arrangement is not directly correlated with the kinetics of base excision (the initial step in the repair of this mis-pair), although it does depend on the sequence context of the mismatch (Sanchez et al., 2003). An A·G mismatch in the sequence d(GAG) with *anti/anti* glycosidic angle conformations had a base cleavage rate that is threefold faster than with a sequence context of d(TAG), having *syn/anti* glycosidic angle conformations.

A comprehensive molecular dynamics simulation study (Rossetti et al., 2015) has examined all possible mismatches in three different sequence environments. In accordance with the structural data outlined above, the simulations have shown that DNA duplexes are able to accommodate mismatches with ease due to the flexibility of the backbone and base pairs, and without altering overall helicity. An NMR study (Ghosh et al., 2014) of an oligonucleotide containing a G·G mis-pair followed by C·C also found that overall B-DNA helicity was retained (PDB id 2MJX), but at the cost of

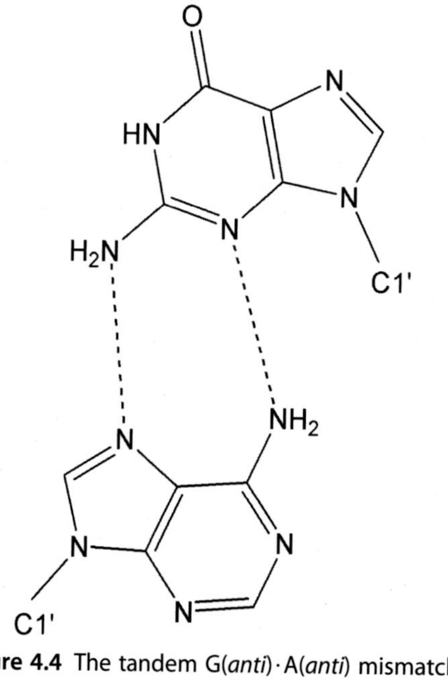

Figure 4.4 The tandem G(*anti*)·A(*anti*) mismatch.

some backbone distortion arising from the two mis-matches. The simulations also suggested that changes in, for example, groove widths, which can occur, do have implications for recognition by mismatch repair proteins.

4.1.3 Alkylation mismatches

Methylation at the O6 position of a guanine base in duplex DNA induces the resulting destabilising G(OMe)·C base pair to undergo a transition mutation to a G(OMe)·T base pair when the DNA is replicated. Methylation of the O6 carbonyl group changes the C6—O6 bond from a double to a single bond, and hence alters the overall pattern of tautomerism in the guanine ring system, with atom N1 no longer having an attached hydrogen atom. NMR studies on the G(OMe)·C mis-pair in an oligomer duplex (Kalnik et al., 1989) have indicated that both bases have *anti* glycosidic angles, in an overall B type helix, with "wobble" hydrogen bonding solely between N2(G) and O2(C), and with the O6 methyl group oriented towards the cytosine (Fig. 4.5a).

The crystal structure of a (Z-DNA) oligonucleotide duplex containing this mis-pair (Ginell et al., 1990) shows an arrangement that is more compatible with standard Watson—Crick hydrogen bonding — distances O6(G) … N4(C) and N1(G) … N3(C) are normal hydrogen-bonding ones (Fig. 4.5b). This suggests that even though the crystals were grown at pH 7.0, atoms N1(G) or N3(C) have acquired a proton so that partial Watson—Crick hydrogen bonding can occur, which is still less stable than a standard G·C base pair. The various possible arrangements for A·G mismatches have been described in the previous section. When the guanine in this mismatch is methylated, the number of possible hydrogen-bonded base pair arrangements becomes even greater (Ginell et al., 1994), although sequence context and/or a requirement for adenine protonation will exclude the majority from consideration in specific instances. The crystal structure (Ginell et al., 1994) of a dodecanucleotide duplex with two such A·G(OMe) mis-pairs shows an arrangement with two hydrogen bonds and A(*syn*)·G(*anti*) glycosidic angles (Fig. 4.5c). A common theme in this and many other mis-pair crystal structures is the location of water molecules that are hydrogen bonded to (and often bridging between) the bases involved in the mismatches.

The G(OMe)·T base pair (Table 4.2), which also destabilises a duplex, has been observed to have Watson—Crick-type hydrogen bonding in two B-DNA dodecamer crystal structures (Leonard et al., 1990; Vojtechovsky et al., 1995), with two hydrogen bonds assigned between the two bases (Fig. 4.5d). By contrast, a G(OMe)·T mis-pair with only a single hydrogen bond was inferred from an NMR study (Patel et al., 1986) of the sequence d(CGTGAATTCG(OMe)CG), with the mismatch in a distinct sequence context from the crystal structures. The overall shape of the G(OMe)·T base pair in both the NMR and crystallographic structures is close to that of a standard G·C one, whereas the G(OMe)·C one is rather less so, especially in the so-called wobble arrangement (Fig. 4.5a). This difference does not explain why there is only a weak preference for G(OMe)·T base pairs to be incorporated into DNA during replication and not to be readily recognised by repair enzymes.

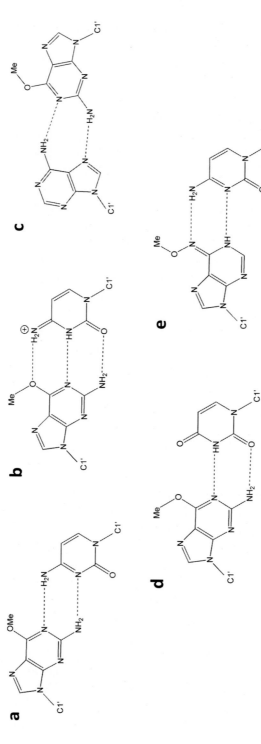

Figure 4.5 (a) The G(OMe)·C wobble base pair. (b) The G(OMe)·C Watson—Crick base pair. (c) The A·G(OMe) base pair. (d) The G(OMe)·T base pair. (e) The A(OMe)·C Watson—Crick base pair.

Table 4.2 Selected DNA crystal structures incorporating purine·pyrimidine or alkylation mis-matches.

Base pair mismatch	PDB id
G·T	113D
A·C	1D99
A(OMe)·C	456D
A(OMe)·T	457D
G(OMe)·T	218D
G·A(8-oxo)	1D75
C·G(8-oxo)	183D
G(OMe)·C	1D24

Mutagens such as hydroxylamine can methoxylate the N6 position of adenine. The crystal structure (Chatake, Ono, et al., 1999) of a dodecamer with two methoxyadenine residues shows that these modified bases form analogous Watson—Crick-like base pairs (Fig. 4.5e) with a cytidine on the opposite strand. In order for this to occur, the modified adenines must be in the imino form, as shown, enabling it to mimic a guanine, with the result that cytidine rather than thymine is mis-incorporated during replication. Thymine opposite this adenine lesion also results in Watson—Crick-like base pairing (Chatake, Hikima, et al., 1999).

Oxidative damage to bases can occur by free-radical attack, especially at guanine sites, where the primary product is 7,8-dihydro-8-oxo-guanine. The crystal structure of a duplex containing this lesion (Lipscomb et al., 1995) shows that the standard Watson—Crick G·C hydrogen bonding arrangements are unimpaired, and alternative G(*syn*) conformations are not involved, in accord with NMR studies (Kouchakdjian et al., 1991). This base change positions the 8-oxo group in the major groove, where it can be available for recognition by DNA repair enzymes.

A diverse range of chemical carcinogens can bind covalently to DNA, typically as chemically reactive metabolites. These are discussed in Chapter 5, as part of the wider topic of small-molecule DNA recognition.

4.1.4 Alkylation at Watson—Crick base pairs

5-Methylcytosine is the product of enzymatic methylation by methyltransferases of cytosine at the 5-position of the base. This modified base is involved in epigenetic gene regulation in eukaryotic species and its effects on DNA structure have been extensively studied (Table 4.3). A high-resolution crystal structure (Theruvathu et al., 2013), PDB id 4MKW, of the Dickerson—Drew dodecamer with 5-methylcytosine at the 3-position has an overall RMSD with the native structure of 0.22 Å. This close similarity was confirmed by a solution NMR study and by other crystallographic studies (Table 4.3), such as that by Renciuk et al., 2013.

Hydroxylation of the methyl group to form 5-hydroxymethylcytosine can occur during epigenetic demethylation back to unmodified cytosine, and further oxidation can

Table 4.3 Crystal structures of DNA sequences with modified cytosines. Resolution limits of the crystal structure analyses are shown.

Alkylation	PDB id	Helix type	Resolution (Å)
5-Me-C	4GJU	B	1.41
5-Me-C	6JV5	B	1.40
5-Me-C	4MKW	B	1.22
5-Me-C	4GLG	B	1.72
5-carboxyl-C	5ZAT	B	1.06
5-carboxyl-C	4PWM	B	1.95
5-hydroxy-Me-C	419V	B	1.02
5-formyl-C	5ZAS	A	1.56
5-formyl-C	4QC7	B	1.90
5-formyl-C	4QKK	F (A)	1.4
5-formyl-C	5MVU	A	2.3

result in 5-formylcytosine and 5-carboxylcytosine (Gruber et al., 2018). The consequences of incorporation of these modifications at position 9 of the Dickerson—Drew dodecamer have been studied by crystallographic and biophysical methods (Szulik et al., 2015). The three structures (Table 4.4; PDB ids 419V, 4QC7 and 4PMW) are all closely similar to the high-resolution native crystal structure (Tereshko et al., 1999), with RMSD values of 0.67, 0.46 and 0.49 Å, respectively, and have standard B-type double helices. Watson—Crick base pairs were intact in all the modified sequences and the substituents all protruded into the major groove, with no evidence of imino tautomerization of the modified cytosines, in agreement with NMR data. The biophysical data showed enhanced stabilisation produced by 5-carboxylcytosine substitution, which was not evident in any observable structural changes and therefore may be attributable to electronic effects. Crystallographic studies have been reported with 5-methylcytosine, 5-formylcytosine, 5-hydroxymethylcytosine and 5-carboxylcytosine modifications to the self-complementary decamer sequence d(CCAGXGCTGG) (Fu et al., 2019) at position X. These sequences crystallise in a variety of space groups — $C2$, $P6_1$, $C2$ and $P2_1$ — whereas the modified Dickerson—Drew dodecamers all tend to crystallise in the same orthorhombic space group, $P2_12_12_1$. The 5-formylcytosine-modified decamer crystallises as an A-DNA helix (Fig. 4.6), while all the others are in the B-DNA genus. The bases at the 5-formylcytosine sites have an ~2 Å shift displacement relative to the adjacent base pairs, resulting in a 3—5 Å increase in major groove width and a 2—3 Å decrease of minor groove width. These changes were not observed with the other cytosine modifications in this study.

Table 4.4 Crystal and NMR structures of selected nucleic acid triple helices.

Triple helix type	Crystal or NMR	PDB id
DNA·DNA-DNA	NMR	136D
PNA·DNA-PNA	X-ray at 2.5 Å	1PNN
DNA·DNA-DNA	X-ray at 1.8 Å	1D3R
DNA·DNA-LNA	NMR	1W86
RNA·RNA-RNA	X-ray at 2.5 Å	6SVS

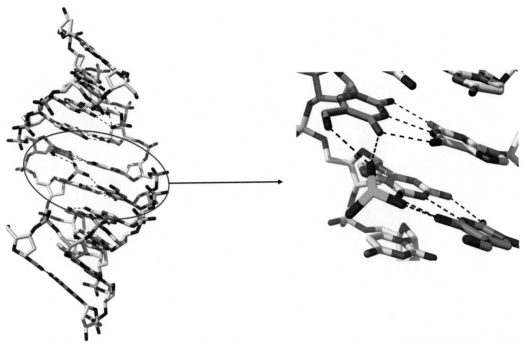

Figure 4.6 View of the crystal structure (PDB id 5XAS) of the DNA decamer d(CCAGXGCTGG) with 5-formylcytosine at position X on each strand (Fu et al., 2019). The enlarged view on the right-hand side shows the carbon atoms of the 5-formylcytosine nucleosides coloured magenta, together with the Watson—Crick hydrogen bonds with the guanine bases. A carbonate ion arising from the DNA purification (with its carbon atom coloured green) is seen bridging the two formyl groups with three hydrogen bonds.

The effects of the 5-formylcytosine group on DNA structure have been reported to vary with sequence, and possibly with crystallisation conditions. Thus, a retention of overall B-DNA type has been reported for a dodecamer structure (Szulik et al., 2015), both in the crystalline state and in solution. A crystallographic analysis (Raiber et al., 2015) of the self-complementary dodecamer d(CTAXGXGXGTAG), i.e. with three 5-formylcytosine (X) sites on each strand forming an alternating six base-pair sequence, has revealed what has been reported to be a remarkably contrasting helix (Fig. 4.7a). This is underwound compared to either A- or B-DNA, with 13 base pairs per turn, and has been termed the "F-helix". The structure has an extensive water network in the major groove, with the formyl groups playing a pivotal role in being the focus for the water contacts (Fig. 4.7b), which together with the many distortions in base-pair morphology is possibly responsible for the "F-helix". A second independent structure determination (Hardwick et al., 2017) on the identical sequence has challenged this interpretation, concluding that this structure (PDB id 5MVU) has an A-type helix in the crystal, and a B-helix in solution (from a parallel NMR study). The RMSD value for the central eight base-pair sequence in this crystal structure compared to the same sequence in the 4QKK structure (Raiber et al., 2015) is 1.21 Å (and this has an RMSD of 1.08 Å from canonical A-DNA). Moreover, comparisons of both structures with the A-DNA crystal structure (PDB id 5MVK) of the unmodified sequence (Hardwick et al., 2017) suggest that crystal

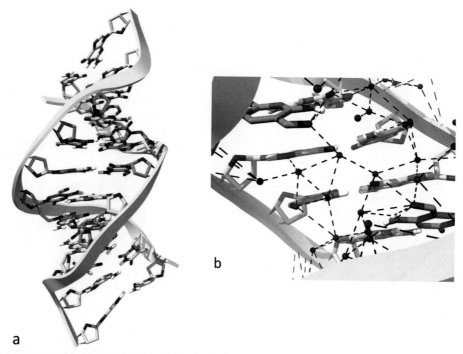

Figure 4.7 (a) The structure of the F-helix dodecamer (PDB id 4QKK; Raiber et al., 2015). The carbon atoms of the 5-formylcytosine nucleosides have been coloured magenta. (b) A magnified view of part of the major groove highlighting water molecules and hydrogen bonds involving the carbonyl atoms of the formyl groups.

packing factors are responsible for the minor differences observed, but that all are of A-DNA type in the crystal. The interested reader will find that viewing the F-helix (PDB id 4QKK) down the helix axis shows its A-DNA morphology, with a characteristic void in the centre of the helix.

4.2 DNA triple helices

4.2.1 Introduction

The formation of a triple helix by two pyrimidine strands and one purine strand was discovered (Felsenfeld et al., 1957) soon after the structure of the double helix itself was determined, when solutions of poly(A) and poly(U) were mixed in appropriate proportions, forming a 1:1:1 three-stranded polynucleotide complex, poly (U·AU). A molecular model for this novel helix was proposed from analysis of fibre diffraction data (Arnott et al., 1976) on both this ribo-polynucleotide and the analogous deoxy-polynucleotide. These studies indicated that the triple helix formed is right-handed. The fibre diffraction data are consistent with an (adenine) purine strand Watson—Crick hydrogen bonding to a (thymine or uracil) pyrimidine one, and a third (thymine or uracil) pyrimidine strand Hoogsteen hydrogen bonding to the purine strand and parallel to this strand (Figs. 4.8 and 4.9). This hydrogen-bonding arrangement is termed a base

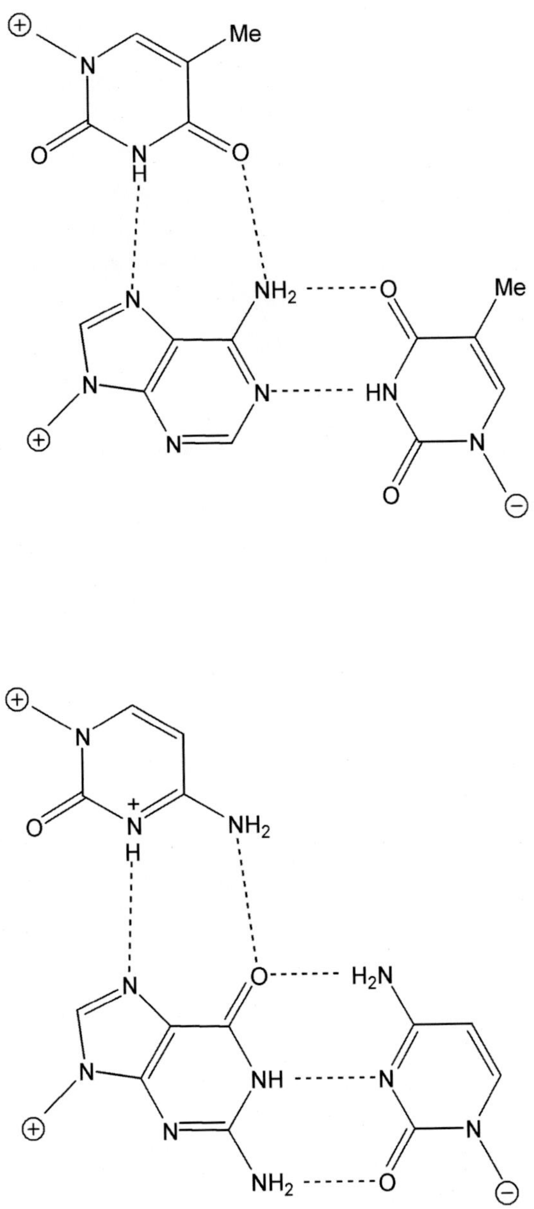

Figure 4.8 The T·AT (top) and C⁺·GC (bottom) triplex hydrogen-bonding arrangements in a Py-Pu-Py triple helix.

triplet and the triple helix itself is often termed a triplex. We adopt the convention here for the triplet X·YZ, that YZ represents the Watson–Crick hydrogen-bonded base pair, with base X being in the third strand and hydrogen bonded in some way to base Y. Subsequent fibre diffraction studies on both poly (U·AU) and its deoxy analogue poly d(T·AT) have more precisely defined the overall structural parameters of these triple helices (Chandrasekaran et al., 2000a,b). The 12-fold DNA triple helix has the sugars in all three strands adopting a C2′-*endo* conformation, and has a pitch of 38.4 Å. The poly (U·AU) ribonucleotide structure has an 11-fold triplex helix and a more complex set

Figure 4.9 Cartoon view of a DNA parallel triple helix, with the third (pyrimidine) strand shown shaded in blue.

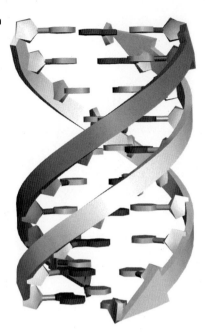

of sugar conformations, with the ribonucleosides in the three strands having C3′-*endo*, C2′-*endo* and C2′-*endo* puckers, respectively. Overall though both triplexes are rather similar, having an A-DNA-like appearance.

A guanine and two cytosine-containing strands can also form a parallel triple helix, which therefore has the same polarity of strands as the T·AT triplex. However, now the third-strand cytosine is required to be protonated at the N3 position in order for two hydrogen bonds to be formed and so enabling effective Hoogsteen hydrogen bonding to take place (Fig. 4.8). The $C^+ \cdot GC$ triplex is isosteric with the T·AT one. Protonation of a cytosine base at N3 can only take place at about pH 5.0, well below physiological pH. This implies that there is a potentially severe practical limitation on $C^+ \cdot GC$ triplex formation in biological systems, and there has been very considerable effort to circumvent this problem (see below). Substitution of a methyl group or bromine at the 5-position of cytosine results in some stabilisation of triplex formation for $C^+ \cdot GC$-containing sequences at ca pH 7 (i.e. around physiological pH). The precise reasons for this are not clear. It is possible that there is an increased hydrophobic contribution to third-strand binding when there are these substituents at the 5-position of cytosine, rather than there being a shift in the pK_a of the N3 atom of cytosine. Triple helices can be stable under some circumstances with both cytosines and thymines in the third strand, i.e. with mixed pyrimidine sequences. These are further discussed below.

The triple helix phenomenon was for many years no more than a laboratory curiosity, until it was found that stretches of triple helix could be formed by oligonucleotides of

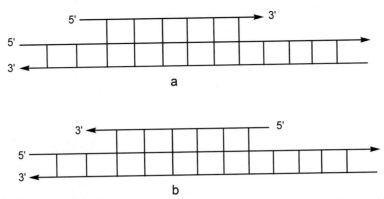

Figure 4.10 Schematic view of the two categories of triple helix formed by an oligonucleotide binding to a long stretch of duplex DNA, with strand directions indicated by *arrows*. The third strand is shown (a) in a parallel orientation and (b) in an anti-parallel orientation to the first duplex strand, which is conventionally a purine one.

appropriate sequence hybridising to much longer duplex DNA molecules (Fig. 4.10a). For example, the oligonucleotide sequence d(TTCTTTTCTTCTTTCTTTTT), with a covalently attached strand cleavage agent, has been used to bind to and cleave a unique site on a yeast chromosome (Strobel & Dervan, 1990), demonstrating the exceptionally high specificity for a target duplex sequence that triplex formation can achieve. However, in spite of this specificity, triple helices are inherently readily disrupted as a consequence of the proximity of the third strand and its negatively charged phosphate groups to the two duplex strands.

4.2.2 Structural studies

The model derived from the original fibre diffraction studies (Arnott et al., 1976) suggested that the third strand in this ribonucleotide triple helix occupies the major groove of the purine-pyrimidine double helix. This key feature has been retained in all subsequent models and ultimately has been verified by both NMR and single-crystal studies. The parallel triple helix in the fibre diffraction model has some (though not all) features of classic A-DNA (Fig. 4.11). It has bases significantly (~3.5 Å) displaced from the helix axis and an average helical twist of ~30°. The duplex minor groove width of 10.7 Å is almost identical to the A-DNA one of 11.0 Å. However, its wide major groove width, of 9.8 Å, which is necessary to accommodate the third strand, is much greater than in standard duplex A-DNA. All deoxyribose sugars in this model have C3′-*endo* pucker.

Triple helices can be formed by relatively short oligonucleotides in solution, such as d(A)$_{12}$·2d(T)$_{12}$, and several NMR studies have been performed on them (see, for example, de los Santos et al., 1989; Macaya et al., 1992; Pasternack et al., 2002; Rajagopal

Figure 4.11 The A-form parallel triple helix, as suggested from fibre diffraction studies (Arnott et al., 1976), shown in van der Waals space-filling representation. Phosphorus atoms are coloured yellow, and the third Hoogsteen strand is at the front of this view, nestling in the enlarged major groove.

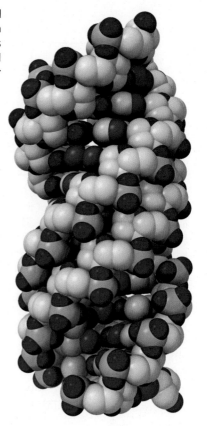

& Feigon, 1989; Radhakrishnan & Patel, 1994; Radhakrishnan et al., 1992). Intramolecular triplexes formed by a single strand of appropriate sequence folding back on itself tend to be more stable and have been especially well studied. The sugar residues in parallel triplexes tend to have B-DNA type C2′-*endo* sugar puckers, whereas the helical twists are closer to those of A-DNA. Overall the parallel triple helices derived from NMR data have morphology closest to that of B-type duplex DNA, with the differences in, for example, x-displacement, being due to the need to expand the size of the major groove to accommodate the third strand (Fig. 4.9). The third DNA strand in a parallel triplex can be replaced by an RNA one, with some increase in stability. This hybrid triple helix still retains much of its B-like character (Gotfredsen et al., 1998), in spite of the presence of the RNA strand, with sugar puckers of the DNA component in the C2′-*endo* range and an x-displacement of −0.7 Å.

Several molecular mechanics and dynamics simulation studies have been performed on triple-helical DNAs. The early studies did not produce a consistent single structural model, but rather showed a diversity of structural types, with differing helix morphologies and sugar puckers (Cheng & Pettitt, 1992; Laughton & Neidle, 1992). This is a reflection of several factors that have been common to many early nucleic acid simulations but are probably more marked for triplexes because of the challenge of being able to adequately treat the electrostatics of three negatively charged DNA strands in close proximity. This can be addressed more adequately with current simulation methodology, for example, by using the particle-mesh Ewald method. Differences in sugar pucker are in large part a result of the earlier use of different force-fields with subtly different pucker preferences. Thus, the increasing ability to undertake long-time-scale stable simulations, together with improved electrostatic treatment, has been able to address both issues, resulted in simulated parallel triplex B-type structures that have many features in common with experimental data (Shields et al., 1997).

Triplexes associated with the neurodegenerative disease Friedreich's ataxia have been analysed in several simulation studies. Boehm et al. (2018) have used the AMBER parm99 force field (Chapter 1), to show that the purine third strand anti-parallel triplex formed from the GAA·TTC duplex associated with this genetic disease is highly distorted and significantly less stable than the pyrimidine third strand in the parallel triplex, in agreement with experimental data. All possible triplexes have been systematically simulated (Zhang et al., 2020) using the AMBER package, however, with the conclusion that both parallel and antiparallel triplexes formed from these disease sequences are stable. These differences may be due to differences in parameter sets, reflecting the reality that a consistent parameterisation for triplexes is not yet available.

Very few single-crystal DNA triplex structures have been determined, in spite of much effort, which has often produced at best crystals with excellent external morphology but little long-range internal order (Liu et al., 1996). However, several crystal structures of duplex DNA oligonucleotides with overhanging ends have been determined. These have revealed the details of triplet arrangements of bases in their crystal lattices. For example, the structure of d(GCGAATTCG) involves Hoogsteen G·GC base triplets formed by interactions between adjacent molecules in the crystal (Van Meervelt et al., 1995). However, the stretches of triplets in these structures are too short to provide information about the morphology of true triple helices. A significantly longer triple-stranded arrangement has been studied in the structure

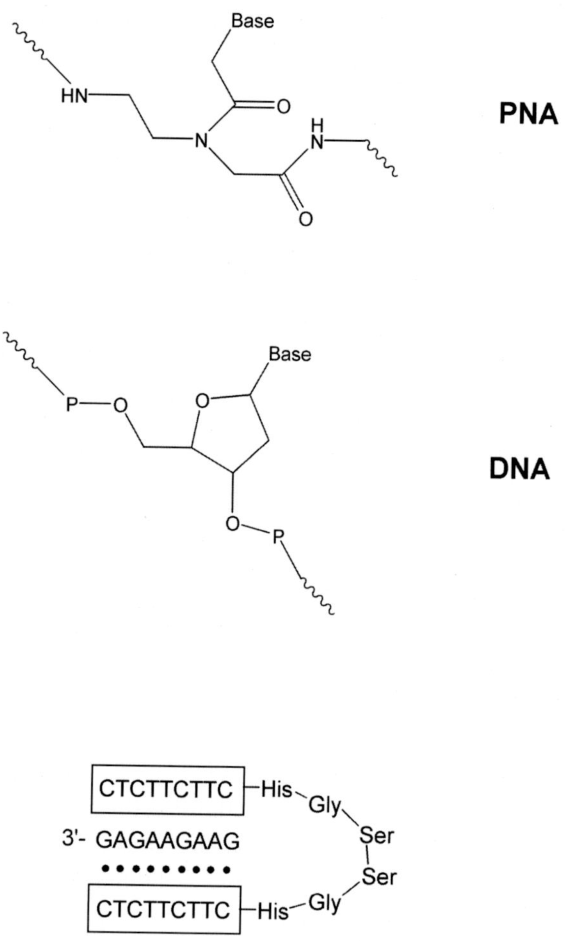

Figure 4.12 The molecular structure of the peptide nucleic acid (PNA) repeating unit, compared to a DNA backbone. Also shown are the sequences of DNA and PNA strands in the crystal structure of the foldback PNA–DNA triple helix (Betts et al., 1995). The PNA strands are shown highlighted in boxes.

(Betts et al., 1995) of a complex between an 18-mer oligopyrimidine foldback peptide nucleic acid (PNA) sequence, and a 9-mer oligopurine sequence (Fig. 4.12). PNA molecules are nucleic acid mimics in which the sugar-phosphate linkage between successive bases has been replaced by an uncharged 2-aminoethylglycine unit with closely similar spacial properties to the natural linker (Fig. 4.12). They form very stable hybridisation complexes both with other PNA molecules and with DNA oligonucleotides (Nielsen et al., 1991; Uhlmann, 1998). The PNA–DNA triplex crystal structure (Fig. 4.13) has features distinct from those of either A- or B-DNA, with 16 bases per turn and a large x-displacement, of 6.8 Å (Betts et al., 1995). All sugars of the DNA strand have C3′-*endo* puckers. All 10 T·AT and 8 C$^+$·GC base triplets in this structure have the expected Hoogsteen geometry. The overall morphology of this triplex, which has been termed the

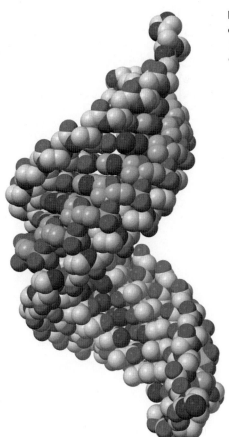

Figure 4.13 van der Waals representation of the 2.5 Å crystal structure of the PNA–DNA triple helix (PDB id 1PNN), with the carbon atoms of the DNA strand coloured light blue (Betts et al., 1995).

P-form, is governed by the requirements of the PNA strands, which are known to prefer to adopt A-type geometries. A subsequent PNA crystal structure (Petersson et al., 2005) has a complex lattice arrangement of left- and right-handed duplex structures, which interact together to form a triplex.

A true DNA triple helix has been crystallised (Rhee et al., 1999), using a construct of three overlapping oligonucleotide sequences, such that there is a central symmetric segment of six base triplets forming a short stable triplex region, flanked by a hexamer duplex at each end (Fig. 4.14). Crystals were grown at pH 5.0 since the dominance of $C^+ \cdot GC$ triplets in the sequence requires a low pH for optimum stability. The resulting crystal structure (Fig. 4.15), at 1.8 Å resolution, confirms all the general features of triplexes inferred from earlier modelling and NMR studies. In particular, the short distances (2.7 Å) between N3 of a cytosine and N7 of a guanine in a $C^+ \cdot GC$ triplet confirm the

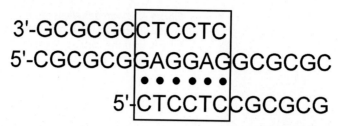

Figure 4.14 The sequences used to co-crystallise a parallel DNA triple helix (Rhee et al., 1999), with the triple helical segment highlighted.

Figure 4.15 The 1.8 Å crystal structure (PDB id 1D3R) of the parallel DNA triple helix (Rhee et al., 1999).

existence of a second hydrogen bond in this C·G Hoogsteen base pair, and thus of cytosine N3 protonation (Fig. 4.16). Most of the sugars have C2′-*endo* pucker. This feature, combined with the low base inclination of 6° in the triplex region and the x-displacement of −2 Å, suggests that the triplex part of the structure could be categorised

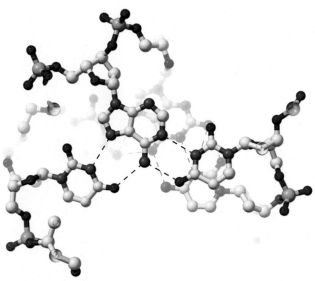

Figure 4.16 View onto two successive base triplets in the triple-helix crystal structure PDB id 1D3R (Rhee et al., 1999), with the adenine in the centre of the triplets.

closer to the B-type family than A-type. However, as with all triplex structures, other features do not accord with this classification, and the structure is best not described in such terms.

Locked nucleic acids (see below) can form stable triple helices, and one such structure has been solved by NMR methods (Sorensen et al., 2004). The third (pyrimidine) strand has alternating DNA and LNA nucleotides, but with sequences chosen to be identical to those previously studied as standard DNA.DNA·RNA and DNA.DNA·DNA triple helices (Gotfredsen et al., 1998; Tarköy et al., 1998). The Watson—Crick duplex part of the structure is intermediate between A- and B-forms, with a high mean propeller twist of $21°$, a mean inclination to the helix axis of $10°$, and sugar puckers in the $C2'$-*endo* region. The sugar puckers in the third strand are alternating between $C2'$-*endo* and $C3'$-*endo* forms. The large propeller twists between the Hoogsteen bases, as well as the Watson—Crick ones results in highly unusual (albeit weak) additional hydrogen bonding between pyrimidines, which form a spiral running along the whole of the structure. However, overall, the morphology is surprisingly similar with that of other (natural) triple helices (Fig. 4.17).

An all-RNA triplex-containing crystal structure has been determined (Ruszkowska et al., 2020), formed from the folding of a 79-mer RNA sequence and which contains 11 consecutive base triplets (PDB id 6SVS). This structure will be discussed in detail in Chapter 6.

Figure 4.17 The ensemble of NMR structures (PDB id 1W86) determined for a locked nucleic acid DNA·DNA-LNA hybrid (Sorensen et al., 2004).

4.2.3 Anti-parallel triplexes and non-standard base pairings

Triple helices can also be formed with the third strand consisting solely of purine residues, so that the resulting helix has Pu·PuPy triplets of bases (Beal & Dervan, 1991). Such a triple helix forms most readily with G·GC base triplets, having a G·G hydrogen-bonding arrangement that is likely to be as shown in Fig. 4.18a, with the third, purine strand being anti-parallel to the purine strand of the duplex. This arrangement is pH in-dependent, contrasting with the C^{+}·GC triple helix. The overall stability of anti-parallel triplexes is frequently greater than that of parallel ones. The guanosine nucleoside on the third strand of the G·GC base triplet can have either *anti* or *syn* arrangements (Fig. 4.18a or b). Both NMR (Radhakrishnan & Patel, 1994; Radhakrishnan et al., 1991) and molecular simulation studies (Laughton & Neidle, 1992) suggest that the *anti* one is actu-ally preferred, though no crystal structures of anti-parallel triplexes have been reported to date.

There have been numerous attempts to extend the sequence restrictions on triple-strand recognition (Gowers & Fox, 1999) using natural DNA bases. There are two over-riding challenges that any such an approach must overcome:

- the need to maximise base … base triplet hydrogen bonding

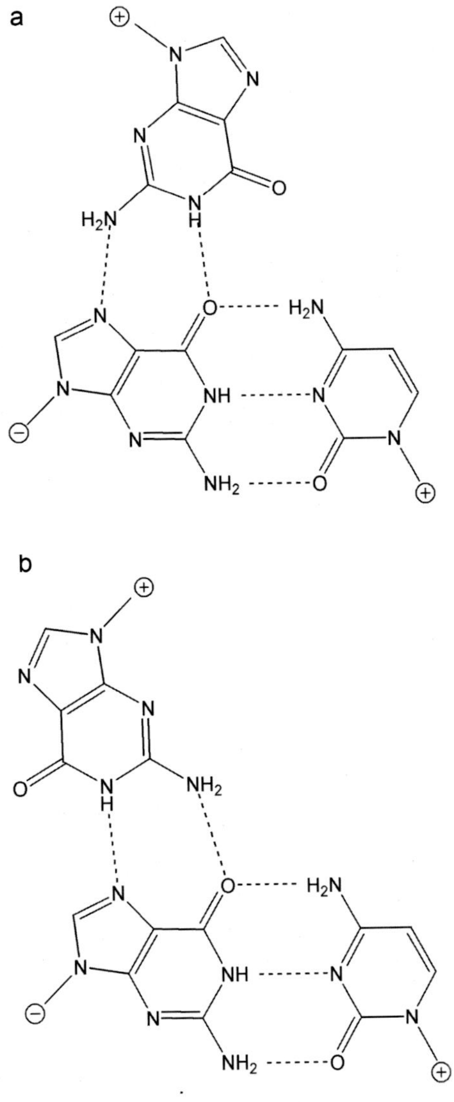

Figure 4.18 (a) Anti-parallel and (b) parallel G·GC base triplet hydrogen-bonding arrangements.

- the distortions in DNA structure that are introduced when inserting one or a few purines in an oligopyrimidine third strand (and vice versa). These distortions reduce intrastrand base stacking and are probably a major contributor to the instability of many possible base triplets

    However, some base combinations have been shown not to significantly destabilise the resulting structures. For example, a thymine base in a pyrimidine third strand can

Figure 4.19 (a) The G·TA base triplet. (b) The T·CG base triplet.

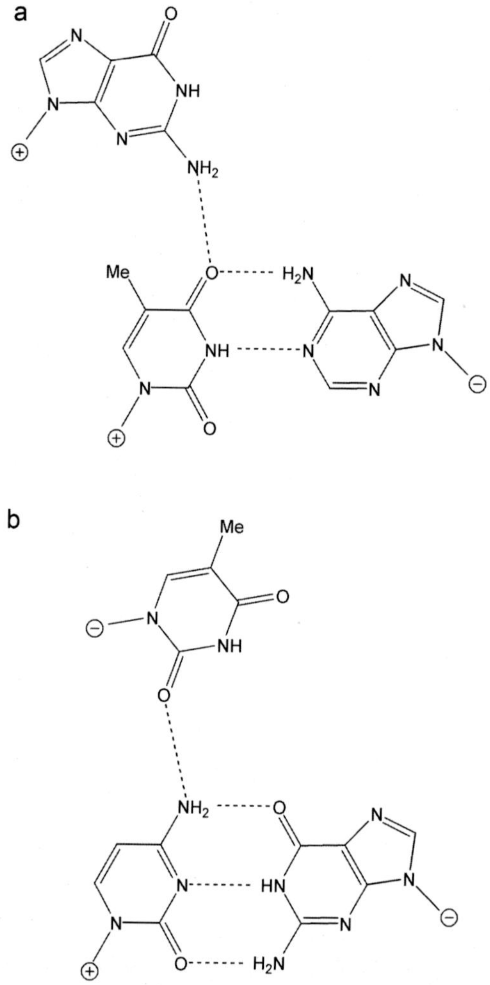

be replaced by a guanine, with the resultant G·TA base triplet (Fig. 4.19a) being relatively stable (Griffin & Dervan, 1989). However, the distortions that this is likely to produce in the DNA backbone suggest a limit on the number of such triplets that can be tolerated within a sequence. Almost all other possible triplet mismatches are highly destabilising to triple helix formation (Kiessling et al., 1992; Mergny et al., 1991) such that triplex formation for an otherwise stable sequence can be abolished by even a single such mismatch within it. The nature of the flanking sequences is generally important for stability of all types of non-standard triplet sequence. There are several mismatches, such as T·CG (Fig. 4.19b), that may be stable in some flanking sequence circumstances, but clearly do not work in others. Table 4.5 details the "code" for triple-strand recognition.

Table 4.5 Relative stabilities of all 16 possible base triplets.

Triplet	Stability	Triplet	Stability
$^+$C·GC	++++	C·CG	+
$^+$A·GC	++	A·CG	−
G·GC	+	G·CG	+
T·GC	++	T·CG	++
C·AT	+	C·TA	−
A·AT	+	A·TA	−
G·AT	+	G·TA	+++
T·AT	++++	T·TA	+

Adopted from Soyfer and Potaman (1996).

More general mixed sequence recognition of all four Watson—Crick base pairs in a duplex (i.e. C·G, G·C, A·T and T·A) is almost certainly only achievable with non-standard, unnatural "designer" bases (Griffin et al., 1992). These may be capable of being incorporated into a general-sequence oligonucleotide and hybridising with a target duplex with high affinity. The use of analogues of thymine with a potential positive charge at physiological pH (for example, 5-(1-propargylamino)-2′-deoxyuridine; Fig. 4.20) results in a 100-increase in binding affinity (Bijapur et al., 1999). This is possibly due to favourable electrostatic interactions with adjacent phosphate groups. Adding a second amino substituent results in, for example, 2′-aminoethoxy-5-(3-aminoprop-1-ynyl) uridine, which produces a significant further improvement in triplex stability, even at pH 7.0 (Osborne et al., 2004). Active recognition of all four bases in a target duplex can be achieved by extending this concept with a third strand containing four different synthetic nucleotides (Rusling et al., 2005). The TA step in a sequence can be as effectively recognised as the AT step is recognised by T, by using a synthetic thiazolyl-alanine deoxynucleoside (Wang et al., 2005).

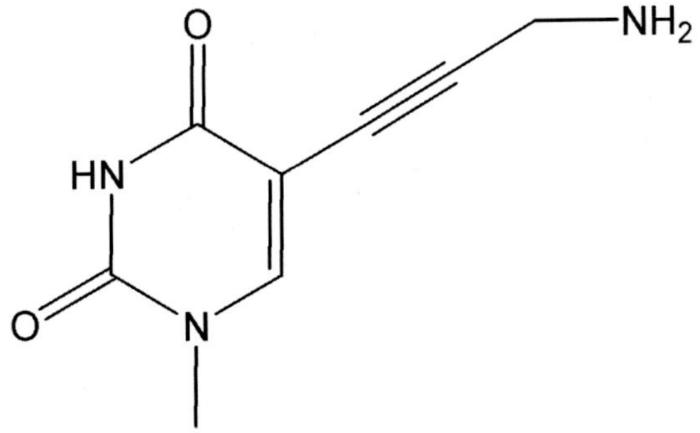

Figure 4.20 The structure of 5-(1-propargylamino)-2′-deoxyuridine (Bijapur et al., 1999).

The requirement for the $C^+ \cdot GC$ triplet to be protonated is a major limitation when selecting potential genomic targets for triplex recognition (see below). This is especially constraining with contiguous $C^+ \cdot GC$ triplets. There have been numerous attempts to devise non–natural cytosine mimics that provide two appropriate hydrogen bonds to the Hoogsteen face of the guanine. One involves the C–nucleoside analogue 2-aminopyridine, which has a pK_a of 6.86 compared to that of 4.3 for deoxycytidine itself, and will thus be largely protonated at neutral pH, so ensuring that two hydrogen bonds are likely to be formed (Fig. 4.21). This has been borne out by the formation of a stable triplex at pH 7 with 2-aminopyridine-containing third strands (Bates et al., 1996; Hildebrand et al., 1997).

The relative weakness of the third strand association in triple helices can be enhanced by triplex-selective ligands. These are typically molecules that can intercalate between base triplets, in a manner analogous to conventional duplex intercalation (see Chapter 5). Molecules such as acridines can be covalently attached to either the 5′ or 3′ end of the third strand oligonucleotide, and rational design considerations have enabled the optimisation of a variety of ligands to be undertaken (Escudé et al., 1998; Keppler et al., 2001). Attachment of a ligand to covalently cross-link the target duplex provides yet greater enhancement of binding. Psoralen has been a well-studied bifunctional cross-linking agent, binding at the sequence 5′-TpA, thereby placing a restriction on its general applicability for triplexes. In addition, psoralen has a requirement for UV activation in order to form covalent cross-links. An alternative approach is to use backbone-modified

Figure 4.21 The triplet formed by 2-aminopyridine with a G·C base pair (Bates et al., 1986; Hildebrand et al., 1997).

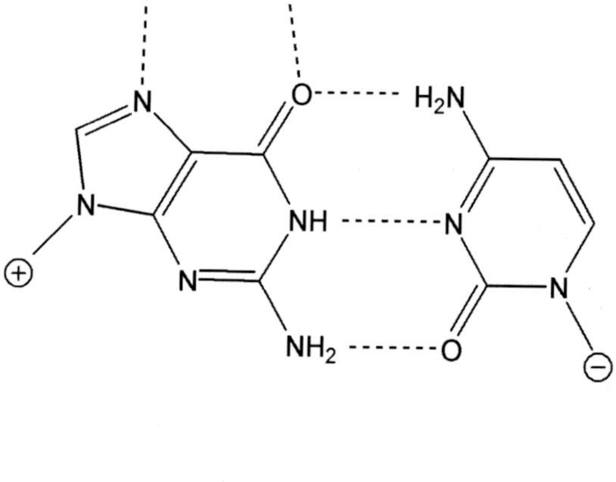

oligonucleotides for enhancement of stabilisation. The use of N3'-P phosphoramidite-modified oligonucleotides is of especial promise, since they hybridise to duplex sequences with significantly higher affinity than unmodified ones, and they are stable in cellular environments — this backbone modification is resistant to nuclease degradation (Escudé et al., 1996). It has been shown (Rusling et al., 2009; Sayoh et al., 2020) that the stability of a triplex sequence depends on the underlying stability of the embedded duplex.

4.2.4 Triplex applications

Much of the interest in the triplex phenomenon arose from the findings that triplex formation at a desired gene locus using an appropriate third strand oligonucleotide can specifically inhibit transcription of a particular gene (Grigoriev et al., 1993), for example, by binding to a promotor region that contains a suitable triplex-forming sequence. This has been termed the "antigene" approach to artificial gene regulation, and complements attempts to read DNA sequences with purely synthetic molecules targeted against the minor groove (see Chapter 5). Both methods require a target site of ca 16–20 bases for uniqueness in the human genome, which can readily be achieved by a triplex-forming oligonucleotide provided that the target site has an uninterrupted run of purines or pyrimidines. Since the stability of a triplex is length dependent, an even longer sequence is to be preferred. The (unsurprising) relative rarity of perfect triplex-forming sequences has encouraged the studies that have been outlined above, which are aimed at extending the in-built sequence restrictions on triplex formation. The goal of such studies is triplex recognition of a regulatory or coding sequence for any gene that is a suitable therapeutic target, and thus in principle is not confined to those with perfect oligopyrimidine or oligopurine triplex sites.

Both parallel and anti-parallel triplex-forming oligonucleotides have been used in a wide range of in vitro and cell-based experiments to demonstrate the viability of the triplex approach, especially its ability to down-regulate endogenous genes such as the oncogene c-*MYC* (Catapano et al., 2000). This goal has been conclusively demonstrated against an HIV polypurine tract in cell culture (Faria et al., 2000), using N3'-P phosphoramidite-modified oligonucleotides to optimise triplex stability and produce inhibition of transcription elongation. Extension of the triplex approach to in vivo situations has not to date generally advanced beyond early stages of development (Giovannangeli & Hélène, 2000), not least because of the difficulties involved in ensuring that oligonucleotides are stabilised against metabolism and are effectively transported into the cell nucleus.

A web-based search engine is available at http://spi.mdanderson.org/tfo/ to search for potential triplex-forming sequences throughout the human and other genomes (Gaddis et al., 2006).

4.3 Guanine quadruplexes

4.3.1 Introduction

It has long been known that guanosine monophosphate (guanylic acid) can self-associate in solution to form gels (Bang, 1910). This aggregation has also been observed with guanosine-rich oligo- and polynucleotides, provided a monovalent cation such as potassium or sodium is present to provide stabilisation. These observations were given a structural explanation half a century after Bang's original study, when diffraction patterns from fibres formed by guanylic acid were interpreted as arising from a novel four-stranded helix (Gellert et al., 1962). The same conclusion was subsequently reached with fibre diffraction patterns of guanosine-containing polynucleotides (Arnott et al., 1974; Zimmerman et al., 1975), with four guanine bases, one from each strand, being involved in a planar four-stranded (tetrad) arrangement and having G ... G base pairing between them (Fig. 4.22a). This very stable arrangement is termed a guanine (G)-quartet, or G-tetrad. It involves a total of eight hydrogen bonds between the "Watson—Crick" face of one guanine base and the "Hoogsteen" major groove face of another, in a manner analogous to that observed in some G·G mismatched duplexes. Thus, the G-tetrad is the four-strand equivalent of Watson—Crick base pairing for two-stranded helices. The four-fold G-helix as determined from fibre diffraction studies is right-handed and has a 23_2 repeat (i.e. 11 tetrads per complete helical turn), with a helical twist of 31.3° and a rise per tetrad of 3.41 Å (Arnott et al., 1974; Zimmerman et al., 1975). A representative G-helical arrangement is shown in Fig. 4.22b.

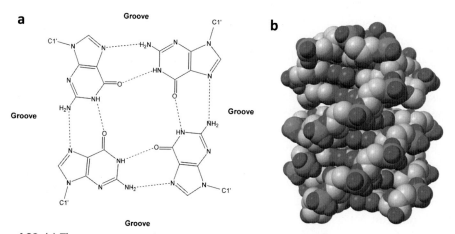

Figure 4.22 (a) The arrangement of guanine bases and Hoogsteen hydrogen bonds in the guanine tetrad. (b) van der Waals representation of a G-helix with seven G-tetrads.

There has been a dramatic increase in interest in higher-order nucleic acids, especially those containing the G-tetrad motif over the past 20 years, not least because guanine-rich sequences have wide-spread (but not random) prevalence in all eukaryotic and many prokaryotic genomes, and have been identified, for example, in eukaryotic promoter and immunoglobulin switch regions, in recombination hot-spots (Mani et al., 2009; Simonsson, 2001) and more recently at sites of genomic instability (Cheloshkina & Poptsova, 2021; Lejault et al., 2021; Lerner & Sale, 2019; McKinney et al., 2020). The occurrence of these sequences in the DNA at the ends of eukaryotic chromosomes (known as telomeres) has been well-characterised. Telomeric DNA comprises tandem repeats of short guanine tracts, sometimes including adenines/thymines, such as G_nT_n, $G_nT_nG_n$, G_nA_n or d(TTAGGG)$_n$, together with associated telomeric proteins (de Lange, 2005). The nature of the telomere repeat is species dependent with for example the sequence d(TTAGGG)$_n$ being conserved in vertebrates (Meyne et al., 1989). The function of telomeres is to protect chromosomal ends from unwanted recombination and nuclease damage. Human telomeric DNA in somatic cells is typically 8—10 kilobases, of duplex DNA. However, the terminal 100—200 nucleotides at the 3' end are single-stranded (de Lange, 2018; Wright et al., 1997), and are therefore in principle conformationally unconstrained, although in cells this single-strand overhang is associated with telomeric proteins such as the single-strand binding protein POT1 (Lei et al., 2004), which together with the associated TPP1 protein serves to protect the ends (Xu et al., 2021). Single-stranded telomeric DNA in the absence of protein can fold back and then dimerise to form four-stranded hairpin loops that can be stabilised (Sen & Gilbert, 1988; Sundquist & Klug, 1989) by the formation of the guanine tetrad motif outlined above. It can also form intramolecular structures by repeated fold-backs, as well as from telomeric sequence oligonucleotides such as d(TTAGGG)$_4$. These G-tetrad-containing arrangements are termed quadruplexes or tetraplexes (Gilbert & Feigon, 1999; Henderson et al., 1987), and can also be formed by short-length oligonucleotides of appropriate sequence, typified, for a single-strand unimolecular quadruplex, as:

$$G_m \ X_n \ G_m \ X_o \ G_m \ X_p \ G_m$$

where m is the number of G residues in each short G-tract. These are usually (but not necessarily all) directly involved in G-tetrad interactions. X_n, X_o and X_p can be any combination of nucleotide residues, including G, forming loops between the G-tetrads.

Some quadruplex sequences have been shown to possess aptamer properties when folded into four-stranded arrangements and have been used to target a range of cellular proteins important in human disease, such as thrombin and VEGF (for example, Roxo et al., 2019; Riccardi et al., 2021). One quadruplex aptamer AS1411 has remarkable anti-cancer properties (Bates et al., 2017; McMicken et al., 2003), and has been evaluated in clinical trials for human cancers; its mechanism of action, though not yet fully elucidated, may involve the protein nucleolin and its role in RNA processing. The structure of an aptamer closely related to the AS1411 quadruplex has been determined (Do et al., 2017) and is discussed in a subsequent section.

The evidence for the existence of stable quadruplex structures in vivo was controversial until the advent of specific probes directly demonstrating their presence in cells, initially being antibody-based for fixed cells, but more recently with small molecules that can be used in live cells. Immunofluorescence detection using the quadruplex-specific engineered antibodies BG4 (Biffi et al., 2013) and 1H6 (Henderson et al., 2014) in a variety of cell types has shown the presence of quadruplexes in cell nuclei and at telomeric ends of chromosomes. Both antibodies are available commercially and have been widely adopted as tools in quadruplex cell biology studies. BG4 has been used to demonstrate elevated quadruplex levels in human stomach and liver cancer tissues (Biffi et al., 2014) and in diffuse large B-cell lymphoma patient-derived tissues (Xu et al., 2020). Several small-molecule quadruplex probes have been reported that enable the existence and location of quadruplexes in live cells to be examined in detail. For example, the fluorescent triangulenium derivative DAOTA-M2 has been used to image dynamic processes (DNA damage response, small-molecule displacement, helicase action) within live cells since the imaging microscopy used is based on fluorescent lifetime rather than fluorescent yields changes (Summers et al., 2021). These live cell probes have much promise for the future study of quadruplex function and its exploitation for therapeutic goals.

Numerous proteins with quadruplex-binding functionality have been identified, notably helicases with quadruplex-unwinding abilities, such as FancJ, the Bloom's helicase, DEAH, DHX36 and RHAU helicase (Estep et al., 2019; Lansdorp & van Wietmarschen, 2019; Schult & Paeschke, 2020). So overall there is now considerable data (for example, Eddy & Maizels, 2006; Spiegel et al., 2020) supporting the concept not only that quadruplex structures can form or be induced to form, within coding and non-coding regions of human and many other genomes, as well as in telomeric sequences, but that they have a variety of functions depending on their location.

The location and number of putative G-tract sequences with quadruplex-forming potential have been examined in the human genome using bioinformatics approaches (see, for example, Huppert & Balasubramanian, 2005; Puig Lombardi & Londoño-Vallejo, 2020; Rawal et al., 2006; Todd et al., 2005). Occurrence is in many regions within the genome (notably in promoter sequences (Huppert & Balasubramanian, 2007; Vannutelli et al., 2020)), as well as in many bacterial, viral and other genomes (for example, Ruggiero & Richter, 2020; Wu et al., 2021). It is though important to realise that the occurrence of a particular quadruplex motif sequence does not necessarily indicate that it forms a stable quadruplex structure, either within its genomic environment, or even as an isolated sequence. In addition, none of the current algorithmic approaches can reliably predict all quadruplex sequences above and beyond those represented by the generalised search motif $G_m X_n G_m X_o G_m X_p G_m$, which assumes that loop length varies between one and seven nucleotides. There are quadruplex examples in the literature where, for example, guanines from the "loop" regions become inserted into core tetrads, where bulges occur, where two-tetrad quadruplexes can form or where stable quadruplexes exist with loops extended in length beyond the widely assumed limit of 1-7 nucleotides (Guédin et al., 2010). Little is known about quadruplexes formed by through-space interactions from G-tracts at two distant sites.

Several web-based tools for locating quadruplex sequences within genes are available, including:

G4Hunter (Brázda et al., 2019) at http://bioinformatics.ibp.cz/#/
G4CatchAll (Doluca, 2019) at http://homes.ieu.edu.tr/odoluca/G4Catchall/
QuadBase2 (Dhapola & Chowdhury, 2016) at http://quadbase.igib.res.in/

A comprehensive list is available (Puig Lombardi & Londoño-Vallejo, 2020), which also surveys the many studies in this area, together with the advantages/drawbacks of each individual computational approach. None are perfect, which is unsurprising given the current (2021) incomplete knowledge about quadruplex folding.

A standard approach to begin answering the question of whether a given sequence actually forms a stable quadruplex is to use a combination of biophysical probe methods such as NMR and circular dichroism (CD), the monitoring of melting behaviour (Mergny et al., 1998), together with, if possible, structure determination by crystallography/NMR. Bioinformatics can sometimes mislead with false positives by suggesting that a given sequence can or even does form a quadruplex, yet the supposition always needs to be validated by experiment. A study of the genome of the Parvovirus B19 illustrates this point (Bua et al., 2020). The small genome of this single-stranded DNA virus is 5.6 kb in length and an informatics search found several putative quadruplex-forming sequences, with one having a high quadruplex-forming score:

d(**GGG**ACTTCCGGAATTA**GGG**TTGGCTCT**GGG**CCAGCTTGCTT**GGGG**)

The four most likely G-tracts that might form a G-tetrad core in the sequence are highlighted in bold. CD and melting studies did not reveal any stable quadruplex signatures, possibly on account of the length of loops determining a preference for alternative hairpin structures rather than folding into a quadruplex.

The early bioinformatics studies on quadruplex sequence occurrence (Huppert & Balasubramanian, 2005; Todd et al., 2005) indicated a total of ca 350,000 quadruplex sequences in the human genome. More recent analyses (summarised in Puig Lombardi & Londoño-Vallejo, 2020) have suggested higher numbers. However, many quadruplexes in cells may have only a transient existence, especially if they are constrained within nucleosomes. An experimental approach has used antibody-based G4 chromatin immunoprecipitation and high-throughput sequencing methodology "G4 ChIP-seq", enabling the mapping of quadruplex locations in an immortalised cell line (Hänsel-Hertsch et al., 2016). The resulting ca. 10,000 quadruplex structures occur mostly in nucleosome-depleted regulatory regions at transcriptionally active sites, with over-representation in the promoters of many cancer-related genes, such as *SRC* and *c-MYC*. By contrast, the number of quadruplex structures in the "normal" counterpart of the immortalised cell line was found to be much lower, ca. 1,500. This result emphasises the differences between normal and immortalised (i.e. pre-cancerous) cells and provides further evidence for the suitability of quadruplexes as therapeutic targets in human cancers (see Chapter 5). This is given added significance by the recent finding that quadruplexes in chromatin can recruit transcription factors, especially at promoter sites of highly expressed genes, but that appropriate quadruplex-binding small molecules can effectively compete at these sites, thereby halting transcription (Spiegel et al., 2021).

The concept that quadruplex sequences are over-represented in promoter sequences (Huppert & Balasubramanian, 2007) has been well validated by the findings of quadruplex-forming sequences and stable quadruplex structures in the promoter regions of numerous oncogenes and cancer-relevant genes. Examples include *K-RAS* (Cogio & Xodo, 2006), c-*KIT* (with two separate quadruplex sequences occurring upstream of the transcription start site: Rankin et al., 2005; Fernando et al., 2006), *BCL2* (Dexheimer et al., 2006), *VEGF* (Guo et al., 2008), h*TERT* (Palumbo et al., 2012), *RET* (Tong et al., 2011) and *HIF* (De Armond et al., 2005). These quadruplexes have been characterised by biophysical as well as structural methods (see Section 4.2.5 below). Much attention has focused on the several G-rich sequences that form quadruplexes in the NHE III nuclease hypersensitive region of the c-*MYC* oncogene (Siddiqui-Jain et al., 2002; Simonsson, 2001), and these exceptionally stable quadruplexes have been well characterised by biophysical methods, with several NMR structures as well as a crystal structure, being available (see Section 4.2.5 below). As a result of this over-representation it has been proposed (Siddiqui-Jain et al., 2002) that induction of quadruplex structures by quadruplex-specific small molecules could be used to down-regulate the expression of such genes, since the formation of a quadruplex—ligand complex could impede the progression of RNA polymerase (Estep et al., 2019). Many small-molecule ligands, starting with the porphyrin TMPyP4, have been studied for this purpose (Asamitsu et al., 2019; Hurley et al., 2006; Neidle, 2016; Siddiqui-Jain et al., 2002), often with the goal of eventual therapeutic utility. Quadruplexes in viruses, bacteria and human neurological diseases are also increasingly becoming the focus of drug discovery efforts and several such quadruplexes have been structurally characterised (see below). The structural features of quadruplex—small molecule complexes are discussed further in Chapter 5.

4.3.2 Overall structural features of quadruplex DNA

As of August 2021, a total of 386 curated quadruplex structures are listed in the CSSR database, at http://g4.x3dna.org/.

The fundamental building block of all quadruplex structures is the G-tetrad. These are stacked one on another in a quadruplex structure, with a minimum of two being required for structural stability. Metal (or ammonium) ions, preferably $K^+ > Na^+ > NH_4^+$, are necessary for stabilisation and they are situated in the central cavity between or in the plane of each tetrad (see below). The tetrads tend to have a (quasi) helical twist of ca 36° between successive tetrad repeats, giving the exterior of some quadruplexes an appearance in two dimensions superficially similar to a stretch of B-DNA double helix (Fig. 4.22b) The number of guanines in each individual G-tract is sometimes (though not invariably) constant, and thus can be related to the number of G-tetrads formed in the folded quadruplex. Thus, in vertebrate telomeric DNA, where the repeat sequence is d(TTAGGG), almost all the quadruplexes formed by either two or four repeats have three stacked G-tetrads. There are several ways of forming a quadruplex structure from a given DNA (or RNA) quadruplex sequence, depending in large part on the number of short G-tracts. Four separate strands can associate together in an intermolecular (tetramolecular) manner (Fig. 4.23) or by a single-strand folding back on itself in an intramolecular structure (Fig. 4.24a–f). The five distinct example topologies shown in Fig. 4.24 differ in the orientations of the phosphodiester chain within the various strands of the complex. Whether the strands are parallel or anti-parallel to each other depends in part on the nature and length of the sequence, on the length and nature of the loops and sometimes on the counter-ion used.

The four guanosine nucleosides in an individual tetrad can in principle exist in either *anti* or *syn* glycosidic angle conformations, and thus there are 16 possible combinations

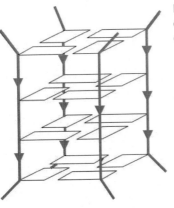

Figure 4.23 Schematic of an all-parallel tetramolecular G-quadruplex with four G-tetrads. The arrows represent strand directions.

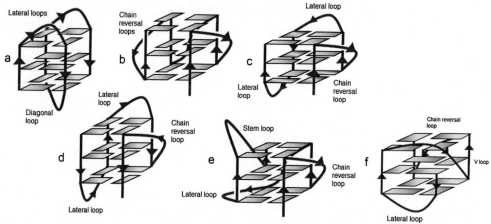

Figure 4.24 Schematic views of six diverse types of intramolecular folds in G-quadruplexes, emphasising the different types of loops. In each case the backbone is shown as a red line or curve and its directionality is indicated by *arrows*. The various types of loops are indicated. (a) The basket-type antiparallel arrangement formed from the human telomeric sequence d[AGGG(TTAGGG)₃] in sodium ion solution, as determined by NMR (Wang & Patel, 1993), PDB id 143D. (b) The parallel topology with three chain reversal loops, as found in the crystal structure of the d[AGGG(TTAGGG)₃] sequence, obtained from potassium-containing solution (Parkinson et al., 2002), PDB id 1KF1. (c) The (3 + 1) form-1 hybrid topology, with two lateral loops and one chain reversal loop, found by NMR for 26-mer human telomeric DNA sequences in potassium ion solution (Dai, Carver, et al., 2007; Luu et al., 2006), PDB ids 2GKU and 2HY9, respectively. (d) The (3 + 1) form-2 hybrid topology, also with two lateral loops and one chain reversal loop, and with the chain direction reversed compared to the form-2 hybrid structure. This was also determined by NMR, for a 26-mer human telomeric DNA sequences in potassium ion solution (Phan, Kuryavyi, Luu et al., 2007), PDB id 2JPZ. (e) The parallel topology of a 22-mer c-*KIT* promoter quadruplex, determined by both crystallography and NMR, with two chain-reversal loops (one of which is at the rear of this view), together with a lateral and a long stem loop (Phan, Kuryavyi, Burge, et al., 2007; Wei et al., 2015), with PDB ids 3QXR and 2O3M, respectively. (f) The anti-parallel topology of the *chl1* intronic G-quadruplex, determined in potassium solution by NMR (Kuryavyi & Patel, 2010), with PDB id 2KPR.

(Fig. 4.25). The mutual orientation of individual strands in a quadruplex has consequences for the glycosidic angles, and vice versa. Thus, an all–parallel orientation for all four strands, as in the tetramolecular quadruplex shown schematically in Fig. 4.23, requires all glycosidic angles to adopt an *anti* conformation. Anti-parallel quadruplexes can have a variety of orientations for the four backbones, whether these are from four separate strands (tetramolecular), two strands (bimolecular) or from a single (i.e. intramolecular) folded strand. Almost all naturally occurring genomic and telomeric quadruplexes are unimolecular. These can have *syn* and *anti* guanosine glycosidic angles, arranged in a way that is particular for a given topology and set of strand orientations. Thus, an all–parallel stranded intramolecular quadruplex will have all *anti* glycosidic

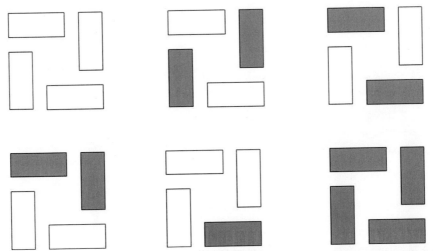

Figure 4.25 Schematic showing 6 of the 16 possible arrangements for the glycosidic angles of individual guanosines in an individual G-tetrad. Shaded rectangles represent guanosine nucleotides in *syn* conformations, and unshaded rectangles represent *anti* arrangements. This shading convention is commonly used in the literature.

angles (or rarely, all *syn*). A (3 + 1) form-1 hybrid quadruplex (Fig. 4.24c), with three parallel strands and one anti-parallel, has a pattern of *syn-anti-anti, syn-anti-anti, syn-syn-anti* and *syn-anti-anti* angles for the four strands, starting from the 5' direction. The (3 + 1) form-2 (Fig. 4.24d) has a pattern of *syn-anti-anti, syn-syn-anti, syn-anti-anti* and *syn-anti-anti* angles. These glycosidic angles thus depend on the nature and position of the loops and ensure correct strand directionality and maximum stacking of G-tetrads on one another. Both types of hybrid structures have the diagonally opposite strands running in opposing directions.

The counter-ions at the centre of the tetrad play a key role in maintaining the integrity of quadruplex structures. They do this by forming ionic interactions with the O6 atoms of the tetrad guanines. At one extreme an ion can be positioned symmetrically in the plane of four guanines in a tetrad. Potassium ions tend to be situated symmetrically between two consecutive tetrads so that they are octahedrally coordinated, to all eight O6 atoms from both tetrads (Fig. 4.26). This arrangement is favoured by ammonium as well as potassium ions. The smaller sodium ion can coordinate all four O6 atoms in a single tetrad while being in the plane of all four guanines, although it can also be accommodated midway between tetrads. A continuum between both extreme positions can be seen in several crystal structures (for example, Horvath & Schultz, 2001; Schultze et al., 1999), with the formation of a continuous channel of positively charged counter-ions along the inner axis of the quadruplex (Fig. 4.26). Thus a poly(dG) or poly(GMP) G-helix will have in effect a continuous G-wire of metal ions running through its centre.

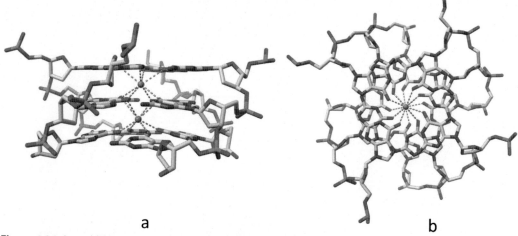

a   b

Figure 4.26 (a and b) Two orthogonal views of the octahedral coordination of potassium ions, symmetrically positioned between G-tetrad planes. The dashed lines indicate ionic interactions with O6 atoms of the guanines. The structure shown is the intramolecular G-quadruplex formed from the human telomeric sequence d[AGGG(TTAGGG)$_3$] as found in the crystal structure obtained from potassium-containing solution (Parkinson et al., 2002).

The variable sequences between the individual short G-tracts in a quadruplex generally form extra-helical loops, of which there are four types (Fig. 4.24a–f). Quadruplexes can also have at least one of the four strands running anti-parallel to the others. These topologies are found in the majority of bimolecular and intramolecular quadruplex structures determined to date.

1. Adjacent linked parallel strands require a connecting loop to link the bottom G-tetrad with the top G-tetrad, leading to propeller-type loops (also termed strand-reversal or chain-reversal loops). These loops can be short, down to a single nucleotide in length and straddle an edge of the stacked tetrad core. A quadruplex can have between one and three such loops, which has all strands being necessarily parallel to each other.

2. Lateral loops (also termed edgewise loops) join adjacent anti-parallel strands and are positioned over a terminal tetrad face of a quadruplex. Two of these loops can be located either on the same or opposite ends, corresponding to head-to-head or head-to-tail linking, respectively, when in bimolecular quadruplexes.

3. The second type of anti-parallel strand connecting loop, the diagonal loop, joins opposite strands. These loops are also positioned over a terminal tetrad. In this instance the directionalities of adjacent strands must alternate between parallel and anti-parallel.

4. The rare V-shaped loop (Fig. 4.24f) (Zhang et al., 2001) connects two adjacent anti-parallel strands and straddles a stacked tetrad edge, analogous to the strand-reversal loop. The start or end of the V-loop is in the middle of the tetrad stack.

Loop size was defined in the early bioinformatics searches as being between one and seven nucleotides. Subsequent biophysical studies (see, for example, the study of a Parvovirus B19 sequence (Bua et al., 2020)) have indicated that longer loops tend to impart destabilisation, and most well-characterised quadruplexes have short (1-7 nucleotide) loops. A survey of loop sequences and their frequency of occurrence (Todd et al., 2005) found that in the human genome, quadruplexes with at least one single-nucleotide loop are over-represented, with a T or an A loop being the most prevalent of all. This is broadly in accord with the frequent finding of single-nucleotide loops in the sequences of many subsequently discovered quadruplexes. Interestingly an experimental study examining 80 model sequences (Guédin et al., 2010) has shown that a single long loop together with two very short loops may be well tolerated. This is likely to be a consequence of the effects of secondary structure stabilisation in the long loop, as has been suggested for the quadruplex encoded in the promoter of the h*TERT* gene (Palumbo et al., 2012). One consequence of the common occurrence of single nucleotide loops, which are sterically constrained to be of the strand-reversal type, is that these tend to lock quadruplexes into a single parallel topology, especially when there is more than one such loop in a structure. This is because a single nucleotide is sterically unable to stretch across the distance between either two adjacent tetrads (a lateral loop) or between two opposite tetrads (in a diagonal loop). The majority of genomic quadruplexes identified to date have at least one single-nucleotide strand-reversal loop and tend to have all-parallel topologies (Hazel et al., 2004), even when there is added complexity, for example, with a long stem-loop. Quadruplex structural polymorphism is then restricted to those quadruplexes with at least two loops with two or more nucleotides, exemplified by human telomeric quadruplexes (see below). When the quadruplex sequence contains more than four G-tracts, an equilibrium between competing structures can occur, as with the 27-nucleotide nuclease-sensitive site NHE III$_1$ upstream of the c-*MYC* transcription start site, which contains six G-tracts (shown in bold):

$$5' - (\text{T}\textbf{GGGG}\text{A}\textbf{GGG}\text{T}\textbf{GGGG}\text{A}\textbf{GGG}\text{T}\textbf{GGGG}\text{AA}\textbf{GG})$$

Several discrete quadruplexes have been identified in this sequence, although not all have equal relevance to the transcriptional regulation of the c-*MYC* gene, or its targeting by small molecules (González & Hurley, 2010).

All right-handed quadruplex structures have normally at least four grooves, in contrast with the two in a double helix. These are defined by the cavities bounded by the phosphodiester backbones. Groove dimensions are variable and depend on strand direction, as well as overall topology, the nature of the loops and glycosidic angles. Grooves in quadruplexes with only lateral or diagonal loops are structurally simple, and the walls of these grooves are bounded by monotonic sugar phosphodiester groups. By contrast, grooves that incorporate strand-reversal loops have more complex structural features that reflect the insertion of these loops into the grooves. In general, two adjacent parallel strands generate a medium width groove; and two adjacent anti-parallel strands can generate either narrow or wide grooves (Li et al., 2021). In a narrow groove, all phosphates point into the groove; thus in an adjacent groove all phosphates on the shared strand between the two grooves necessarily point away from the groove making it either wide or medium but not narrow. Two adjacent grooves cannot both be narrow. A wide groove has all phosphates oriented away from the groove. All quadruplexes have the large planar surface of a tetrad at the 5′ and 3′ ends of their core of stacked tetrads, which in some instances have terminal non-G bases stacked on them, such as in the (3 + 1) hybrid structures (see below). Fig. 4.27 illustrates some of these groove features, and shows that

1. In a parallel quadruplex with all strand-reversal loops the overall structure is more disc-like than globular, there are more than four loop/groove cavities around the tetrad core and all these grooves are short. The accessible surface areas on the terminal tetrad planar surfaces are maximised.

2. In an anti-parallel quadruplex grooves can be elongated, especially when the two strands are anti-parallel. The accessible surface areas on the terminal tetrad surfaces are reduced since the diagonal and/or lateral loops cover part of the tetrad surface.

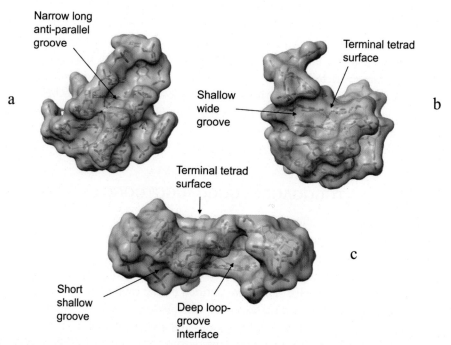

Figure 4.27 Views of the solvent-accessible surface of two human telomeric quadruplex structures, showing grooves and tetrad platforms at the ends of the tetrad cores. (a), (b) Two views of the anti-parallel quadruplex PDB id 6JKN. (c) View of the parallel quadruplex PDB id 1KF1.

Some general patterns of the relationships between strand polarity and base pairing have emerged from studies on triplexes and quadruplexes, as well as on Watson–Crick, left-handed and mispaired duplexes (Lavery et al., 1992; Westhof, 1992). For example, the pattern of glycosidic angles for a base pair is a consequence of the relationship between strand directions in quadruplexes, with *syn* angles being produced by anti-parallel strands and *anti* angles by parallel ones. This has led to a general classification scheme for nucleic acid structural types (Lavery et al., 1992), based on these polarity and base pairing criteria. Prediction of some new structural types can emerge naturally from the scheme. The variety of DNA structures that have been established over the past 20 years and described in this chapter suggests that many more remain to be experimentally determined, especially for quadruplex nucleic acids.

There are therefore several structural variables, notably number of G-tetrads: loop type (sequence and length), strand polarity and guanosine glycosidic angle, that together result in quadruplexes as a class having high topological and structural variety. There are few general rules that can give one confidence in topology prediction, not least because there are still insufficient experimental structures available (Tables 4.6 and 4.7) for prediction to be consistently reliable. Loop size and sequence are probably the most important determinants of topology. Very short loop sequences with ≤2 nucleotides are more likely to form strand-reversal loops (for purely steric reasons since lateral and diagonal loops require at least two nucleotides in order to bridge their attachment points on the tetrads), and thus impose a parallel strand topology on that part of a quadruplex (Bugaut & Balasubramanian, 2008; Hazel et al., 2004). Longer sequences are more likely to prefer lateral or diagonal loops. In these instances, groove width may become enlarged due to the presence of the bulky loops. Even apparently minor changes in loop size (or in flanking 5′ or 3′ sequences) can have dramatic consequences for overall topology, as is illustrated below. An approach to potential topology prediction has been made (da Silva, 2007; Karsisiotis et al., 2013) by considering *syn* vs *anti* glycosidic angles and sugar puckers for the individual nucleotides in a strand. Loop permutation, i.e. the position of a given loop within a quadruplex, is also a factor in governing topology (Cheng et al., 2018). However, we are still far from being able to confidentially predict the folds of many quadruplexes, especially when nucleotides in loops become actively involved in stabilising the G-tetrad cores, for example, by donating a loop guanosine to become part of a tetrad.

4.3.3 Telomeric quadruplex structures

The sequences d(TGGGGT) (Aboul-ela et al., 1994) from the telomeric repeat sequence found in *Tetrahymena* and d(TTGGGG) (Wang & Patel, 1992) form parallel-stranded quadruplex structures in solution with all-*anti* glycosidic angles (Fig. 4.23). Others such as d(GGTTTTCGG) (Wang et al., 1991) form anti-parallel arrangements with

Table 4.6 Crystal and NMR structures of selected telomeric DNA quadruplexes. All are potassium-associated, unless otherwise specified.

Quadruplex sequence	Crystal or NMR	PDB id	Structure type	Organism
d(TGGGGT)	Crystal	352D	Parallel	*Tetrahymena*
d(GGGGTTTGGGG)	Crystal	2AVH	Dimeric antiparallel	*Oxytricha nova*
d(GGGGTTTGGGG)	NMR	1U64	Dimeric antiparallel	*Oxytricha nova*
d(GGGGTTTGGG)	NMR	1LVS	Dimeric antiparallel	*Oxytricha nova*
d(GGGGTTTGGGG)	Crystal	1JPQ	Dimeric antiparallel	*Oxytricha nova*
d(GGGGTTTGGGG) + Na⁺	NMR	156D	Dimeric antiparallel	*Oxytricha nova*
d(GGGGTTTGGGG) + telomere end-binding protein + Na⁺	Crystal	1JB7	Dimeric antiparallel	*Oxytricha nova*
d[AGGG(TTAGGG)₃]	Crystal	1KF1	Parallel	*Homo sapiens*
		6IP3		
d[AGGG(TTAGGG)₃] + Na⁺	NMR	143D	Anti-parallel basket	*Homo sapiens*
d[TTGGG(TTAGGG)₃A]	NMR	2GKU	(3 + 1) hybrid fold-1	*Homo sapiens*
d[AAAGGG(TTAGGG)₃AA]	NMR	2HY9	(3 + 1) hybrid fold-1	*Homo sapiens*
d[TTAGGG(TTAGGG)₃TT]	NMR	2JPZ	(3 + 1) hybrid fold-2	*Homo sapiens*
d[(GGGTTA)₃GGGT]	NMR	2KF8	2-Tetrad basket	*Homo sapiens*
d[TTAGGG)₃TTA(ᴮʳG)GGTTA] + Na⁺	NMR	2MBJ	(2 + 2) Anti-parallel	*Homo sapiens*
d[GGGTTAGᴮʳGGTTAGGGTTAGᴮʳGG]	Crystal	6JKN	Anti-parallel chair	*Homo sapiens*

Table 4.7 Crystal and NMR structures of selected promoter and viral DNA quadruplexes.

Gene	Quadruplex sequence	Crystal or NMR	Organism	PDB id
BCL2	d(GGGCGCGGGAGGAATTGGGCGGG)	NMR	Homo sapiens	2F8U
cKIT1	d(AGGGAGGGCGCTGGGAGGAGGG)	NMR	Homo sapiens	203M
cKIT1	d(AGGGAGGGCGC(^{Br}U)GGGAGGAGGG)	Crystal	Homo sapiens	3QXR
cKIT2	d(CGGGCGGGCGCTAGGGAGGGT)	NMR	Homo sapiens	2KYP
cMYC	d(TGAGGGTGGGTAGGGTGGGTAA)	NMR	Homo sapiens	1XAV
cMYC	d(TAGGGAGGGTAGGGAGGGT)	NMR	Homo sapiens	2LBY
cMYC	d(TGAGGGTGGGTAGGGTGGGTAA)	Crystal	Homo sapiens	6AU4
kRAS	d(AGGGCGGTGTGGAATAGGGAA)	Crystal	Homo sapiens	6N65
kRAS	d(AGGGCGGTGTGGAATAGGGAA)	NMR	Homo sapiens	5I2V
kRAS	d(GGAG$_5$AGAAG$_3$AGAAG$_3$TGTGGCGGGA)	NMR	Homo sapiens	6T2G
VEGF	d(CGGGGCGGGCCTTGGGCGGGGT)	NMR	Homo sapiens	2M27
PIM	d(GCGGGGAGGGCGCGCCAGCGGGGTCGGG)	NMR	Homo sapiens	7CV3
PARP1	d(TGGGTCCCAGGGCGGGGCTTGGG)	NMR	Homo sapiens	6AC7
RET	d(GGGGCGGGGCGGGGGCGGGGT)	NMR	Homo sapiens	2L88
B-RAF	d(GGGCGGGGAGGGGAAGGGA)	Crystal	Homo sapiens	4H29
WNT	d(GGGCCACCGGGCAGTGGGCGGG)	NMR	Homo sapiens	6L92
hTERT	d(AGGGIAGGGGCTGGGAGGGC)	NMR	Homo sapiens	2KZD
hTERT	d(AIGGGAGGGGICTGGGAGGGC)	NMR	Homo sapiens	2KZE
CEB25	d(AAGGGTGGGTGTAAGTGTGGGTGGGT)	NMR	Homo sapiens	2LPW
Promoter	d(GGGGGAGGGGTACAGGGGTACAGGGG)	Crystal	Dictyostelium discoideum	6FTU
U3 LTR	d(GGGAGGCCGTGGCCTGGGCGGGACTGGGG)	NMR	HIV	6H1K
_	d(GGGTAGGGCAGGGGACACAGGGT)	NMR	Papillomavirus	5O4D
KCNN4 mini-satellite	d(GGTCTGAGGGAGAGGGGCTGGGT)	NMR	Homo sapiens	7ATZ

alternating *syn-anti* glycosidic bonds. The parallel d(TGGGGT) quadruplex structure, stabilised by sodium ions, has also been studied in detail by X-ray crystallography (Phillips et al., 1997). The structure, which contains four independent quadruplexes in the crystallographic asymmetric unit, has been solved to exceptionally high resolution (0.95 Å), enabling many details to be precisely defined. Two quadruplexes are stacked together in the crystal, with a channel of seven sodium ions running through the centre of the structure. These occupy a range of positions for these ions (and hence of ion—ion distances) with respect to the tetrad planes, although there is a clear pattern of increasing co-planarity with the tetrads towards the ends of the tetraplex dimer. Most of the sugar rings are in the normal B-DNA C3′-*endo* conformation. However, the distributions of values for the backbone torsion angles δ and ε are outside the normal duplex ranges, suggesting that the quadruplex backbone is strained, possibly to enable the hydrogen bonding of the tetrads to occur.

The sequence d(GGGGTTTTGGGG), from the telomere of the *Oxytricha nova* organism, has been extensively studied by structural methods. The general arrangement is of a four-stranded, bimolecular folded quadruplex with two strands associating together as a hairpin structure. There are four stacked hydrogen-bonded G-tetrads (Fig. 4.28), sandwiched on top of each other at a separation of 3.4 Å. Several crystallographic analyses have been reported for this sequence, with sodium ions and co-crystallised within the

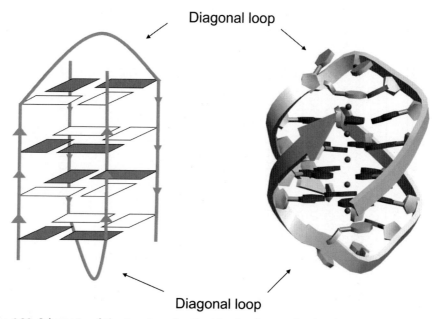

Diagonal loop

Diagonal loop

Figure 4.28 Schematic of the topology for the bimolecular quadruplex formed by two strands of d(GGGGTTTTGGGG), as found in the crystal structure of the *Oxytricha nova* quadruplex (Haider, Neidle and Parkinson, 2003). The rectangles, representing guanosine nucleosides, are shaded as detailed in Figure 4.25.

structure of a protein-single-stranded telomere complex (Horvath & Schultz, 2001), and also crystallised as the native structure in the presence of potassium ions (Haider et al., 2002). Both structures show an identical bimolecular quadruplex topology with adjacent strands anti-parallel to each other and the loops joining diagonal strands. There are alternating *syn* and *anti* glycosidic conformations along any one strand (Fig. 4.28), and the arrangement of strands results in a G-tetrad arrangement with a *syn-syn-anti-anti* pattern of glycosidic angles and diagonal loops. This topology is fully in accord with that determined in a series of NMR studies (Schultze et al., 1999; Smith & Feigon, 1992) on this sequence with sodium, potassium and ammonium ions. The overall fold is thus retained in all these environments. There are some minor differences between the NMR and crystal structures that are mostly concerned with the conformations of thymines in the loops, which are highly ordered in the crystal structures with extensive stacking between them (Fig. 4.28).

Apparently minor changes to the *Oxytricha* sequence d(GGGGTTTTGGGG) result in quadruplexes with major structural differences from the symmetric bimolecular quadruplex described above. Two distinct bimolecular quadruplexes (PDB ids 2AVH and 2AVJ) are formed in the crystal by d($G_4T_3G_4$) and a brominated analogue d[G_4(^{Br}U)T_2G_4], with one being a head-to-tail lateral loop dimer in which all adjacent strands are anti-parallel, and the other is a head-to-head hairpin quadruplex with one adjacent strand parallel and the other anti-parallel (Hazel et al., 2006). Decreasing the size of one G-tract in the native *Oxytricha* sequence by one guanosine at the 5′-end, to form d(GGGTTTTGGGG), results in an asymmetric bimolecular quadruplex with just three tetrad planes, and having lateral, diagonal and an edge-contacting loop (Šket et al., 2003). Again, the fold is not dependent on the nature of the metal ion. This structure is one of the few known examples of a bimolecular quadruplex with an unequal number of parallel (three) and anti-parallel (one) strands (Fig. 4.29).

Figure 4.29 Cartoon representation of one of the ensemble of solution structures (PDB id 1U64) for the asymmetric bimolecular quadruplex formed by two strands of d(GGGTTTTGGGG) (Šket et al., 2003).

4.3.4 Human telomeric quadruplex structures

A complex pattern of folding has been found in the intramolecular fold-back quadruplexes (Figs. 4.24 and 4.30) formed by the sequence d[AGGG(TTAGGG)₃] and its many close variants, each having small changes in sequence, mostly at the 5′ and/or 3′ ends. Almost all consist of four repeats of the vertebrate telomeric sequence d(TTAGGG). The original NMR structure in sodium solution (Wang & Patel, 1993) of d[AGGG(TTAGGG)₃] shows each strand flanked by one parallel and one antiparallel neighbour (Figs. 4.24a and 4.30a). There is a core of three guanine tetrads flanked by two lateral and one diagonal TTA loop. The asymmetry of the loops results in significant differences between the widths of the four grooves. For the NMR structure in the ensemble with the lowest RMSD, two grooves have intermediate widths of 11.5—13.5 Å, and are flanked by a narrow (<10 Å) and a wide groove (>15 Å).

The crystal structure of the same sequence, with crystals grown from potassium-containing solution, shows a remarkably different topology (Figs. 4.24b, 4.26 and 4.30b), with all strands parallel and with all three TTA loops arranged in a strand reversal topology (Parkinson et al., 2002). This structure determination was at 2.1 Å resolution: more recent re-determinations at 1.40 Å (Nuthanakanti et al., 2019) and at 1.35 Å (PDB id 7KLP) (Li et al., 2021) show closely similar structures, with the higher resolution enabling a greater number of associated water molecules and networks to be identified.

The conditions used to form these crystals have some similarities to those in the cell nucleus, with typical potassium ion concentrations of ca 100 mM, so it has been suggested that this topology is relevant to the folded state of the 3′-single-stranded telomeric DNA overhang in cells (Renciuk et al., 2009). This is in accord with CD studies at high potassium ion (Rujan et al., 2005) and high quadruplex concentrations (Kejnovská et al., 2014). An NMR structural study under molecular crowding conditions using polyethylene glycol (PEG) with four variants of the intramolecular quadruplex sequence found that they each formed a distinct structure in more dilute conditions, but as crowding (i.e. PEG concentration) increased, all four structures converted to the parallel form (Heddi & Phan, 2011): PDB id 2LD8. The topology found in the crystal structure is one low-energy form that is favoured by G-quadruplex/G-tetrad stacking, and perhaps by the sheer crowded environment of the high-packing density in a crystal structure.

However, it is also apparent from several biophysical studies (for example, Le et al., 2014; Li et al., 2005; Ying et al., 2003) that human telomeric quadruplex sequences in more dilute potassium solution can fold into several conformers, whose nature depends on the precise sequence involved. Individual structures can be identified by subtly altering the precise sequence used so that a particular species can dominate and so its structure can be determined. Several other topologies have been observed in NMR studies (Table 4.6 and Figs. 4.24 and 4.30) using sequences related to the 22-mer, often with short 3′ and/or 5′ flanking sequences that serve to stabilise particular structures, such

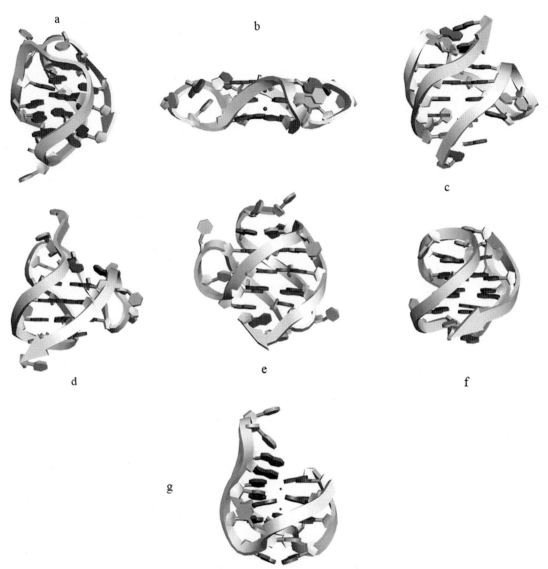

Figure 4.30 Cartoon representations of the various structures determined of unimolecular human telomeric DNA sequences. Sequences are given in Table 4.6. For the NMR-determined structures an individual structure has been taken from the deposited ensemble of structures (see Table 4.6). (a) Basket-type anti-parallel structure in Na^+ solution, PDB id 143D. (b) Parallel structure, PDB id 1KF1. (c) (3 + 1) form-1 hybrid structure, PDB ids 2GKU and 2HY9. (d) (3 + 1) form-2 hybrid structure, PDB id 2JPZ. (e) (2 + 2) anti-parallel structure in Na^+ solution, PDB id 2MBJ. (f) Two-tetrad basket structure, PDB id 2KF8. (g) Anti-parallel chair structure, PDB id 6JKN.

as ones with $(3 + 1)$ loops (Ambrus et al., 2006; Dai et al., 2007; Luu et al., 2006; Lim et al., 2009, 2013; Phan and Patel, 2003), where bases of the flanking sequences can stack on top of $5'$ or $3'$ end tetrads. The two forms of the $(3 + 1)$ fold have unusual topologies, with three parallel and one anti-parallel strand, so that there are one propeller and two lateral loops (Fig. 4.30c and d). Thus, we now have structural views of most species present in the K^+-solution of human intramolecular telomeric quadruplex sequences. The consensus is that most sequences in dilute solution form hybrid and anti-parallel structures.

The question of how these mono-quadruplex structures relate to the protein-free folded structure of the 100+ nucleotide single-stranded human telomeric DNA overhang has not yet been answered by high-resolution structure determination. The overhang length suggests that it is capable of folding into several consecutive multimer quadruplex structures. Biophysical and molecular modelling/simulation studies have produced several alternative models. These are of interest as potential drug targets, possibly providing unique selectivity at the interface between individual quadruplex units compared to single quadruplex targets. A modelling and molecular dynamics study (Haider et al., 2008) based on the intramolecular parallel quadruplex with three strand-reversal loops concluded that the individual units in such a multimer could effectively stack on top on one another (Fig. 4.31a), with four quadruplexes, having d(TTA) linkers, comprising 93 nucleotides. The effective stacking is facilitated by the large accessible surface area of the terminal tetrads of each individual quadruplex. An integrated simulation and biophysical study (Monsen et al., 2021) has concluded that multimers in solution comprise a 25:75% mixture of $(3 + 1)$ form-1 and form-2 individual units (Fig. 4.31b). The simulations indicate that both types of models can accommodate small-molecule binding between individual units as there appears to be some flexibility for the linkers. In cells vertebrate telomeric DNA is protected by copies of the POT1/TPP1 protein heterodimer, which can be displaced by quadruplex multimer formation: this occurs at higher local concentrations than those typically used in biophysical studies so as with the telomeric quadruplex monomer it may well be that different species predominate under different crowding conditions.

4.3.5 Promoter and other quadruplexes

The findings that quadruplex sequences are over-represented in many cancer genes have generated much interest in them as therapeutic targets, and in their structural characterisation. Table 4.7 lists the majority that have been so characterised to date. There are few crystal structures in this list, reflecting the continuing challenges of obtaining diffraction-quality quadruplex crystals. By contrast with the human telomeric quadruplexes, when

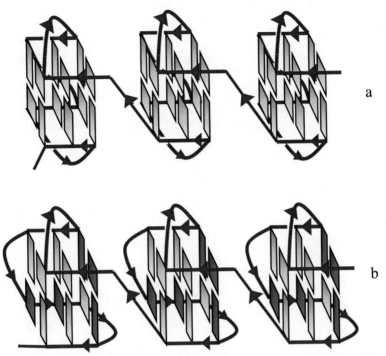

Figure 4.31 Cartoon views of folded telomeric DNA comprising human telomeric quadruplex multi-mers. (a) Constructed from all-parallel individual quadruplex units. (b) Constructed from (3 + 1) hybrid units.

comparisons can be made, solution NMR and crystal structures normally concur on topology, sometimes with minor differences in loop conformations. Most of these structures have at least one single-nucleotide (strand-reversal) loop, and many have all-parallel topology. However, added complexity with, for example, insertion of a guanine loop into a tetrad, as observed in the c-*KIT* structures, is not uncommon, as are stem loops with their own secondary structures.

The intramolecular quadruplex structures available to the G-rich strand of the nuclease-hypersensitive NHE III1 promoter sequence in the *c-MYC* oncogene have been extensively characterised by NMR methods. There are six short G-tracts overall in the natural 27-mer sequence:

$$5' - (\text{T}\textbf{GGGG}\text{A}\textbf{GGG}\text{T}\textbf{GGGG}\text{A}\textbf{GGG}\text{T}\textbf{GGGG}\text{AA}\textbf{GG})$$

Structural studies have been undertaken on quadruplexes with either four or five such tracts. The overall folds are closely similar, with all-parallel topologies and with all loops constrained to be in strand-reversal conformations (Ambrus et al., 2005; Phan et al., 2004; Phan et al., 2005). The major and most physiologically important form includes the four

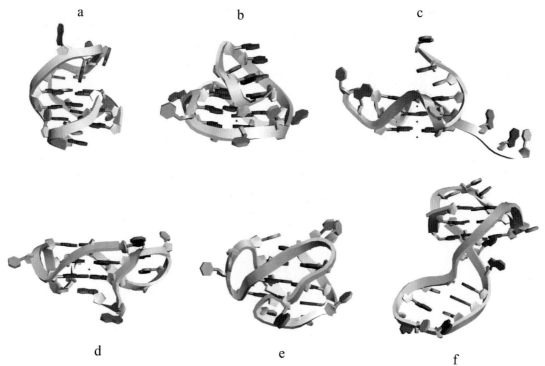

Figure 4.32 Cartoon views of several promoter and other quadruplex structures. (a) A *BCL2* promoter quadruplex, with three parallel and one anti-parallel strand, showing a structure selected from the ensemble of NMR structures (Dai et al., 2006), PDB id 2F8U. (b) A *c-KIT* quadruplex, with all strands parallel, from the crystal structure (Wei et al., 2012), PDB id 3QXR. There is a short 5-nucleotide structured stem loop positioned over the top tetrad, with two A·G base pairs in this unique quadruplex fold. (c) The crystal structure of a *c-MYC* quadruplex (Stump et al., 2018), PDB id 6AU4. (d) The crystal structure of the *K-RAS* quadruplex (Ou et al., 2020), PDB id 6N65. (e) A structure taken from the ensemble of NMR structures for the *PARP1* promoter (Sengar et al., 2019), PDB id 6AC7. (f) One of the ensemble of NMR structures of the quadruplex found in the U3 promoter region of the HIV-1 long terminal repeat (Butovskaya et al., 2018), PDB id 6H1K. This unique arrangement has a (3 + 1) hybrid fold, with three tetrads, a 12-nucleotide diagonal loop containing a conserved duplex stem, a 3-nucleotide lateral loop, a single-nucleotide strand-reversal loop and a V-shaped loop.

G–tracts at the 3′ end of the sequence (Ambrus et al., 2005). A crystallographic analysis of this 22–mer (Stump et al., 2018) has found the same topology as the earlier NMR studies (Fig. 4.32c). Both NMR and crystallographic studies have used sequences with conservative mutations to obtain interpretable spectra (NMR) or useable crystals (crystallography). There are two structurally closely similar independent quadruplexes in the c-*MYC* quadruplex crystallographic asymmetric unit, providing further support for the topology being the physiologically important one. A principal difference with the NMR structure is that the terminal 5′ d(TGA) is stacked onto the terminal tetrad in the NMR structures, whereas it is in an extended conformation in the crystal, with the adjacent tetrad surface being more exposed.

A recurrent theme in both NMR and crystal structures (*BCL2, c-KIT, K-RAS, PARP1*) is the existence of a terminal base or bases platform that stacks onto a terminal tetrad plane, providing additional stabilisation (Fig. 4.32). Related platforms have also been found in several of the (3 + 1) telomeric quadruplexes, where they are formed by the flanking sequences, and perform the same stabilising function. An unusual quadruplex fold with such a feature occurs in a sequence in the c–*KIT* kinase promoter region (Rankin et al., 2005). This structure, conserved in solution and in the crystal (Phan et al., 2007; Wei et al., 2012) (Fig. 4.32b), has an isolated guanine, previously assumed to be part of a loop tract, being involved in G-tetrad core formation, despite the presence of four three-guanine tracts. There are four loops: two single-residue strand-reversal loops, a two-residue loop and a five-residue stem-loop at one end of the core quadruplex, which contain two A·G base pairs. The sole differences in the NMR and crystallographic structures (Wei et al., 2012) are in groove dimensions.

The potassium channel *KCNN4* gene contains a G-rich minisatellite 304 bp downstream of the transcription start site, which can form a G-quadruplex. The solution structure of this quadruplex has been determined by NMR methods (Vianney & Weisz, 2021), and has revealed the presence of a highly unusual combination of two lateral loops, a zero–nucleotide V-shaped loop and a strand-reversal loop (Fig. 4.33), which may well provide suitable for selective drug targeting since the *KCNN4* gene is over-expressed in some human cancers.

Figure 4.33 One of the ensemble of NMR structures for the quadruplex from a mini-satellite region of the KCNN4 gene (Vianney & Weisz, 2021), PDB id 7ATZ.

4.3.6 Quadruplex–duplex junctions

The structural features of quadruplexes embedded within a long duplex DNA environment, for example, within a promoter sequence, are not yet established. However, some structural aspects of quadruplex–duplex junctions have been examined by NMR (with five representative constructs: Lim & Phan, 2013) and crystallographic studies (Russo-Krauss, 2016). These demonstrate that attachment of a B-DNA duplex to several types of quadruplex fold is sterically feasible, although all the structures have a duplex on only one side of a quadruplex. The structures have some features in common with quadruplexes having long stem loops/hairpins. One of the crystal structures (Fig. 4.34a) shows a parallel stranded quadruplex with single-nucleotide strand reversal loops and a long 3′ end sequence, which is hybridised to a complementary 8-mer sequence, forming a short, connected duplex at this end. There is a T·A·T base triplet at the quadruplex–duplex interface, positioned between the terminal tetrad and the first (T·A) base pair of the duplex, so ensuring a continuous stack of bases between the two nucleic acid moieties. The NMR structures comprise in each case a single strand (for example, a 34-mer, PDB id 2M92), with the duplex as in effect a hairpin (Fig. 4.34b). Modelling studies have demonstrated the feasibility of attaching more than one hairpin to the quadruplex, with several attachment points being possible.

The inference from these structures is that a quadruplex can be induced within a genomic duplex sequence without long-range perturbations of duplex structure. The

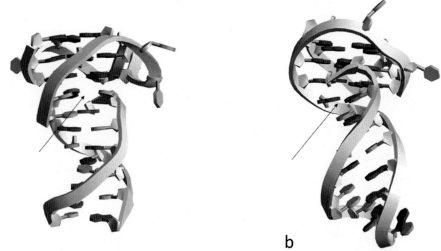

a   b

Figure 4.34 (a) Crystal structure of a quadruplex–duplex junction (Russo Krauss et al., 2016), PDB id 5DWW. The parallel stranded quadruplex has one single-nucleotide and two two-nucleotide strand reversal loops. The arrow indicates the T·A·T base triad at the duplex–quadruplex interface. (b) One of the ensemble of NMR structures for an intramolecular duplex–quadruplex junction (Lim & Phan, 2013), PDB id 2M92, showing an eight base-pair duplex linked to the quadruplex. An A·G base pair at the junction, highlighted by an arrow, is wedged between the tetrads and the base pairs, which are inclined by ca 40°.

observations of local mispairing at the junction, either by a small-molecule binding or by protein-induced effects, may though be sufficient for a DNA damage response to occur in cells (Rodriguez et al., 2012).

4.3.7 Left-handed quadruplexes

The quadruplex aptamer AS1411 has previously been mentioned as having anti-proliferative properties. In solution it is a mixture of multiple quadruplex species. An NMR study (Do et al., 2017) of a closely related sequence (AT11), d(TGGTGGTGGTTGTTGTGGTGGTGGTGGT) which forms a single species in solution, has revealed an unusual topology (Fig. 4.35a) with two two-tetrad subunits, connected by a single thymine nucleotide. The stacking between the two subunits is 3′ to 5′. A further sequence, with a single G (underlined) difference compared to AT11 d(TGGTGGTGGTG̲GTTGTGGTGGTGGTGTT) remarkably forms a left-handed quadruplex, which was observed in solution (by NMR) and in a crystal structure (Fig. 4.35b) (Chung et al., 2015). The structure closely resembles that of AT11, apart from the inversion of strand direction, in that it retains the two-tetrad subunit arrangement and the prominent central irregular zig-zag backbones and groove running parallel to the tetrad planes and encircling around most of each structure. The left-handedness was confirmed by the circular dichroism (CD) spectrum, which is inverted compared

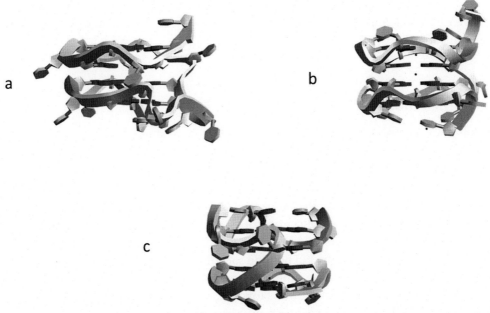

Figure 4.35 Left-handed and related quadruplex structures. (a) Crystal structure of the right-handed anti-proliferative aptamer AT11 (Do et al., 2017), PDB id 2N3M. (b) Crystal structure of a left-handed quadruplex (Chung et al., 2015), PDB id 4U5M. (c) Crystal structure of a quadruplex showing both left- and right-handed parallel folds (Winnerdy et al., 2019), PDB id 6QJO.

to the normal right-handed one of AT11. Thus, the positive peak at ca 270 nm and the negative one at ca 250 nm for a normal right-handed parallel quadruplex becomes a negative peak and positive peak, respectively. The backbone and groove irregularity is an inherent consequence of the two-tetrad subunit arrangement in both left- and right-handed structures. It has subsequently been possible to design quadruplex sequences with both left- and right-handed features (Winnerdy et al., 2019). A modular approach has been used, fusing a right-handed quadruplex motif (from the thrombin aptamer) and a left-handed motif, linked for the crystal structure, with a single thymine:

$$d(GGTTGGTGTGGTTGTGG - T - GTGGTGGTGGTG)$$

The left-handed motif alters the thrombin quadruplex topology from right-handed antiparallel to right-handed parallel (Fig. 4.35c). The two tetrad subunits have opposite strand polarity. This structure confirms the strong preferences observed in other left-handed quadruplexes, for single-nucleotide loops. Subtle changes in G/T position have revealed a second minimal left-handed quadruplex motif, d(GGTGGTGGTGTG), in addition to the initial one, d(GTGGTGGTGGTG), which can also form a stable structure, as revealed by NMR and crystallography (Das et al., 2021).

Do such sequences occur naturally in genomes and if so, what is their prevalence and possible function? A very small number of hits have been found in the human genome, of which two are within genes (Schmitt et al., 2015). However, their function and prevalence in other genomes remain to be investigated. The short history of quadruplex structures does though indicate that other unusual folds could well exist: of the 10,000 envisaged in the human genome (Hänsel-Hertsch et al., 2016), we only have structural information on <1% to date.

4.4 The i-motif

Diffraction studies on poly (C) fibres produced at acidic pH were interpreted (Langridge & Rich, 1963) in terms of a parallel-stranded duplex with $C \cdot C^+$ mismatch base pairs (Fig. 4.36). As with four-stranded G-helices, this type of arrangement was largely forgotten for many years. A subsequent NMR study on the cytosine-rich oligonucleotide d(TCCCC) showed that this sequence formed, not a duplex, but instead a novel four-stranded arrangement, distinct from the G-quadruplex (Gehring et al., 1993), which was termed the i-motif. The basic unit is the $C \cdot C^+$ mismatch base pair, with one of the two cytosine protonated at the N3 position such that three hydrogen bonds are formed. The protonation implies that the i-motif is most stable at slightly acidic pH, which is the case for most simple i-motif structures. However, stability at physiological pH can occur (Day et al., 2014), especially under molecular crowding or negative superhelicity conditions (Sun & Hurley, 2009).

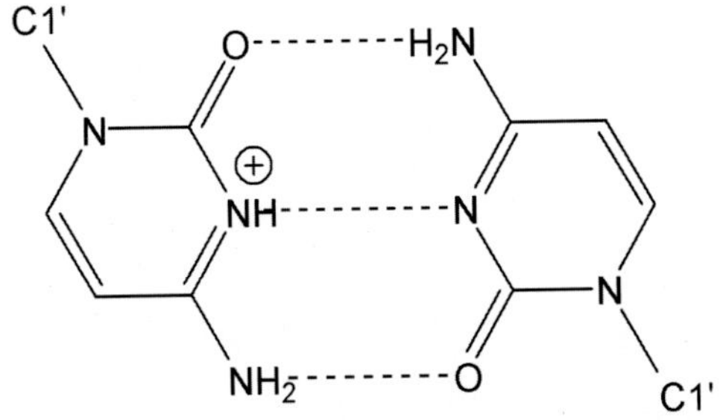

Figure 4.36 The hydrogen-bonding arrangement in a C...C$^+$ base pair.

The arrangement of the i-motif bears some resemblance to the original fibre diffraction model for poly (C) in that there the two strands connected by the C·C$^+$ base pairs are in a parallel orientation. Two of these duplexes are intercalated into each other, resulting in sets of consecutive C·C$^+$ base pairs oriented approximately at right angles to each other (Figs. 4.37 and 4.38a). The DNA strand complementary to a guanine-rich double-stranded centromeric or telomeric sequence is, by definition, cytosine-rich. Several crystal structures of such sequences have been reported (Table 4.8), all of which have this structural motif (Berger et al., 1995; Cai et al., 1998; Kang et al., 1994, 1995). These structures also show that the grooves in the i-motif structures are highly asymmetric, with two broad, shallow major grooves, of width typically 16 Å. Each is flanked by a very narrow minor groove (Fig. 4.38a and b), of width ca 6 Å. The helical twist between consecutive C·C$^+$ base pairs in either duplex tends to be small, between 12 and 20°.

The i-motif sequence motif is assumed to occur naturally in duplex telomeric DNA, when the strands separate, as the complement to the quadruplex-forming G-rich strand. An NMR structure (Phan et al., 2000) has been determined for the human telomeric i-motif sequence d(CCCTAA^{5me}CCCTAACCCUAACCCT). This forms an intramolecular i-motif structure with three lateral loops joining the strands (Fig. 4.38b), so that there are two narrow and two wide grooves. Importantly it has been demonstrated that the i-motif can co-exist with other DNA structural motifs, as found in the crystal structure of a hybrid structure (Fig. 4.38c), which has a two G-tetrad central region flanked on both sides by an A·A·T base triplet then three C·C$^+$ base pairs. The crystal structure of a DNA containing a d(CCA) sequence (Chen et al., 2014) is of interest in that this triplet repeat is associated with the neurological disorder fragile X syndrome. The structure comprises (Fig. 4.38d) a four-stranded region formed from the two d(CCG) moieties each folding back to form a hairpin and hydrogen bonded to each other. This region has three C·C$^+$ base pairs; an additional one occurs in the parallel duplex stem, together with G·G and A·A base pairs. These and other structures reveal the versatility of higher-order DNA (G-quadruplex and i-motifs), especially when combined with non-standard base pairs and triplets.

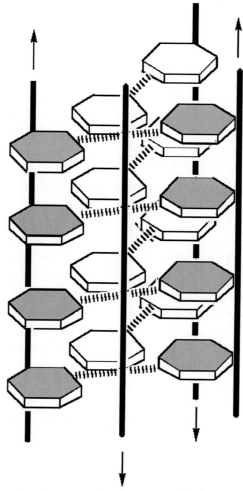

Figure 4.37 Schematic view of the arrangement of base pairs in an i-motif four-stranded structure. The C·C⁺ base pairs shown shaded form one double helix, which is intercalated into that formed by the unshaded C·C⁺ base pairs.

The role of the i-motif in cells has assumed increasing significance since the realisation that the arrangement can be stable under some cellular conditions, and especially since the direct observations of its existence in human (Zeraati et al., 2018) and other cell types (Tang et al., 2020) using antibodies specifically raised against the i-motif. The motif has over-representation in promoter regions of human genes (Wright et al., 2017; Zeraati et al., 2018), which is unsurprising since i-motif sequences (and structures) would complement quadruplex arrangements at these loci, even though i-motif structures are likely to be more dynamic. There is accumulating evidence that these play a role in the control of gene expression for the *BCL2* and c-*MYC* genes (see, for example, Kang et al., 2014). Moreover, quadruplexes and i-motifs at a particular gene locus appear to have complementary roles in gene expression that can be exploitable since quadruplex-stabilising

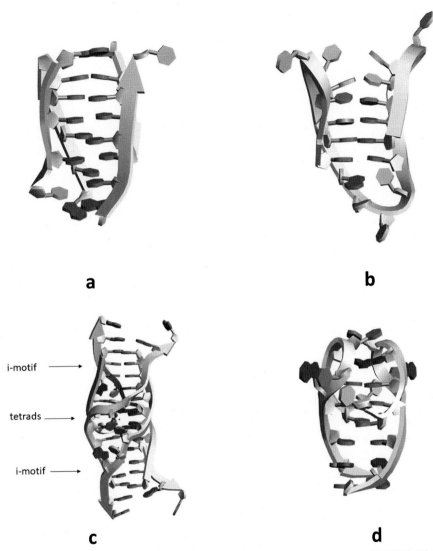

Figure 4.38 Structures of i-motif sequences. (a) Crystal structure of sequence d(ACCCT) (Weil et al., 1999), PDB id 1BQJ. (b) One of the ensemble of NMR structures (Phan et al., 2000), for the human telomeric i-motif sequence d(CCCTAA^{5me}CCCTAACCCUAACCCT), PDB id 1A83. (c) Crystal structure of a quadruplex i-motif hybrid (Chu et al., 2019), PDB id 6TZQ. (d) Crystal structure of a CCG triplet repeat containing an i-motif (Chen et al., 2014), PDB id 4PZQ.

small molecules destabilise i-motif structures, and vice versa (King et al., 2020). Although the study of i-motifs as therapeutic targets is still in its infancy compared to quadruplexes (Abou Assi et al., 2018; Brown & Kendrick, 2021), the diversity of folds in those few i-motif structures for which there is experimental data suggests that the strategy is well founded.

Table 4.8 Selected crystal and NMR structures of nucleic acids containing i-motifs.

Sequence	NMR or crystal	PDB code no.
d(CCCC)	Crystal	190D
d(CCCT)	Crystal	191D
d(ACCCT)	Crystal	1BQJ
d(TAACCC)	Crystal	200D
d(CCCAAT)	Crystal	241D
d(AACCCC)	Crystal	294D
d(CCTTTCCTTTACCTTTCC)	NMR	1A83
d(CCCTAA^{5me}CCCTAACCCUAACCCT)	NMR	1EL2
d(TAACCCTAA)	Crystal	6TQI
d(CCAGGCTGCAA)	Crystal	6TZQ
d(TCCGCCGCCGA)	Crystal	4PZQ
d(TCCTTTTCCA)	NMR	2MRZ

4.5 DNA junctions

4.5.1 Holliday junction structures

The process of homologous genetic recombination involves the crossing-over of strands from two DNA sequences. The central intermediate in this process is believed to be a four-way-branched junction-type structure (Fig. 4.39), termed a Holliday junction. It consists of four strands, each participating in four anti-parallel base-paired arms. A wide variety of biophysical techniques, such as gel electrophoresis and fluorescence energy transfer, have been used in attempts to define the three-dimensional structure of these junctions (Lilley & Norman, 1999). Data from these studies (Duckett et al., 1995), together with molecular modelling (von Kitzing et al., 1990), resulted in the proposal of the X-structure, with two pairs of almost parallel helices and a region in the centre where the DNA strands from the two starting duplexes cross over.

There have been numerous attempts in the past to crystallise junction-type DNA structures (Lilley & Norman, 1999), mostly involving long DNA sequences. Success was finally achieved independently in two laboratories, very surprisingly with two short closely similar decanucleotide sequences. The first, d(CCGGGACCGG), was crystallised (Ortiz-Lombardia et al., 1999) with the intent to study contiguous G·A mismatches in duplex DNA. The crystal structures of the fully inverted-repeat sequences d(CCGGTACCGG) and d(CCGCTAGCGG) show a closely similar arrangement (Eichman et al., 2000), albeit without any possible perturbing effects of the G·A mismatch. All three structures, in common with subsequent Holliday junction structures, consist of two pairs of arms of the junction stacked on each other to form two continuous anti-parallel duplexes that have totally normal linear B-DNA conformations (Figs. 4.40 and 4.41). All bases are in base-paired arrangements in the duplex regions. The junction itself is formed by a few changes in backbone conformations, chiefly in the two

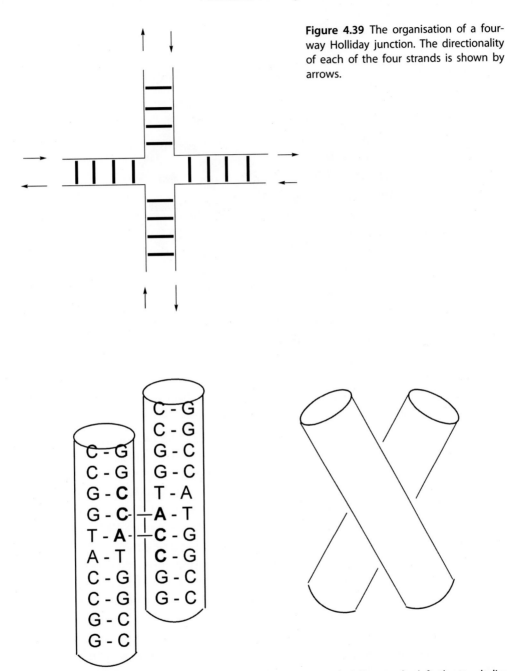

Figure 4.39 The organisation of a four-way Holliday junction. The directionality of each of the four strands is shown by arrows.

Figure 4.40 Schematic views of a Holliday junction structure, showing, on the left, the two helical arms and their connection points via the two central cross-over sequences. The relative angle of the two helical arms is shown on the right.

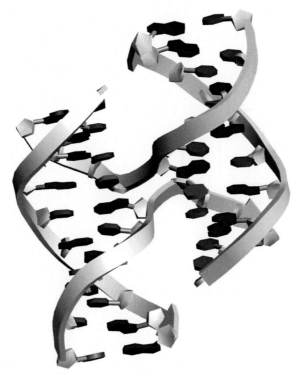

Figure 4.41 Crystal structure of the Holliday junction formed by four decamer strands (Vander Zanden et al., 2016), PDB id 5DSA.

cytosine-7 nucleotides on the two inner-facing strands. The two arms are right-hand twisted with an angle between them of ~40°, in accord with the stacked X-junction model — it is remarkable how close the crystal structure matches the conclusions from the earlier biophysical and modelling studies. The 5′-ACC sequence is at the core of the junction and may be important for its stabilisation. The closeness of four phosphate groups at the junction requires compensating cationic charge, and a sodium ion has been located in a buried position within the junction, bridging two adenosine-6 phosphate oxygen atoms and situated on the local twofold axis of the structure. Crystal structures of the Holliday junction structure formed by the sequence d(TCGGTACCGA) with sodium, calcium and strontium ions (Thorpe et al., 2003) have provided detailed views of the roles played by ions, not only at the junction itself, but also in the helical stems.

Comparative analysis of several junction structures (Eichman et al., 2002; Ho, 2017) has shown that forming the junction has little impact on the sequence-dependent B-form structure of the duplex DNA arms, or the pattern of hydration in the duplex grooves. A general methodology for analysing and comparing the geometry of Holliday junction structures has been proposed, that provides descriptors for the relative orientation, rotation and translation of the duplex helical arms (Watson et al., 2004).

The effects of epigenetic base modification on Holliday junction structure have been studied in several structures (van der Zanden et al., 2016). For example, 5-methylcytosine substituted for cytosine in the cross-over region (Table 4.9) does not affect the overall structure (Fig. 4.41), with the methyl group being accommodated in the central trinucleotide region without distortion. Surprisingly the duplex arms are more B-like than in the native junction structures, so that the methyl group has a stabilising influence. A Holliday junction crystal structure built from extended d(TTAGGG) telomeric DNA repeats (Haider et al., 2018) has two junctions in close proximity (Fig. 4.42), providing a model for strand exchange during homologous recombination.

4.5.2 DNA enzyme structures

The well-established ability of RNAs to fold into complex arrangements (ribozymes) possessing catalytic activities against a range of substrates (see Chapter 6) leads to the question of whether quadruplex (and other more complexes) folded DNAs can also possess catalytic activity. Quadruplexes can show activity (Yum et al., 2019), when complexed with a suitable group such as a haem and this occurs more often and with greater catalytic variety than was initially envisaged (see Yum et al., 2019 for further details). More complex catalysis has been found in the crystal structures of "DNA enzymes", deoxyribozymes, which are junction structures with specialised sequences responsible for catalytic activity. These structures were discovered using in vitro selection procedures that certain DNA sequences can act as enzymes, for example, being able to cleave phosphodiester bonds (Breaker & Joyce, 1994). To date no natural functions for deoxyribozymes have been reported.

The "10–23" motif, which catalyses the cleavage of RNA sequences, has been co-crystallised (Nowakowski, Shim, Prasad, et al., 1999) with an inactivated RNA substrate. The resulting crystal structure consists of a dimer of the DNA:RNA complex in which there are two DNA and two RNA strands. These form five double-helical regions (Figs. 4.43 and 4.44a). A four-way junction is produced in this structure by the confluence of the hexanucleotide loop with three stem regions. As in the Holliday junction

Table 4.9 Crystal structures of nucleic acid junction arrangement.

DNA sequence	Structure type	PDB code no.
d(CCGGTACCGG)	Holliday junction	1DCW
d(CCGGGACCGG)	Holliday junction	467D
d(TCGGTACCGA)	Holliday junction	1NQS
d(CCGGCG^{5Me}CCGG)	Holliday junction with 5-MeC	5DSA
d(GTTAGGGTTCTTGAACCCTTG GGTTACTTGTTACCCTAACC)	Telomeric holliday junction	6GDN
10-23 DNA enzyme + RNA	Deoxyribozyme	1BR3
10-23 DNA enzyme + RNA	Deoxyribozyme	1EGK
d(ATCCGATGGATCATACGGTCGGAGG GGTTTGCCGTTTAAGTGCC) + RNA	Deoxyribozyme	5CKK

Figure 4.42 Crystal structure of the double Holliday junction structure formed from telomeric DNA repeats (Haider et al., 2018), PDB id 6GDN.

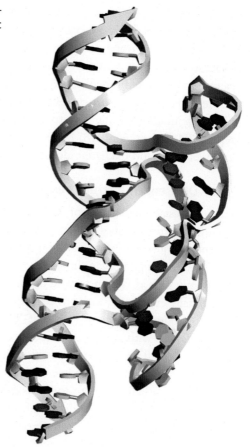

structures, all bases are base paired and stacked. All these junction structures have an overall X-form, and there are few differences between them, apart from the A-form nature of two stems in the 10–23 junction (which arises from the involvement of RNA). By contrast, a 108-nucleotide DNA–RNA complex (Nowakowski, Shim, Stout et al. 1999) adopts an X-junction structure but with a left-handed orientation of the arms.

Other deoxyribozymes that have been subsequently found display a wide range of catalytic activities. These include not only RNA ligation and DNA hydrolysis but also a number of organic chemistry reactions such as Diels–Alder, phosphorylation and glycolysis. Over 800 deoxyribozymes have been identified to date (Ponce-Salvatierra et al., 2020) and information on them is available from the website https://genesilico.pl/DNAmoreDB/. Only a total of eight crystal structures have been reported to date. That of a 44-mer RNA-ligating DNA, PDB id 9DB1, complexed to a 15-mer RNA (Ponce-Salvatierra et al., 2016) is notable for capturing the structure after the RNA ligation has occurred. The RNA forms two short heteroduplexes with the 5′ and 3′ ends of

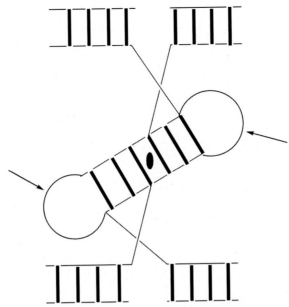

Figure 4.43 Schematic view of the connectivity of the helices and loop regions is found in the crystal structure of the 10–23 DNA enzymes (Nowakowski, Shim, Prasad, et al., 1999), PDB id 1BR3.

the DNA (Fig. 4.44b), which are performing a stapling function, holding the RNA in place. The active site is between these two duplexes (at an angle of 120° to each other), where the RNA is in close proximity. The catalytic domain involves extensive tertiary interactions between four layers of hydrogen-bonded bases, some of which involve long-range interactions that serve to fold the structure into a double pseudoknot. This ensures that the two RNA nucleotides r(AG) are in direct contact with a T·T base pair at the heart of the active site.

4.6 Unnatural DNA structures

A very large number of chemically modified DNAs have been reported in the literature, many of which have been constructed for use as antisense probes (McKenzie et al., 2021; Ochoa & Milam, 2020). There is a significant crystallographic literature on some of these DNA analogues (Egli & Pallan, 2007, 2010). Modifications that have been studied cover all possible functional groupings in an oligonucleotide, ranging from backbone changes through to deoxyribose and base alterations. We focus here on some of the more structurally studied ones, of which the PNA backbone is of particular importance as a biological probe (Nielsen, 2004; Nielsen et al., 1991; Saarbach et al., 2019: see also Section 4.2.2).

Two crystal structures of PNAs within triplex structures have been reported (Section 4.2.2), as well as a small number of crystal and NMR structures of PNA-containing duplexes. One such crystal structure (Rasmussen et al., 1997) is of a hexamer duplex with the base sequence CGTACG and with pure PNA backbones in both strands. The helix (Fig. 4.45) is exceptionally open, with a 28 Å width and an 18 Å pitch; although the base pairs are perpendicular to the helix axis, this is far from a classic B-form. A more

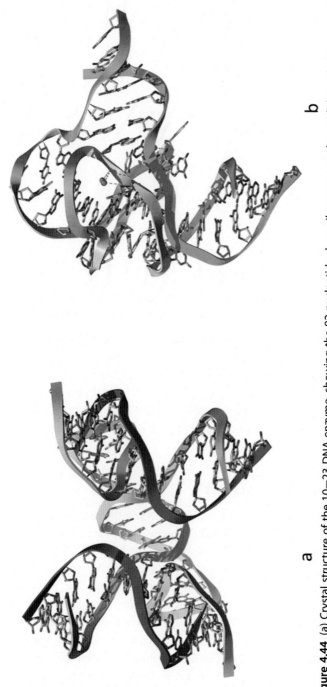

Figure 4.44 (a) Crystal structure of the 10—23 DNA enzyme, showing the 82-nucleotide deoxyribozyme together with bound RNA molecules (with orange cartoon backbones) (Nowakowski, Shim, Prasad, et al., 1999), PDB id 1BR3. (b) Crystal structure of a deoxyribozyme (Ponce-Salvatierra et al., 2016) with bound RNA substrate, coloured purple, PDB id 5CKK.

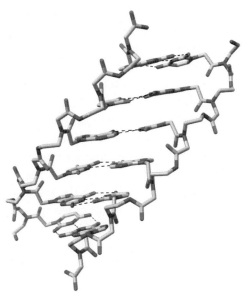

Figure 4.45 Crystal structure of a PNA duplex (Rasmussen et al., 1997), PDB id 1PUP.

complex decamer duplex PNA—DNA structure (Menchise et al., 2003) has one conventional DNA strand and a complementary anti-parallel strand incorporating three chiral D-lysine monomers (Fig. 4.46). This helix also has a wide, deep major groove and a correspondingly shallow and narrow minor groove, with the base pairs perpendicular to the helix axis. The helix type adopted by PNAs is conformationally more restricted than with

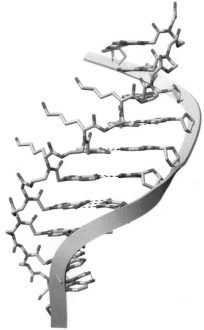

Figure 4.46 Crystal structure of a decamer duplex PNA—DNA (Menchise et al., 2003), PDB id 1NRB. The backbone of the DNA strand is shown in ribbon representation, and that of the PNA in stick form.

standard backbones, and has been termed the P-form, by contrast with the standard A- and B-forms. The complexity of PNA structures has also been illustrated by the formation of both left- and right-handed helices in the same PNA crystal structure (Petersson et al., 2005). An extension to the traditional PNA backbone, with added rigidity and duplex stabilisation, has been the development of a cyclopentane PNA (Zheng et al., 2021). The crystal structure of a fully modified cyclopentyl PNA with a DNA decamer (Fig. 4.47) shows an antiparallel right-handed, slightly irregular helix with a helical twist of 27Å, a rise of 3.4 Å and a pitch of ca 13 base pairs per complete helical turn. As with other PNA—DNA duplexes, the base—base hydrogen bonding is of the Watson—Crick type.

Replacement of natural DNA bases with synthetic analogues has long been an active area of study. Expanded-size bases have been explored as the basis for an alternative genetic system, which can be recognised by the standard replication apparatus of the cell. Base extension with the addition of a benzene ring (see Fig. 4.48) raises the question of whether a DNA constructed from such "x" base pairs can still form a conventional double helix, a prerequisite for biological functioning. The NMR structure of an eight-base xDNA double helix (Lynch et al., 2006) has been described. This sequence (Table 4.10) has been designed so that as many as possible nearest-neighbour effects for the xDNA bases are sampled. The resulting structure (Fig. 4.49) is a right-handed double helix with all glycosidic angles in the *anti* range and all sugars having C2'-*endo* pucker. Although the helix has some resemblance to the B-form, there are significant differences in local and global morphology. The average helical twist is 30°, giving 12 residues per helical turn, and

Figure 4.47 Crystal structure of a DNA—PNA duplex in which the standard PNA residues have all been replaced by cyclopentyl-PNA ones (Zheng et al., 2021), PDB id 7KZL. The backbone of the DNA strand is shown in ribbon representation, and that of the PNA in stick form.

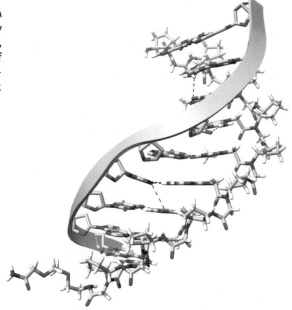

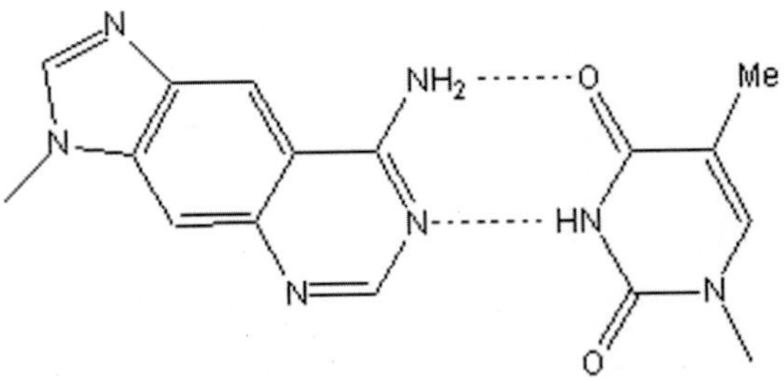

Figure 4.48 xA·T and xG·C base pairs, showing the purines extended with a benzene ring to form xDNA bases.

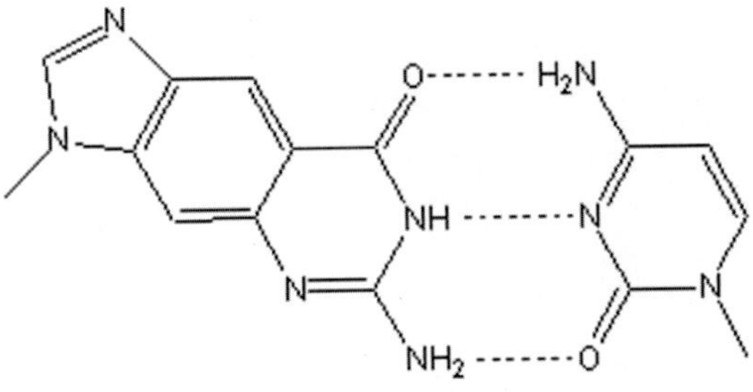

xA:T

xG:C

Table 4.10 Crystal and NMR structures of unnatural nucleic acids.

Sequence	PDB code no.
$d(C_{pn}G_{pn}T_{pn}A_{pn}C_{pn}G_{pn})$	1PUP
[a]$d(pAGTGATCTAC)$	1NR8
$+ d(G_{pn}T_{pn}A_{pn}GATCA_{pn}C_{pn}T)$	
[b]$d(_xTG_xTA_xC_xGC_xA_xGT)$	2ICZ
$+ d(A_xCA_xTGC_xGTC_xA)$	
[c]$d(C_hG_hA_hA_hT_hT_hC_hG)$	2H9S
[d]$d(GCGTA_LACGC)$	1I5W
[e]$d(G_{cp}A_{cp}T_{cp}G_{cp}T_{cp}G_{cp}A_{cp}T_{cp}A_{cp})$	7KZL
$+ d(TATDACATC)$	

[a]pn indicates a peptide nucleic acid (PNA) backbone, replacing a standard backbone.
[b]x indicates a benzylated base, replacing a standard base.
[c]h indicates a dideoxy-β-D-glucopyranose sugar, replacing a standard deoxypentose sugar.
[d]L indicates a locked nucleic acid sugar, replacing a standard deoxypentose sugar.
[e]cp indicates a cyclopentyl PNA.

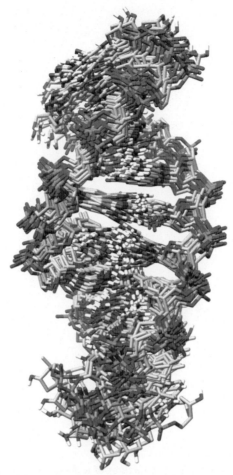

Figure 4.49 The ensemble of NMR structures of xDNA (Lynch et al., 2006), PDB id 2ICX.

the average base pair inclination to the helix axis is 24°. Consequently, the minor groove is over 7 Å wider in xDNA than in B-DNA. The structure analysis suggests that even if DNA polymerase has difficulties in coping with the large size of xDNA bases, they are not an impediment to double helix formation, and a subsequent in vivo study using *E. coli* cells has shown that xDNA can encode natural DNA sequences during replication, resulting in the correct replication partners (Delaney et al., 2009).

We conclude this chapter with a structure that addresses the question of why Nature has chosen five-membered pentose sugar rings as part of the building blocks in nucleic acids rather than the more naturally abundant hexose motif. The crystal structure of a homo-DNA octamer duplex with all nucleosides having deoxyhexose rings replaced by deoxypentose (Fig. 4.50) has been determined (Egli et al., 2006). The structure determination (itself a crystallographic tour-de-force) shows that although the sequence forms an anti-parallel duplex helix (Fig. 4.51), the arrangement is highly irregular (and bears

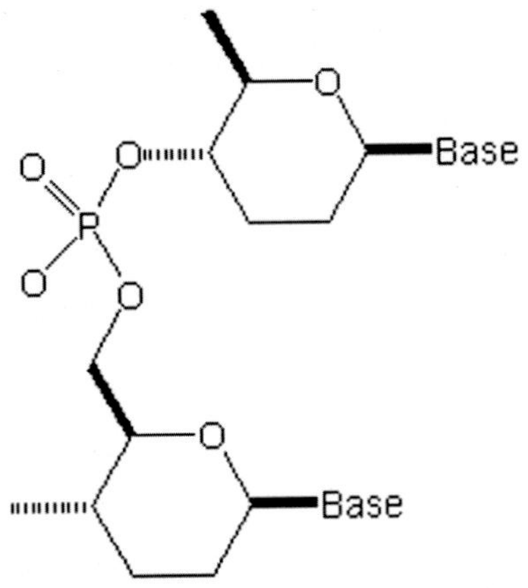

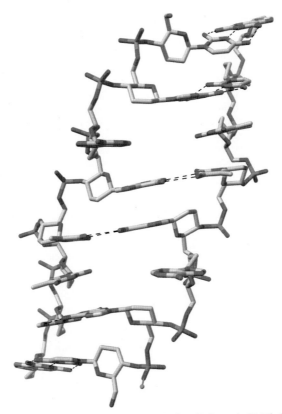

Figure 4.50 The dideoxyhexose arrangement in homo DNA, replacing the deoxyribose in conventional DNA.

Figure 4.51 The crystal structure of the homo-DNA duplex (Egli et al., 2006), PDB id 2H9S. Note the irregular backbone and the four non-hydrogen-bonded bases.

little resemblance to previous predictions of homo-DNA structure). Four bases are non-paired and protrude out from the structure, and there is almost no base—base stacking within each strand. Base-step geometry is also highly irregular along the sequence. Overall, the structure provides a rationalisation of why homo-DNAs are unable to form stable pairing arrangements, and ultimately why Nature did not choose them.

References

Abou Assi, H., Garavís, M., González, C., & Damha, M. J. (2018). i-Motif DNA: structural features and significance to cell biology. *Nucleic Acids Research, 46*, 8038—8056.

Aboul-ela, F., Murchie, A. I. H., Norman, D. G., & Lilley, D. M. J. (1994). Solution structure of a parallel-stranded tetraplex formed by d(TG₄T) in the presence of sodium ions by nuclear magnetic resonance spectroscopy. *Journal of Molecular Biology, 253*, 458—471.

Ambrus, A., Chen, D., Dai, J., Bialis, T., Jones, R. A., & Yang, D. (2006). Human telomeric sequence forms a hybrid-type intramolecular G-quadruplex structure with mixed parallel/antiparallel strands in potassium solution. *Nucleic Acids Research, 34*, 2723—2735.

Ambrus, A., Chen, D., Dai, J., Jones, R. A., & Yang, D. (2005). Solution structure of the biologically relevant G-quadruplex element in the human c-MYC promoter. Implications for G-quadruplex stabilization. *Biochemistry, 44*, 2048—2058.

Arnott, S., Bond, P. J., Selsing, E., & Smith, P. J. C. (1976). Models of triple-stranded polynucleotides with optimised stereochemistry. *Nucleic Acids Research, 10*, 2459—2470.

Arnott, S., Chandrasekaran, R., & Marttila, C. M. (1974). Structures for polyinosinic acid and polyguanylic acid. *Biochemical Journal, 141*, 537—543.

Asamitsu, S., Obata, S., Yu, Z., Bando, T., & Sugiyama, H. (2019). Recent progress of targeted G-quadruplex-preferred ligands toward cancer therapy. *Molecules, 24*, 429.

Bang, I (1910). Untersuchungen über die Guanylsäure. *Biochemische Zeitschrift, 26*, 293—311.

Bates, P. J., Laughton, C. A., Jenkins, T. C., Capaldi, D. C., Roselt, P. D., Reese, C. B., & Neidle, S. (1996). Efficient triple helix formation by oligodeoxyribonucleotides containing α- or β-2-amino-5-(2-deoxy-D-ribofuranosyl) pyridine residues. *Nucleic Acids Research, 24*, 4176—4184.

Bates, P. J., Reyes-Reyes, E. M., Malik, M. T., Murphy, E. M., O'Toole, M. G., & Trent, J. O. (2017). G-quadruplex oligonucleotide AS1411 as a cancer-targeting agent: Uses and mechanisms. *Biochimica et Biophysica Acta (BBA) - General Subjects, 1861*, 1414—1428.

Beal, P. A., & Dervan, P. B. (1991). Second structural motif for recognition of DNA by oligonucleotide-directed triple-helix formation. *Science, 251*, 1360—1363.

Berger, I., Kang, C., Fredian, A., Lockshin, C., Ratliff, R., Moysis, R., & Rich, A. (1995). Extension of the four-stranded intercalated cytosine motif by adenine adenine base pairing in the crystal structure of d(CCCAAT). *Nature Structural Biology, 2*, 416—425.

Betts, L., Josey, J. A., Veal, J. M., & Jordan, S. R. (1995). A nucleic acid triple helix formed by a peptide nucleic acid-DNA complex. *Science, 270*, 1838—18341.

Biffi, G., Tannahill, D., McCafferty, J., & Balasubramanian, S. (2013). Quantitative visualization of DNA G-quadruplex structures in human cells. *Nature Chemistry, 5*, 182—186.

Biffi, G., Tannahill, D., Miller, J., Howat, W. J., & Balasubramanian, S. (2014). Elevated levels of G-quadruplex formation in human stomach and liver cancer tissues. *PloS One, 9*, e102711.

Bijapur, J., Keppler, M. D., Bergqvist, S., Brown, T., & Fox, K. R. (1999). 5-(1-propargylamino)-2'-deoxyuridine (UP): A novel thymidine analogue for generating DNA triplexes with increased stability. *Nucleic Acids Research, 27*, 1802—1809.

Boehm, B. J., Whidborne, C., Button, A. L., Pukala, T. L., & Huang, D. M. (2018). DNA triplex structure, thermodynamics, and destabilisation: Insight from molecular simulations. *Physical Chemistry Chemical Physics, 20*, 14013—14023.

Bohon, J., & de los Santos, C. R. (2003). Structural effect of the anticancer agent 6-thioguanine on duplex DNA. *Nucleic Acids Research, 31*, 1331—1338.

Bohon, J., & de los Santos, C. R. (2005). Effect of 6-thioguanine on the stability of duplex DNA. *Nucleic Acids Research, 33*, 2880—2886.

Brázda, V., Kolomazník, J., Lýsek, J., Bartas, M., Fojta, M., Šťastný, J., & Mergny, J. L. (2019). G4Hunter web application: A web server for G-quadruplex prediction. *Bioinformatics, 35*, 3493—3495.

Breaker, R. R., & Joyce, G. F. (1994). A DNA enzyme that cleaves RNA. *Chemical Biology, 1*, 223—229.

Brown, T., Hunter, W. N., Kneale, G., & Kennard, O. (1986). Molecular structure of the G· A base pair in DNA and its implications for the mechanism of transversion mutations. *Proceedings of the National Academy of Sciences of the United States of America, 83*, 2402—2406.

Brown, S. L., & Kendrick, S. (2021). The i-motif as a molecular target: More than a complementary DNA secondary structure. *Pharmaceuticals (Basel), 14*, 96.

Bua, G., Tedesco, D., Conti, I., Reggiani, A., Bartolini, M., & Gallinella, G. (2020). No G-quadruplex structures in the DNA of Parvovirus B19: Experimental evidence versus bioinformatic predictions. *Viruses, 12*, 935.

Bugaut, A., & Balasubramanian, S. (2008). A sequence-independent study of the influence of short loop lengths on the stability and topology of intramolecular DNA G-quadruplexes. *Biochemistry, 47*, 689—697.

Cai, L., Chen, L., Raghavan, S., Ratliff, R., Moysis, R., & Rich, A. (1998). Intercalated cytosine motif and novel adenine clusters in the crystal structure of the *Tetrahymena* telomere. *Nucleic Acids Research, 26*, 4696—4705.

Catapano, C. V., McGuffie, E. M., Pacheco, D., & Carbone, G. M. R. (2000). Inhibition of gene expression and cell proliferation by triple helix-forming oligonucleotides directed to the *c-myc* gene. *Biochemistry, 39*, 5126—5138.

Chandrasekaran, R., Giacometti, A., & Arnott, S. (2000a). Structure of poly (U).poly (A).poly (U). *Journal of Biomolecular Structure & Dynamics, 17*, 1023—1034.

Chandrasekaran, R., Giacometti, A., & Arnott, S. (2000b). Structure of poly (dT).poly (dA).poly (dT). *Journal of Biomolecular Structure & Dynamics, 17*, 1011—1022.

Chatake, T., Ono, A., Ueno, Y., Matsuda, A., & Takénaka, A. (1999). Crystallographic studies on damaged DNAs. I. An N(6)-methoxyadenine residue forms a Watson-Crick pair with a cytosine residue in a B-DNA duplex. *Journal of Molecular Biology, 294*, 1215—1222.

Chatake, T., Hikima, T., Ono, A., Ueno, Y., Matsuda, A., & Takénaka, A. (1999). Crystallographic studies on damaged DNAs. II. N(6)-methoxyadenine can present two alternate faces for Watson-Crick base-pairing, leading to pyrimidine transition mutagenesis. *Journal of Molecular Biology, 294*, 1223—1230.

Cheloshkina, K., & Poptsova, M. (2021). Comprehensive analysis of cancer breakpoints reveals signatures of genetic and epigenetic contribution to cancer genome rearrangements. *PLoS Computational Biology, 17*, e1008749.

Chen, Y. W., Jhan, C. R., Neidle, S., & Hou, M. H. (2014). Structural basis for the identification of an i-motif tetraplex core with a parallel-duplex junction as a structural motif in CCG triplet repeats. *Angewandte Chemie International Edition in English, 53*, 10682—10686.

Cheng, M., Cheng, Y., Hao, J., Jia, G., Zhou, J., Mergny, J. L., & Li, C. (2018). Loop permutation affects the topology and stability of G-quadruplexes. *Nucleic Acids Research, 46*, 9264—9275.

Cheng, Y.-K., & Pettitt, B. M. (1992). Hoogsteen versus reversed-hoogsteen base pairing: DNA triple helixes. *Journal of the American Chemical Society, 114*, 4465—4474.

Chou, S.-H., Cheng, J.-W., & Reid, B. R. (1992). Solution structure of [d(ATGAGCGAATA)]$_2$: Adjacent G: A mismatches stabilized by cross-strand base-stacking and B$_{II}$ phosphate groups. *Journal of Molecular Biology, 228*, 138—155.

Chou, S.-H., Chin, K.-H., & Wang, A. H.-J. (2003). Unusual DNA duplex and hairpin motifs. *Nucleic Acids Research, 31*, 2461—2474.

Chou, S.-H., Zhu, L., & Reid, B. R. (1997). Sheared purine x purine pairing in biology. *Journal of Molecular Biology, 267*, 1055—1067.

Chu, B., Zhang, D., & Paukstelis, P. J. (2019). A DNA G-quadruplex/i-motif hybrid. *Nucleic Acids Research, 47*, 11921—11930.

Chung, W. J., Heddi, B., Schmitt, E., Lim, K. W., Mechulam, Y., & Phan, A. T. (2015). Structure of a left-handed DNA G-quadruplex. *Proceedings of the National Academy of Sciences of the United States of America, 112*, 2729–2733.

Cogio, S., & Xodo, L. E. (2006). G-quadruplex formation within the promoter of the KRAS proto-oncogene and its effect on transcription. *Nucleic Acids Research, 34*, 2536–2549.

da Silva, M. W. (2007). Geometric formalism for DNA quadruplex folding. *Chemistry, 13*, 9738–9745.

Dai, J., Carver, M., Punchihewa, C., Jones, R. A., & Yang, D. (2007). Structure of the hybrid-2 type intramolecular human telomeric G-quadruplex in K+ solution: Insights into structure polymorphism of the human telomeric sequence. *Nucleic Acids Research, 35*, 4927–4940.

Dai, J., Chen, D., Jones, R. A., Hurley, L. H., & Yang, D. (2006). NMR solution structure of the major G-quadruplex structure formed in the human BCL2 promoter region. *Nucleic Acids Research, 34*, 5133–5144.

Dai, J., Punchihewa, C., Ambrus, A., Chen, D., Jones, R. A., & Yang, D. (2007b). Structure of the intramolecular human telomeric G-quadruplex in potassium solution: A novel adenine triple formation. *Nucleic Acids Research, 35*, 2440–2450.

Das, P., Ngo, K. H., Winnerdy, F. R., Maity, A., Bakalar, B., Mechulam, Y., Schmitt, E., & Phan, A. T. (2021). Bulges in left-handed G-quadruplexes. *Nucleic Acids Research, 49*, 1724–1736.

Das, P., Winnerdy, F. R., Maity, A., Mechulam, Y., & Phan, A. T. (2021). A novel minimal motif for left-handed G-quadruplex formation. *Chemical Communications, 57*, 2527–2530.

Day, H. A., Pavlou, P., & Waller, Z. A. E. (2014). i-Motif DNA: Structure, stability and targeting with ligands. *Bioorganic & Medicinal Chemistry, 22*, 4407–4418.

De Armond, R., Wood, S., Sun, D., Hurley, L. H., & Ebbinghaus, S. W. (2005). Evidence for the presence of a guanine quadruplex forming region within a polypurine tract of the hypoxia inducible factor 1alpha promoter. *Biochemistry, 44*, 16341–16350.

Delaney, J. C., Gao, J., Liu, H., Shrivastav, N., Essigmann, J. M., & Kool, E. T. (2009). Efficient replication bypass of size-expanded DNA base pairs in bacterial cells. *Angewandte Chemie International Edition in English, 48*, 4524–4527.

de Lange, T. (2005). Shelterin: The protein complex that shapes and safeguards human telomeres. *Genes & Development, 19*, 2100–2110.

de Lange, T. (2018). Shelterin-mediated telomere protection. *Annual Review of Genetics, 52*, 223–247.

de los Santos, C., Rosen, M., & Patel, D. J. (1989). NMR studies of DNA $(R^+)_n \cdot (Y^-)_n \cdot (Y^+)_n$ triple helices in solution: Imino and amino proton markers of $T \cdot A \cdot T$ and $C \cdot G \cdot C^+$ base-triple formation. *Biochemistry, 28*, 7282–7289.

Dexheimer, T. S., Sun, D., & Hurley, L. H. (2006). Deconvoluting the structural and drug-recognition complexity of the G-quadruplex-forming region upstream of the bcl-2 P1 promoter. *Journal of the American Chemical Society, 128*, 5404–5415.

Dhapola, P., & Chowdhury, S. (2016). QuadBase2: Web server for multiplexed guanine quadruplex mining and visualization. *Nucleic Acids Research, 44*, W277–W283.

Do, N. Q., Chung, W. J., Truong, T. H. A., Heddi, B., & Phan, A. T. (2017). G-quadruplex structure of an anti-proliferative DNA sequence. *Nucleic Acids Research, 45*, 7487–7493.

Doluca, O. (2019). G4Catchall: A G-quadruplex prediction approach considering atypical features. *Journal of Theoretical Biology, 463*, 92–98.

Duckett, D. R., Murchie, A. I. H., Giraud-Panis, M.-J. E., Pöhler, J. R., & Lilley, D. M. J. (1995). Structure of the four-way DNA junction and its interaction with proteins. *Philosophical Transactions of the Royal Society of London - A, B347*, 27–36.

Eddy, J., & Maizels, N. (2006). Gene function correlates with potential for G4 DNA formation in the human genome. *Nucleic Acids Research, 34*, 3887–3896.

Egli, M., & Pallan, P. S. (2007). Insights from crystallographic studies into the structural and pairing properties of nucleic acid analogs and chemically modified DNA and RNA oligonucleotides. *Annual Review of Biophysics and Biomolecular Structure, 36*, 281–305.

Egli, M., & Pallan, P. S. (2010). Crystallographic studies of chemically modified nucleic acids: A backward glance. *Chemistry and Biodiversity, 7*, 60–89.

Egli, M., Pallan, P. S., Pattanayek, R., Wilds, C. J., Lubini, P., Minasov, G., Dobler, M., Leumann, C. J., & Eschenmoser, A. (2006). Crystal structure of homo-DNA and nature's choice of pentose over hexose in the genetic system. *Journal of the American Chemical Society, 128*, 10847—10856.

Eichman, B. F., Ortiz-Lombardia, M., Aymami, J., Coll, M., & Ho, P. S. (2002). The inherent properties of DNA four-way junctions: Comparing the crystal structures of Holliday junctions. *Journal of Molecular Biology, 320*, 1037—1051.

Eichman, B. F., Vargason, J. M., Mooers, B. H. M., & Ho, P. S. (2000). The Holliday junction in an inverted repeat DNA sequence: Sequence effects on the structure of four-way junctions. *Proceedings of the National Academy of Sciences of the United States of America, 97*, 3971—3976.

Escudé, C., Nguyen, C. H., Kukreti, S., Janin, Y., Sun, J.-S., Bisagni, E., Garestier, T., & Hélène, C. (1998). Rational design of a triple helix-specific intercalating ligand. *Proceedings of the National Academy of Sciences of the United States of America, 95*, 3591—3596.

Escudé, C., Giovannangeli, C., Sun, J.-S., Lloyd, D. H., Chen, J. K., Gryaznov, S., Garestier, T., & Hélène, C. (1996). Stable triple helices formed by oligonucleotide N3'—>P5' phosphoramidates inhibit transcription elongation. *Proceedings of the National Academy of Sciences of the United States of America, 93*, 4365—4369.

Estep, K. N., Butler, T. J., Ding, J., & Brosh, R. M. (2019). G4-interacting DNA helicases and polymerases: Potential therapeutic targets. *Current Medicinal Chemistry, 26*, 2881—2897.

Faria, M., Wood, C. D., Perrouault, L., Nelson, J. S., Winter, A., White, M. R. H., Hélène, C., & Giovannangeli, C. (2000). Targeted inhibition of transcription elongation in cells mediated by triplex-forming oligonucleotides. *Proceedings of the National Academy of Sciences of the United States of America, 97*, 3862—3867.

Felsenfeld, G., Davies, D. R., & Rich, A. (1957). Formation of a three-stranded polynucleotide molecule. *Journal of the American Chemical Society, 79*, 2023—2024.

Fernando, H., Reszka, A. P., Huppert, J., Ladame, S., Rankin, S., Venkitaraman, A. R., Neidle, S., & Balasubramanian, S. (2006). A conserved quadruplex motif located in a transcription activation site of the human c-kit oncogene. *Biochemistry, 45*, 7854—7860.

Fu, T., Liu, L., Yang, Q. L., Wang, Y., Xu, P., Zhang, L., Liu, S., Dai, Q., Ji, Q., Xu, G. L., He, C., Luo, C., & Zhang, L. (2019). Thymine DNA glycosylase recognizes the geometry alteration of minor grooves induced by 5-formylcytosine and 5-carboxylcytosine. *Chemical Science, 10*, 7407—7417.

Gaddis, S. S., Wu, O., Thames, H. D., DiGiovanni, J., Walborg, E. F., macLeod, M. C., & Vasquez, K. M. (2006). A web-based search engine for triplex-forming oligonucleotide target sequences. *Oligonucleotides, 16*, 196—201.

Gao, X., & Patel, D. J. (1988). G(syn)·A(anti) mismatch formation in DNA dodecamers at acidic pH: pH-dependent conformational transition of G·A mispairs detected by proton NMR. *Journal of the American Chemical Society, 110*, 5178—5182.

Gao, Y.-G., Robinson, H., Sanishvili, R., Joachimiak, A., & Wang, A. H.-J. (1999). Structure and recognition of sheared tandem G x A base pairs associated with human centromere DNA sequence at atomic resolution. *Biochemistry, 38*, 16452—16460.

Gehring, K., Leroy, J.-L., & Guéron, M. (1993). A tetrameric DNA structure with protonated cytosine·cytosine base pairs. *Nature, 363*, 561—565.

Gellert, M., Lipsett, M. N., & Davies, D. R. (1962). Helix formation by guanylic acid. *Proceedings of the National Academy of Sciences of the United States of America, 48*, 2013—2018.

Ghosh, A., Kar, R. K., Krishnamoorthy, J., Chatterjee, S., & Bhunia, A. (2014). Double GC: GC mismatch in dsDNA enhances local dynamics retaining the DNA footprint: A high-resolution NMR study. *ChemMedChem, 9*, 2059—2064.

Gilbert, D. E., & Feigon, J. (1999). Multistranded DNA structures. *Current Opinion in Structural Biology, 9*, 305—314.

Ginell, S. L., Kuzmich, S., Jones, R. A., & Berman, H. M. (1990). Crystal and molecular structure of a DNA duplex containing the carcinogenic lesion O6-methylguanine. *Biochemistry, 29*, 10461—10465.

Ginell, S. L., Vojtechovsky, J., Gaffney, B., Jones, R. A., & Berman, H. M. (1994). Crystal structure of a mispaired dodecamer, d(CGAGAATTC(O6^{Me})GCG)$_2$, containing a carcinogenic O6-methylguanine. *Biochemistry, 33*, 3487—3493.

Giovannangeli, C., & Hélène, C. (2000). Triplex technology takes off. *Nature Biotechnology, 18*, 1245–1246.

González, V., & Hurley, L. H. (2010). The c-MYC NHE III1: Function and regulation. *Annual Review of Pharmacology and Toxicology, 50*, 111–129.

Gotfredsen, C. H., Schultze, P., & Feigon, J. (1998). Solution structure of an intramolecular pyrimidine-purine-pyrimidine triplex containing an RNA third strand. *Journal of the American Chemical Society, 120*, 4281–4289.

Gowers, D. M., & Fox, K. R. (1999). Towards mixed sequence recognition by triple helix formation. *Nucleic Acids Research, 27*, 1569–1577.

Greene, K. L., Jones, R. L., Li, Y., Robinson, H., Wang, A. H.-J., & Wilson, W. D. (1994). Solution structure of a GA mismatch DNA sequence, d(CCATGAATGG)₂, determined by 2D NMR and structural refinement methods. *Biochemistry, 33*, 1053–1062.

Griffin, L. C., & Dervan, P. B. (1989). Recognition of thymine·adenine base pairs by guanine in a pyrimidine triple helix motif. *Science, 245*, 967–971.

Griffin, L. C., Kiessling, L. L., Beal, P. A., Gillespie, P., & Dervan, P. B. (1992). Recognition of all four base pairs of double-helical DNA by triple-helix formation: Design of nonnatural deoxyribonucleosides for pyrimidine·purine base pair binding. *Journal of the American Chemical Society, 114*, 7976–7982.

Grigoriev, M., Praseuth, D., Guieysse, A. L., Robin, P., Thoung, N. T., Hélène, C., & Harel-Bellan, A. (1993). Inhibition of gene expression by triple helix-directed DNA cross-linking at specific sites. *Proceedings of the National Academy of Sciences of the United States of America, 90*, 3501–3505.

Gruber, D. R., Toner, J. J., Miears, H. L., Shernyukov, A. V., Kiryutin, A. S., Lomzov, A. A., Endutkin, A. V., Grin, I. R., Petrova, D. V., Kupryushkin, M. S., Yurkovskaya, A. V., Johnson, E. C., Okon, M., Bagryanskaya, E. G., Zharkov, D. O., & Smirnov, S. L. (2018). Oxidative damage to epigenetically methylated sites affects DNA stability, dynamics and enzymatic demethylation. *Nucleic Acids Research, 46*, 10827–10839.

Guédin, A., Gros, J., Alberti, P., & Mergny, J. L. (2010). How long is too long? Effects of loop size on G-quadruplex stability. *Nucleic Acids Research, 38*, 7858–7868.

Guo, K., Gokhale, V., Hurley, L. H., & Sun, D. (2008). Intramolecularly folded G-quadruplex and i-motif structures in the proximal promoter of the vascular endothelial growth factor gene. *Nucleic Acids Research, 36*, 4598–4608.

Haider, S., Li, P., Khiali, S., Munnur, D., Ramanathan, A., & Parkinson, G. N. (2018). Holliday junctions formed from human telomeric DNA. *Journal of the American Chemical Society, 140*, 15366–15374.

Haider, S., Parkinson, G. N., & Neidle, S. (2002). Crystal structure of the potassium form of an *Oxytricha nova* G-quadruplex. *Journal of Molecular Biology, 320*, 189–200.

Haider, S., Parkinson, G. N., & Neidle, S. (2008). Molecular dynamics and principal components analysis of human telomeric quadruplex multimers. *Biophysical Journal, 95*, 296–311.

Hansel-Hertsch, R., Beraldi, D., Lensing, S. V., Marsico, G., Zyner, K., Parry, A., Di Antonio, M., Pike, J., Kimura, H., Narita, M., Tannahill, D., & Balasubramanian, S. (2016). G-quadruplex structures mark human regulatory chromatin. *Nature Genetics, 48*, 1267–1272.

Hardwick, J. S., Ptchelkine, D., El-Sagheer, A. H., Tear, I., Singleton, D., Phillips, S. E. V., Lane, A. N., & Brown, T. (2017). 5-Formylcytosine does not change the global structure of DNA. *Nature Structural & Molecular Biology, 24*, 544–552.

Hazel, P., Huppert, J., Balasubramanian, S., & Neidle, S. (2004). Loop-length-dependent folding of G-quadruplexes. *Journal of the American Chemical Society, 1286*, 16405–16415.

Hazel, P., Parkinson, G. N., & Neidle, S. (2006). Topology variation and loop structural homology in crystal and simulated structures of a bimolecular DNA quadruplex. *Journal of the American Chemical Society, 128*, 5480–5487.

Heddi, B., & Phan, A. T. (2011). Structure of human telomeric DNA in crowded solution. *Journal of the American Chemical Society, 133*, 9824–9833.

Henderson, E., Hardin, C. C., Walk, S. K., Tinoco, I., Jr., & Blackburn, E. H. (1987). Telomeric DNA oligonucleotides form novel intramolecular structures containing guanine-guanine base pairs. *Cell, 51*, 899–908.

Henderson, A., Wu, Y., Huang, Y. C., Chavez, E. A., Platt, J., Johnson, F. B., Brosh, R. M., Sen, D., & Lansdorp, P. M. (2014). Detection of G-quadruplex DNA in mammalian cells. *Nucleic Acids Research, 42*, 860—869.

Hildebrand, Blaser, A., Parel, S., & Leumann, C. J. (1997). 5-Substituted 2-aminopyridine *c*-nucleosides as protonated cytidine equivalents: Increasing efficiency and selectivity in DNA triple-helix formation. *Journal of the American Chemical Society, 119*, 5499—5511.

Ho, P. S. (2017). Structure of the Holliday junction: Applications beyond recombination. *Biochemical Society Transactions, 45*, 1149—1158.

Horvath, M. P., & Schultz, S. C. (2001). DNA G-quartets in a 1.86 Å resolution structure of an *Oxytricha nova* telomeric protein-DNA complex. *Journal of Molecular Biology, 310*, 367—377.

Huppert, J. L., & Balasubramanian, S. (2005). Prevalence of quadruplexes in the human genome. *Nucleic Acids Research, 33*, 2901—2907.

Huppert, J. L., & Balasubramanian, S. (2007). G-quadruplexes in promoters throughout the human genome. *Nucleic Acids Research, 35*, 406—413.

Hurley, L. H., von Hoff, D. D., Siddiqui-Jain, A., & Yang, D. (2006). Drug targeting of the c-MYC promoter to repress gene expression via a G-quadruplex silencer element. *Seminars in Oncology, 33*, 498—512.

Kalnik, M. W., Li, B. F. L., Swann, P. F., & Patel, D. J. (1989). O6-ethylguanine carcinogenic lesions in DNA: An NMR study of O6^{et}G·C pairing in dodecanucleotide duplexes. *Biochemistry, 28*, 6182—6192.

Kang, C., Berger, I., Lockshin, C., Ratliff, R., Moysis, R., & Rich, A. (1994). Crystal structure of intercalated four-stranded d(C3T) at 1.4 A resolution. *Proceedings of the National Academy of Sciences of the United States of America, 91*, 11636—11640.

Kang, C., Berger, I., Lockshin, C., Ratliff, R., Moysis, R., & Rich, A. (1995). Stable loop in the crystal structure of the intercalated four-stranded cytosine-rich metazoan telomere. *Proceedings of the National Academy of Sciences of the United States of America, 92*, 3874—3878.

Kang, H.-J., Kendrick, S., Hecht, S. M., & Hurley, L. H. (2014). The transcriptional complex between the BCL2i-Motif and hnRNP LL is a molecular switch for control of gene expression that can be modulated by small molecules. *Journal of the American Chemical Society, 136*, 4172—4185.

Karsisiotis, A. I., O'Kane, C., & da Silva, M. W. (2013). DNA quadruplex folding formalism—a tutorial on quadruplex topologies. *Methods, 64*, 28—35.

Kejnovská, I., Vorlíčková, M., Brázdová, M., & Sagi, J. (2014). Stability of human telomere quadruplexes at high DNA concentrations. *Biopolymers, 101*, 428—438.

Keppler, M. D., Neidle, S., & Fox, K. R. (2001). Stabilisation of TG- and AG-containing antiparallel DNA triplexes by triplex-binding ligands. *Nucleic Acids Research, 29*, 1935—1942.

Kiessling, L. L., Griffin, L. C., & Dervan, P. B. (1992). Flanking sequence effects within the pyrimidine triple-helix motif characterized by affinity cleaving. *Biochemistry, 31*, 2829—2834.

King, J. J., Irving, K. L., Evans, C. W., Chikhale, R. V., Becker, R., Morris, C. J., Pena Martinez, C. D., Schofield, P., Christ, D., Hurley, L. H., Waller, Z. A. E., Iyer, K. S., & Smith, N. M. (2020). DNA G-quadruplex and i-motif structure formation is interdependent in human cells. *Journal of the American Chemical Society, 142*, 20600—20604.

Kouchakdjian, M., Bodepudi, V., Shibutani, S., Eisenberg, M., Johnson, F., Grollman, A. P., & Patel, D. J. (1991). NMR structural studies of the ionizing radiation adduct 7-hydro-8-oxodeoxyguanosine (8-oxo-7H-dG) opposite deoxyadenosine in a DNA duplex. 8-Oxo-7H-dG(syn).dA(anti) alignment at lesion site. *Biochemistry, 30*, 1403—1412.

Kuryavyi, V., & Patel, D. J. (2010). Solution structure of a unique G-quadruplex scaffold adopted by a guanosine-rich human intronic sequence. *Structure, 18*, 73—82.

Lane, A. N., Jenkins, T. C., Brown, D. J. S., & Brown, T. (1991). N.m.r. determination of the solution conformation and dynamics of the A·G mismatch in the d(CGCAAATTGGCG)$_2$ dodecamer. *Biochemical Journal, 279*, 269—281.

Langridge, R., & Rich, A. (1963). Molecular structure of helical polycytidylic acid. *Nature, 198*, 725—728.

Lansdorp, P., & van Wietmarschen, N. (2019). Helicases FANCJ, RTEL1 and BLM act on guanine quadruplex DNA in vivo. *Genes (Basel), 10*, 870.

Laughton, C. A., & Neidle, S. (1992a). Molecular dynamics simulation of the DNA triplex d(TC)$_5$·d(GA)$_5$·d(C+T)$_5$. *Journal of Molecular Biology, 223*, 519–529.

Laughton, C. A., & Neidle, S. (1992b). Prediction of the structure of the Y$^+$·R$^-$·R($^+$)-type DNA triple helix by molecular modelling. *Nucleic Acids Research, 20*, 6535–6541.

Lavery, R., Zakrzewska, K., Sun, J.-S., & Harvey, S. C. (1992). A comprehensive classification of nucleic acid structural families based on strand direction and base pairing. *Nucleic Acids Research, 20*, 5011–5016.

Le, H. T., Dean, W. L., Buscaglia, R., Chaires, J. B., & Trent, J. O. (2014). An investigation of G-quadruplex structural polymorphism in the human telomere using a combined approach of hydrodynamic bead modeling and molecular dynamics simulation. *Journal of Physical Chemistry B, 118*, 5390–5405.

Lei, M., Podell, E. M., & Cech, T. R. (2004). Structure of human POT1 bound to telomeric single-stranded DNA provides a model for chromosome end-protection. *Nature Structural & Molecular Biology, 11*, 1223–1229.

Lejault, P., Mitteaux, J., Sperti, F. R., & Monchaud, D. (2021). How to untie G-quadruplex knots and why? *Cell Chemical Biology, 15*, S2451–S9456.

Leonard, G. A., Booth, E. D., & Brown, T. (1990a). Structural and thermodynamic studies on the adenine.guanine mismatch in B-DNA. *Nucleic Acids Research, 18*, 5617–5623.

Leonard, G. A., Thomson, J., Watson, W. P., & Brown, T. (1990b). High-resolution structure of a mutagenic lesion in DNA. *Proceedings of the National Academy of Sciences of the United States of America, 87*, 9573–9576.

Lerner, L. K., & Sale, J. E. (2019). Replication of G quadruplex DNA. *Genes (Basel), 10*, 95.

Li, J., Correia, J. J., Wang, L., Trent, J. O., & Chaires, J. B. (2005). Not so crystal clear: The structure of the human telomere G-quadruplex in solution differs from that present in a crystal. *Nucleic Acids Research, 33*, 4649–4659.

Li, K., Yatsunyk, L., & Neidle, S. (2021). Water spines and networks in G-quadruplex structures. *Nucleic Acids Research, 49*, 519–528.

Li, Y., Zon, G., & Wilson, W. D. (1991a). NMR and molecular modeling evidence for a G.A mismatch base pair in a purine-rich DNA duplex. *Proceedings of the National Academy of Sciences of the United States of America, 88*, 26–30.

Li, Y., Zon, G., & Wilson, W. D. (1991b). Thermodynamics of DNA duplexes with adjacent G.A mismatches. *Biochemistry, 30*, 7566–7572.

Lilley, D. M. J., & Norman, D. G. (1999). The Holliday junction is finally seen with crystal clarity. *Nature Structural Biology, 6*, 897–899.

Lim, K. W., Amrane, S., Bouaziz, S., Xu, W., Mu, Y., Patel, D. J., Luu, K. N., & Phan, A. T. (2009). Structure of the human telomere in K+ solution: A stable basket-type G-quadruplex with only two G-tetrad layers. *Journal of the American Chemical Society, 131*, 4301–4309.

Lim, K. W., Ng, V. C., Martin-Pintado, N., Heddi, B., & Phan, A. T. (2013). Structure of the human telomere in Na$^+$ solution: An antiparallel (2+2) G-quadruplex scaffold reveals additional diversity. *Nucleic Acids Research, 41*, 10556–10562.

Lim, K. W., & Phan, A. T. (2013). Structural basis of DNA quadruplex-duplex junction formation. *Angewandte Chemie International Edition in English, 52*, 8566–8569.

Lipscomb, L. A., Peek, M. E., Morningstar, M. L., Verghis, S. M., Miller, E. M., Rich, A., Essigmann, J. M., & Williams, L. D. (1995). X-Ray structure of a DNA decamer containing 7,8-dihydro-8-oxoguanine. *Proceedings of the National Academy of Sciences of the United States of America, 92*, 719–723.

Liu, K., Sasisekharan, V., Miles, H. T., & Raghunathan, G. (1996). Structure of Py·Pu·Py DNA triple helices. Fourier transforms of fiber-type x-ray diffraction of single crystals. *Biopolymers, 39*, 573–589.

Luu, K. N., Phan, A. T., Kuryavyi, V., Lacroix, L., & Patel, D. J. (2006). Structure of the human telomere in K$^+$ solution: An intramolecular (3 + 1) G-quadruplex scaffold. *Journal of the American Chemical Society, 128*, 9963–9970.

Lynch, R. L., Liu, H., Gao, J., & Kool, E. T. (2006). Toward a designed, functioning genetic system with expanded-size base pairs: Solution structure of the eight-base xDNA double helix. *Journal of the American Chemical Society, 128*, 14704–14711.

Macaya, R., Wang, E., Schultze, P., Sklenár, V., & Feigon, J. (1992). Proton nuclear magnetic resonance assignments and structural characterization of an intramolecular DNA triplex. *Journal of Molecular Biology, 225*, 755—773.

Mani, P., Yadav, V. K., Das, S. K., & Chowdhury, S. (2009). Genome-wide analyses of recombination prone regions predict role of DNA structural motif in recombination. *PloS One, 4*, e4399.

McKenzie, L. K., El-Khoury, R., Thorpe, J. D., Damha, M. J., & Hollenstein, M. (2021). Recent progress in non-native nucleic acid modifications. *Chemical Society Reviews*. https://doi.org/10.1039/d0cs01430

McKinney, J. A., Wang, G., Mukherjee, A., Christensen, L., Subramanian, S. H. S., Zhao, J., & Vasquez, K. M. (2020). Distinct DNA repair pathways cause genomic instability at alternative DNA structures. *Nature Communications, 11*, 236.

Monsen, R. C., Chakravarthy, S., Dean, W. L., Chaires, J. B., & Trent, J. O. (2021). The solution structures of higher-order human telomere G-quadruplex multimers. *Nucleic Acids Research, 49*, 1749—1768.

McMicken, H. W., Bates, P. J., & Chen, Y. (2003). Antiproliferative activity of G-quartet-containing oligonucleotides generated by a novel single-stranded DNA expression system. *Cancer Gene Therapy, 10*, 867—869.

Menchise, V., De Simone, G., Tedeschi, T., Corradini, R., Sforza, S., Marchelli, R., Capasso, D., Saviano, M., & Pedone, C. (2003). Insights into peptide nucleic acid (PNA) structural features: The crystal structure of a D-lysine-based chiral PNA-DNA duplex. *Proceedings of the National Academy of Sciences of the United States of America, 100*, 12021—12026.

Mergny, J.-L., Phan, A. T., & Lacroix, L. (1998). Following G-quartet formation by UV-spectroscopy. *FEBS Letters, 435*, 74—78.

Mergny, J.-L., Sun, J.-S., Rougée, M., Montenay-Garestier, T., Barcelo, F., Chomilier, J., & Hélène, C. (1991). Sequence specificity in triple helix formation: Experimental and theoretical studies of the effect of mismatches on triplex stability. *Biochemistry, 30*, 9791—9798.

Meyne, J., Ratliff, R. L., & Moyzis, R. K. (1989). Conservation of the human telomere sequence (TTAGGG)n among vertebrates. *Proceedings of the National Academy of Sciences of the United States of America, 86*, 7049—7053.

Neidle, S. (2016). Quadruplex nucleic acids as novel therapeutic targets. *Journal of Medicinal Chemistry, 59*, 5987—6011.

Nielsen, P. E. (2004). PNA technology. *Molecular Biotechnology, 26*, 233—248.

Nielsen, P. E., Egholm, M., Berg, R. H., & Buchardt, O. (1991). Sequence-selective recognition of DNA by strand displacement with a thymine-substituted polyamide. *Science, 254*, 1497—1500.

Nowakowski, J., Shim, P. J., Prasad, G. S., Stout, C. D., & Joyce, G. F. (1999). Crystal structure of an 82-nucleotide RNA-DNA complex formed by the 10-23 DNA enzyme. *Nature Structural Biology, 6*, 151—156.

Nowakowski, J., Shim, P. J., Stout, C. D., & Joyce, G. F. (1999). Alternative conformations of a nucleic acid four-way junction. *Journal of Molecular Biology, 300*, 93—102.

Nuthanakanti, A., Ahmed, I., Khatik, S. Y., Saikrishnan, K., & Srivatsan, S. G. (2019). Probing G-quadruplex topologies and recognition concurrently in real time and 3D using a dual-app nucleoside probe. *Nucleic Acids Research, 47*, 6059—6072.

Ochoa, S., & Milam, V. T. (2020). Modified nucleic acids: Expanding the capabilities of functional oligonucleotides. *Molecules, 25*, 4659.

Ortiz-Lombardia, M., González, A., Eritja, R., Aymami, J., Azorin, F., & Coll, M. (1999). Crystal structure of a DNA Holliday junction. *Nature Structural Biology, 6*, 913—917.

Osborne, S. D., Powers, V. E. C., Rusling, D. A., Lack, O., Fox, K. R., & Brown, T. (2004). Selectivity and affinity of triplex-forming oligonucleotides containing 2'-aminoethoxy-5-(3-aminoprop-1-ynyl)uridine for recognizing AT base pairs in duplex DNA. *Nucleic Acids Research, 32*, 4439—4447.

Ou, A., Schmidberger, J. W., Wilson, K. A., Evans, C. W., Hargreaves, J. A., Grigg, M., O'Mara, M. L., Iyer, K. S., Bond, C. S., & Smith, N. M. (2020). High resolution crystal structure of a KRAS promoter G-quadruplex reveals a dimer with extensive poly-A pi-stacking interactions for small-molecule recognition. *Nucleic Acids Research, 48*, 5766—5776.

Palumbo, S. L., Ebbinghaus, S. W., & Hurley, L. H. (2012). Formation of a unique end-to-end stacked pair of G-quadruplexes in the hTERT core promoter with implications for inhibition of telomerase by G-quadruplex-interactive ligands. *Journal of the American Chemical Society, 131,* 10678–10691.

Parkinson, G. N., Lee, M. P. H., & Neidle, S. (2002). Crystal structure of parallel quadruplexes from human telomeric DNA. *Nature, 417,* 876–880.

Pasternack, L. B., Lin, S.-B., Chin, T.-M., Lin, W.-C., Huang, D.-H., & Kan, L.-S. (2002). Proton NMR studies of 5'-d-(TC)₃ (CT)₃ (AG)₃-3' — a paperclip triplex: The structural relevance of turns. *Biophysical Journal, 82,* 3170–3180.

Patel, D. J. (1992). Covalent carcinogenic guanine-modified DNA lesions: Solution structures of adducts and crosslinks. *Current Opinion in Structural Biology, 2,* 345–353.

Patel, D. J., Bouaziz, S., Kettant, A., & Wang, Y. (1999). In S. Neidle (Ed.), *Oxford handbook of nucleic acid structure* (pp. 389–453). Oxford.

Patel, D. J., Shapiro, L., Kozlowski, S. A., Gaffney, B. L., & Jones, R. A. (1986). Structural studies of the O6meG.T interaction in the d(C-G-T-G-A-A-T-T-C-O6meG-C-G) duplex. *Biochemistry, 25,* 1036–1042.

Petersson, B., Nielsen, B. B., Rasmussen, H., Larsen, I. K., Gajhede, M., Nielsen, P. E., & Kastrup, J. S. (2005). Crystal structure of a partly self-complementary peptide nucleic acid (PNA) oligomer showing a duplex-triplex network. *Journal of the American Chemical Society, 127,* 1424–1430.

Phan, A. T., Guéron, M., & Leroy, J.-L. (2000). The solution structure and internal motions of a fragment of the cytidine-rich strand of the human telomere. *Journal of Molecular Biology, 299,* 123–144.

Phan, A. T., Kuryavyi, V., Luu, K. N., & Patel, D. J. (2007). Structure of two intramolecular G-quadruplexes formed by natural human telomere sequences in K+ solution. *Nucleic Acids Research, 35,* 6517–6525.

Phan, A. T., Kuryavyi, V., Burge, S. E., Neidle, S., & Patel, D. J. (2007). Structure of an unprecedented G-quadruplex scaffold in the human c-kit promoter. *Journal of the American Chemical Society, 129,* 4386–4392.

Phan, A. T., Kuryavyi, V., Gaw, H. Y., & Patel, D. J. (2005). Small-molecule interaction with a five-guanine-tract G-quadruplex structure from the human MYC promoter. *Nature Chemical Biology, 1,* 167–173.

Phan, A. T., Modi, Y. S., & Patel, D. J. (2004). Propeller-type parallel-stranded G-quadruplexes in the human c-myc promoter. *Journal of the American Chemical Society, 126,* 8710–8716.

Phan, A. T., & Patel, D. J. (2003). Two-repeat human telomeric d(TAGGGTTAGGGT) sequence forms interconverting parallel and antiparallel G-quadruplexes in solution: Distinct topologies, thermodynamic properties, and folding/unfolding kinetics. *Journal of the American Chemical Society, 125,* 15021–15027.

Phillips, K., Dauter, Z., Murchie, A. I. H., Lilley, D. M. J., & Luisi, B. J. (1997). The crystal structure of a parallel-stranded guanine tetraplex at 0.95 Å resolution. *Journal of Molecular Biology, 273,* 171–182.

Ponce-Salvatierra, A., Boccaletto, P., & Bujnicki, J. M. (2020). DNAmoreDB, a database of DNAzymes. *Nucleic Acids Research, 49,* D76–D81.

Ponce-Salvatierra, A., Wawrzyniak-Turek, K., Steuerwald, U., Höbartner, C., & Pena, V. (2016). Crystal structure of a DNA catalyst. *Nature, 529,* 231–234.

Privé, G. G., Heinemann, U., Chandrasegaran, S., Kan, L.-S., Kopka, M. L., & Dickerson, R. E. (1987). Helix geometry, hydration, and G·A mismatch in a B-DNA decamer. *Science, 238,* 498–504.

Puig Lombardi, E., & Londoño-Vallejo, A. (2020). A guide to computational methods for G-quadruplex prediction. *Nucleic Acids Research, 48,* 1–15.

Radhakrishnan, I., Patel, D. J., Veal, J. M., & Gao, X. L. (1992). Solution conformation of a G·TA [guanosine·thymidine-5'-adenylic acid] triple in an intramolecular pyrimidine·purine·pyrimidine DNA triplex. *Journal of the American Chemical Society, 114,* 6913–6915.

Radhakrishnan, I., & Patel, D. J. (1994). DNA triplexes: Solution structures, hydration sites, energetics, interactions, and function. *Biochemistry, 33,* 11405–11416.

Radhakrishnan, I., de los Santos, C., & Patel, D. J. (1991). Nuclear magnetic resonance structural studies of intramolecular purine·purine·pyrimidine DNA triplexes in solution. Base triple pairing alignments and strand direction. *Journal of Molecular Biology, 221,* 1403–1418.

Raiber, E. A., Murat, P., Chirgadze, D. Y., Beraldi, D., Luisi, B. F., & Balasubramanian, S. (2015). 5-Formylcytosine alters the structure of the DNA double helix. *Nature Structural & Molecular Biology, 22*, 44—49.

Rajagopal, P., & Feigon, J. (1989). Triple-strand formation in the homopurine:homopyrimidine DNA oligonucleotides d(G-A)$_4$ and d(T-C)$_4$. *Nature, 339*, 637—640.

Rankin, S., Reszka, A. P., Huppert, J., Zloh, M., Parkinson, G. N., Todd, A. K., Ladame, S., Balasubramanian, S., & Neidle, S. (2005). Putative DNA quadruplex formation within the human c-kit oncogene. *Journal of the American Chemical Society, 127*, 10584—10589.

Rasmussen, H., Kastrup, J. H., Nielsen, J. N., Nielsen, J. M., & Nielsen, P. E. (1997). Crystal structure of a peptide nucleic acid (PNA) duplex at 1.7 Å resolution. *Nature Structural Biology, 4*, 98—101.

Rawal, P., Kummarasetti, V. B., Ravindran, J., Kumar, N., Halder, K., Sharma, R., Mukerji, M., Das, S. K., & Chowdhury, S. (2006). Genome-wide prediction of G4 DNA as regulatory motifs: Role in *Escherichia coli* global regulation. *Genome Research, 16*, 644—655.

Renciuk, D., Blacque, O., Vorlickova, M., & Spingler, B. (2013). Crystal structures of B-DNA dodecamer containing the epigenetic modifications 5-hydroxymethylcytosine or 5-methylcytosine. *Nucleic Acids Research, 41*, 9891—9900.

Renciuk, D., Kejnovská, I., Skoláková, P., Bednárová, K., Motlová, J., & Vorlícková, M. (2009). Arrangements of human telomere DNA quadruplex in physiologically relevant K+ solutions. *Nucleic Acids Research, 37*, 6625—6634.

Rhee, S., Han, Z., Liu, K., Miles, H. T., & Davies, D. R. (1999). Structure of a triple helical DNA with a triplex-duplex junction. *Biochemistry, 38*, 16810—16815.

Riccardi, C., Napolitano, E., Platella, C., Musumeci, D., & Montesarchio, D. (2021). G-quadruplex-based aptamers targeting human thrombin: Discovery, chemical modifications and antithrombotic effects. *Pharmacology & Therapeutics, 217*, 107649.

Rodriguez, R., Miller, K. M., Forment, J. V., Bradshaw, C. R., Nikan, M., Britton, S., Oelschlaegel, T., Xhemalce, B., Balasubramanian, S., & Jackson, S. P. (2012). Small-molecule-induced DNA damage identifies alternative DNA structures in human genes. *Nature Chemical Biology, 8*, 301—310.

Rossetti, G., Dans, P. D., Gomez-Pinto, I., Ivani, I., Gonzalez, C., & Orozco, M. (2015). The structural impact of DNA mismatches. *Nucleic Acids Research, 43*, 4309—4321.

Roxo, C., Kotkowiak, W., & Pasternak, A. (2019). G-quadruplex-forming aptamers-characteristics, applications, and perspectives. *Molecules, 24*, 3781.

Ruggiero, E., & Richter, S. N. (2020). Viral G-quadruplexes: New frontiers in virus pathogenesis and antiviral therapy. *Annual Reports in Medicinal Chemistry, 54*, 101—131.

Rujan, I. N., Meleney, J. C., & Bolton, P. H. (2005). Vertebrate telomere repeat DNAs favor external loop propeller quadruplex structures in the presence of high concentrations of potassium. *Nucleic Acids Research, 33*, 2022—2031.

Rusling, D. A., Powers, V. E. C., Ranasinghe, R. T., Wang, Y., Osborne, S. D., Brown, T., & Fox, K. R. (2005). Four base recognition by triplex-forming oligonucleotides at physiological pH. *Nucleic Acids Research, 33*, 3025—3032.

Rusling, D. A., Rachwal, P. A., Brown, T., & Fox, K. R. (2009). The stability of triplex DNA is affected by the stability of the underlying duplex. *Biophysical Chemistry, 145*, 105—110.

Russo Krauss, I., Ramaswamy, S., Neidle, S., Haider, S., & Parkinson, G. N. (2016). Structural insights into the quadruplex-duplex 3' interface formed from a telomeric repeat: A potential molecular target. *Journal of the American Chemical Society, 138*, 1226—1233.

Ruszkowska, A., Ruszkowski, M., Hulewicz, J. P., Dauter, Z., & Brown, J. A. (2020). Molecular structure of a U·A-U-rich RNA triple helix with 11 consecutive base triples. *Nucleic Acids Research, 48*, 3304—3314.

Saarbach, J., Sabale, P. M., & Winssinger, N. (2019). Peptide nucleic acid (PNA) and its applications in chemical biology, diagnostics, and therapeutics. *Current Opinion in Chemical Biology, 52*, 112—124.

Sanchez, A. M., Volk, D. E., Gorenstein, D. C., & Lloyd, R. S. (2003). Initiation of repair of A/G mismatches is modulated by sequence context. *DNA Repair, 2*, 863—878.

Sayoh, I., Rusling, D. A., Brown, T., & Fox, K. R. (2020). DNA structural changes induced by intermolecular triple helix formation. *ACS Omega, 5*, 1679—1687.

Schult, P., & Paeschke, K. (2020). The DEAH helicase DHX36 and its role in G-quadruplex-dependent processes. *Biological Chemistry, 402,* 581–591.

Schultze, P., Hud, N. V., Smith, F. W., & Feigon, J. (1999). The effect of sodium, potassium and ammonium ions on the conformation of the dimeric quadruplex formed by the *Oxytricha nova* telomere repeat oligonucleotide d(G(4)T(4)G(4)). *Nucleic Acids Research, 27,* 3018–3028.

Sen, D., & Gilbert, W. (1988). Formation of parallel four-stranded complexes by guanine-rich motifs in DNA and its implications for meiosis. *Nature, 334,* 364–366.

Sengar, A., Vandana, J. J., Chambers, V. S., Di Antonio, M., Winnerdy, F. R., Balasubramanian, S., & Phan, A. T. (2019). Structure of a (3+1) hybrid G-quadruplex in the PARP1 promoter. *Nucleic Acids Research, 47,* 1564–1572.

Šket, P., Črnugelj, M., & Plavec, J. (2003). Small change in a G-rich sequence, a dramatic change in topology: New dimeric G-quadruplex folding motif with unique loop orientations. *Journal of the American Chemical Society, 125,* 7866–7871.

Spiegel, J., Adhikari, S., & Balasubramanian, S. (2020). The structure and function of DNA G-Quadruplexes. *Trends in Chemistry, 2,* 123–136.

Spiegel, J., Cuesta, S. M., Adhikari, S., Hänsel-Hertsch, R., Tannahill, D., & Balasubramanian, S. (2021). G-quadruplexes are transcription factor binding hubs in human chromatin. *Genome Biology, 22,* 117.

Summers, P. A., Lewis, B. W., Gonzalez-Garcia, J., Porreca, R. M., Lim, A. H. M., Cadinu, P., Martin-Pintado, N., Mann, D. J., Edel, J. B., Vannier, J. B., Kuimova, M. K., & Vilar, R. (2021). Visualising G-quadruplex DNA dynamics in live cells by fluorescence lifetime imaging microscopy. *Nature Communications, 12,* 162.

Shields, G. C., Laughton, C. A., & Orozco, M. (1997). Molecular dynamics simulations of the d(T·A·T) triple helix. *Journal of the American Chemical Society, 119,* 7463–7469.

Siddiqui-Jain, A., Grand, C. L., Bearss, D. J., & Hurley, L. H. (2002). Direct evidence for a G-quadruplex in a promoter region and its targeting with a small molecule to repress c-*MYC* transcription. *Proceedings of the National Academy of Sciences of the United States of America, 99,* 11593–11598.

Simonsson, T. (2001). G-quadruplex DNA structures - variations on a theme. *Biological Chemistry, 382,* 621–628.

Skelly, J. V., Edwards, K. J., Jenkins, T. C., & Neidle, S. (1993). Crystal structure of an oligonucleotide duplex containing G·G base pairs: Influence of mispairing on DNA backbone conformation. *Proceedings of the National Academy of Sciences of the United States of America, 89,* 804–808.

Smith, F. W., & Feigon, J. (1992). Quadruplex structure of *Oxytricha* telomeric DNA oligonucleotides. *Nature, 356,* 164–168.

Somerville, L., Krynetski, E. Y., Krynetskaia, N. F., Beger, R. D., Zhang, W., Marhefka, C. A., Evans, W. E., & Kriwacki, R. W. (2003). Structure and dynamics of thioguanine-modified duplex DNA. *Journal of Biological Chemistry, 278,* 1005–1011.

Sorensen, J. J., Nielsen, J. T., & Petersen, M. (2004). Solution structure of a dsDNA: LNA triplex. *Nucleic Acids Research, 32,* 6078–6085.

Soyfer, V. N., & Potaman, V. N. (1996). *Triple helical nucleic acids.* New York: Springer-Verlag.

Strobel, S. A., & Dervan, P. B. (1990). Site-specific cleavage of a yeast chromosome by oligonucleotide-directed triple-helix formation. *Science, 249,* 73–75.

Stump, S., Mou, T. C., Sprang, S. R., Natale, N. R., & Beall, H. D. (2018). Crystal structure of the major quadruplex formed in the promoter region of the human c-MYC oncogene. *PloS One, 13.* e0205584.

Sun, D., & Hurley, L. H. (2009). The importance of negative superhelicity in inducing the formation of G-quadruplex and i-motif structures in the c-myc promoter: Implications for drug targeting and control of gene expression. *Journal of Medicinal Chemistry, 52,* 2863–2874.

Sundquist, W. I., & Klug, A. (1989a). Telomeric DNA dimerizes by formation of guanine tetrads between hairpin loops. *Nature, 342,* 825–829.

Szulik, M. W., Pallan, P. S., Nocek, B., Voehler, M., Banerjee, S., Brooks, S., Joachimiak, A., Egli, M., Eichman, B. F., & Stone, M. P. (2015). Differential stabilities and sequence-dependent base pair opening dynamics of Watson-Crick base pairs with 5-hydroxymethylcytosine, 5-formylcytosine, or 5-carboxylcytosine. *Biochemistry, 54,* 1294–1305.

Tang, W., Niu, K., Yu, G., Jin, Y., Zhang, X., Peng, Y., Chen, S., Deng, H., Li, S., Wang, J., Song, Q., & Feng, Q. (2020). In vivo visualization of the i-motif DNA secondary structure in the *Bombyx mori* testis. *Epigenetics & Chromatin, 13*, 12.

Tarköy, M., Phipps, A. K., Schultze, P., & Feigon, J. (1998). Solution structure of an intramolecular DNA triplex linked by hexakis (ethyleneglycol) units: d(AGAGAGAA-(EG)$_6$-TTCTCTCT-(EG)$_6$-TCTCTCTT. *Biochemistry, 37*, 5810—5819.

Tereshko, V., Minasov, G., & Egli, M. (1999). The Dickerson—Drew B-DNA dodecamer revisited at atomic resolution *J. Amer. Chem. Forestry and Society, 121*, 470—471.

Theruvathu, J. A., Yin, Y. W., Pettitt, B. M., & Sowers, L. C. (2013). Comparison of the structural and dynamic effects of 5-methylcytosine and 5-chlorocytosine in a CpG dinucleotide sequence. *Biochemistry, 52*, 8590—8598.

Thorpe, J. H., Gale, B. C., Teixeira, S. C., & Cardin, C. J. (2003). Conformational and hydration effects of site-selective sodium, calcium and strontium ion binding to the DNA Holliday junction structure d(TCGGTACCGA)$_4$. *Journal of Molecular Biology, 327*, 97—109.

Todd, A. K., Johnston, M., & Neidle, S. (2005). Highly prevalent putative quadruplex sequence motifs in human DNA. *Nucleic Acids Research, 33*, 2901—2907.

Tong, X., Lan, W., Zhang, X., Wu, H., Liu, M., & Cao, M. (2011). Solution structure of all parallel G-quadruplex formed by the oncogene RET promoter sequence. *Nucleic Acids Research, 39*, 6753—6763.

Uhlmann, E. (1998). Peptide nucleic acids (PNA) and PNA-DNA chimeras: From high binding affinity towards biological function. *Biological Chemistry, 379*, 1045—1052.

vander Zanden, C. M., Rowe, R. K., Broad, A. J., Robertson, A. B., & Ho, P. S. (2016). Effect of hydroxymethylcytosine on the structure and stability of Holliday junctions. *Biochemistry, 55*, 5781—5789.

van Meervelt, L., Vlieghe, D., Dautant, A., Gallois, B., Précigoux, G., & Kennard, O. (1995). High-resolution structure of a DNA helix forming (C·G)*G base triplets. *Nature, 374*, 742—744.

Vannutelli, A., Belhamiti, S., Garant, J. M., Ouangraoua, A., & Perreault, J. P. (2020). Where are G-quadruplexes located in the human transcriptome? *NAR Genom. Bioinformation, 2*, lqaa035.

Vianney, Y. M., & Weisz, K. (2021). First tandem repeat of a potassium channel KCNN4 minisatellite folds into a V-Loop G-quadruplex structure. *Biochemistry.* https://doi.org/10.1021/acs.biochem.1c00043

Vojtechovsky, J., Eaton, M. D., Gaffney, B., Jones, R. A., & Berman, H. M. (1995). Structure of a new crystal form of a DNA dodecamer containing T·(O6Me)G base pairs. *Biochemistry, 34*, 16632—16640.

von Kitzing, E., Lilley, D. M. J., & Diekmann, S. (1990). The stereochemistry of a four-way DNA junction: A theoretical study. *Nucleic Acids Research, 18*, 2671—2683.

Wang, Y., & Patel, D. J. (1992). Guanine residues in d(T$_2$AG$_3$) and d(T$_2$G$_4$) form parallel-stranded potassium cation stabilized G-quadruplexes with anti glycosidic torsion angles in solution. *Biochemistry, 31*, 8112—8119.

Wang, Y., de los Santos, C., Gao, X., Greene, K., Live, D., & Patel, D. J. (1991). Multinuclear nuclear magnetic resonance studies of Na cation-stabilized complex formed by d(G-G-T-T-T-T-C-G-G) in solution: Implications for G-tetrad structures. *Journal of Molecular Biology, 222*, 819—832.

Wang, Y., & Patel, D. J. (1993). Solution structure of the human telomeric repeat d[AG$_3$(T$_2$AG$_3$)$_3$] G-tetraplex. *Structure, 1*, 263—282.

Wang, Y., Rusling, D. A., Powers, V. E. C., Lack, O., Osborne, S. D., Fox, K. R., & Brown, T. (2005). Stable recognition of TA interruptions by triplex forming oligonucleotides containing a novel nucleoside. *Biochemistry, 19*, 5884—5892.

Watson, J., Hays, F. A., & Ho, P. S. (2004). Definitions and analysis of DNA Holliday junction geometry. *Nucleic Acids Research, 32*, 3017—3027.

Webster, G. D., Sanderson, M. R., Skelly, J. V., Neidle, S., Swann, P. F., Li, B. F., & Tickle, I. J. (1990). Crystal structure and sequence-dependent conformation of the A·G mispaired oligonucleotide d(CGCAAGCTGGCG). *Proceedings of the National Academy of Sciences of the United States of America, 87*, 6693—6697.

Wei, D., Husby, J., & Neidle, S. (2015). Flexibility and structural conservation in a c-KIT G-quadruplex. *Nucleic Acids Research, 43*, 629—644.

Wei, D., Parkinson, G. N., Reszka, A. P., & Neidle, S. (2012). Crystal structure of a c-kit promoter quadruplex reveals the structural role of metal ions and water molecules in maintaining loop conformation. *Nucleic Acids Research, 40*, 4691–4700.

Weil, J., Min, T., Yang, C., Wang, S., Sutherland, C., Sinha, N., & Kang, C. (1999). Stabilization of the i-motif by intramolecular adenine-adenine-thymine base triple in the structure of d(ACCCT). *Acta Crystallographica, D55*, 422–429.

Westhof, E. (1992). Westhof's rule. *Nature, 358*, 459–460.

Wright, E. P., Huppert, J. L., & Waller, Z. A. E. (2017). Identification of multiple genomic DNA sequences which form i-motif structures at neutral pH. *Nucleic Acids Research, 45*, 2951–2959.

Wright, W. E., Tesmer, V. M., Huffman, K. E., Levene, S. D., & Shay, J. W. (1997). Normal human chromosomes have long G-rich telomeric overhangs at one end. *Genes & Development, 11*, 2801–2809.

Winnerdy, F. R., Bakalar, B., Maity, A., Vandana, J. J., Mechulam, Y., Schmitt, E., & Phan, A. T. (2019). NMR solution and X-ray crystal structures of a DNA molecule containing both right- and left-handed parallel-stranded G-quadruplexes. *Nucleic Acids Research, 47*, 8272–8281.

Wu, F., Niu, K., Cui, Y., Li, C., Lyu, M., Ren, Y., Chen, Y., Deng, H., Huang, L., Zheng, S., Liu, L., Wang, J., Song, Q., Xiang, H., & Feng, Q. (2021). Genome-wide analysis of DNA G-quadruplex motifs across 37 species provides insights into G4 evolution. *Communications Biology, 4*, 98.

Xu, M., Axhemi, A., Malgowska, M., Chen, Y., Leonard, D., Srinivasan, S., Jankowsky, E., & Taylor, D. J. (2021). Active and passive destabilization of G-quadruplex DNA by the telomere POT1-TPP1 complex. *Journal of Molecular Biology, 433*, 166846.

Xu, Y.-Z., Jenjaroenpun, P., Wongsurawat, T., Byrum, S. D., Shponka, V., Tannahill, D., Chavez, E. A., Hung, S. S., Steidl, C., Balasubramanian, S., Rimsza, L. M., & Kendrick, S. (2020). Activation-induced cytidine deaminase localizes to G-quadruplex motifs at mutation hotspots in lymphoma. *NAR Cancer, 2*, zcaa029.

Yang, W. (2006). Poor base stacking at DNA lesions may initiate recognition by many repair proteins. *DNA Repair, 5*, 654–666.

Ying, L., Green, J. J., Li, H., Klenerman, D., & Balasubramanian, S. (2003). Studies on the structure and dynamics of the human telomeric G quadruplex by single-molecule fluorescence resonance energy transfer. *Proceedings of the National Academy of Sciences of the United States of America, 100*, 14629–14634.

Yum, J. H., Park, S., & Sugiyama, H. (2019). G-quadruplexes as versatile scaffolds for catalysis. *Organic and Biomolecular Chemistry, 17*, 9547–9561.

Zeraati, M., Langley, D. B., Schofield, P., Moye, A. L., Rouet, R., Hughes, W. E., Bryan, T. M., Dinger, M. E., & Christ, D. (2018). i-motif DNA structures are formed in the nuclei of human cells. *Nature Chemistry, 10*, 631–637.

Zhang, Y., El Omari, K., Duman, R., Liu, S., Haider, S., Wagner, A., Parkinson, G. N., & Wei, D. (2020a). Native *de novo* structural determinations of non-canonical nucleic acid motifs by X-ray crystallography at long wavelengths. *Nucleic Acids Research, 48*, 9886–9898.

Zhang, J., Fakharzadeh, A., Pan, F., Roland, C., & Sagui, C. (2020b). Atypical structures of GAA/TTC trinucleotide repeats underlying Friedreich's ataxia: DNA triplexes and RNA/DNA hybrids. *Nucleic Acids Research, 48*, 9899–9917.

Zhang, N., Gorin, A., Majumdar, A., Kettani, A., Chernichenko, N., Skripkin, E., & Patel, D. J. (2001). V-shaped scaffold: A new architectural motif identified in an A x (G x G x G x G) pentad-containing dimeric DNA quadruplex involving stacked G(anti) x G(anti) x G(anti) x G(syn) tetrads. *Journal of Molecular Biology, 311*, 1063–1079.

Zheng, H., Botos, I., Clausse, V., Nikolayevskiy, H., Rastede, E. E., Fouz, M. F., Mazur, S. J., & Appella, D. H. (2021). Conformational constraints of cyclopentane peptide nucleic acids facilitate tunable binding to DNA. *Nucleic Acids Research, 49*, 713–725.

Zimmerman, S. B., Cohen, G. H., & Davies, D. R. (1975). X-ray fiber diffraction and model-building study of polyguanylic acid and polyinosinic acid. *Journal of Molecular Biology, 92*, 181–192.

Further reading

Mispairing

Brown, T., & Kennard, O. (1992). Structural basis of DNA mutagenesis. *Current Opinion in Structural Biology,* *2,* 354—360.

Hunter, W. N. (1992). Crystallographic studies of DNA containing mismatches, modified and unpaired bases. *Methods in Enzymology, 211,* 221—231.

Modrich, P. (1987). DNA mismatch correction. *Annual Review of Biochemistry, 56,* 435—466.

Modrich, P. (2006). Mechanisms in eukaryotic mismatch repair. *Journal of Biological Chemistry, 281,* 30305—30309.

Singer, B., & Grunberger, D. (1983). *Molecular biology of mutagens and carcinogens.* New York: Plenum Press.

Triplexes

Bacolla, A., Wang, G., & Vasquez, K. M. (2015). New perspectives on DNA and RNA triplexes as effectors of biological activity. *PLoS Genetics, 11,* e1005696.

Chandrasekaran, A. R., & Rusling, D. A. (2017). Triplex-forming oligonucleotides: A third strand for DNA nanotechnology. *Nucleic Acids Research, 46,* 1021—1037.

Fox, K. R. (2000). Targeting DNA with triplexes. *Current Medicinal Chemistry, 7,* 17—37.

Fox, K. R., & Brown, T. (2005). An extra dimension in nucleic acid sequence recognition. *Quarterly Reviews of Biophysics, 38,* 311—320.

Fox, K. R., & Brown, T. (2011). Formation of stable DNA triplexes. *Biochemical Society Transactions, 39,* 629—634.

Jain, A., Wang, G., & Vasquez, K. M. (2008). DNA triple helices: Biological consequences and therapeutic potential. *Biochemie, 90,* 117—1130.

Moser, H., & Dervan, P. B. (1987). Sequence-specific cleavage of double helical DNA by triple helix formation. *Science, 238,* 645—650.

Neidle, S. (1997). Recent developments in triple-helix regulation of gene expression. *Anti-Cancer Drug Design, 12,* 433—442.

Thuong, N. T., & Hélène, C. (1993). Sequence-specific recognition and modification of double-helical DNA by oligonucleotides. *Angewandte Chemie International Edition, 32,* 666—690.

Telomeres and quadruplexes

Bryan, T. M. (2020). G-quadruplexes at telomeres: Friend or Foe? *Molecules, 25,* 3686.

Burge, S., Parkinson, G. N., Hazel, P., Todd, A. K., & Neidle, S. (2006). Quadruplex DNA: Sequence, topology and structure. *Nucleic Acids Research, 34,* 5402—5415.

Cech, T. R. (2004). Beginning to understand the end of the chromosome. *Cell, 116,* 273—279.

Chaires, J. B., & Graves, D. (Eds.). (2013). *Quadruplex nucleic acids.* Berlin, Germany: Springer-Verlag.

Davies, J. T. (2004). G-tetrads 40 years later: From 5'-GMP to molecular biology and supramolecular chemistry. *Angewandte Chemie International Edition, 43,* 668—698.

del Villar-Guerra, R., Trent, J. O., & Chaires, J. B. (2018). G-quadruplex secondary structure from circular dichroism spectroscopy. *Angewandte Chemie International Edition in English, 57,* 7171—7175.

Kosiol, N., Juranek, S., Brossart, P., Heine, A., & Paeschke, K. (2021). G-quadruplexes: A promising target for cancer therapy. *Molecular Cancer, 20,* 40.

Krauskopf, A., & Blackburn, E. H. (2000). Telomeres and their control. *Annual Review of Genetics, 34,* 331—358.

Lim, C. J., & Cech, T. R. (2021). Shaping human telomeres: From shelterin and CST complexes to telomeric chromatin organization. *Nature Reviews Molecular Cell Biology.* https://doi.org/10.1038/s41580-021-00328-y

Neidle, S., & Balasubramanian, S. (2006). In *Quadruplex nucleic acids.* Cambridge, UK: Royal Society of Chemistry.

Rigo, R., Palumbo, M., & Sissi, C. (2017). G-Quadruplexes in human promoters: A challenge for therapeutic applications. *Biochimica et Biophysica Acta, 1861,* 1399–1413.

Seimiya, H. (2020). Crossroads of telomere biology and anticancer drug discovery. *Cancer Science, 111,* 3089–3099.

Spiegel, J., Adhikari, S., & Balasubramanian, S. (2020). The structure and function of DNA G-quadruplexes. *Trends in Chemistry, 2,* 123–136.

Varshney, D., Spiegel, J., Zyner, K., Tannahill, D., & Balasubramanian, S. (2020). The regulation and functions of DNA and RNA G-quadruplexes. *Nature Reviews Molecular Cell Biology, 21,* 459–474.

Williamson, J. R. (1994). G-quartet structures in telomeric DNA. *Annual Review of Biophysics, 23,* 703–730.

Winnerdy, F. R., & Phan, A. T. (2020). Quadruplex structure and diversity. *Annual Reports in Medicinal Chemistry, 54,* 45–73.

DNA junctions

Hays, F. A., Watson, J., & Ho, P. S. (2003). Caution! DNA crossing: Crystal structures of Holliday junctions. *Journal of Biological Chemistry, 278,* 49663–49666.

Ho, P. S., & Eichman, B. F. (2001). The crystal structures of DNA Holliday junctions. *Current Opinion in Structural Biology, 11,* 302–308.

Lilley, D. M. J. (2000). Structures of helical junctions in nucleic acids. *Quarterly Reviews of Biophysics, 33,* 109–159.

CHAPTER 5

Principles of small molecule—DNA recognition

5.1 Introduction

The recognition of a DNA molecule by a small molecule or protein can be non-specific, not choosing between any individual nucleotide, or nucleotide sequence. At the other extreme, it can be highly specific, solely recognising a defined sequence within a gene, a genome or even an organism such as a bacterium or a virus. Selective recognition can be performed very effectively by DNA itself, as has been seen in Chapter 4, with DNA—DNA recognition being governed in large part by base—base hydrogen bonding. A single strand of oligonucleotide can recognise

- sequences within the strand itself, folding back to form an intramolecular duplex, triplex or quadruplex;
- another single strand to form a duplex or a bimolecular quadruplex;
- a duplex to form a triplex; ora
- even three other strands to form a tetramolecular quadruplex.

There is an increasingly stringent sequence requirement ongoing from duplex to triplex to quadruplex, as detailed in Chapter 4. All these utilise the hydrogen-bonding ability of DNA bases over and above Watson—Crick hydrogen bonding. This is also the manner by which direct readout of DNA sequence information can be achieved using small molecules and proteins.

The task of DNA sequence recognition has as its pre-requisite knowledge of the frequency of occurrence of a given DNA sequence within a defined length of DNA. This can be directly obtained from a DNA databank if the entire target sequence is known, which is the case for the human genome, and for genomes in many other species, eukaryotic and prokaryotic (see, for example, the genome websites at www.sanger.ac.uk and www.genome.gov).

The frequency of occurrence of a given sequence can be readily estimated by assuming a random distribution of the four nucleotides, and of sequences generally. It is then possible to calculate the frequency of occurrence for a given size of a DNA recognition site. The reciprocal of the former is the number of unique sites for a given site size. The data in Table 5.1 suggest that the human genome, comprising 3.097 billion nucleotides, requires a recognition site of 16—18 nucleotides in order to be reasonably certain of uniqueness, for example, if a single sequence within a gene of therapeutic interest is to be targeted. There are over 40,000 genes coded by the entire human genome, corresponding to about 1%—2% of the human genome length. As of May 2021 this

Principles of Nucleic Acid Structure
ISBN 978-0-12-819677-9, https://doi.org/10.1016/B978-0-12-819677-9.00005-6

Table 5.1 The statistics of DNA sequence selectivity, based on a random distribution of nucleotides.

Site size	Frequency n, as 1:n
2	10
3	32
4	136
5	512
6	2080
7	8196
10	524,800
12	8,390,656
15	536,870,912
16	2,147,516,416

corresponds to 20,442 known coding and 23,982 non-coding genes (http://www.ensembl.org/Homo_sapiens). An important caveat is that alternative splicing variants may increase this number two- or threefold, and there is now much evidence of significant variations between individuals. Thus, the sequence recognition problem suggests that the size of site for unique recognition may be to the lower end of the 16—18 nucleotide range. However, since human and many other gene sequences are now freely available, genome-wide searches should be made at the outset if sequence recognition targeting is being considered, to ascertain the actual uniqueness of a target site. Software such as that available at the Ensembl browser (http://www.ensembl.org) enables such searches to be performed. Importantly, many types of sequence pattern are not randomly distributed, and may turn out to be under- or over-represented, typified by the G-tract feature of G-quadruplexes (Chapter 4).

For direct readout to occur, an incoming molecule must access the base pairs of a target sequence via a DNA groove. The sequence itself should be accessible and free from tightly packaged chromatin organisation, i.e. the gene should be actively transcribing. The orientation of this incoming group is important since hydrogen bonding to and from bases, and hence direct readout, are predominantly directional interactions. The edges of the bases in Watson—Crick duplex DNA have hydrogen-bonding potential (Fig. 5.1) that is only partially satisfied by base pairing itself (Seeman et al., 1976), and therefore is available for recognition by ligands with hydrogen-bonding complementarity. This pattern of hydrogen-bond donation and acceptance has built-in sequence dependency. Thus, in the major groove, a C•G Watson—Crick base pair has a pattern of (donor, acceptor, acceptor) hydrogen bonding that is distinct from that for the reversed G•C base pair. The A•T and T•A base pairs have identical major groove donor/acceptor patterns, but the presence of the methyl group on thymine introduces an asymmetry in the groove that can, at least in principle, enable effective discrimination between these

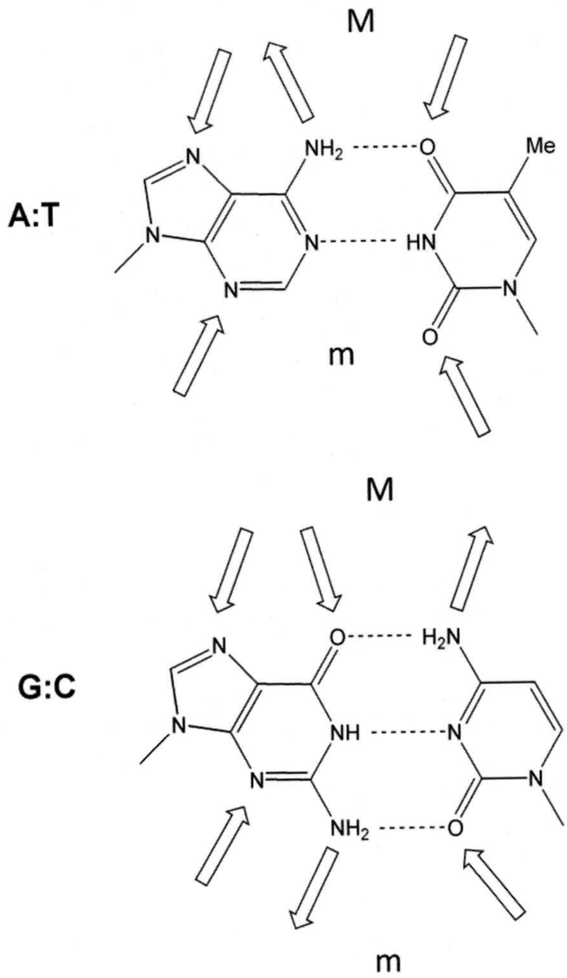

Figure 5.1 The pattern of hydrogen-bond donors and acceptors in A•T and G•C base pairs. The arrows indicate the directionality of the hydrogen bonds to and from the base edges. M and m indicate the major and minor grooves, respectively.

two base-pair sequences. On the other hand, patterns of hydrogen bonding in the minor grooves of each pair of sequences are symmetric, so discrimination on this basis alone by molecules entering the minor groove is not straightforward.

It is widely assumed that the minor groove in B-type DNA sequences is invariably narrow compared to the major groove. This has led to the view that it is unlikely that all three hydrogen bonds at the minor-groove edge of a C•G or G•C base pair or the two in A•T or T•A base pairs are simultaneously available for direct hydrogen-bonding sequence readout, for purely steric reasons. It has been established from the large number of oligonucleotide crystal structures that A/T-rich minor grooves tend to be narrower than average; their widths depend both on the length and the nature of the A/T sequence (see Chapter 3). Thus, accessibility is a consequence of groove width variations, which are themselves sequence dependent in both static and dynamic ways, as discussed in Chapter 3. However, these trends are by no means absolute. A more realistic view is that A/T regions are more flexible than G/C ones, and that circumstances such as the presence of the spine of hydration often result in narrowed minor grooves in A/T tracts of sequence. Other circumstances where the energetics of interaction over-rides the stability of structured water in a groove can readily alter this feature and markedly widen the groove. Consequently, it is possible to simultaneously access the minor groove edges of both components of the base pairs in a run of A/T sequence with appropriate ligands that can stabilise a widened groove. Other sequence–dependent features of base pairs and base-pair steps, which can be both localised (such as propeller and helical twist) and long range (typified by the bending in B-DNA induced by A-tracts), will affect the geometric relationships between hydrogen bond donors/acceptors on successive bases.

Indirect sequence readout, by definition, involves interaction with more general elements of DNA structure other than the base-pair hydrogen bonds and is consequently inherently less directional. The backbones themselves provide several mechanisms for such readout. Their sequence-dependent structural features have been outlined in Chapters 2 and 3. Advantage can be taken of differences in backbone and phosphate conformations such as B_I or B_{II} phosphate geometry, which can act as recognition signals. Perhaps more important are differences in distances between phosphate groups that necessarily result from backbone conformational differences, which can be recognised by hydrogen bonding/electrostatic interactions with basic amino acid side chains. The B-DNA minor groove has extended regions of apolar character. Carbon and hydrogen atoms of sugar residues, notably C1′, H1′, C4′, H4′, C5′ and H5′, form a significant part of the van der Waals and solvent-accessible surface walls of this groove (Fig. 5.2). The apolar groove surface is interspersed at regular intervals by highly polar phosphate groups.

Figure 5.2 A surface representation of a canonical B-DNA duplex (PDB id 1BNA), looking onto the major and minor grooves, and with the hydrogen atoms at the surface of the grooves coloured white (oxygen atoms are coloured red). Note that the hydrogen atoms area prominent part of the walls of the minor groove surface.

The surface of the groove floor comprises some carbon/hydrogen atoms along the base edges, together with the polar base hydrogen bond donor/acceptor atoms. Groove depth is shallower in G/C regions on account of the exocyclic N2 substituent of guanine.

There are also significant differences in electronic character between A•T and G•C base pairs, with the latter being more electron rich. Thus, electron-deficient planar molecules that can stack with individual base pairs (such as intercalating drug molecules, or phenylalanine/tryptophan side chains) will tend to bind preferentially to G•C base pairs. Such preferences are by their nature localised to individual base-pair sites. There is a tendency for reduced stacking between bases in B-DNA duplexes for pyrimidine-3′,5′-purine dinucleoside steps compared to purine-3′,5′-pyrimidine ones (Fig. 5.3). The consequence of these differences is that the dinucleoside sequences CpG, TpG, CpA and TpA (of pyrimidine-3′,5′-purine type) are more readily unstacked than their counterparts GpC, GpT, ApC and ApT. The cost in energetic terms between them is only a few kcal/mole. However, this is sufficient for there to be a measurable preference for pyrimidine-3′,5′-purine base steps in the first group to be more readily unstacked and separated so that planar molecules of appropriate size can be inserted and bind intercalatively between them.

The methyl group of thymine can play an important role in major-groove sequence readout, by being able to make stable van der Waals non-bonded contacts with hydrophobic groups or side chains such as those of leucine, isoleucine and alanine. Negative indirect readout can also take place when the presence of a particular group, substituent or side chain would result in the occurrence of steric clashes with base substituents, and consequent destabilisation of the resulting DNA—ligand complex. Examples include (i) the bulky methyl group of thymine driving major groove recognition to a G/C region and (ii) the bulky exocyclic amino group at the 2-position of guanine, whose presence in the minor groove acts as a deterrent to many minor groove ligands. They then bind more readily in A/T regions.

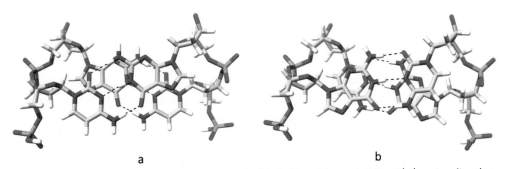

a   b

Figure 5.3 Base stacking in a canonical B-DNA double helix at (a) a pyrimidine-3′,5′-purine dinucleoside step and (b) a purine-3′,5′-pyrimidine step.

The electronic characteristics of an individual base pair produce an electric field and potential in its vicinity. These electronic features are distinct for the various classical DNA structural types, as well as for changes in oligonucleotide sequences (Otero-Navas & Seminario, 2012; Pullman & Pullman, 1981). Of especial significance is the finding that for B-DNA an electrostatic potential minimum occurs in the major groove of oligo (dG)•(dC) sequences, and in the minor groove of oligo (dA)•(dT) sequences such as the 5'-AATT sequence in the Dickerson—Drew dodecamer. These all have greater negative electrostatic potential than around the phosphate groups, which explains why cationic ligands and basic protein side chains tend to cluster in the grooves rather than around the phosphates. These electrostatic potentials and fields are important in directing an incoming ligand to a region of sequence; small local variations in net charge and field around an individual base pair then could assume some importance, especially for electrophilically driven covalent bonding.

5.2 DNA—water interactions

Water molecules play a pivotal role in biological systems, in large part by use of the unique hydrogen-bonding ability of an individual water molecule. It can simultaneously act as acceptor and donor for up to a total of four hydrogen bonds, often forming extended networks, even in bulk water. The hydration of DNA is essential for the maintenance of its structural integrity, as was recognised in the early days of fibre diffraction studies (Franklin & Gosling, 1953; Fuller et al., 2004). At least 12—15 first-shell water molecules are associated with each nucleotide in the canonical B-form. Historically, it has been presumed that some of these are clustered around phosphate groups, together with cationic counter-ions such as sodium, potassium or ammonium for charge neutralisation. One would expect other water molecules to be associated with the bases, possibly taking advantage of the hydrogen-bonding facility (Fig. 5.1) provided by the edges of the bases (Seeman et al., 1976).

Fibre diffraction analyses have successfully located the positions of some counter-ions and water molecules (Fuller et al., 2004), although complete details of structured water arrangements are unavailable from this method. However, only with the advent of medium and subsequently high-resolution single-crystal analyses of oligonucleotides has it been possible to define the positions of the major proportion of the first- and second-shell water molecules associated with a given DNA sequence. NMR methods can provide valuable complementary information, by locating some of the least mobile water molecules, such as the minor groove spine of hydration in the dodecamer d(CGCGAATTCGCG) (Halle & Denisov, 1998; Kuninec & Wemmer, 1992; Liepinsh et al., 1992). These methods can also provide data on residence times, which are typically in the low nanosecond range (Phan et al., 1999). Classic crystallography, by contrast, provides a picture of hydration averaged over the time scale of the X-ray experiment, which even for a high-flux synchrotron radiation source is many orders of magnitude greater than these times. The application of femtosecond x-ray free-electron laser sources to this and other nucleic acid systems promises to provide structural information on the time scale for dynamic events; the technique has already been successfully applied to visualise the reaction states of a riboswitch (Stagno et al., 2017). Molecular dynamics simulations, which are now increasingly performed in the millisecond range, can realistically model water motions around oligonucleotides. Several simulations have found excellent correlations between the positions for statistical averaged waters from the simulations, and those found from crystallography (see, for example, Duan et al., 1997; Feig & Pettitt, 1999; Rueda et al., 2004).

In general, the number of water molecules located by crystallography is highly dependent on the resolution of the structure and the thermal motion of individual water molecules (as well as on the quality of the raw X-ray diffraction data). It is necessary to achieve high resolution (≥ 1.5 Å) in a crystal structure analysis to find more than a fraction of the water molecules beyond the first hydration shell. The almost-universal use of cryo-crystallographic methods to collect low-temperature diffraction data, especially at high-intensity synchrotron X-ray sources, has enabled a significant proportion of these associated water molecules to be located, since the freezing of crystals at liquid nitrogen temperatures enables all molecular motions to be decreased, and also diminishes radiation damage. Motions are also reduced for quasi-bulk water molecules, which are involved in extensive water—water interactions (Pal et al., 2003). In favourable cases complete second- and even third-shell water molecules have been observed (Table 5.2). However, some water molecules are undoubtedly disordered, and only partially occupy discrete positions. These are harder to locate and differentiate from background noise in electron-density maps. The availability of database search and comparison methods has enabled systematic studies of water clustering to be made using the large number of oligonucleotide structures in the Nucleic Acid Database (Schneider & Berman, 1995; Schneider et al., 1998). In reality, water arrangements surrounding nucleic acids are complex and can comprise networks of water molecules hydrogen bonded together in diverse ways. This has been graphically illustrated by a combined X-ray and neutron diffraction study (Arai et al., 2005) of the hydration pattern surrounding the self-complementary decanucleotide d(CCATTAATGG). The ability of neutron diffraction to reliably locate hydrogen atom positions enabled the orientation of many water molecules to be unambiguously determined, such as those involved in intrastrand bridges between bases in base pairs. Orientations were also determined for waters in the minor groove, the "spine of hydration" (Table 5.2). It was concluded that the hydration is built on an irregular hexagonal motif, although the analysis of waters beyond the first and second shells was constrained by the moderate resolution of the neutron data (3 Å).

The nature and extent of DNA hydration also provides an indication of accessibility to ligands other than water itself, which are borne out by calculations of the

Table 5.2 Some oligonucleotide B-DNA crystal structures showing hydration networks and, in some instances, localised metal ions in the grooves.

Sequence	Feature	Resolution Å	PDB id
d(CGCGAATTCGCG)	Spine of hydration	1.90	1BNA
d(CGCGAATTCGCG)	Spine of hydration	1.30	4C64
d(CGCGAATTCGCG)	Spine of hydration	1.10	436D
d(CGCGAATTCGCG)	Spine of sodium ions	1.40	355D
d(CGCGAATTCGCG)	Potassium ions	1.75	428D
		1.20	1FQ2
d(CGCGAATTCGCG)	Calcium ions	3.00	426D
d(CCAGTACTGG)	Hydration at ultra-high resolution	0.74	1D8G
d(CCATTAATGG)	Neutron diffraction study of hydration	3.00	1WQZ

solvent-accessible surface characteristics of DNA. These have shown (Alden & Kim, 1979) that there are some differences between A and B helical types in terms of available surface area in the grooves, as well as between A•T and G•C base pairs. Sequence-dependent backbone conformational variability will also alter accessibility — when the minor groove in A/T regions is narrow its width is sufficient for the binding of only one water molecule per base pair in the first hydration shell. The first and second shells together produce a network of water molecules, the spine of hydration (see Figs. 5.6 and 5.7), which is two-dimensional in form.

The hydration patterns around phosphate groups in oligonucleotide crystal structures are dependent on whether the helix is an A, B or Z type (Saenger et al., 1986; Schneider et al., 1992). Detailed bridging hydration arrangements around phosphate groups have also been inferred from fibre diffraction studies of DNA polymers (Fuller et al., 2004). Analysis of water molecules around phosphate groups in single crystals of A-, B- and Z-DNA duplexes has found that each of the two charged oxygen atoms in a phosphate group can accommodate up to three water molecules of hydration (Schneider et al., 1998). The ester oxygen atoms, unsurprisingly, have little preference for hydration. In the A-form, successive phosphate groups along the phosphodiester backbone are close enough together for water molecules to be able to form bridges between them, as indeed is observed in many crystal structures. Such bridging patterns are not found in B-form oligonucleotides, with their more separated phosphate groups. These water molecules prefer to hydrogen bond to the anionic phosphate oxygen atoms, whereas the ester oxygen atoms O3′ and O5′ are rarely hydrated. By contrast the deoxy sugar ring oxygen atom O4′ frequently participates in water networks, in both A- and B-form and in quadruplex DNA structures. Several A-DNA octamer high-resolution structures have extended major groove networks, which can form hexagonal, pentagonal or ribbon-linked water molecules (Egli et al., 1998; Kennard et al., 1986). The coordination of these motifs to base edges appears to be a major factor in maintaining the integrity of A-form oligonucleotide structures.

DNA-bound water molecules are frequently associated with metal ions, especially sodium, calcium, potassium or magnesium (see, for example, Minasov et al., 1999). These ions, together with ammonium, are required for neutralisation of the negatively charged phosphate groups, although only a small proportion of such ions are typically located in crystal structures. It has been suggested that sodium ions can occupy part of the minor groove spine of hydration in B-DNA, possibly in fast exchange with water molecules. This is discussed in detail below. Magnesium ions are normally hexa-hydrated and have been well located in many oligonucleotide crystal structures. Typically, their role is to bridge phosphate groups (Fig. 5.4). They can also lie in the major groove, contacting hydrogen-bond acceptors at the base edges, in both instances with hydrogen bonds from the waters of hydration associated with the magnesium ion.

5.2.1 Hydration in the grooves in detail

Systematic examination of the observed distribution of water molecules in known oligo-nucleotide crystal structures has shown that the dependence of individual base hydration on the helical type (A, B or Z) is ultimately due to the size and electronic characteristics of the grooves in each category. For example, the B-DNA major groove is dominated by large, hydrophobic thymine methyl groups in A/T regions. The width of this major groove is such that water–DNA networks would have to be three-dimensional, even

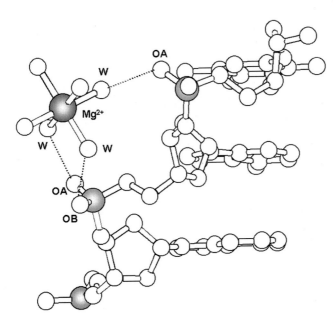

Figure 5.4 Part of the crystal structure of a typical B-DNA oligo-nucleotide, showing a hydrated magnesium ion coordinated to and bridging phosphate groups. Water molecules are labelled W and the non-diester phosphate oxygen atoms are labelled OA and OB. Hydrogen bonds are shown as dotted lines.

for the first shell of hydration. These networks would ideally need to straddle between and contact the two backbones as well as contacting base edge donor/acceptors. It appears that such arrangements are looser than minor groove ones, and associated water molecules are more mobile in the major groove. It is thus unsurprising that water molecules have only been observed to form extensive major groove networks in B-DNA oligonucleotide structures studies at high resolution and undertaken at low temperatures. Instead, the major groove base hydration that has been observed in crystal structures is much more localised (Schneider & Berman, 1995). Water molecules tend to form discrete arrangements around particular bases, which sometimes extend to one or two bases in either direction. Purine bases frequently have water molecules hydrogen bonded to them in the clustered manner shown in Fig. 5.5. Pyrimidine bases, by contrast, usually have a single associated water molecule, which can then hydrogen bond to another base, often via a second water molecule.

The hydration of the B-form minor groove has been the subject of much study and debate. The initial location (Drew & Dickerson, 1981) of the spine of hydration in the A/T region of the crystal structure of the sodium salt of the Dickerson−Drew duplex sequence d(CGCGAATTCGCG)$_2$ was at a resolution of 1.9 Å (see Chapter 3), which is only moderate by current standards. Nonetheless, the pronounced network of solvent molecules observed in this analysis is still accepted as essentially correct. This network involves a linear array of hydrogen-bonded water molecules extending over the central AATT region of the minor groove, with every other solvent molecule also hydrogen bonded to a pair of base edge hydrogen-bond acceptors (Fig. 5.6). These thus bridge between adjacent base pairs and are held together by waters that do not directly contact the DNA. The base-pair bridging waters also make hydrogen-bond contacts with O4' deoxyribose ring atoms, although in this structure the contacts are mostly with one strand. More recent high-resolution re-determinations of the structure (for example, Chiu et al., 1999; Lercher et al., 2014; Tereshko et al., 1999) have confirmed the spine's existence, and have also shown, with greater precision, a larger number of associated water molecules. Thus, in a 1.30 Å analysis (Lercher et al., 2014), not only is the spine itself more extensively defined, but higher-shell water molecules linked to the spine were also located (Fig. 5.7).

The assignment of all these solvent atoms as water molecules has been re-examined in several low-temperature determinations of this structure, which also used crystallographic refinement methods based on more precise geometric parameters than were available to the original study. An analysis, extending to a resolution of 1.38 Å, has suggested that the primary shell of solvent (which directly contacts base-pair edges) is partially occupied by sodium ions, and partly by water molecules (Shui et al., 1998). These assignments are based on the behaviour of the solvent during the refinement, together with valence calculations. However, subsequent crystal structure determinations, one at yet higher resolution (1.1 Å) using synchrotron data, have not found unequivocal evidence for sodium

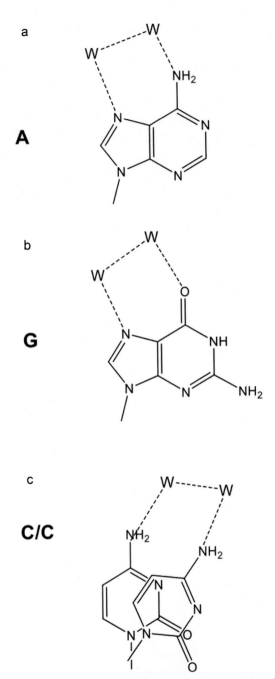

Figure 5.5 Typical patterns of water clusters in a B-DNA major groove. (a) and (b) are around individual purine bases. (c) Water molecules bridging between two consecutive pyrimidines in the same strand.

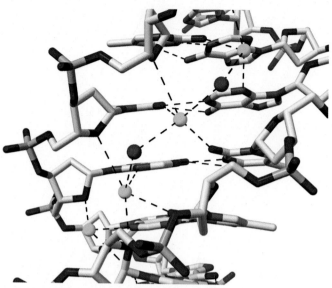

Figure 5.6 A view of the spine of water molecules in the original Dickerson—Drew dodecanucleotide crystal structure (Drew & Dickerson, 1981), PDB id 1BNA. The water molecules are shown as shaded spheres, and hydrogen bonds as dashed lines. The cyan-coloured water molecules are hydrogen bonding to the edges of bases forming the floor of the minor groove: the mauve-coloured waters are bridging between them.

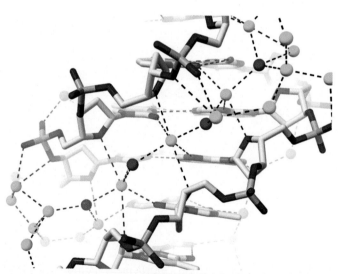

Figure 5.7 The spine of hydration, as observed in a high-resolution 1.30 Å crystal structure of the Dickerson—Drew dodecanucleotide (Lercher et al., 2014), PDB id 4C64. The spine of hydration corresponds to shells 1 and 2. Water molecules are coloured with the convention used in Fig. 5.6.

ion participation in the spine's structure (Chiu et al., 1999; Tereshko et al., 1999). The structure determinations of more recently discovered crystal forms of the dodecamer also concur with this view (Johansson et al., 2000; Liu & Subirana, 1999; Minasov et al., 1999). These analyses also confirm that minor groove hydration in other DNA structures extends well beyond the first shell and second shell of the spine itself, with, for example, each group of three water molecules forming the nucleus of a hexagon of higher-shell water molecules. These eventually form hydrogen bonds to phosphate groups. The higher-shell solvent molecules in these hexagons are highly mobile, with large thermal factors. It is thus unsurprising that they have not been observed in the earlier lower-resolution studies.

Why has the unequivocal assignment of solvent molecules as waters or sodium ions been so challenging? Since X-rays are diffracted by electrons in atoms, there are obvious problems of differentiating one type of atom from another when the number of electrons in each is closely similar. An oxygen atom has 8 electrons, and a sodium cation has 10. Even very high-resolution diffraction data have difficulty in distinguishing one from another. The problem is compounded by the mobility of solvent molecules, and by the probability that some have partial occupancies. On the other hand, discrimination between water molecules and potassium ions should be more straightforward with diffraction data extending to at least 2 Å. This is indeed the case (Shui et al., 1998). X-ray crystallography, unless at ultra-high resolution, cannot hope to differentiate water molecules from ammonium ions. NMR approaches using isotopically labelled ^{15}N ammonium ions have been used to show that they also partially occupy some minor groove sites, in fast exchange with water molecules (Hud et al., 1999). The overall picture that emerges is one in which small solvent molecules continually aggregate in and dissociate from various buried sites (the grooves) and around exterior sites (the phosphate groups), on the average in geometrically defined ways. Water molecules play the major role, but ammonium and alkali metal ions can also come into play. Both small molecule and protein–nucleic acid complexes are also dependent on solvent for their stability. Water molecules can play an active role in the recognition processes in these complexes even though they still have short residence times (Billiter et al., 1996; Laage et al., 2017). Ions can play a significant role, in maintaining both local and large-scale conformational features such as A-tract bending (McConnell & Beveridge, 2000). A recent cryo neutron crystallographic study shows the promise of this technique, provided suitably large crystals can be grown, in this instance having a volume of ca 1 mm^3. This analysis at 1.5 Å resolution, of the classic Z-DNA duplex formed from the sequence d(CGCGCG) in deuterated water and containing deuterated ammonium sulphate in solution, has shown that this technique is sufficiently powerful to even determine the orientation of the deuterium atoms for a number of the water molecules (Harp et al., 2021). This is the most detailed view to date of DNA solvation in crystals, undistorted by metal ions or polyamines, which were not included in the crystallisation mix.

The spine of hydration concept is a compelling one that has stood well the test of time. However, it is based on crystallographic observations at cryogenic temperatures, where water molecules become necessarily more immobilised. More recently, it has been demonstrated using chiral non-linear vibrational spectroscopy with two DNA sequences (McDermott et al., 2017), that the spine exists at ambient temperatures, as a chiral superstructure. Water spines have also been observed in high-resolution DNA quadruplex crystal structures (Li et al., 2021), although these differ from the spines in duplex DNA. Firstly, their extent and appearance depend on the width and length of the grooves, which themselves depend on quadruplex and loop topology. Secondly, it is common to have waters almost exclusively hydrogen bonding to phosphate groups, since these are oriented into the groove and not away from it, as shown in Fig. 5.8. The spine in this structure has just two contacts with guanine $-NH_2$ groups at the floor of the groove. The wider grooves/loops in DNA quadruplexes tend to have three-dimensional clusters of water molecules.

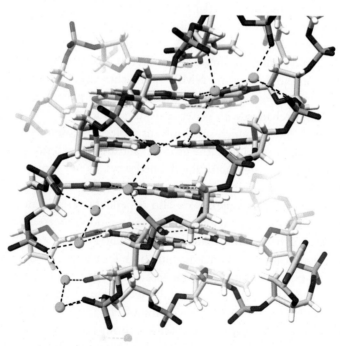

Figure 5.8 The spine of hydration in the narrow groove of the DNA quadruplex d($GT_2(G_4T_2)_3G_4$), PDB id 6XT7, consisting of seven connected water molecules, all which hydrogen bond to phosphate groups and none contact base edges.

5.3 General features of DNA–drug and small-molecule recognition

Many clinically important drugs and antibiotics are believed to exert their primary biological action by means of non-covalent interaction with DNA and subsequent inhibition of the template functions of transcription and replication. Activity has been shown for many of these agents against many human and animal diseases, ranging from cancer, bacterial infections and a variety of viral diseases to parasitic diseases such as malaria and trypanosomiasis. Cancer chemotherapy has traditionally been in large part based on DNA-interacting cytotoxic drugs, several of which are covalently binding agents, and many of these continue to be in clinical use today, even though they have limitations on account of their toxicity and limited cancer cell selectivity. Their effects are directly on DNA template function. Thus, selective inhibition of DNA-directed RNA synthesis is produced by the antitumour antibiotic actinomycin D, whose non-covalent DNA interactions have been especially well studied at the molecular level. For many of these molecules, drug–DNA interaction in vivo involves concomitant interactions with the enzymes DNA topoisomerase I or II to form ternary complexes. These interactions can lead to lethal DNA double-strand breaks, and eventually to selective tumour cell death, often via activation of the p53 tumour suppressor protein apoptosis pathways. However, cytotoxic drugs generally have high toxicity to both normal and cancer cells, and their therapeutic index (the ratio between a therapeutic and a toxic dose) is usually low. So unsurprisingly there has been a desire to rationally design new, improved ones, as well as to understand the molecular basis for the action of existing drugs. This has traditionally been the impetus for many of the large number of structural studies in this field. These have mostly been on binary drug–DNA complexes. More recently information has become available at the atomic level on ternary complexes with protein partners (see Chapter 8). This is from crystallographic studies since the proteins involved, notably the topoisomerases, are generally too large for NMR structural analysis.

Present-day cancer (and anti-infective) therapeutics is mostly concerned with molecular targeted agents, directed against proteins, enzymes and pathways that are selective to a particular cancer or infective agent. However, DNA-interactive agents, even though they may appear to have the disadvantage of non-selectivity, in practice are often clinically useful because they target protein–DNA complexes that may be over-expressed or mutated in the disease state. For example, the utility of many DNA-reactive drugs in cancer is related to the over-expression of DNA topoisomerase enzymes associated with higher levels of proliferation in some cancers. The potential of DNA-interactive agents targeted against specific genes has yet to be fully realised, although the recent interest in targeting quadruplex DNA and RNA in individual over-expressed genes is currently an active area of investigation.

DNA-interactive small molecules can be conveniently categorised into two major classes based on their mode of interaction — non-covalent binding and covalent bonding. There are sub-divisions within each class, which frequently overlap. Electrostatic-dominated non-covalent binding involves sequence-neutral non-selective interactions between a cationic group and negatively charged phosphates on DNA. Examples of such molecules include the natural polyamines such as spermine and spermidine. They are frequently used in nucleic acid crystallisation trials, to shield the highly negatively charged individual DNA or RNA molecules from each other. It is rare for these poly-amines to be observed in electron density maps. This suggests that, in common with alkali metal ions, their positions are mobile. The majority of non-covalent binding drugs are formally positively charged, so electrostatic contributions are always a significant component of ligand binding.

The rationalisation of experimental drug—DNA structures by computer modelling methods has long been an active field, especially when such studies are accompanied by predictions of drug modifications with enhanced DNA affinity or selectivity. As with all theoretical studies on nucleic acids, there has been the problem of adequate force field paramerisations, so that, for example, reliable quantitation of binding energies has been a challenge. Encouragingly, a recent extended molecular dynamics study of ethidium bromide intercalation complexes with two different DNA oligonucleotides has shown that the current DNA force field parameterisations in the AMBER system (see Chapter 1) can result in reliable simulations consistent with experimental data (Galindo-Murillo & Cheatham, 2021). Molecular dynamics simulations and free energy calculations can thus be used with confidence to predict ranking order and rationalise experimental binding geometries, provided the simulations are run for sufficient times and that high-quality force fields are used. Examples include the correct prediction of su-perior DNA binding of netropsin compared to distamycin (Dolenc et al., 2005) and the correct prediction of binding site size of aminoacridine dimers (Rowell et al., 2020). Pat-terns of hydration in the minor groove are modified or displaced by drug binding, and their analyses have been used to derive a set of potentials that have found use in quanti-tative docking studies (Ge et al., 2005).

The ability to discriminate between different sites (i.e. sequences) and different ligands has assumed increasing importance with the widespread use of in silico methods to rapidly screen very large libraries of compounds in order to find leads for subsequent me-dicinal chemistry and pharmaceutical development (Luo et al., 2019). A widely used source of compounds is the ZINC database (http://zinc.docking.org/), a free database of over 750 million commercially available compounds set up for virtual screening (Sterling & Irwin, 2015). DOCK is a popular and well-validated computer program for library screening of ligands, often used for fitting to DNA (see the website at http://dock.compbio.ucsf.edu/). The current version is DOCK 6.9 and an algorithm for flexible ligand docking is available within the package. A scoring function based on an implementation of AMBER molecular dynamics has also been incorporated into DOCK. Calculation of ligand–DNA binding energies within DOCK using the generalised Born model for taking account of solvent-mediated electrostatic energetics is now well validated (Allen et al., 2015; Kang et al., 2003), and can enable a series of DNA-binding ligands to be ranked. However, scoring functions, even for the well-constrained ligand–minor groove complexes, are sensitive to minor changes in geometry, and cannot always be relied upon to consistently predict structures that reproduce the crystal structures of these complexes (Evans & Neidle, 2006). Docking algorithms aim to achieve a balance between accuracy and speed of calculation. If just one or a very few ligands are to be examined, then extensive molecular dynamics calculations can be employed (see, for example, Cieplak et al., 1990 for an early study on distamycin, and Dolenc et al., 2005). These enable conformational flexibility to be sampled to a greater extent than is possible during in silico library screening. Monte Carlo methods have also been used (Rohs et al., 2005) in trial docking studies since they also offer these advantages over screening algorithms. However, neither method enables multi-million compound libraries to be rapidly screened since they are still highly compute-intensive even for an individual drug–DNA system. None of these methods can yet reliably predict absolute binding or free energies in a routine manner; the best that can be hoped for is accurate prediction of relative energies for a series of related molecules (Luo et al., 2019; Wang & Laughton, 2007).

In silico screening has been widely used in the quadruplex field to find new leads and new chemotypes for subsequent pharmaceutical development (Alcaro & Musetti, 2013; D'Aria et al., 2020; Monsen & Trent, 2018). Finding selective and high-affinity quadruplexes has its own set of challenges, not least the open nature of most ligand-binding sites on quadruplexes, at the terminal tetrads of a quadruplex core.

5.4 Intercalative binding

Many antibiotics, antibacterial and antitumour drugs are characterised by the possession of an extended planar heteroaromatic ring system (chromophore). This essential structural feature is typically a chromophore of three fused six-membered rings (Fig. 5.9),

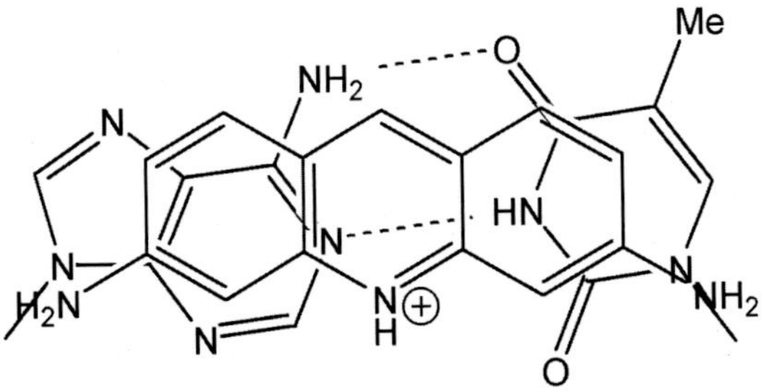

Figure 5.9 The proflavine molecule has an acridine ring, a typical DNA intercalative chemotype. Proflavine is shown superimposed onto a base pair, highlighting their similar dimensions.

which is approximately the same size as a base pair itself. The intercalation hypothesis, originally proposed by Leonard Lerman (Lerman, 1961), suggests that the planar chromophore of the drug molecule becoming inserted in between adjacent base pairs in an intercalative manner (Fig. 5.10). The drug chromophore is stabilised by polarisation forces and van der Waals dispersion interactions between its planar group and the base pairs surrounding it. Optimal attractive interactions occur when the intercalating chromophore has polarised character, incorporating one or more heteroatoms such as nitrogen. Electron deficiency is also important, and a formal cationic charge is common (e.g. in the acridines). Intercalation results in an extension of the double helix by 3.4 Å per bound drug molecule, together with changes in helical twist (unwinding) for the base pairs at and adjacent to each binding site (Waring, 1970). The base pairs either side of an intercalated drug thus increase in separation from 3.4 Å to 6.8 Å. The maximum

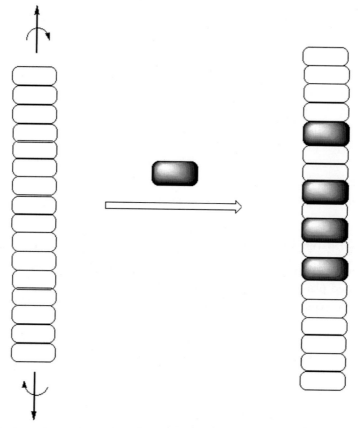

Figure 5.10 A schematic representation of planar drug molecules binding intercalatively into a DNA double helix, causing base-pair extension and unwinding. Base pairs and drug molecules are shown respectively as open and shaded rectangles.

number of drug molecules which can be bound to a DNA duplex is one for every three base pairs, i.e. every other potential site becomes occupied at drug saturation. So, the immediate base pair step either side of a binding site cannot bind a drug molecule. This is the neighbour-exclusion principle, which originates from the interpretation of solution binding data, but is fundamentally a consequence of the structural constraints forced on nucleic acid duplex backbone geometry by intercalation, so that two intercalation sites, each with 6.8 Å separated base pairs, cannot exist next to each other in a duplex without invoking severe steric clashes.

Much of the interest in intercalating drugs has focused on their potential for antitumour activity. Some, such as the anthracycline antibiotic doxorubicin (also known as Adriamycin) and its synthetic mimetic mitoxantrone, are still of major clinical value in the treatment of many of the common solid tumours as well as leukaemias. For many of these drugs and their derivatives, DNA-binding abilities correlate approximately with biological activity. However, in spite of much activity in this field, the past 30 years have produced very few new DNA intercalative therapeutic agents with a superior antitumour activity/toxicity profile compared to the long-established intercalative drugs such as doxorubicin. Drugs such as the anthracyclines form ternary complexes involving the topoisomerase family of proteins, which are often over-expressed in tumours. As a consequence DNA strand breaks are produced, which can damage both tumour and normal cell DNA.

5.4.1 Simple intercalators

DNA-intercalator recognition itself is in large part sequence-neutral, although, as described in Section 5.1, base pair stacking/unstacking requirements usually result in a small preference for a pyrimidine-3′,5′-purine sequence step at the binding site. "Simple intercalators" consist solely or primarily of an intercalating chromophore, often carrying a positive charge on the ring system — for example, proflavine and ethidium bromide (Fig. 5.11). The py-pu sequence preference at the drug-binding sites is generally obeyed both for these simple intercalators and for the complex ones such as daunomycin and its analogues. For the former, crystal structures of their complexes are restricted to DNA and RNA dinucleoside monophosphate "mini" duplexes, provided the sequences are pyrimidine-3′,5′-purine (see, for example, Neidle et al., 1977; Shieh et al., 1980), although an (as yet unpublished) entry of a high-resolution crystal structure of a proflavine—hexanucleotide complex is available in the Protein Data Bank (PDB id 3FT6). The more complex intercalators, especially daunomycin and its analogues, have been mostly studied structurally (Fig. 5.11 and Table 5.3) as hexanucleotide self-complementary duplex complexes, with again drugs binding at the py-pu sites (see, for example, Wang et al., 1987). Typically, as in Fig. 5.12 the sequence d(CGTACG) has two drug

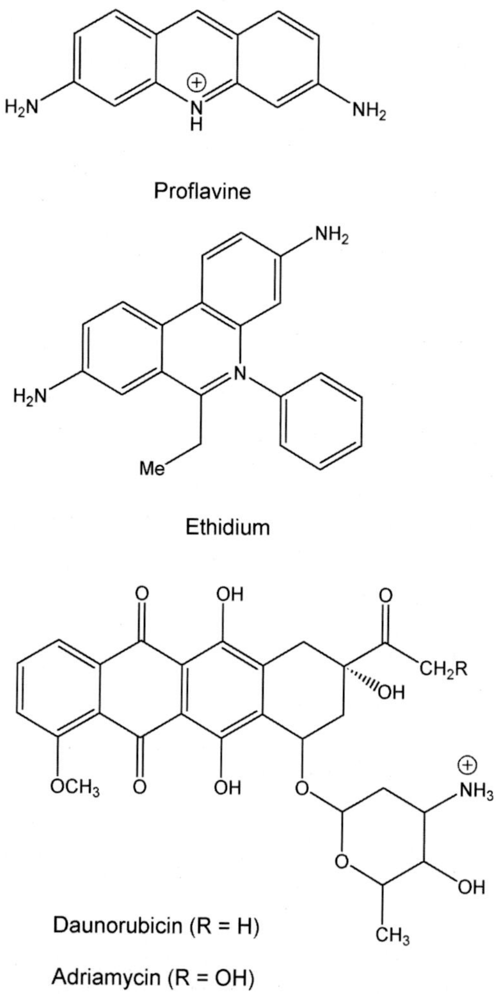

Figure 5.11 The structures of some representative intercalating molecules. Adriamycin is a trade name for doxorubicin.

Proflavine

Ethidium

Daunorubicin (R = H)

Adriamycin (R = OH)

molecules bound, one at each CG site. Analysis of conformational changes in intercalated dinucleoside duplex structures and the observation of "mixed" sugar puckers (C3'-*endo*-3',5'-C2'-*endo*) led initially to the suggestion that these are a necessary consequence of the intercalation process, and would explain the neighbour exclusion effect. This feature was not observed in several crystal structures involving the acridine drug proflavine (Neidle et al., 1977; Shieh et al., 1980), and is no longer considered to be indicative of any fundamental structural property. Final confirmation of this has come with analysis of the structure of a rhodium ligand—DNA complex.

The degree of helix unwinding produced by intercalation is dependent on the nature of the bound drug. It can be measured experimentally (Waring, 1970) using covalently

Table 5.3 Crystal structures of some drug–oligonucleotide intercalation complexes.

Drug	Sequence	Resolution in Å	PDB id code
Ditercalinium	d(CGCG)	1.70	1D32
Ellipticine	d(CGTACG)	1.50	1Z3F
Daunomycin	d(CGTACG)	1.90	1D11
Daunomycin	d(CGATCG)	1.50	1DA0
Doxorubicin	d(CGATCG)	1.70	1D12
4′-Epi-doxorubicin	d(TGATCA)	1.40	1D54
Bis-daunomycin	d(CGATCG)	2.20	1AGL
Nogalamycin	d(CGTACG)	1.70	1D21
Nogalamycin	d(CCCGGG)	2.40	282D
MAR70 disaccharide	d(CGTACG)	1.20	1R68
9-Amino-DACA	d(CGTACG)	1.60	465D
Triostin A	d(GCGTACGC)	2.00	1VS2
Echinomycin	d(ACGTACGT)	1.60	2ADW
Echinomycin	d(ACGTCGT)	1.55	5YTZ
Actinomycin	d(GAAGCTTC)	3.00	173D
Actinomycin	d(ATGCGGCAT)	2.60	4HIV
Rhodium complex	d(GUTGCAAC)	1.20	454D
Rhodium complex	d(CGGAAATTCCCG)	1.10	2O1I
Cu-TMPy	d(CGTACG)	2.40	231D
Ni-TMPy	d(CCTAGG)	0.90	1EM0
Psoralen	d(CCGCTAGCGG)	2.20	1FHY
Bis-acridine	d(CGTACG)	1.70	2GB9

closed-circular DNA, with reference to the standard value of 26° per bound drug molecule, for ethidium bromide. Crystallographic analyses of intercalation complexes have shown qualitative unwinding; however, the helical twist angle for the two base pairs immediately surrounding an intercalated drug is not necessarily equivalent to this angle. The ribodinucleoside (CpG) duplex complex with proflavine (Neidle et al., 1977) and the hexamer duplex d(CGTACG) with bound daunomycin (Wang et al., 1987) both have normal B-DNA helical twist angles of 36° for the base pairs flanking the drug. In the latter case, there are changes in helical twist at adjacent base-pair steps. The total unwinding angle can be defined as the sum of cumulative changes in helical twist over all affected base pairs. The process of base-pair separation to produce an intercalation site also results in other changes, especially in backbone conformation, and sometimes also in the details of base pair and base step morphology. The crystal structure (Canals et al., 2005) of the anticancer drug ellipticine (which has an experimentally determined unwinding angle of 22°) complexed with the hexanucleotide duplex formed by d(CGTACG) shows that the two drug-binding sites each have helical twist angles of 21−22°, with some further

Figure 5.12 The crystal structure (Wang et al., 1987), PDB id 1D11, of an intercalation complex involving the anticancer drug daunomycin bound at each of the two terminal CpG steps of the hexanucleotide duplex d(CGTACG). The two drug molecules are shown in ball and stick form with mauve shaded carbon atoms, and the daunosamine substituent sugar rings are seen to reside in the minor groove.

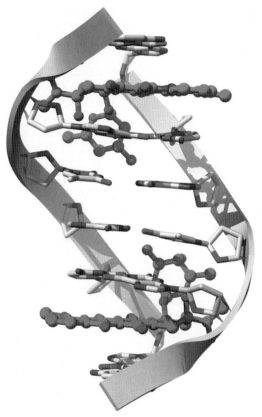

unwinding being apparent at adjacent sites. However, this in total is still less than the experimental value, indicating the limitations of these crystallographic models for quantitatively mirroring intercalation into long double-helical DNA sequences.

5.4.2 Complex intercalators

More complex intercalator molecules than ethidium or proflavine have attached groups such as side chains, sugar rings or peptide units — for example, the antitumour drugs actinomycin D and daunomycin (Figs. 5.11 and 5.13). These pendant groups normally reside in the minor groove of B-type DNA, where they are stabilised by van der Waals interactions, and can hydrogen bond to adjacent bases, providing sequence-specific direct readout. For example, the threonine residue in the cyclic pentapeptide part of the actinomycin molecule hydrogen bonds to the N2 and N3 atoms of a guanosine nucleoside, which is then structurally constrained to be on the 5'-side of the intercalating drug

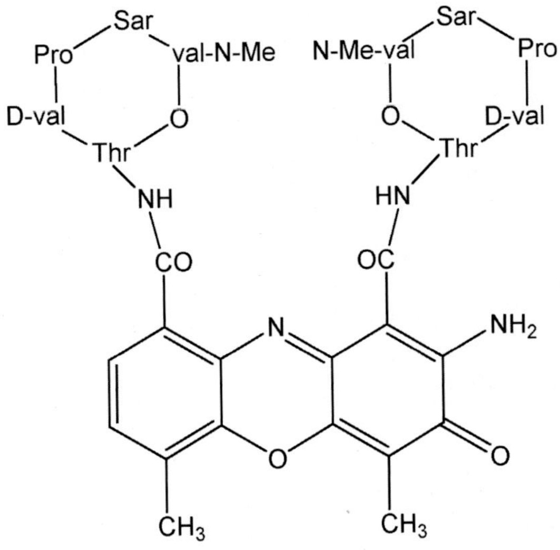

Figure 5.13 The structure of actinomycin.

Actinomycin D

chromophore, giving a binding site requirement for the sequence GpX. This has been shown by crystallographic analyses of actinomycin bound to several relevant sequences such as d(GAAGCTTC) (Kamitori & Takusagawa, 1992) and d(ATGCTGCAT) (Hou et al., 2002). The structures, exemplified by the complex with the latter sequence, show that the phenoxazone chromophore of the drug is intercalated at the GpC sites (Fig. 5.14). The two pentapeptide groups are situated in the minor groove of the DNA helix, in agreement with an early NMR study of an actinomycin complex with the sequence d(AAAGCTTT) (Liu et al., 1991). The crystal structure of actinomycin bound to the sequence d(ATGCGGCAT), containing a CGG repeat (important in several neurodegenerative diseases) flanking a G•G mismatch, has shown (Lo et al., 2013) that the drug at these sequences induces several major changes in DNA conformation, including base flipping, sharp bending and a left-handed twist (Fig. 5.15).

Numerous crystallographic and NMR analyses have been reported of hexanucleotide complexes involving members of the anthracycline family of anticancer drugs and their analogues, in large part because of their continuing clinical importance. Most adopt structures analogous to that first observed in the original daunomycin complex (Wang et al., 1987). More complex substituents than the monosaccharide of daunomycin and doxorubicin still reside in the minor groove, as shown in the crystal structure (Fig. 5.16) of the d(CGTACG) complex with the MAR70 analogue (Temperini et al., 2005), which has two linked sugar group attached to the anthracycline chromophore. The crystal structure shows that these penetrate further into the minor groove on binding.

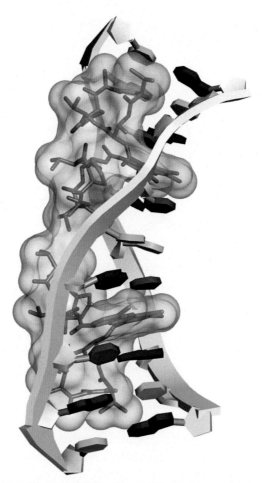

Figure 5.14 Crystal structure (Hou et al., 2002), PDB id 1MNV, of a complex between actinomycin and the octanucleotide duplex formed by d(ATGCTGCAT). The two bound drug molecules are shown in transparent solvent-accessible mode, with the drug molecules also shown in stick form. The DNA minor groove is on the left and the intercalation of both actinomycin chromophores between d(GpC) steps can be clearly seen.

NMR and crystallographic studies on oligonucleotide complexes of the antitumour antibiotic nogalamycin (Fig. 5.17) have revealed a novel binding mode for this drug, with the two attached groups residing, one in each groove; the drug chromophore itself is intercalated as expected (Egli et al., 1991; Smith et al., 1995; Williams & Searle, 1999). The aglycone group is held in the minor groove and the amino sugar in the major groove (Fig. 5.18), with drug–oligonucleotide hydrogen bonding to N2 and O6 atoms of guanines, thereby providing simultaneous major and minor groove direct sequence

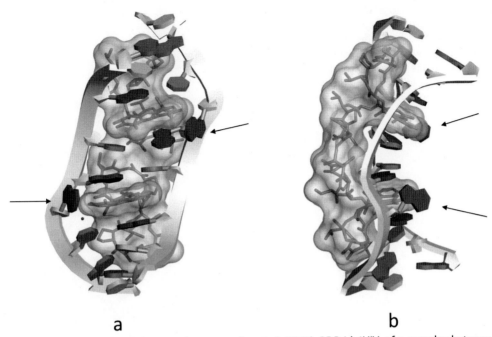

a

b

Figure 5.15 Two views of the crystal structure (Lo et al., 2013), PDB id 4HIV, of a complex between actinomycin and the sequence d(ATGCGGCAT). (a) A view into the minor groove, showing the base flipping and distortions to base pairs indicated by arrows. (b) Rotated by 90°, showing the curvature induced in the structure by drug binding.

recognition, which agrees well with DNA footprinting data (Fox & Waring, 1986). In contrast with the simpler intercalators, the mechanism of nogalamycin intercalation into duplex DNA is not straightforward. The two attached groups, one at each end of the molecule, are sufficiently bulky to ensure that the drug chromophore cannot become intercalated without one or other of these groups effectively passing through or opening-up the intercalation site. This implies that drug—DNA association and dissociation will both be exceptionally slow processes as a direct result of the structural constraints forced on the complex, as has been found experimentally.

5.4.3 Major groove intercalation

Most "complex" intercalating molecules have their attached groups bound in the minor groove, with the DNA duplex retaining B-like character. These drugs are sometimes natural products and the pendant groups are then sugar residues or modified peptides, often with hydrophobic character which complement the hydrophobic areas on the DNA groove walls. By contrast, several classes of intercalating molecules have been synthesised which subsequent structural studies have shown to have their attached groups bound in

Figure 5.16 The crystal structure (Temperini et al., 2005) of the anthracycline derivative MAR70 bound to the sequence d(CGTACG), PDB id 1R68, showing the two saccharide rings of each drug molecule residing in the major groove, with carbon atoms of the drug coloured mauve.

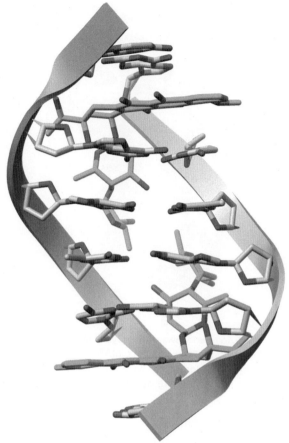

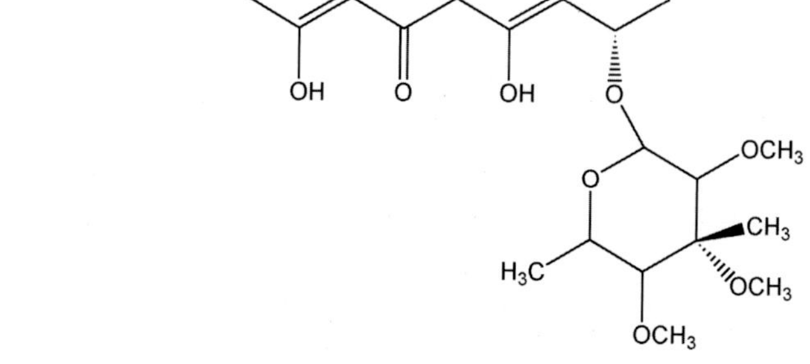

Figure 5.17 The structure of the antitumour antibiotic nogalamycin.

Nogalamycin

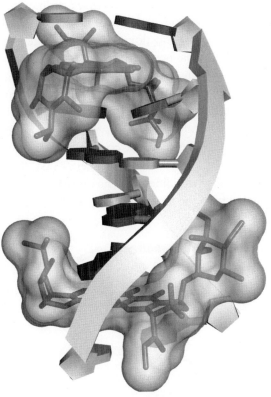

Figure 5.18 The crystal structure (Smith et al., 1995), PDB id 182D, of the complex between nogalamycin and the duplex sequence d(TGATCA), showing a transparent view of the solvent-accessible surface of the two bound drug molecules, through which stick representations can be seen.

the major groove. Major groove binding has been demonstrated for a family of synthetic anticancer drugs based on the acridine carboxamide skeleton. For example, the crystal structure has been determined (Adams et al., 2000a,b; Todd et al., 1999) of a complex between the d(CGTACG)$_2$ duplex and the 9-amino derivative of the parent drug in this series of acridine derivatives, termed "DACA" (Fig. 5.19). This shows that the cationic end of the dimethylaminoethyl side chain forms hydrogen bonds with O6

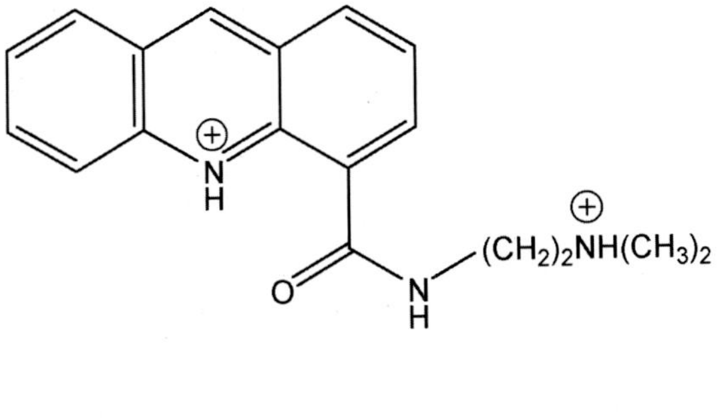

Figure 5.19 The structure of the acridine antitumour drug DACA.

and N7 guanine base edge atoms in the major groove (Fig. 5.20). These acridine-based drugs are also active against the enzyme DNA topoisomerase II, as are the minor groove intercalators such as the anthracyclines daunorubicin and doxorubicin. A related bis-acridine, also a 4-carboxamide derivative (Fig. 5.21) has been found, at least in the crystal lattice (Hopcroft et al., 2006) to form a novel non-covalent inter-helix cross-linked arrangement (Fig. 5.22), in which the two halves of the ligand each intercalate into a CG site, but in two separate symmetry-related duplexes.

Numerous metal-containing intercalating molecules have been synthesised as probes of DNA structure, and several are photo-active. The rhodium complex (Fig. 5.23) is typical. It was designed to recognise the major groove of the sequence 5'-TGCA, i.e. unusually intercalating at the 5'-GC purine-3',5'-pyrimidine site. The remarkable crystal structure (Kielkopf et al., 2000b) of a complex with the recognition sequence embedded

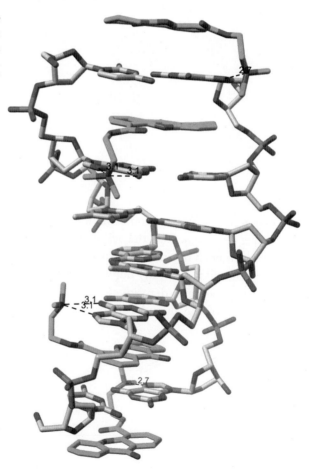

Figure 5.20 The crystal structure (Todd et al., 1999), PDB id 456D, of a complex between 9-amino-DACA and the duplex sequence d(CGTACG), showing the side chains in the major groove of the DNA and its hydrogen bonding to base edges. The carbon atoms of the 9-amino-DACA are coloured mauve.

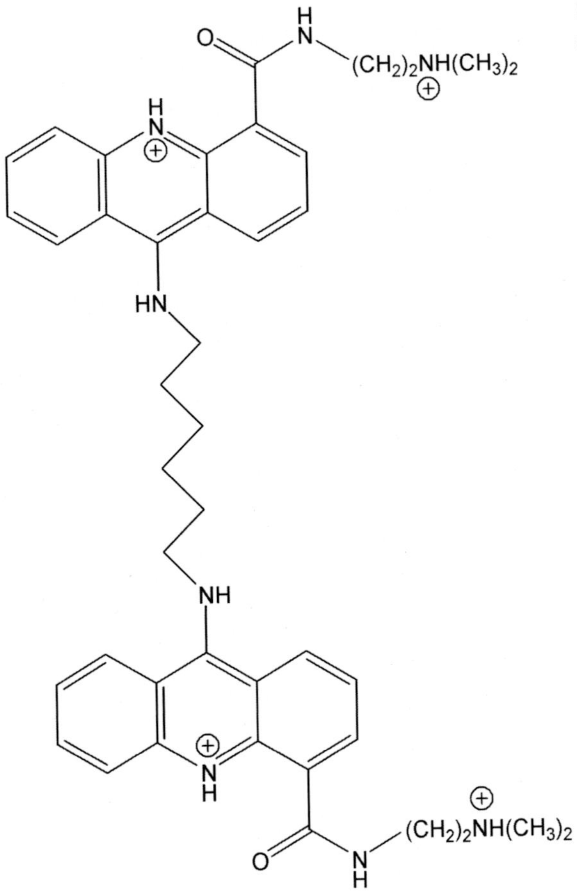

Figure 5.21 The structure of a bis-(4-carboxamide-acridine) molecule.

into an octanucleotide duplex has five independent complex moieties in the crystallographic asymmetric unit. They all show consistent structural features and conformations, within each of the planar group inserted from the major groove direction, and the bulky non-planar part of the ligand fitting snugly against the walls and floor of the major groove (Fig. 5.24). This high-resolution structure is notable for an intercalation complex since the ligand is embedded at the centre of a significant length of DNA sequence, thus eliminating possible end-effects on intercalation conformation. It provides a view of the molecular features of intercalation that are relevant to the situation in bulk DNA. The long-standing controversy surrounding the nature of nucleoside sugar puckers at the intercalation site (see Section 5.4.1) has been resolved by this structure determination, with the observation of consistent C2′-*endo* puckers at the ligand-binding site.

Figure 5.22 View of part of the crystal structure of a bis-acridine complex, showing the cross-linking arrangement involving the bis-acridine molecule that brings two duplexes close together (Hopcroft et al., 2006), PDB id 2GB9.

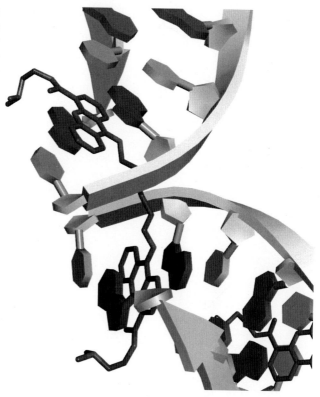

A series of rhodium complexes that have been designed to target mis-matched DNA sequences also interact by intercalative-type binding, as has been shown (Pierre et al., 2007) in the crystal structure of a complex with a duplex formed by the sequence d(CGGAAATTCCCG). In this structure the bulky rhodium complex binds at two distinct sites. At a normal base pair site, in the centre of the helix, the ligand binds in a conventional intercalative manner (Fig. 5.25) and the two A•T base pairs separate by the standard 6.8 Å distance. However, at the two A•C mis-match sites the ligand insertion results in the ejection of the adenine and cytosine bases, into the minor and major grooves, respectively.

A highly unusual and unexpected intercalation motif has been observed with the hexamer sequence d(CGTACG) in complexes with several diverse drugs and ligands — an acridine-4-carboxamide (Adams et al., 1999; Thorpe et al., 2000), a bis-acridine and a synthetic daunomycin analogue, ametantrone (Yang et al., 2000). In each case, conventional major groove intercalation into a duplex structure was expected. Instead, four duplexes in the crystal lattice come together to form an intercalation "platform" with

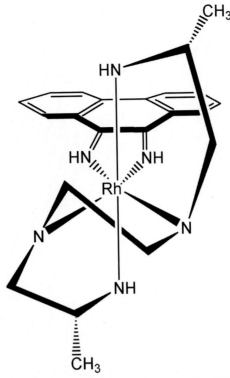

Figure 5.23 Structure of an intercalating rhodium complex (Kielkopf et al., 2000b).

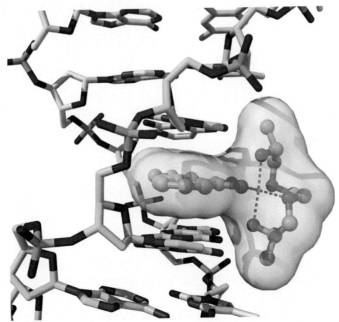

Figure 5.24 Crystal structure of one duplex in the intercalation complex (Kielkopf et al., 2000b), PDB id 454D, between a rhodium compound and an octanucleotide duplex, with part of the solvent accessible surface of the rhodium complex shown snugly intercalated between the central base pairs.

Figure 5.25 Crystal structure of a rhodium complex (shown in van der Waals space filling representation), bound to a mis-matched DNA oligonucleotide (Pierre et al., 2007), PDB id 2O1I.

terminal cytosines displaced out from one end of each duplex — this feature has been found in several native and protein-bound oligonucleotide crystal structures (see Chapters 3 and 7). Each displaced cytosine then effectively strand-invades another symmetry-related duplex so that a C•G base pair is formed. A pseudo-intercalation cavity is formed by two such adjacent duplexes, which has some resemblance to four-way junctions, and to ligand-binding sites in DNA quadruplexes. Whether this type of structure has biological relevance, or is solely an artifact of crystal lattice packing, remains to be determined. The crystal structure of a true four-way junction complex has been determined with a covalently bound derivative of the largely planar molecule psoralen (Eichman et al., 2001). Junction—ligand complexes are further discussed in Section 5.5.

Among more complex multi-groove-binding molecules are the porphyrins such as tetra-N-methyl-pyridyl-porphyrin (TMPyP4) (Fig. 5.26). Molecular modelling studies with this ligand have suggested that it could fit into a B-DNA intercalating site with two of its substituent N-methyl-pyridyl groups in each groove, although its observed sequence preferences have been interpreted in terms of minor groove binding at A/T regions and intercalative in G/C ones. Intercalation has been inferred by NMR methods on an oligonucleotide complex, to take place at the sequence CpG, in accordance with the pyrimidine-3′,5′-purine sequence preference rule for intercalators (Marzilli et al., 1986). This would necessarily place two of the non-planar N-methylpyridine substituents in each groove, major and minor. However, the crystal structure of a nickel complex with the hexanucleotide duplex d(CGATCG)$_2$ (Bennett et al., 2000) and a copper complex of this porphyrin does not show full intercalation (Lipscomb et al., 1996). Instead, the porphyrin is best described as being semi-intercalated at the two terminal CpG sites, with the end cytosine in each case being swung out of the helix (Fig. 5.27). Thus, the

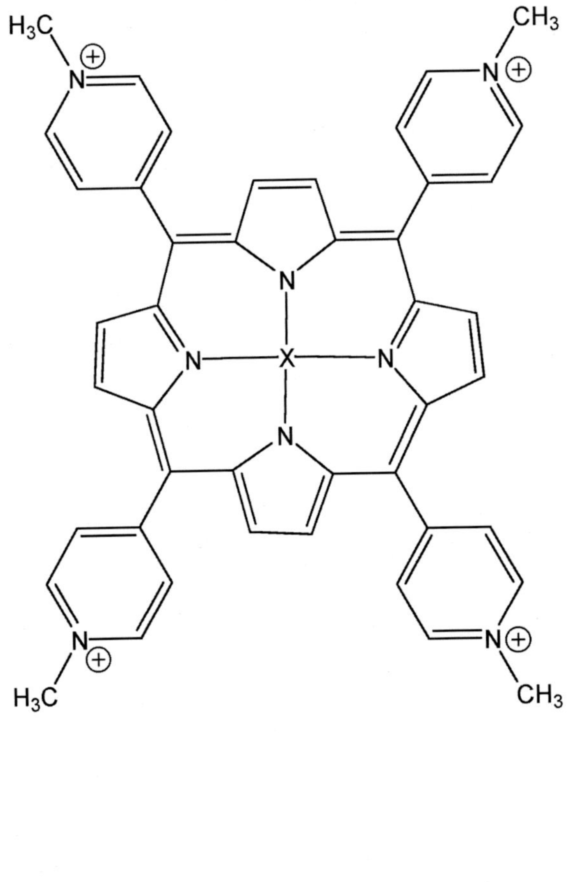

Figure 5.26 Structure of tetra-N-methylpyridine-porphyrin, TMPyP4. Atom X can be any one of several metal ions, such as nickel, copper or zinc. The native porphyrin molecule has hydrogen atoms attached to two of the four inner-facing nitrogen atoms.

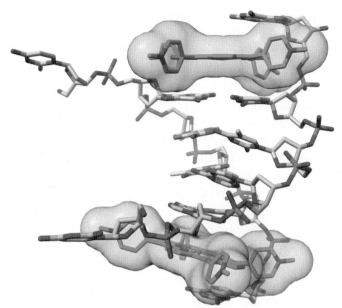

Figure 5.27 The crystal structure (Lipscomb et al., 1996), PDB id 231D, of the copper complex of TMPyP4 bound to the sequence d(CGTACG). The two TMpyP4 molecules are shown as solvent-accessible surfaces with the stick form shown in the interior. The central coordinated copper atom, shown as a sphere, is visible in the lower porphyrin molecule.

overall amount of porphyrin-base stacking is less than expected for full intercalation. The nickel complex of TMPyP4 forms a quite distinct complex (Bennett et al., 2000) with the d(CCTAGG)$_2$ duplex sequence. This hexanucleotide forms a conventional B-DNA duplex. However, the porphyrin molecule, which is significantly buckled out of planarity, is not intercalated, but stacked at the ends of a pseudo-dodecanucleotide formed by two end-to-end hexanucleotide duplexes. The crystal packing of this structure has one such dodecanucleotide duplex positioned with a porphyrin molecule capping the ends and also positioned against the widened minor groove of another duplex, thereby providing a visualisation of the groove-binding mode of this ligand.

5.4.4 Bis intercalators

Yet more complex are the drug molecules typified by the echinomycin and triostin family of antitumour natural product antibiotics (Fig. 5.28), which have two intercalating chromophores linked together by cyclic oligopeptides. These molecules intercalate such that the two chromophores, whose planes are separated by 10.2 Å, can simultaneously bind at sites separated by two intervening base pairs (Fig. 5.29). This mode of binding, termed bis-intercalation, has been validated by several crystallographic studies on echinomycin and triostin A bound to oligonucleotide sequences (Quigley et al.,

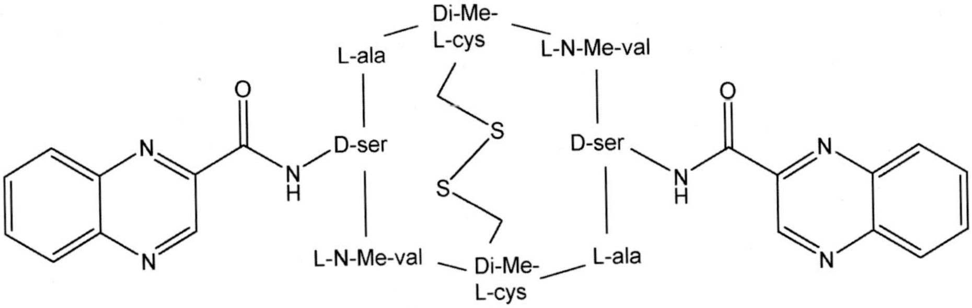

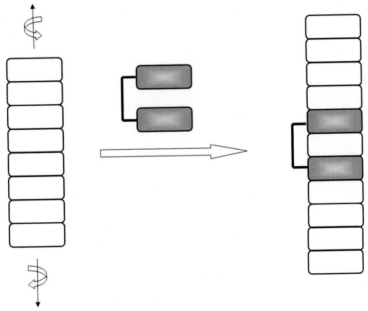

Triostin A

Figure 5.28 The structure of triostin A.

Figure 5.29 A schematic view of bis-intercalation.

1986; Wang et al., 1984) and NMR (Gao & Patel, 1988; Gilbert & Feigon, 1992). Bis-intercalators can also be characterised by unwinding experiments on closed-circular DNA, since they produce approximately twice the unwinding per bound drug molecule compared to a mono-intercalator. Echinomycin itself has a sequence requirement for the dinucleoside 5′-d(CpG), with two C•G base pairs being flanked by the two quinoxaline chromophores. The peptide ring system sits in the DNA minor groove, with specific hydrogen bonding to the 2-amino group of the guanine.

An unexpected finding in the crystal structures (Fig. 5.30) of these drug complexes has been the occurrence of Hoogsteen hydrogen bonding for the base pairs "externally" flanking the bound drug — the two base pairs between the chromophores are always in a standard Watson—Crick arrangement. NMR studies indicate that this major structural change from standard B-DNA can occur with these drugs and certain short oligonucleotide sequences in solution. However, chemical probe experiments (McLean et al.,

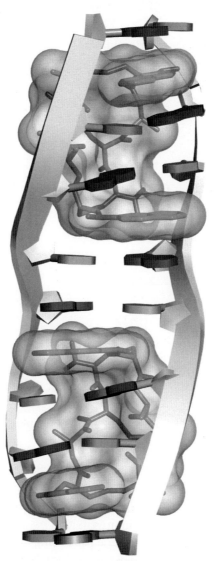

Figure 5.30 The crystal structure (Wang et al., 1984), PDB id 1VS2, of a complex between an octanucleotide duplex and two triostin molecules, each bound in a bis-intercalative manner.

1989) with much longer (160 base pair) sequences that are more truly representative of biological DNA strongly suggest that Hoogsteen base pairing does not occur in these sequences. It appears then that Hoogsteen pairing does not always need to be formed as part of the structural changes in a DNA duplex consequent to bis-intercalative recognition, and that these changes may be dependent on sequence length. A subsequent crystal structure analysis at higher resolution (1.6 Å) has two crystallographically independent echinomycin complexes, each bis-intercalated into a duplex formed by the sequence d(ACGTACGT). This shows Watson–Crick base pairing for three of the eight base pairs flanking the drug chromophore (Cuesta-Seijo et al., 2006), whereas crystallographic studies on both these sequences in a different crystal form, and with a d(GCGTACGC) complex (Cuesta-Seijo & Sheldrick, 2005), showed a more consistent pattern of Hoogsteen base pairing. This suggest that Hoogsteen pairing can occur, but that the energy barrier to the Watson–Crick arrangement is low and can even be overcome by crystal packing forces. The preference of echinomycin for 5′-d(CpG) sites is enhanced when a T•T mismatch separates two such sites (Wu et al., 2018). This effect has been ascribed to small changes in local DNA structure on binding the first echinomycin molecule, resulting in a tight minor groove pocket for binding of the second drug molecule (Fig. 5.31).

A novel series of synthetic bis-intercalating analogues of daunomycin have been developed using structure-based drug design principles (Chaires et al., 1997), based on two assumptions:

- such molecules would be superior DNA binders compared to daunomycin and its analogues since their two chromophores would together produce greater stacking interaction energy.
- superior DNA affinity would result in superior biological activity, since it has been found in the anthracycline series of compounds that intercalative DNA-binding affinity can correlate with cellular cytotoxicity. This in turn can indicate trends in anti-tumour activity. Thus, it was anticipated that bis-daunomycins would show enhanced anti-cancer activity compared to their mono-analogues.

The well-established hexanucleotide–daunomycin crystal structures (Wang et al., 1987), as described above, were taken as a starting-point, with two drug molecules bound in the hexanucleotide duplex. The two minor groove pendant sugar groups were replaced by a p-xylene linker group through the exocyclic amino groups on each sugar ring, thus generating a single bis-intercalating molecule. The resulting molecule "WP631" (Fig. 5.32) has been shown to exhibit all the enhanced DNA-binding properties predicted for a bis-intercalator. Both NMR and crystallographic studies (Fig. 5.33) have verified this binding mode (Hu et al., 1997). The DNA affinity of WP631, at 10^{11} mol^{-1}, approaches that of many regulatory proteins. However, this is

Figure 5.31 The crystal structure of the complex between echinomycin and d(ACGTCGT), PDB id 5YPZ, showing a central T•T base mismatch pair (Wu et al., 2018).

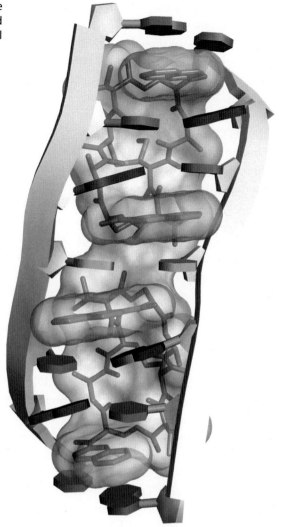

not reflected in an equivalent leap in biological response since its antitumour activity is not greatly superior to that of the native daunomycin. So, in spite of the elegance of the drug design process in this instance, there appears to be an inherent limitation to the therapeutic index and thus the efficacy of intercalating molecules as antitumour agents. This is unsurprising, since even molecules such as WP631 will have high toxicity to normal cells and are by themselves not capable of targeting specific oncogene

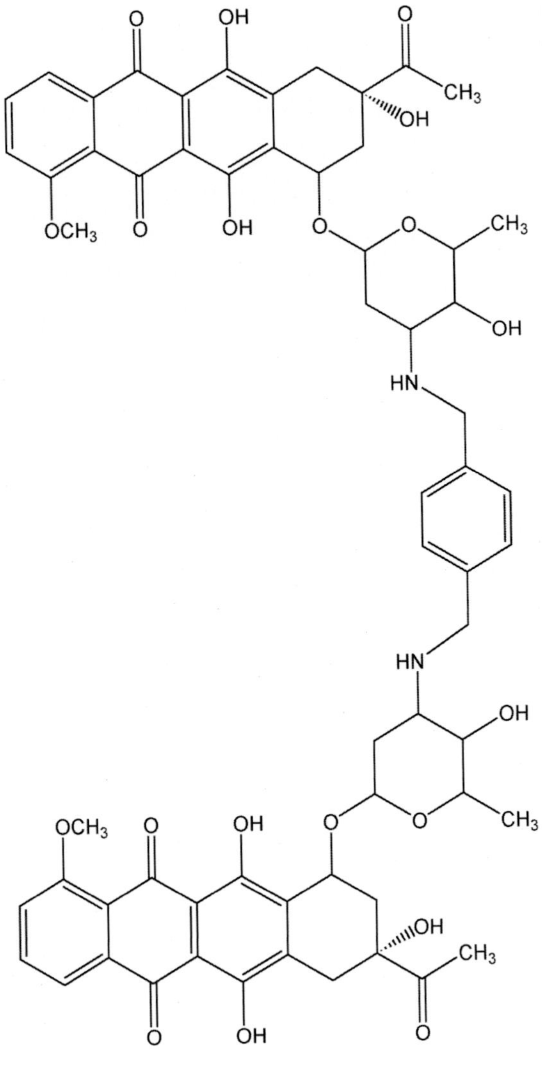

Figure 5.32 The structure of the bis-daunomycin molecule, WP631 (Chaires et al., 1997).

sequences without at the same time binding to many other genes involved in normal cell functions.

Bis-intercalation from the major groove direction is rare. It has been observed for bis-acridines (Hopcroft et al., 2006; see above), and (Peek et al., 1994) in a molecule comprising two pyridocarbazole units, linked by a flexible aliphatic chain, which has been co-crystallised with the short duplex sequence d(CGCG). Each planar group of the ligand is intercalated at the CpG step, and the linker is positioned in the major groove. There is a marked bend in the DNA, towards the minor groove, which may be an innate feature induced by this type of bis-intercalating molecule.

Figure 5.33 The crystal structure (Hu et al., 1997), PDB id 1AGL, of the bis-daunomycin molecule WP631 bound to the duplex sequence d(CGTACG), viewed towards the minor groove and showing the linker group covering four base pairs.

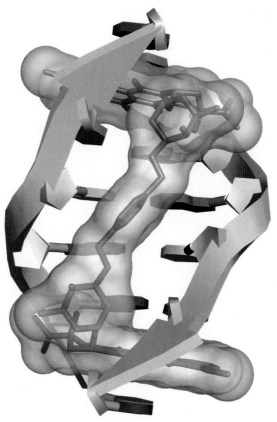

5.5 Intercalative-type binding to higher-order DNAs

5.5.1 Triplex DNA–ligand interactions

Binding of small molecules to triplex DNA is a well-established approach for stabilising tracts of triple-stranded sequences (Cassidy et al., 1994; Escude et al., 1998; Fox et al., 1995; Mergny et al., 1992; Silver et al., 1997). The effective molecules in use have the common features of planar aromatic chromophores containing several fused electron-deficient rings, and it is presumed that they intercalate into the triplex at some point. Several show preferential binding to triplexes over duplex DNA. Effective stabilising ligands include benzo[e]pyridoindole, the alkaloid coralyne and several disubstituted amido-anthraquinones. There is little direct structural data on these complexes, although molecular modelling studies (for example, with amido-anthraquinones; Keppler et al., 1999) have successfully rationalised binding behaviour in terms of intercalative-type models. Selectivity for triple-stranded DNA arises from a combination of base-chromophore stacking and substituent binding in triplex grooves.

5.5.2 Ligand binding to quadruplex DNAs

The first report of small molecule binding to a DNA G-quadruplex involved the simple intercalator molecule ethidium (Guo et al., 1992), although the mode of binding was not defined. This report lied dormant in the literature for some years, since when there has been an explosion of interest in such four-stranded structures formed from telomeric and other DNA sequences, and their complexes with a wide range of small molecule compounds (see, for example, Han & Hurley, 2000; Mergny & Hélène, 1998; Monchaud & Teulade-Fichou, 2008; Neidle, 2016; Organisian & Bryan, 2007; Savva & Georgiades, 2021). This activity initially follows on from the findings that telomeric DNA is progressively shortened in normal cells but is stabilised in length in most tumour cells by a specialised reverse transcriptase enzyme complex, telomerase (Cech, 2000; McEachern et al., 2000), which is over-expressed in most cancer cells but not normal cell. This enzyme requires the telomere strand, acting as primer, to be single stranded to hybridise with the RNA template component of the telomerase complex. Folding the primer strand into a quadruplex arrangement effectively inhibits telomerase function (Zahler et al., 1991). The folding process can be promoted by a range of intercalator-type molecules. Many such compounds, including substituted anthraquinones, acridines and the tetra-N-methyl-pyridyl porphyrin TMPyP4, have been subsequently developed, and analogues typically show correlations between quadruplex binding and telomerase inhibition (see, for example, Read et al., 1999; Riou et al., 2002; Sun et al., 1997; Wheelhouse et al., 1998). Several of the first-generation of these molecules, notably TMPyP4 itself, are also effective duplex-binding agents, and thus show only moderate selectivity for G-quadruplexes. Listings of all quadruplex- (and i-motif) binding small molecules can be found in www.G4LDB.com, which up to August 2021 includes over 3100 compounds.

That it is possible to generate selective G-quadruplex-interactive molecules has been shown by many subsequent studies (for example, Asamitsu, et al., 2019; Islam et al., 2016; Savva & Georgiades, 2021). Optimising quadruplex affinity has been an equally important and associated goal. Whereas ethidium itself binds only weakly to quadruplexes, its derivatives with appropriate side-chain functionality have been shown to bind with high affinity, and to be potent telomerase inhibitors, as have, for example, several dibenzophenanthroline derivatives (Koeppel et al., 2001; Mergny et al., 2001), the natural product telomestatin (Kim et al., 2002) and trisubstituted acridine compounds. These latter molecules have been successfully designed to exploit the groove differences between quadruplexes and duplexes and are also effective telomerase inhibitors, exemplified by the trisubstituted acridine compound BRACO19 (Read et al., 2001). This and the pentacyclic acridinium salt compound RHPS4 (Salvati et al., 2007) have demonstrated in vivo antitumour activity that appears to be related to their telomere binding and quadruplex induction. BRACO19 and RHPS4 were optimised to more active analogues but toxicity issues have to date precluded any further development of them as antitumour drugs (Neidle, 2020).

Crystallographic and NMR structural information together with biophysical (and sometimes biological data) is available on several small molecule—telomeric quadruplex complexes (Table 5.4 and Fig. 5.34). These have overwhelmingly concluded that intercalative-type binding does not occur between stacked G-tetrads, but rather ligands are almost exclusively externally stacked at the ends of quadruplexes (Fedoroff et al., 1998; Gavathiotis et al., 2003; Read & Neidle, 2000), with both 1:1 and 2:1 complexes being characterised, depending on the nature of the ligand. The crystal and NMR structures show that the planar aromatic or heteroaromatic moiety is stacked on to one or both ends of the G-tetrad core, and crucially has π-π overlap with one or more of the guanines in the tetrad (Figs. 5.34c and 5.35c and e). Many of the ligands also have side-chain substituents, which tend to lie in the quadruplex grooves, often with a cationic charged atom close to or sometimes in contact with a backbone phosphate group (see, for example, Collie et al., 2012).

High-affinity quadruplex binding small-molecule compounds, all of which are believed to bind by an end-stacking mode, are in one of three broad chemotypes comprising:

(i) An extended planar or nearly planar extended heteroaromatic chromophore such as the porphyrin TMPyP4 (Parkinson et al., 2007), which can itself carry a formal positive charge, together with one or more pendant substituents (side chains), which also often contain a terminal cationic charge. The majority of quadruplex-binding compounds are of this type.

(ii) Two or more linked aromatic groupings, such that the complete moiety is still coplanar, albeit with some flexibility, as in the bis-quinolinium compound Phen-DC(3). These compounds may overlap with a greater G-quartet surface area than a conventional extended chromophore (Le et al., 2015; Ohnmacht et al., 2014).

(iii) Macrocyclic ring systems comprising simple five- and/or six-membered rings linked in such a way that the complete assembly is planar or nearly planar (Iida & Nagasawa, 2013; Nielsen & Ulven, 2010). The classic compound of this type is the natural product telomostatin, a macrocyclic polyoxazole with potent telomerase inhibitory activity.

The diversity of possible quadruplex-type arrangements for G-rich sequences provides a complex set of geometries and targets to be explored for drug action. The classic end-binding mode is found in the crystal structure (Haider et al., 2003) of a disubstituted amido acridine compound bound to a bimolecular quadruplex formed by two molecules of the sequence d(GGGGTTTTGGGG). The structure (Fig. 5.34a) shows a single-bound acridine molecule, at the end of four-stacked G-tetrads constituting the quadruplex core. The ligand is also held within the tetra-thymine loop, with direct in-plane hydrogen bonding between the acridine chromophore and a thymine base. The 3,6-bis-[3-pyrrolidino-propionamide] side chains project outwards and are too short to fully

Table 5.4 Crystal and NMR structures of some small-molecule complexes with DNA quadruplexes.

Ligand	Sequence	Quadruplex	Crystal or NMR	PDB id
Bis-amido-acridine	d(GGGGTTTTGGGG)	*Oxytricha* telomere	Crystal	1L1H
BRACO19	d(TAGGGTTAGGGT)	Hu telomere	Crystal	3CE5
TMPyP4	d(TAGGGTTAGGG)	Hu telomere	Crystal	2HRI
TMPyP4	d(TGAG$_3$TGGIGAG$_3$TG$_4$AAGG)	c-MYC promoter	NMR	2A5R
Berberine	d[TAG$_3$(TTAG$_3$)$_3$]	Hu telomere	Crystal	3R6R
N-methyl mesoporphyrin IX	d[(GGGTT)$_3$GGG]	Synthetic	Crystal	6PNK
BMVC	d(TGAG$_3$TG$_3$TAG$_3$TG$_3$TAA)	c-MYC promoter	NMR	6O21, 6JJ0
Epiberberine	d[(TTAG$_3$)$_4$TT]	Hu telomere	NMR	6CCW
Platinum tripod compound	d[AAA(GGGTTA)$_3$ GGGAA]	Hu telomere	NMR	5Z80
Triangulenium DAOTA-M2	d(TAG$_3$AG$_3$TAG$_3$AG$_3$)	c-MYC promoter	NMR	5LIG
Colchicine	d(G$_4$CG$_4$CG$_4$CG$_4$T)	RET promoter	NMR	6JWD
Naphthalene diimide compound MM41	d[AGGG(TTAGGG)$_3$	Hu telomere	Crystal	3UYH
Bis-quinolinium compound Phen–DC(3)	d(TGAG$_3$TG$_2$TGAG$_3$TG$_4$AAG$_2$)	c-MYC promoter	NMR	2MGN
Telomestatin derivative	d(TTG$_3$TTAG$_3$TTAG$_3$TTAG$_3$A)	Hu telomere	NMR	2MB3
Quindoline	d(TGAG$_3$TG$_3$TAG$_3$TG$_3$TAA)	c-MYC promoter	NMR	2L7V
DC-34 (a substituted 1-benzofuran-3-carboxamide)	d(TGAG$_3$TG$_3$TAG$_3$TG$_3$TAA)	c-MYC promoter	NMR	5W77

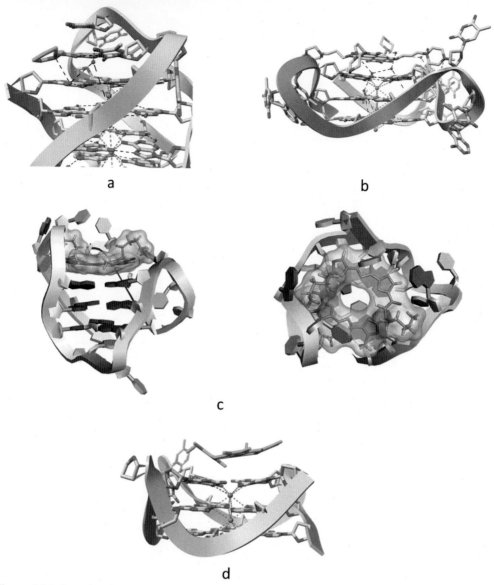

a

b

c

d

Figure 5.34 Crystal and NMR structures of quadruplex–ligand complexes involving telomeric and related sequences. (a) Crystal structure (Haider et al., 2003), PDB id 1L1H of a di-substituted aminoalkylamido acridine compound complexed with the bimolecular quadruplex formed by the sequence d(GGGGTTTTGGGG). The ligand is shown in stick representation and the small spheres show the potassium ions in the central channel of the quadruplex structure. (b) Crystal structure of a complex between a parallel topology human telomeric quadruplex and the tetra-substituted naphthalene diimide compound MM41 (Micco et al., 2013), PDB id 3UYH. Water molecules involved in ligand–quadruplex binding are shown. Two of the substituent groups have their cationic ring nitrogen atoms hydrogen bonding to water molecules, which in turn hydrogen bond to quadruplex phosphate groups. (c) Two views of one of the ensemble of NMR structure of an intramolecular (3 + 1) human telomeric hybrid G-quadruplex bound to a telomestatin derivative (Chung et al., 2013), PDB id 2MB3. The right-hand view, onto the G-tetrad planes, show the extensive overlap between the macrocycle and the terminal G-tetrad bases. (d) The crystal structure of a complex between N-methyl mesoporphyrin IX and a parallel quadruplex formed from the sequence d[(GGGTT)$_3$GGG] (Lin et al., 2020), PDB 6PNK, showing the non-planarity of the ligand and also that of the G-tetrad on which it is stacked, highlighting the ability of a G-tetrad to accommodate ligands that do not necessarily follow the general rule of heteroaromatic planarity.

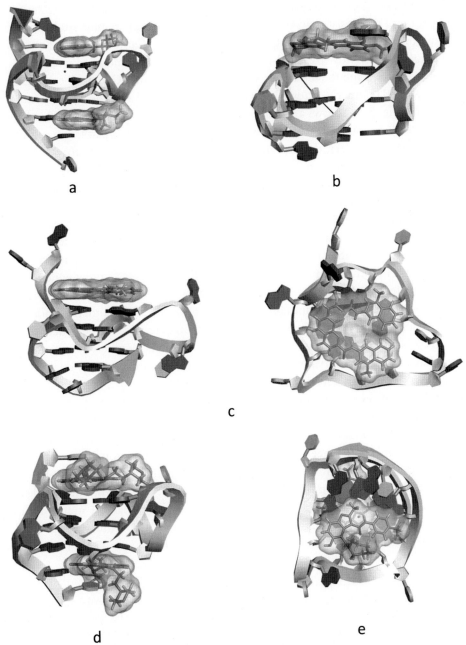

Figure 5.35 Crystal and NMR structures of quadruplex—ligand complexes involving promoter quadruplex sequences. (a) One of the ensemble of NMR structures of a quindoline complex with a c-*MYC* promoter quadruplex (Dai et al., 2011), PDB id 2L7V. Two quindoline molecules are bound, one at each end of the G-tetrad core, and are stacked between them and 5′ and 3′ flanking bases. (b) One of the ensemble of NMR structures of a complex between the *RET* promoter quadruplex and the alkaloid

fill the quadruple groove. Instead they are pinned in place by water molecules, which hydrogen bond to the cationic end-groups. A further crystallographic study, in total of seven disubstituted amido acridine quadruplex complexes (Campbell et al., 2009), found that a range of differently sized end-groups could all be accommodated within the large space of the groove and the tetraloop. The co-crystal structure (Campbell et al., 2008) of a human parallel telomeric quadruplex with the trisubstituted acridine experimental drug BRACO19 (Fig. 5.36) shows a more complex arrangement, with the two flexible pyrrolidine side chains situated in open regions of the grooves, such that larger substituents could also fit in. However, the slim anilino group is firmly held in a narrow cavity. As in the ether acridine complex crystal structures, hydrogen bonding involving water molecules play a role in cementing ligand to quadruplex phosphate groups and base edges. The acridines have cationic charge on the central ring nitrogen atom of the chromophore, which is normally found to be aligned with the potassium channel in the centre of a quadruplex. A typical arrangement for ligand binding to a parallel topology, in this instance a tetrasubstituted naphthalene-diimide compound, is shown in Fig. 5.34b, with the drug substituents in the grooves. Again, since this was a crystallographic study, water molecules have been located, mediating between side chain and DNA. The potent natural product telomestatin binds preferentially to non-parallel telomeric quadruplexes, and the structure of a complex involving a telomestatin derivative has been determined in solution by NMR methods (Fig. 5.34c). The individual five-membered rings of the macrocycle have extensive overlap with the tetrad bases, in accordance with the exceptional affinity for this telomeric quadruplex.

G-rich sequences in the promoter regions of several genes, especially those involved with cellular proliferation and cancer, are widely studied as potential quadruplex targets for therapy, in cancer and increasingly in other diseases where the function of a gene containing a quadruplex sequence can be implicated in disease initiation and/or progression. The principal quadruplex in the promoter region of the ubiquitous c-MYC oncogene is the most studied of these. An NMR structure (Phan et al., 2005) of this quadruplex with bound TMPyP4 shows the ligand stacked externally onto this parallel-stranded quadruplex, in a manner analogous to ligands bound to human telomeric quadruplexes (Fig. 5.34). This confirms and extends the previous NMR data for TMPyP4 complexed

colchicine (Wang et al., 2020), PDB id 6JWD. The quadruplex adopts a classic parallel topology. (c) Two views of one of the ensemble of NMR structures of a complex between the bis-quinolinium compound Phen-DC3 and a quadruplex from the c-MYC promoter sequence (Chung et al., 2014), PDB id 2MGN. The right-hand view is looking down onto the G-tetrad planes (i.e. down the central potassium channel). Two out of the three aromatic groups of the ligand have extensive overlap with the terminal G-quartet. (d) One of the ensemble of NMR structures of a complex between a c-MYC promoter quadruplex and the fluorescence probe triangulenium compound DAOTA-M2 (Kotar et al., 2016), PDB id 5LIG. (e) One of the ensemble of NMR structures of a complex between a c-MYC promoter quadruplex and the synthetic compound DC-34 (Calabrese et al., 2018), PDB id 5W77.

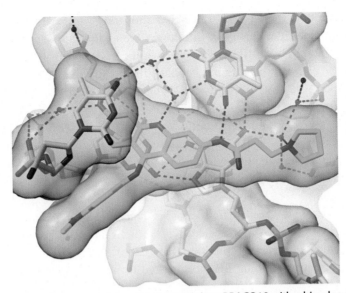

Figure 5.36 A co-crystal structure of the experimental drug BRACO19 with a bimolecular human telomeric quadruplex (Campbell et al., 2008), PDB id 3CE5. The drug is bound onto a terminal G-tetrad of the quadruplex, with a thymine base helping to sandwich the acridine moiety of the drug in position. Several water molecules are seen to mediate between drug and quadruplex, with hydrogen-bond interactions from drug to phosphate groups and nucleobase edges of the quadruplex.

with a G-quadruplex (Wheelhouse et al., 1998), and both have some features in common with the crystal structure of nickel TMPyP4 stacked on the ends of a d(CCTAGG) duplex (Bennett et al., 2000). The structures of several natural products and synthetic compounds bound to the c-*MYC* quadruplex have been determined by NMR in solution (Table 5.5 and Fig. 5.35). The NMR structure of the 2:1 quindoline complex (Fig. 5.35a) shows that this compound is held in place not only by stacking interactions with the G-tetrads but also by interactions with flanking bases, which serve to form a more defined binding site than the large surface area presented by the tetrads alone. On the other hand, the structure of the 1:1 c-*MYC* quadruplex complex with the *bis*-quinolinium compound Phen-DC3 (Fig. 5.35c) demonstrates that ligand size determines

Table 5.5 Crystal structures of some ligand complexes with DNA junctions.

Ligand	Sequence	Resolution in Å	PDB id	Junction type
Bis-acridine	d(TCGGTACCGA)	1.75	2GWA	Holliday
Metallo-helicate	d(CGTACG)	1.70	2ET0	Three-way
Triamino-triazine	d[(^{Br}U) TCTGCTGCTGAA]	1.55	6M4T	Head-to-head four-way

whether one or both tetrad ends can accommodate a ligand. Phen-DC3 is a large crescent-shaped molecule which can fit in only one end of this complex. The synthetic drug-like compound DC-34 (Fig. 5.35e) is a lead compound that has emerged from a systematic drug discovery program targeting this quadruplex. DC-34 downregulates *MYC* transcription in cancer cells by a G4-dependent mechanism. The compound is small enough to bind to both ends of the c-*MYC* quadruplex and the 2:1 NMR solution structure shows several specific ligand—DNA contacts for both bound molecules, which help to rationalise the quadruplex selectivity shown by DC-34. The pursuit of quadruplex selectivity is an important goal in this field, although the very large number of validated quadruplexes (10,000?) means that any conclusions on a given ligand are at best tentative. It is common to hypothesis test selectivity with a panel of quadruplexes representing the best-established one. The alkaloid colchicine has been found using an in silico library screening approach, to bind selectively to the *RET* promoter quadruplex (Wang et al., 2020), which has a parallel topology (Fig. 5.35b). The *RET* gene is over-expressed in several cancer types including ER-positive breast cancer and is therefore a significant cancer target.

The finding (Rangan et al., 2001) of an association between the C-rich strand in the c-*MYC* promoter and the quadruplex-binding TMPyP4 shows that the i-motif may also be of significance as a drug target. A model for this complex has been suggested from NMR data, with the ligand bound externally to an i-motif tetraplex. However, no detailed i-motif complexes with small molecules have yet been reported, although some i-motif selective compounds have been identified (Sedghi Masoud & Nagasawa, 2018).

5.5.3 Ligand binding to junction DNAs

The crystal structures of two DNA four-way Holliday junctions complexed with a derivative of the covalently binding intercalating drug psoralen (Eichman et al., 2001) are remarkable illustrations of the power of certain DNA sequences to form non-duplex structures. One structure, with the sequence d(CCGGTACCGG), is not unexpected, given that this sequence forms a junction structure in the absence of ligand — the complex is nearly identical to the native structure. By contrast, the related sequence d(CCGCTAGCGG) dimerises to a duplex in the absence of psoralen but is induced by the drug to form a related (though non-identical) junction structure. It has been suggested that this is a model to explain the promotion of recombination events by psoralen, leading to repair of the resulting lesions in DNA. These structures have raised the intriguing question of whether other, non-covalently binding drugs can similarly promote and stabilise four-way junctions, since the central junction regions in them can have considerable free space.

An answer to this question has come from the crystal structure of a Holliday junction with a bis-acridine molecule (Fig. 5.21), which is unexpectedly isomorphous with native

Holliday junction structures (Brogden et al., 2007). The drug was observed in clear electron density bound at the central cross-over region (Fig. 5.37); the binding site was created by the movement of two symmetry-related adenine bases out of their stacked positions, creating a cavity for the drug. The challenge posed by this structure is whether selective molecules for junction structures can be devised from it, and if so, whether such ligands can have useful biological or therapeutic functions.

Holliday junctions are frequently used as the defining link motif in artificial supramolecular DNA assemblies (Seeman, 2005), although ligand-directed assembly has not been seriously explored. The unexpected finding that four-way Holliday junctions can be formed by far shorter sequences than was hitherto envisaged has led not only to studies of the ligand complexes outlined above, but also to the realisation that under some circumstances, ligand shape itself can control what type of DNA arrangement is formed. A graphic illustration of this is the formation of a three-way DNA junction by an iron-containing molecule consisting of three bis-pyridylimine groups coordinated to two Fe^{2+} ions. The crystal structure (Oleksi et al., 2006) of this complex bound to the simple hexanucleotide sequence d(CGTACG) shows that a three-way DNA junction has formed around it. Each arm contains two G•C and one A•T base pair. Each of these forms extensive π-π stacks with the phenyl rings of the ligand. The features of the junction, such as

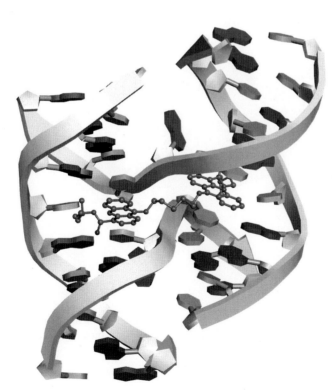

Figure 5.37 A view of the crystal structure (Brogden et al., 2007), PDB id 2GWA, of a complex between a bis-acridine molecule and a Holliday junction. Carbon atoms of the bis-acridine (shown in ball and stick representation) are coloured orange. Each of the acridine chromophores has displaced an adenine base, creating a cavity for itself.

B-DNA arms inclined at an angle of 110°, are closely similar as those observed in the crystal structure of the Cre recombinase complex (Woods et al., 2001).

The expansion of d(CTG) trinucleotide repeats in the 3′-untranslated region of the myotonic dystrophy protein kinase gene is a direct cause of myotonic dystrophy type 1 neurodegenerative disease. The acridine conjugate with the triaminotriazine moiety is a potent experimental agent in this disease that binds strongly to d(CTG) repeat DNA, with selectivity for the T•T mismatches produced by this repeat. A co-crystal structure with a duplex containing three consecutive d(CTG) repeats (Chien et al., 2020) has revealed that the resulting structure comprises a novel four-way junction with a double U-shaped conformation (Fig. 5.38). The two helical parts are at right angles to each other, and each ligand molecule is double-intercalated such that the triaminotriazine is paired with one thymine from a T•T pair, and the other thymine is swung out of the helix.

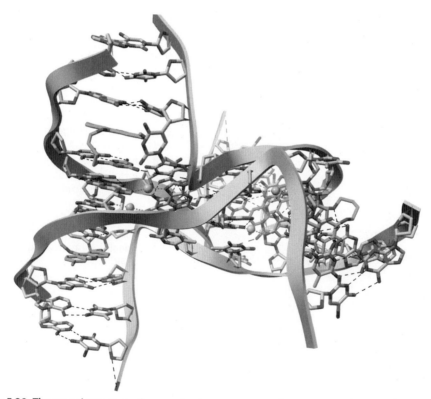

Figure 5.38 The crystal structure of a complex between an acridine-triamminotriazine and a d(CTG)-containing oligonucleotide (Chien et al., 2020), PDB id 6M4T. The bound ligand molecules are coloured magenta. The hydrogen bonding between a thymine base and a triazine group is visible for the ligand at the upper left-hand side.

5.6 Groove-binding molecules

A large and chemically diverse family of compounds can be classified as duplex DNA groove-binders. Many have biological activity, and several of them find medicinal use as anti-parasitic or anti-viral agents. They generally show a preference for binding to A/T regions of DNA (Dervan, 1986; Suckling, 2012; Zimmer & Wähnert, 1986). By contrast with intercalating drugs, they do not significantly perturb DNA structure. They bind exclusively in the minor groove of B-DNA duplexes. They can function as simple blockers of transcription, or as inhibitors of DNA topoisomerase enzymes. This alone is probably insufficient to explain why some minor-groove binders work as drugs. It has been suggested that for those drugs with high therapeutic indices in diseases where an external organism is the causative agent (e.g. in microbial infections), preferential drug binding would occur in extended regions of A/T sequence in the genetic apparatus of the organism. Such regions have been found in the mitochondrial DNA of these organisms and the drugs may therefore be selectively inhibiting their electron transport functions. Groove-binding ligands have also attracted much interest as starting points for the design of sequence-specific molecules capable of the recognition of unique sequences within a genome (see Section 1 in this chapter). This would have general applicability to a wide range of human diseases, although no molecules of this class have as yet progressed to clinical trials.

A notable exception to the A/T sequence selectivity of classic minor groove drugs is illustrated by the anticancer antibiotic chromomycin, comprising five sugar rings linked to a tricyclic, mostly aromatic chromophore; unsurprisingly early biophysical data on DNA binding was interpreted as showing intercalative binding. The drug has high selectivity for a binding site containing at least three G•C base pairs, which is not readily explicable by such a model. The structural basis for the sequence selectivity has been revealed by several structural studies, notably a co-crystal structure (Hou et al., 2004) of a complex with the sequence d(TTGGCCAA). This has confirmed and extended the earlier NMR studies (Gao et al., 1992), and a subsequent analysis (Gochin, 2000) of a complex with the same DNA sequence as in the crystal structure. The crystal structure (Fig. 5.39) shows that chromomycin binds to the widened G/C minor groove of the B-type duplex as a dimer. A magnesium ion plays a crucial role by interacting with the two chromomycin chromophores; the two magnesium-coordinated water molecules also hydrogen bond with various cytosine oxygen atoms in the minor groove. The G/C specificity of chromomycin is a consequence of several hydrogen bonds between oxygen atoms at the chromophore edge and N2 substituent atoms of guanine bases at the floor of the minor groove. The DNA, especially the recognition site itself, is appreciably distorted from B-form ideality, with a pronounced kink, and ca 10° helical unwinding. A crystal structure of chromomycin bound to the related sequence d(TTGGCGAA), which has two G•G mis-matches, has revealed that the combination of drug and mis-match has distorted the DNA into a Z-like right-handed helix (Satange et al., 2019).

Figure 5.39 Crystal structure (Hou et al., 2004), PDB id 1VAQ, of a complex between chromomycin and an octanucleotide duplex. The drug is bound in the minor groove and shown coloured mauve. The magnesium ion that coordinates to two chromomycin molecules is shown at the centre of the structure. Hydrogen bonds between drug and DNA are shown, including those between drug chromophore and guanine N2 atoms.

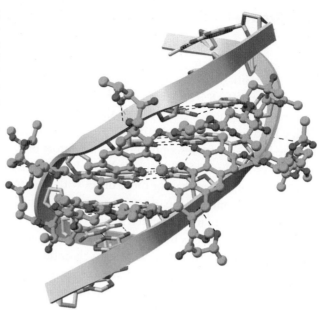

5.6.1 Simple groove-binding molecules

Molecules which have been shown to bind preferential to A/T-rich duplex DNA, but not to duplex RNA or A-form DNA, include synthetic DNA stains typified by Hoechst 33258 and DAPI, anti-trypanocidal agents such as berenil and the drug pentamidine (Fig. 5.40). Biophysical and footprinting studies have demonstrated that these molecules bind with approximately the same affinity to DNA as intercalators (with typical binding affinities of 10^6 mol^{-1}), but do not perturb DNA structure.

The common structural characteristics shared by most "simple" minor groove-binding molecules are:
- positive charge(s)
- linked rather than fused aromatic and/or heteroaromatic rings
- an approximately crescent shape in two dimensions

Many X-ray crystallographic and NMR studies have shown that these molecules bind into the minor groove of B-type DNA duplexes, and that these common features play an important role in the interactions (see, for example, Brown et al., 1992; Kopka et al., 1985; Lane et al., 1991; Patel & Shapiro, 1986).

Most of these structures involve complexes with duplexes formed by dodecanucleotides, most often with the Dickerson—Drew sequence d(CGCGAATTCGCG) and closely related ones such as d(CGCAAATTTGCG). Typically, the ligand is bound in

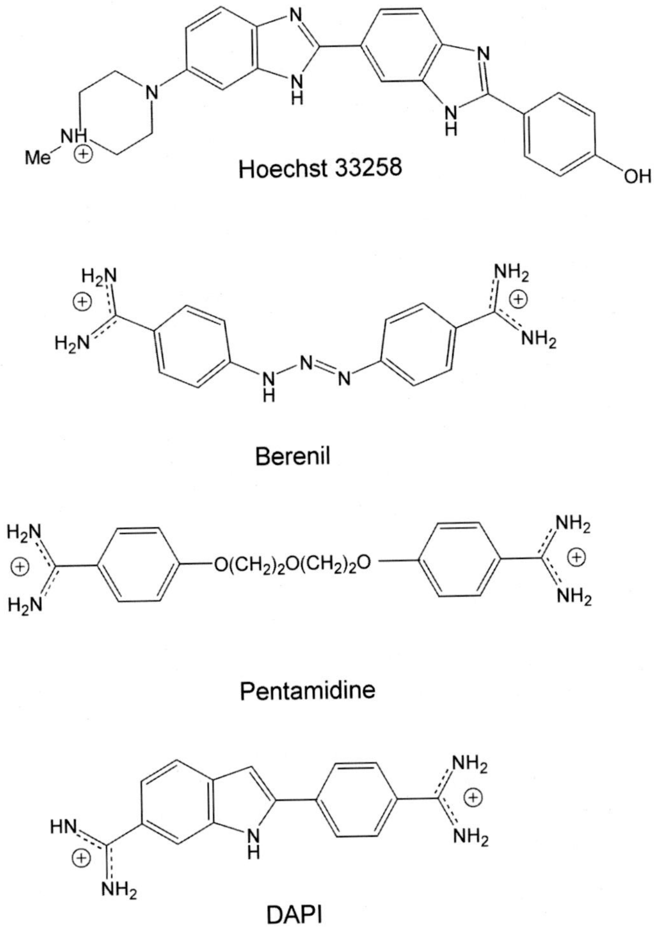

Hoechst 33258

Berenil

Pentamidine

DAPI

Figure 5.40 Some typical groove-binding molecules.

the A/T region, with the aromatic groups of the ligand lying between and parallel to the two sugar-phosphate backbones (Fig. 5.41). There is little change to the local structure of the DNA upon minor groove binding, at least for these ligands. The minor groove tends to be exceptionally narrow in this region, and atoms forming the backbone and groove floor are in close van der Waals contact with the ligand. Much of this contact is hydrophobic in nature, and predominantly involve non-polar atoms C1'/H1', C4'/H4' and C5'/H5' of the backbone, as seen in Fig. 5.2. NMR studies have shown that the chemical shifts induced in these protons on binding can be taken as diagnostic for groove binding (see, for example, Patel & Shapiro, 1986).

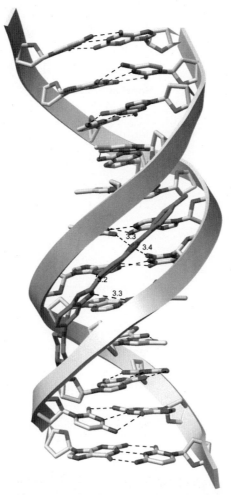

Figure 5.41 A view of the crystal structure (Spink et al., 1994), PDB id 296D, of the complex between Hoechst 33258 and the duplex formed by d(CGCAAATTTGCG), with the atoms of the drug molecule shown shaded mauve. Hydrogen-bond distances between ligand and DNA are shown.

Hydrogen bonding between ligand and DNA is frequently observed, notably between a donor atom on a ligand hetero ring or one in a charged terminal amidinium group (Fig. 5.42). Such hydrogen bonding is to a thymine O2 or an adenine N3 atom, on the floor of the groove. This is termed direct sequence readout and is highly directional. Conversely, sequence readout can also be indirect, with the ligand being able to recognise particular features of backbone geometry, or other sequence-dependent structural features. For A/T sequences indirect readout is a consequence of

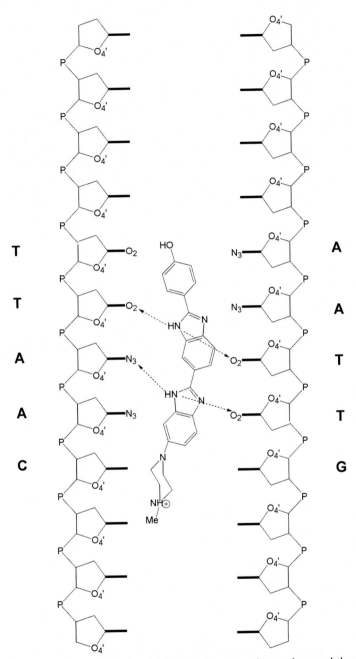

Figure 5.42 Schematic view of the hydrogen bonding between base edges and the benzimidazole groups, in the crystal structures of the Hoechst 33258 molecule complexed with duplex sequences containing the sequence .... AATT ....

(i) the narrow cross-section of groove-binding molecules complementing groove width and (ii) the negative electrostatic potential in the AT minor groove complementing the positive charge(s) on these molecules. Groove interaction also involves at least part of the inner surface of the bound molecule in close contact with the convex surface of the floor of the DNA minor groove itself. Thus, a G/C sequence with an exocyclic NH_2 guanine amino group protruding into the groove will hinder the effective binding of a molecule with a smooth concave inner surface. The observed matching of ligand curvature to that of an A/T minor groove surface for many minor groove-binding molecules has been termed "iso-helicity" (Goodsell & Dickerson, 1986), and involves the shape of the drug inner surface complementing the chiral twist in the curvature of the floor of the minor groove that are a consequence of the right-handed helical nature of the DNA double helix. Iso-helicity (Fig. 5.43) has been a useful concept in the design of many novel

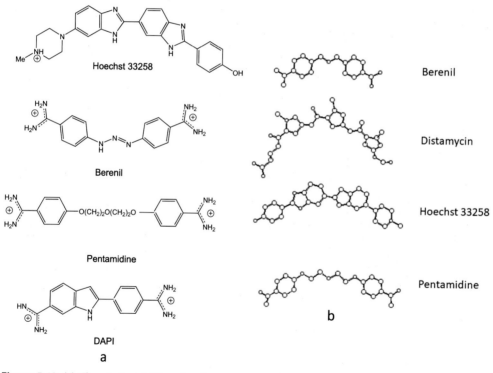

Figure 5.43 (a) Chemical and (b) molecular structures of four minor groove drugs, the latter taken from the crystal structures of their oligonucleotide complexes (see Table 5.6). In each case their concave surface is apparent; the degree of concavity differing from one drug to the other, with distamycin having the greatest curvature.

groove-binding agents, although more recent studies (discussed below) have shown that it is by no means universal.

The relative importance of the various factors contributing to sequence-selective minor groove binding is not always apparent. Some ligands have been shown to have minimal or even no hydrogen bonding to the edges of base pairs, and electrostatic factors are clearly of lesser importance in those instances where the ligands are uncharged. van der Waals and hydrophobic interactions with groove walls are probably the dominant factors in overall stability when ligand is bound in the groove and is driven in part by groove width structure and flexibility. Directed hydrogen bonding, though a smaller contributor to overall ligand stabilisation in its binding site, is responsible for ensuring binding to defined sites and sequences within an overall sequence type. Conversely, ligands that have a general A/T selectivity and do not discriminate between different types of A/T sequence are not involved in significant hydrogen bonding to bases. Hydrogen bonding is a crucial factor that can be exploited in the design of molecules with altered sequence recognition capabilities — see Section 5.6.3.

Another factor to be considered in ligand binding at A/T sequences is the displacement of the highly structured arrangement of water molecules, especially the "spine of hydration" (Section 5.2). The energetic driving force for this displacement is the dominance of the non-polar interactions between ligand and groove walls. Variations in groove width are probably responsible for the greater affinity of molecules such as berenil, netropsin, Hoechst 33258, for sequences containing 5′-AATT compared to 5′-TTAA or 5′-TATA (Abu-daya et al., 1995; Laughton & Luisi, 1999). The chiral nature of the spine of hydration at ambient temperatures (McDermott et al., 2017) also accords with the feature of many groove-binding ligands in having a twist along their length, imparting chirality and three-dimensionality. A conserved cluster of 11 groove-bound water molecules has been found in crystal structures of three minor groove—ligand complexes (Wei et al., 2013), at the boundary of A/T and G/C regions, which may help to maintain the bound ligands in position.

A well-studied sub-class of groove binders comprises molecules with two charged amidinium groups, one at each end. Several compounds of this type have activity against diseases prevalent in tropical countries, notably sleeping sickness, Chagas disease and leishmaniasis (Soeiro et al., 2013). Berenil is typical, with a triazine group linking two phenyl amidinium moieties. Crystallographic and NMR analyses have shown that the molecule covers three to four A•T base pairs (Brown et al., 1992; Lane et al., 1991). The former also find that each amidinium group has an N—H bond facing into the groove, which hydrogen bonds with thymine O2 or adenine N3 base-edge atoms. In the complex with the Dickerson—Drew sequence, a water molecule mediates this interaction at one end of the binding site. Several other bis-amidinium drugs, notably those with a flexible linker such as the drug pentamidine, bind in a similar manner. Pentamidine is of considerable clinical importance, since it has activity against the *Pneumocystis*

carinii pathogen, which is responsible for the opportunistic and life-threatening pneu-
monia that affects many AIDS patients. The drug probably works by selectively inhibiting the topoisomerases of the pathogen, via binding to its A/T–rich genome at points of
topoisomerase selectivity. Although pentamidine is one of the drugs of choice in this disease, its effectiveness is limited, and it produces some toxic side effects. Accordingly, there
have been numerous studies directed to finding more selective and effective analogues,
for which there are good correlations between DNA-binding affinity and biological
efficacy. Dicationic diarylfuran molecules (which are almost isostructural with berenil)
have shown some promise in this regard, and the cyclohexyl derivative (Fig. 5.44) is
100-fold more active than pentamidine itself. A crystallographic study of an oligonucleotide complex with this compound has provided a rationale for its improved DNA-
binding activity, and indirectly for its biological superiority over pentamidine (Boykin
et al., 1995). The bulky cyclohexyl groups fit snugly into the minor groove and make
extensive contacts with the non-polar atoms lining the groove walls, to a greater extent
than analogues with smaller attached groups.

One particular compound in the substituted diarylfuran series, DB289 (Pafurami-
dine), a prodrug of 2,5-bis(4-amidinophenyl)furan (furamidine), is highly active against
trypanosomes, and has been in phase II clinical trials (Wilson et al., 2005), although liver
and kidney toxicities have halted any further clinical development. It is believed to
exert its action by selectively binding to A/T-rich kinetoplast DNA in trypanosomes;
the crystal structure of furamidine bound to the Dickerson–Drew docamer (Laughton
et al., 1996) shows that it forms a conventional iso-helical complex. On the other hand,
a number of ligands have subsequently been discovered with high A/T affinity (and
often corresponding good biological activity against parasitic targets) yet which do
not obey the iso-helicity principle since they have a linear or near-linear shape. One
example is the almost linear meta-substituted diamidine molecule CGP40215A
(Fig. 5.45). The crystal structure of a complex with d(CGCGAATTCGCG)$_2$ shows
that this ligand binds in the minor groove A/T region (Nguyen et al., 2002, 2004),

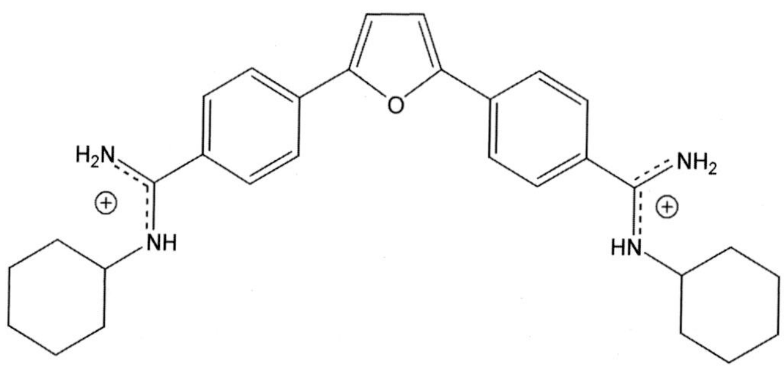

Figure 5.44 The structure of the cyclohexyl derivative of bis-[amidino-phenyl] furan (Boykin et al., 1995).

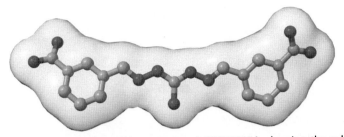

Figure 5.45 The molecular structure of the compound CGP40215A, showing the solvent-accessible surface of the molecule. Its near linearity is apparent when compared to iso-helical groove-binding molecules such as berenil (Fig. 5.43).

in accord with biophysical data showing that it also has high A/T affinity and selectivity. The two central NH groups of the ligand form strong hydrogen bonds to O2 atoms of thymine bases, but the ends of the ligand only make one direct DNA contact in total. This end also makes a short water-mediated contact with base edges while the other end of the ligand is disordered and makes no DNA contacts at all (Fig. 5.46). Complementary molecular dynamics calculations are in accord with conclusions from the crystal structure and indicate that the ligand binds in a see-saw-like manner, pivoted around the central set of ligand—DNA hydrogen bonds.

The linear ligand DB921 has a twisted biphenyl group linked in a linear way to a benzimidazole ring, resulting in a non-curved structure that does not fit the isohelicity concept, yet it also binds effectively to A/T DNA. The crystal structure of its DNA complex (Miao et al., 2005) shows that a water molecule again mediates between one end of the ligand and a DNA base edge (Fig. 5.47), thus effectively creating flexible and dynamic iso-helicity for this otherwise linear drug molecule. How these water molecules relate to the spine of hydration has yet to be fully explored, but it is apparent that they can play a more general role in A/T sequence-selective recognition (Bailly et al., 2003). It is plausible that those waters that are observed in various crystal structures represent the residue left after ligand displaces the bound water network spine.

The bis-benzimidazole compound Hoechst 33258 is widely used as a DNA and chromosomal stain. Footprinting studies have shown that its four to five base pair binding site, although necessarily mostly A/T-containing, does have a G/C pair at one end. Structural studies (Bostock-Smith et al., 1999; Gavathiostis et al., 2000; Spink et al., 1994; Vega et al., 1994) have provided an explanation; the two benzimidazole groups in the molecule form a network of bifurcated hydrogen bonds to three consecutive A•T base pairs (Figs. 5.41 and 5.42). However, the piperidine ring is non-planar and is too bulky to fit into the narrow A/T groove region. Instead, it forces the molecule into a site with a

Figure 5.46 The crystal structure of a complex of CGP40215A with d(CGCGAATTCGCG)$_2$ showing the ligand binding in the minor groove A/T region (Nguyen et al., 2002, 2004), PDB id 1M6F.

widened groove at the $3'$ end, which is best accommodated by a G•C base pair. The head-to-tail arrangement of the two benzimidazole units in Hoechst 33258 has been extended to three such units, as in the molecule shown in Fig. 5.48. This extended ligand (with the trivial name TRIBIZ) binds tightly to a site of 7.5 base pairs (Fig. 5.49). As in the parent Hoechst 22358, each benzimidazole group hydrogen bonds to two consecutive A•T base pairs by means of a pair of bifurcated hydrogen bonds. The TRIBIZ molecule is just about able to maintain all base pairs in hydrogen-bonding register, with some

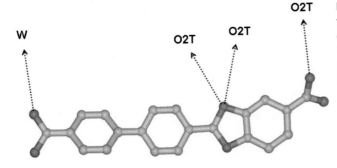

Figure 5.47 Schematic showing the hydrogen bonding to base edges from the linear ligand DB921, as found in the crystal structure of its DNA complex (Miao et al., 2005), PDB id 2B0K.

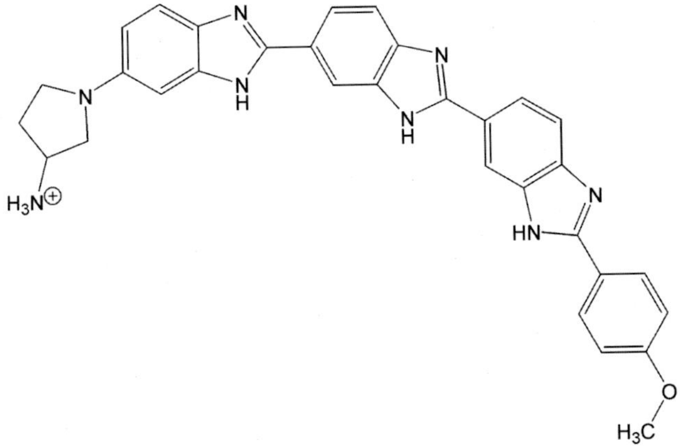

Figure 5.48 The structure of the groove-binding molecule TRIBIZ, containing three linked benzimidazole units (Clark et al., 1996).

changes in local DNA structure being needed to achieve this (Clark et al., 1996; Ji et al., 2001). Further benzimidazole units linked in the same manner would not maintain the correct phasing of hydrogen bonds to successive DNA base pairs. This is a general problem, of "keeping in register" with successive base pairs in a DNA sequence. It is especially significant when designing ligands to recognise upwards of a complete turn of double helix, since even small differences in DNA sequence structure and flexibility can be magnified over longer lengths of DNA. A series of symmetric head-to-head bis-benzimidazole compounds have been devised using structural information gained from the Hoechst—DNA crystal structures (Mann et al., 2001). Several compounds in this series have antibacterial activity against a range of pathogens and one compound (2,2′-Di(4-pyridinyl)-3H,3′H-5,5′-bisbenzimidazole: ridinilazole) has pronounced narrow-spectrum activity against the *Clostridioides difficile* pathogen (Mann et al., 2015). It is currently in phase III clinical trials.

Figure 5.49 A view of the crystal structure (Clark et al., 1996), PDB id 263D, of the complex between the TRIBIZ molecule and the duplex formed by d(CGCAAATTTGCG). The methoxyphenyl group of TRI-BIZ is at the lower end of the binding site in this view.

The consequences of berenil action in vivo, on the yeast genome, have been studied by a microarray method (Eckdahl et al., 2008), and compared with data on short oligo-nucleotides. The A/T preference of berenil was confirmed in this study, and interestingly the major gene expression changes were found in those genes (containing A/T upstream regions) most affected by drug treatment. Although the mechanism for berenil action was not determined in this study, interference with transcription factor binding or induction of chromatin changes are plausible explanations.

5.6.2 Netropsin and distamycin

These two compounds are the classic groove-binding naturally occurring antibiotics (Fig. 5.50) and have been extensively characterised, with their DNA interactions studied by a wide range of biophysical methods (Zimmer & Wähnert, 1986). Both comprise linked N-methyl pyrrole and amide units, with netropsin having two cationic charges compared to the one charge of the larger distamycin molecule. Crystal structures of both drugs complexed with A/T-containing oligonucleotides (Table 5.6) show that the narrow cross-section of both complements the narrow cross-section of the minor groove, again providing an explanation for their observed A/T preferences. Each amide group has its nitrogen atom oriented into the groove, and these participate in (often bifurcated) hydrogen bonding with donor atoms on the edges of consecutive adenine

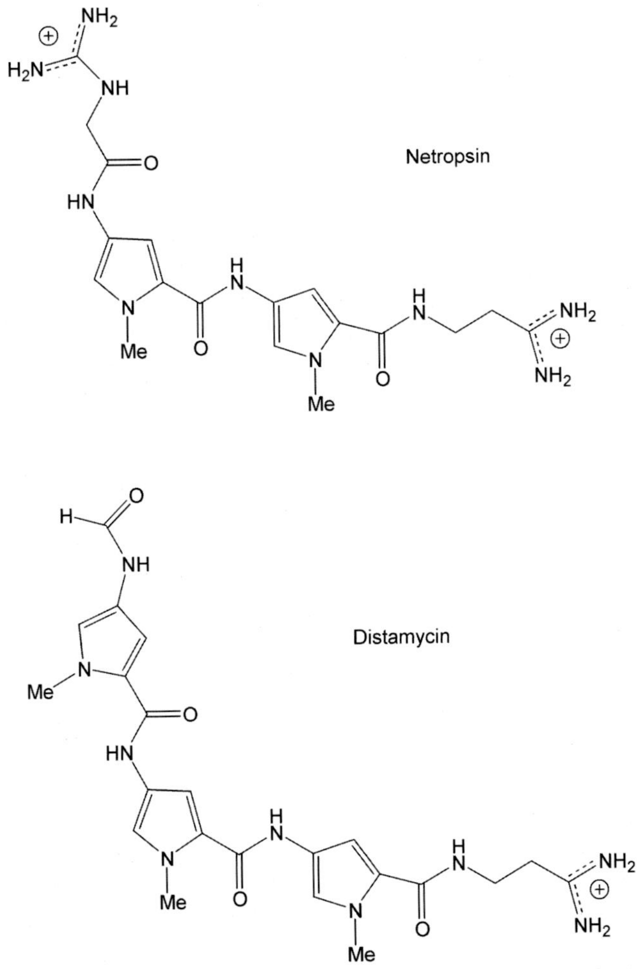

Figure 5.50 The structures of the oligopeptide-like groove-binding drugs netropsin and distamycin.

Table 5.6 Crystal structures of selected drug–oligonucleotide minor groove complexes.

Drug	Sequence	Resolution in Å	PDB id code
Distamycin	d(CGCAAATTTGCG)	2.20	2DND
Netropsin	d(CGCGAATTCGCG)	2.21	6BNA
Hoechst 33258	d(CGCGAATTCGCG)	2.00, 2.00	1D43, 1D44
Netropsin	d(CTTAATTCGAATTAAG)	1.85	1ZTT
Hoechst 33258	d(CGCAAATTTGCG)	2.44, 2.25	264D, 296D
Berenil	d(CGCGAATTCGCG)	2.50	2DBE
Pentamidine	d(CGCGAATTCGCG)	2.10	1D64
Furamidine	d(CGCGAATTCGCG)	2.20	227D
Cyclohexyl-furamidine	d(CGCGAATTCGCG)	1.90	1FMS
CGP40215A	d(CGCGAATTCGCG)	1.78	1M6F
DB921	d(CGCGAATTCGCG)	1.64	2B0K
Meta–OH–Hoechst	d(CGCGAATTCGCG)	2.20, 2.20	302D, 303D
TRIBIZ	d(CGCAAATTTGCG)	2.20	263D
DAPI	d(CGCGAATTCGCG)	2.40	1D30
4,4′-Bis(imidazolinyl amino) diphenylamine CD-27	Moloney murine leukaemia virus + d(CTTAATTCGAATTAAG)	1.75	3FSI
Mono-Im lexitropsin	d(CGCGAATTCGCG))	2.25, 2.25	1LEX, 1LEY
Di-imidazole lexitropsin	d(CATGGCCATG)	1.80	334D
2:1 Distamycin	d(ICICICIC)	1.80	159D
Polyamide Im-Im-Py-Py-β-Dp	d(CCAGGCCTGG)	2.00	365D
Im-Hp-Py-Py-β-Dp	d(CCAGTACTGG)	2.20	407D
Im-Py-Hp-Py	d(CCAGATCTGG)	2.27	1CVX
-[Py-Py-Im-Im]₂-	d(C^{Br}CAGGCCCTGG)	1.50	6M5B
Chromomycin	d(TTGGCCAA)	2.00	1VAQ
Chromomycin	d(TTGGCGAA)	2.50	6J0I
Mithramycin analogue MTM SA-Trp	d(AGAGGCCTCT)	2.00	5JVW

and/or thymine bases (Kopka et al., 1985; Coll et al., 1987). This has been termed Class I binding. Class II binding, which has only been rarely observed, involves disordered drug ends and a consequent reduction in sequence selectivity. Subsequent crystallographic studies on netropsin complexes (see, for example, Nunn et al., 1997; Van Hecke et al., 2005) have confirmed the Class I structural features, although there are sometimes small differences of detail depending on the precise A/T target sequence and on the nature of the flanking sequences.

A powerful approach to determining ligand–DNA structures has used the host–guest technique outlined in Chapter 4, in which the target oligonucleotide is effectively pinned between two host protein molecules, the N-terminal fragment of Moloney murine leukaemia virus reverse transcriptase (Goodwin et al., 2005). The crystal structure of such a complex involving netropsin and the oligonucleotide d(CTTAATTCGAAT-TAAG) illustrates the advantage of this approach, with two AATT drug-binding sites in the one sequence. Isomorphous structures with variants of the target sequence and changes in ligand type can be readily obtained; the method has also been applied to a diamidine derivative of benzimidazole (Goodwin et al., 2006; Tanious et al., 2007) and to a trypanocidal 4,4′-bis(imidazolinylamino)diphenylamine complex (Glass et al., 2009). This compound when unbound is highly twisted due to steric repulsions between the hydrogen atoms on the two phenyl rings. However, when bound to DNA in the minor groove, the compound is forced by the groove shape, to assume a near planar conformation (Fig. 5.51).

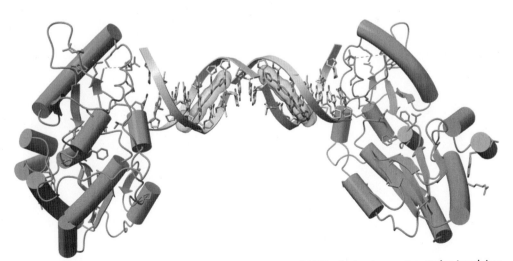

Figure 5.51 The crystal structure (Glass et al., 2009), PDB id 3FSI, of a host–guest complex involving the trypanocidal compound CD27 [4,4′-bis(imidazolinylamino)diphenylamine] (shown coloured mauve and with a semi-transparent solvent-accessible surface) bound in the minor groove of the B-DNA oligonucleotide duplex d(CTTAATTCGAATTAAG), showing the two AATT drug-binding sites in the one sequence. The protein host molecules at the end of the duplex are also shown, in cartoon form.

All of these groove binders show a consistent pattern of interaction with A/T stretches of duplex DNA, with two principal sets of interactions dominating the sequence selectivity: shape complementarity with the narrow groove walls, and either direct or water-mediated hydrogen bonding to base edges at the groove floor. Implied also is a negative factor, that the absence of guanine exocyclic $-NH_2$ substituents on the floor of the groove ensures that the shape and hydrogen-bonding complementarity are optimal at A/T sites. This concept has been used in the design of molecules with potential G/C selectivity (Fig. 5.52). Such molecules, termed "lexitropsins", are based on the proposal

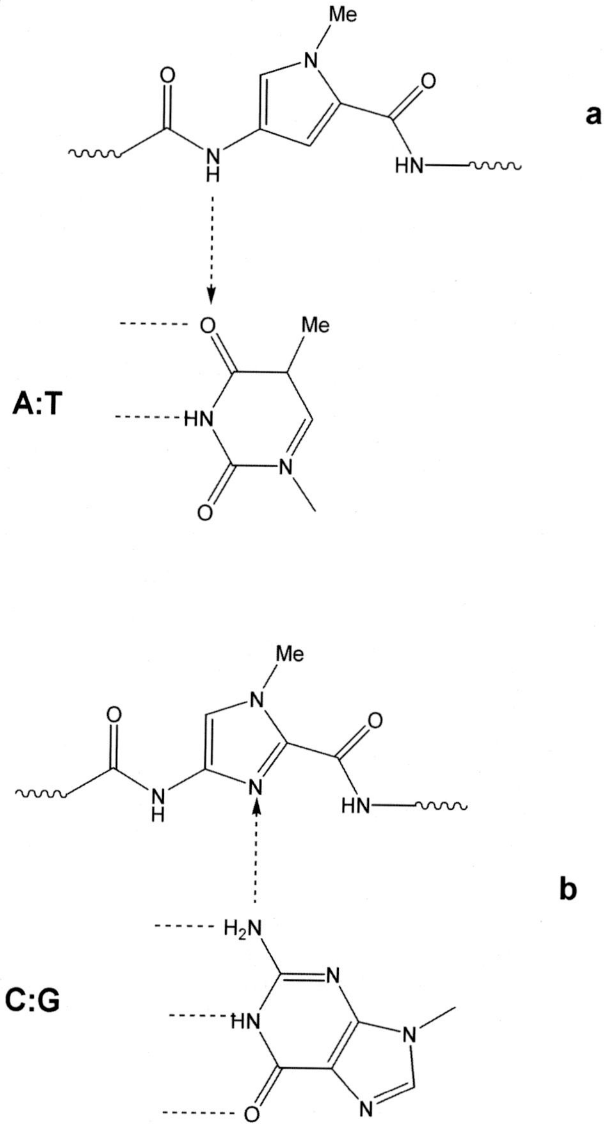

Figure 5.52 (a) The principles of netropsin and distamycin amide — base recognition, showing hydrogen bonding to the thymine of an A•T base pair. (b) The concept of lexitropsin base recognition, with hydrogen bonding from the exocyclic amino substituent of a guanine to the nitrogen atom of an imidazole ring.

(Lown, 1994) of switching hydrogen-bond polarity at the groove floor by means of, for example, an imidazole ring to hydrogen bond to a guanine $-NH_2$ substituent. In practice, lexitropsins have only rarely completely achieved the goal of a total change in sequence specificity. Instead, most bind to both A/T and mixed sequences (Kopka et al., 1997), and tend to show reduced affinities for both types of site. In retrospect this failure can be ascribed to the differences in both groove width and flexibility in G/C compared to A/T regions. The latter tend to be wider and a typical groove-binding molecule with a narrow cross-section will not bind well. The astute reader will also notice that the idealised hydrogen-bonding scheme for lexitropsins implies subtle differences in position for hydrogen-bond donors and acceptors, which are difficult to account for in the design process. In spite of these problems, the lexitropsin concept has proved its value when combined with dimeric groove binding, as discussed in Table 5.6.

An important observation, initially from NMR studies (Pelton & Wemmer, 1990), was that at higher drug:DNA ratios, distamycin can form a 2:1 complex with the duplex sequence d(CGCAAATTTGCG). This has been of major significance for the subsequent design of the polyamide class of sequence-selective ligand. Crystallographic analyses have been reported for a number of other 2:1 distamycin:complexes, with the drug bound to a range of sequences (for example, Chen et al., 1994). All show the drug bound in the minor groove as an anti-parallel head-to-tail manner (Figs. 5.53 and 5.54), in striking contrast to the pattern of simple 1:1 minor groove complexes described above, and to the structure of the 1:1 distamycin-d(CGCAAATTTGCG) complex. The positively charged ends of the two distamycin molecules are far apart in the dimer complexes, suggesting that this arrangement could occur with other groove-binding molecules possessing a single positive charge, such as Hoechst 33258, although no such observations have yet been reported. The NMR and crystallographic structures of the 2:1 complexes have both distamycin molecules involved in close non-bonded contacts with each other, and most importantly, each interacts with just one DNA strand (Fig. 5.55). In the 1:1 complex of distamycin (and those of netropsin and Hoechst 33258), hydrogen bonding to base edges involves sets of three-centre bifurcated hydrogen bonds between an amide group and any two of adenine N3 and thymine O2 atoms. For both distamycin molecules to be accommodated in the 2:1 complex, the width of the narrow minor groove region of the 5'-AAATTT tract must expand from 3.4 to 6.8 Å. Thus, a new picture emerges, so that rather than the minor groove having rigid geometry, it should be considered as being inherently flexible, especially in order to accommodate ligand binding (Bostock-Smith et al., 1999).

Figure 5.53 The crystal structure (Chen et al., 1994), PDB id 159D, of a 2:1 complex between distamycin and the duplex formed by d(ICICI-CIC). The two bound distamycin molecules are seen to form a side-by-side dimer.

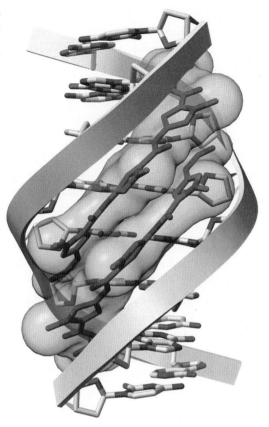

5.6.3 Sequence-specific polyamides

The 2:1 mode of distamycin binding is the basis for many subsequent studies which have made attainable the goal of sequence recognition at the gene level (Dervan & Bürli, 1999; Farkas et al., 2009; Hidaka & Sugiyama, 2020; Mrksich & Dervan, 1993). The key feature demonstrated by the 2:1 complexes is that minor groove recognition can involve simultaneous hydrogen bonding to both bases in a base pair. Thus far greater discrimination is inherently attainable than can be achieved with 1:1 minor groove binders, and more than might be envisaged by solely considering canonical B-DNA minor groove width, which in the absence of a single molecule that can widen it seems incapable of binding a ligand large enough to recognise both bases in a pair. A general approach for switching from A•T to G•C base pair recognition has been to adopt a lexitropsin-type of change, whereby one or more N–methyl pyrrole (Py) groups is changed to an imidazole (Im) one (Figs. 5.52b and 5.55).

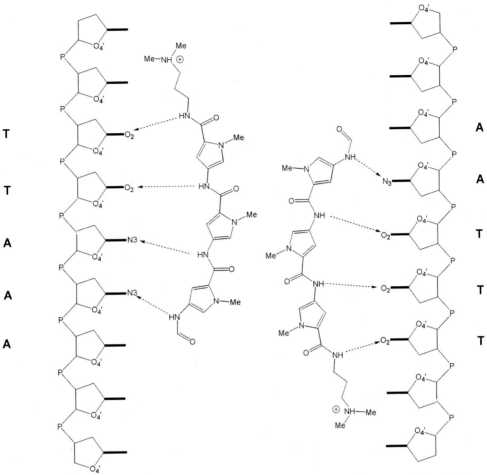

Figure 5.54 A schematic view of the detailed hydrogen bonding to both DNA strands in a 2:1 distamycin dimer, showing the hydrogen bonding to bases in both strands of a five base-pair A/T sequence.

A general recognition code has been established from these studies (Dervan, 2001),which enables a wide range of sequences to be recognised using strings of these Im and Py groups linked by amide groups (Table 5.7a).

The resulting molecules incorporating these rules in their features, termed polyamides, have been shown by footprinting analysis to be able to target, normally as dimers, a wide range of DNA sequences, each with remarkably high selectivity. Their typical specific base interactions are as shown in Fig. 5.55. Other small heterocyclic rings such as

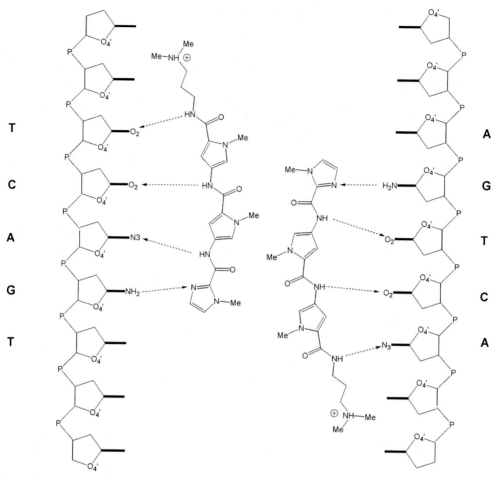

Figure 5.55 A schematic view of the hydrogen bonding to A•T and G•C base pair edges involving the polyamide Im-Py-Py and the target sequence d(… TCAGT …).

Table 5.7 (a) The Im/Py pairing codes for minor groove polyamide recognition.

Pair	C:G	G:C	T:A	A:T
Im/Py	−	+	−	−
Py/Im	+	−	−	−
Hp/Py	−	−	+	−
Py/Hp	−	−	−	+

(b) The benzimidazole pairing codes for minor groove polyamide recognition, involving hydroxybenzimidazole (Hz) and imidazopyridine (Ip).

Pair	C:G	G:C	T:A	A:T
Ip/Py	−	+	−	−
Py/Ip	+	−	−	−
Hz/Py	−	−	+	−
Py/Hz	−	−	−	+

pyrrole (Py) and 3-hydroxypyrrole (Hp) have also been successfully incorporated into polyamides to provide hydrogen bonding to defined bases, and the recognition code has been subsequently extended by the use of benzimidazole (Table 5.7b) as a versatile and drug-like recognition unit (Marques et al., 2004). Validation of these building blocks has been performed in large part using footprinting methods; however, these can only be applied to a small number of sequences at any one time. A high-throughput microarray method has been developed (Warren et al., 2005) that is able to examine all possible sequences available to eight base pairs of DNA, using self-complementary palindromes with a short hairpin sequence. Results from these are in excellent agreement with previous studies, emphasising the future potential of the approach for studying small-molecule DNA sequence and structure specificity.

Greater control of recognition has been achieved by covalently linking two polyamide molecules, which need not have the same sequence of recognition units (Fig. 5.56). This avoids the possibility of slippage between the two unconnected components of a dimer, with potential ambiguity in sequence readout. These hairpin polyamides can achieve exceptionally high selectivity and site affinities, with typical binding constants in the nM range. For example, the molecule Im-Py-Py-Py-γ-Py-Py-Py-Py-β-Dp (where γ is the hairpin linker γ-aminobutyric acid, β is β-alanine and Dp is dimethylaminopropylamide) binds to the sequence d(TGTTAT) with a dissociation constant of 1.1 nM. Several NMR and crystal structures (for example, Chenoweth & Dervan, 2009; de Clairac et al., 1997; Kielkopf et al., 1998a,b; 2000a; Zhang et al., 2003) have been determined for polyamide—oligonucleotide complexes, which have found the predicted patterns of polyamide-base recognition and thus have demonstrated the overall correctness of the approach (Fig. 5.57). Cyclisation of a polyamide, for example, as observed in the crystal structure (Abe et al., 2020), PDB id 6M5B, of the cyclic polyamide -[Py-Py-Im-Im]$_2$- bound to the target sequence d(C^{Br}CAGGCCCTGG), similarly shows high sequence specificity.

An extensive series of studies have shown that polyamides can target DNA response elements which are sites for transcription factor binding, and successfully compete with these regulatory proteins (see, for example, Bremer et al., 2001; Burnett et al., 2006; Dickinson et al., 1999; Gearhart et al., 2005; Gottesefeld et al., 2001, 1997; Mapp et al., 2000; Viger & Dervan, 2006). For example, repression of the 5S RNA gene has been achieved by targeting an eight-ring polyamide to the binding site d(AGTACT) within the TFIIIA transcription factor binding region. This polyamide binds 30-fold more tightly than does TFIIIA to its complete 50 base-pair sequences, and effective inhibition of transcription both in vitro and in cells has been observed. Polyamides can also bind to specialised DNA sequences, as shown by the effects of a β-alanine-pyrrole-imidazole polyamide targeting the extensive GAA•TTC repeats found in the frataxin gene in patients suffering from the neurodegenerative disease Friedreich's ataxia. This polyamide appears to be able to lock the repeats in a B-like conformation, preventing alternative DNA structures from forming and thus aiding fratixin transcription to occur.

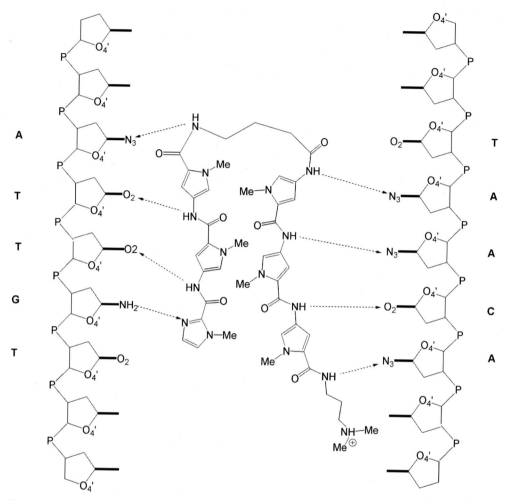

Figure 5.56 A schematic view of a hairpin polyamide, Im-Py-Py-linker-Py-Py-Py, recognising both strands of a G/C-containing sequence.

Polyamides can also be used to up-regulate transcription, by blocking repressor sites. A quite unexpected effect has been found (Henikoff & Vermaak, 2000; Janssen, Cuvier et al., 2000; Janssen, Durussel et al., 2000; Maeshima et al., 2001) with a polyamide targeted to the sequence d(GAGAAGAGAA) in the fruit fly *Drosophila*. This is a satellite DNA sequence and is normally considered to be functionally inert. It was found that the flies had highly specific loss of function in several defined genes, which are presumed to lie close to and thus can be affected by binding to satellite DNA. Overall, these

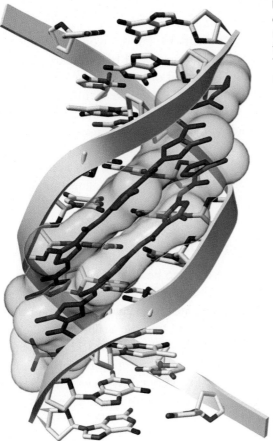

Figure 5.57 The crystal structure (Kielkopf et al., 1998b), PDB id 365D, of the polyamide molecule Im-Im-Py-β-Dp bound to a duplex formed by the sequence d(CCAGGCCTGG). The central part of this sequence, i.e. 5′-AGATCT, is recognised by the polyamide.

remarkable effects have been attributed to the experimental observations of drug-induced opening-up of satellite chromatin and a consequent long-range effect on specific genes. It is tempting to speculate that analogous events may be involved in some of the biological effects observed for the classic A/T-binding minor groove agents and their more recent analogues.

Are polyamides going to deliver the holy grail of single gene discrimination within a human genome in vivo? It is probably premature to be able to definitively answer this, since some major obstacles remain before any desired 16—18 base pairs can be specifically recognised. The strategy does not work equally well for all sequences, and beyond about a length of about seven to nine base pairs, the problem of maintaining polyamide subunits in register sometimes becomes severe. This may be due to differences in flexibility between DNA sequences. Most importantly for therapeutic purposes, polyamides are relatively poorly transported into some cell types compared to conventional minor groove drugs such as Hoechst 33258 or pentamidine. A desirable goal is to devise pharmacophore-like building blocks to mimic polyamides, but this work is still in its infancy. The repertoire of suitable molecules has been extended by the finding (Wang et al., 2000) that an asymmetric aromatic dication based on the diphenyl furan type of molecule (Boykin et al., 1995) can also bind in the minor groove as a dimer. Even so it is remarkable that well-validated results have been obtained with polyamides using in vivo models (for example, with a polyamide targeting the androgen response element in enzalutamide-resistant prostate cancer xenografts; Kurmis & Dervan, 2019; Kurmis et al., 2017), which demonstrate that once polyamides are introduced into a cell nucleus, then sequence-specific effects across a genome both in cells and in animal models are achievable (Finn et al., 2020).

5.7 Small-molecule covalent bonding to DNA

Historically the first DNA-interactive anticancer drugs to be developed were alkylating agents, derived from World War I poison gases. These drugs, the nitrogen mustards, are highly toxic molecules, although their more modern analogues chlorambucil, L-phenyl-alanine mustard (melphalan) and cyclophosphamide still find use against some cancers. Non-specific covalent binding can take place on the phosphodiester backbone or sugar residues. This is often a step in DNA strand scission, as in the case of the bleomycin family of anticancer drugs. Both single- and double-strand breaks can then occur. The latter are usually lethal events to a cell since they are especially difficult to repair. There has been particular emphasis on studies of binding to defined sites on DNA bases by drugs that undertake nucleophilic reactions. Purines are the most susceptible to covalent attack, with guanine being preferred over adenine. Particularly favoured sites are O6 (guanine), N6 (adenine) and N7 in the major groove and N1, N2 (guanine) and N3 in the minor groove. These site preferences are the result of the differing electronic charge distributions in bases, base pairs and in runs of sequence. The propensity for alkylation by nitrogen mustards at N7 of a guanine base embedded within guanine-rich sequences (Hartley et al., 1992; More et al., 2019; Singh et al., 2018) is an inherent property of these sequences, since the same pattern of sequence selectivity has been found both in vitro and in DNA extracted from mustard-treated cells.

Drugs can bind to a single site or to two at once if they have bifunctional capability. This latter cross-linking mode can be either intra- or inter-strand, depending on two principal factors:
- the distance between the two functional groups on the drug
- the nature of the affected DNA sequence — for example, whether two adjacent guanine are on the same or opposite strands

Table 5.8 Oligonucleotide crystal structures with covalently bound drug molecules. In each case the asterix indicates the sites of covalent binding along the DNA sequence.

Sequence	Drug	Resolution in Å	PDB ID code
d(CCTCTG*GTCTCC) + d(GGAGACCAGAGG)	cis-platinum	2.60 1.77	1AIO 3LPV
d(CCTCG*CTCTC) + d(GAGAG*CGAGG)	cis-platinum	1.63	1A2E
d(CCTCTG*GTCTCC) + d(GGAGACCAGAGG)	Oxaliplatin	2.40	1IHH
d(CCTCTG*GTCTCC) + d(GGAGACCACAGG)	Pt(ammine)-cyclohexylamine	2.40	1LU5
d(CCAACGTTGG)	Anthramycin	2.30	274D
Nucleosome with d(GTG) cross-link	Cis-platinum	3.22	3O62
Nucleosome	Oxaliplatin	2.60	3REK
Nucleosome	cis-platinum	3.45	3B6F
Nucleosome	Oxaliplatin	3.45	3B6G

Information on the structures of covalent adducts has to date been obtained from, on the one hand, chemical and biophysical probe experiments, and on the other, from NMR and some X-ray crystallographic studies (Table 5.8). Crystallisation of covalent adducts has proved remarkably difficult, and significant success has mostly been achieved with platinum complexes. In accordance with the theme of this book, we shall focus here on some representative drugs where firm structural information is available, and which are of particular interest at the present time. More extensive accounts of these and the large number of other drugs in this category are available in the reading list at the end of the chapter.

5.7.1 The platinum drugs

The chance discovery and subsequent clinical exploitation of the cytotoxic and antitumour properties of the strikingly simple molecule *cis*-dichlorodiamminoplatinum (II) (cisplatin or *cis*-DDP) is one of the great success stories of cancer chemotherapy. It is a highly toxic drug, yet it and its closely related analogues are the probably most effective single agents still in current use in the cancer clinic, where it is curative in over 90% of testicular cancers, a disease previously hard to treat, and one that had a poor prognosis (Kelland, 2007).

Cisplatin mostly binds to purines, especially guanine at the N7 position. A variety of adducts are formed with DNA in solution, with the major species having intrastrand cross-links (Fig. 5.58) to the sequence d(GG), and to a lesser extent to d(AG). Several more minor species have been characterised, but they are probably of little functional significance. NMR and crystal structures have been determined for oligonucleotides containing cisplatin bound at a variety of d(G) sites. The structure of a platinum-containing

Figure 5.58 A schematic view of the covalent intrastrand cross-linking interactions of the cis-platinum drug with the N7 atoms of two consecutive guanine bases.

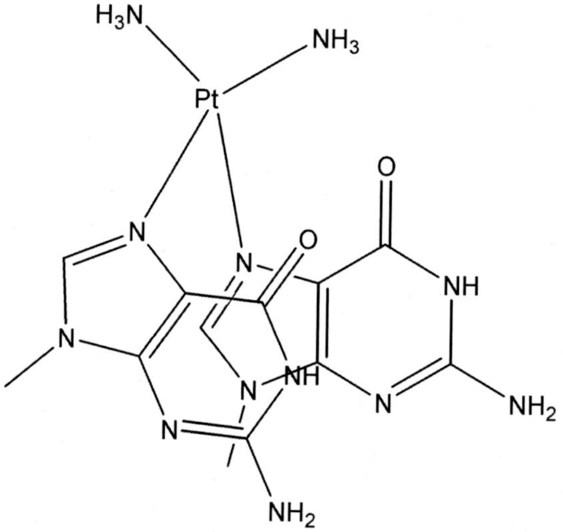

adduct d(CCTCTG*GTCTCC)•d(GGAGACCAGAGG), where G*G represents the intra-strand adduct, has been solved by both techniques (Gelasco & Lippard, 1998; Takahara et al., 1996). The crystal structure has been subsequently re-determined, at significantly higher resolution, 1.77 Å (Todd & Lippard, 2010a). This has revealed some differences in the geometry around the bound platinum atom compared to the earlier analysis (Fig. 5.59), although the DNAs itself in the earlier and more recent structures are closely similar, with a root–mean–square deviation (RMSD) over all atoms of 0.472 Å. Taken together with the NMR study these provide an interesting comparison of results from solution vs the crystal. The length of the platinum–N7 bond, of 2.02 Å, forces the two guanines out of their normal coplanarity, and the conformations of the two adjacent guanosines become highly distorted from native B-DNA. The platinum atom itself has near–normal square-planar geometry for its immediate substituents, whereas in the earlier X-ray study the coordination was ill-defined and distorted. The distortions in the DNA compared to canonical B-DNA are not confined to platinum-bound region of the sequence. The duplex becomes significantly bent, with a large roll towards the major groove at the cisplatin-binding site. The NMR study finds a roll of 49° whereas that from the X-ray analysis (Fig. 5.59a) is less, of 26°. Consequently, the major groove in both structures becomes more compact and the minor groove is wider and shallower than in canonical B-DNA, changing by up to 6 Å in both width and depth. The DNA in the crystal structure has some A-form character on the 5′ side of the bound platinum and is more B-form on the 3′-side. All these features are more exaggerated in the NMR structure, possibly because of the lack of crystal packing constraints.

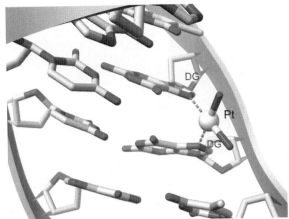

Figure 5.59 Detailed views of the crystal structure of the covalent complex with cis-diamminoplatinum and the sequences d(CCTCTG*GTCTCC) + d(GGAGACCA-GAGG), where G*G represents the intra-strand cross-linked adduct. Note the marked bend of the two halves of the duplex, from the 1.77 Å re-determination (Todd & Lippard, 2010a), PDB id 3LPV.

These structural distortions in DNA are analogous to those produced in DNA by the HMG (high-mobility group) chromosomal proteins such as SRY and LEF-1, which are known to bind to the minor groove of DNA. The crystal structure has been determined of a ternary complex between a 16-base-pair duplex containing a single platinum intra-strand cross-link, and an 89-amino acid domain of the HMG1 protein (Ohndorf et al., 1999). It provides a detailed picture of the interplay between the platinum-induced and protein-induced structural changes in DNA, and how platinum lesions in DNA might be recognised by repair proteins. As in the binary platinum adduct, the DNA is sharply bent towards the major groove, but this time by 61°, and the minor groove in this region resembles that in canonical A-form DNA. The protein-binding site is wholly on the 3′ side of the bound platinum, and the bend itself, whereas the protein–DNA complex alone has protein bound in the centre of the bend. There is more than a passing resemblance between the overall features of the structure and those of the TATA-box-DNA complex. The crystal structures of several nucleosome–platinum drug complexes are now available (Todd & Lippard, 2010b; Wu et al., 2008; Wu et al., 2011). The structure with a d(GTG) cross-link (Todd & Lippard, 2010b) in nucleosome-bound DNA (Fig. 5.60) shows cis-platinum bound to a single site, with one of the two bound guanines forced to be out of the DNA base-pair plane, whereas the other guanine retains its co-planarity, a feature similar to that determined by solution NMR studies on a short DNA sequence containing this adduct (Teuben et al., 1999). The site is inner facing and has little or no structural effect on overall nucleosomal DNA structure. The bending caused by cis-platinum binding is towards the major groove, in accordance with the bending requirements in nucleosomal DNA.

In striking contrast, the solution structure of DNA containing the less abundant G:G interstrand cross-link has been reported to show local unwinding at the lesion such that there is a reversal at this point to left-handedness and an extra-helical arrangement for the

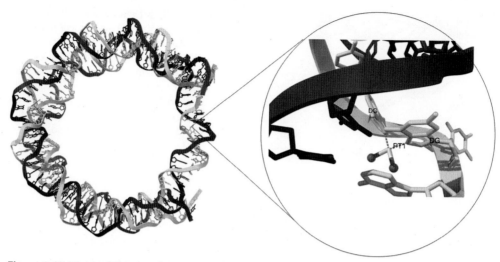

Figure 5.60 Views of the crystal structure of cis-platinum bound to a single d(GTG) site in a nucleo-some (Todd & Lippard, 2010b), PDB id 3O62. The histone proteins have been removed from the nucle-osome core in these views, to enhance clarity.

cytosines paired to the cross-linked guanines (Huang et al., 1995). These structural changes place the platinum in the minor groove. The crystal structure (Coste et al., 1999; PDB id 1A2E) of the same inter-strand cross-link, but now incorporated in the decamer duplex d(CCTCG*CTCTC)•d(GAGAG*CGAGG), also shows these features as well as equivalent minor groove widths, although other aspects such as the degree of unwinding at the cross-link and the degree of bending are strikingly different (Fig. 5.61). This suggests that sequence context may be an important factor in defining the geometry of a cross-link. However, these structural differences may also reflect the real differences between moderate resolution NMR and X-ray methods in their ability in being able to define long-range structural features with precision. In general, as discussed earlier, NMR methods are not as powerful in defining features such as unwinding and bending when the number of NOE distances is not large.

An even greater degree of distortion has been observed in a 1,3 G:G intrastrand cross-linked solution structure, i.e. with d(GTG) (Teuben et al., 1999), with a loss of base pairing around the lesion. Again, a pyrimidine base (here a thymine) is extruded from the helix. These consistent observations of extra-helical bases in platinum adducts suggest that they may be a general phenomenon for certain types of platinum lesion. They may act as signals for subsequent DNA repair proteins, suggesting that the minor adducts are relatively unimportant for the antitumour activity of cisplatin compared to the 1,2-intrastrand cross-link lesion, which is less readily repaired and therefore more persistent.

Figure 5.61 View of the crystal structure of an inter-strand cross-linked adduct of cis-diamminoplatinum (Coste et al., 1999), PDB id 1A2E.

Continuing attempts to improve potency, reduce toxicity and resistance and increase the range of treatable tumours has led to the development of numerous platinum (II) and (IV) derivatives, although only two of them, carboplatin (cis-diammine-1,1-cyclobutanedicarboxylatoplatinum) and oxaliplatin ((1R,2R-diamino-cyclohexane) oxa-latoplatinum (II)), are currently in widespread clinical use. The crystal structure of a cyclohexylamine platinum (IV) adduct with a dodecamer duplex (Silverman et al., 2002) shows that the drug forms an asymmetric 1,2-G:G intrastrand cross-link. The base pairs are kept intact in spite of a large positive roll caused by the cross-linking, and overall the structure has a close resemblance to other platinum—DNA complexes, including that of oxaliplatin, which also has a 1,2-G:G intrastrand cross-link (Spingler et al., 2001). The similar pattern of DNA distortion observed in the various crystal structures contrasts with their diverse biological activity. However, a subsequent NMR structure analysis has compared 1,2-G:G intrastrand cross-link adducts of cis-platin and oxaliplatin in the identical sequence (Wu et al., 2007), which has revealed some differences in geometry that may be reflected in differences in recognition by repair proteins and polymerases. The base-pair steps at and close to the adduct sites show differences in several base morphological parameters especially helical twist and base-pair roll that result in distinct values for the global helical bend. Although crystal packing has been invoked to account for the differences between crystal and solution results, it is likely a consequence of distinct refinement methods combined with the fact that none of these structures approach atomic resolution.

5.7.2 Covalent binding combined with sequence-specific recognition

The anti-tumour antibiotic (+)−CC−1065 (Fig. 5.62) is perhaps the classic sequence-specific DNA recognition compound. It has three pyrrolo–indole units, and its simpler analogue duocarmycin has just one. In both instances, one end of the drug molecule binds covalently to N3 of an adenine via opening of the cyclopropane ring (Hurley et al., 1988; MacMilan & Boger, 2009; Tietze et al., 2009). The rest of the drug lies in the minor groove. CC-1065 covers four base pairs to the 5′ side of this adenine and one base pair on the 3′ side. Specific recognition is to sequences such as 5′-AAAAA; the drug induces bending of 17−22°, comparable to that found in natural A tracts, which enables a close steric fit between drug and DNA to take place. CC-1065 has a specific (and highly stringent) requirement for a sequence that will bend in an appropriate manner. The smaller drug molecule duocarmycin induces fewer structural changes in its target sequences, and the overall conformations remain close to canonical B-DNA (Eis et al., 1997). Bizelesin is a synthetic analogue of CC-1065 with high experimental antitumour activity. It has a chemically reactive chloromethyl group at each end of the molecule and was rationally designed to covalently cross-link to the hexamer sequence 5′-TAATTA. This sequence is intrinsically bent in solution, due to the high propeller twists of the A•T base pairs. Unexpectedly, NMR studies (Thompson & Hurley, 1995) have shown that there are two adduct conformations, in a 40:60 ratio. Both have straightened-out A-tracts but differ in the hydrogen-bonding arrangement of the central A•T base pairs, with one being unpaired and the other having a Hoogsteen arrangement.

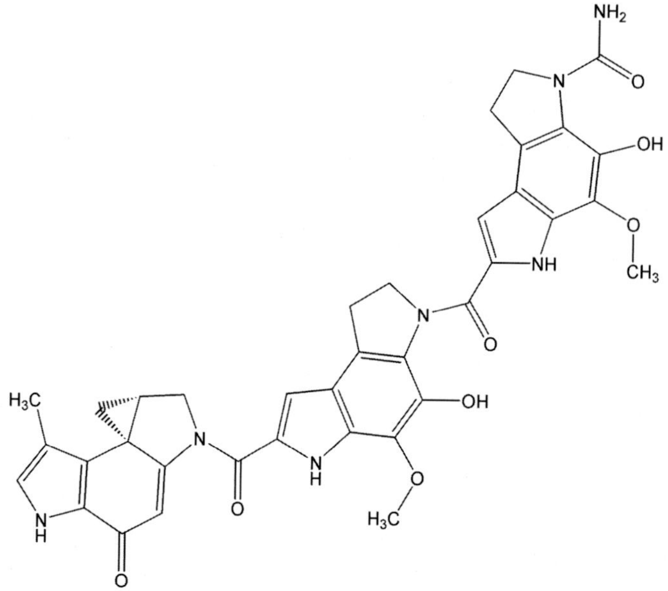

Figure 5.62 The structure of the covalently binding drug CC-1065.

Minor groove alkylation most often occurs at the N2 of guanine. Several compounds that bind at this site have significant experimental anti-tumour activity. Mitomycin, discovered in Japan in fermentation broths over 50 years ago, has found some clinical application in the treatment of solid tumours, although its severe toxicity has precluded extensive use. A few other minor groove alkylators have now progressed to evaluation in clinical trials. Their biological efficacy is undoubtedly a consequence of interference with aspects of gene regulation, although this is not well understood. For at least some compounds such as DSB-120 and its analogues (see paragraph below), their binding to the minor groove of B-form DNA involves little if any perturbation from the canonical structure. This results in a lack of structural lesions for subsequent recognition by a repair protein, so that these adducts tend to be long-lasting, with consequent enhancement of potency.

The pyrrolo[2,1-c][1,4]benzodiazepine (PBD) family of compounds, typified by the naturally occurring antitumour antibiotic anthramycin, binds to the exocyclic N2 atom of guanine bases through the N10—C11 double bond (Fig. 5.63). The crystal structure (Kopka et al., 1994) of an adduct with two bound drug molecules shows that each lies in the minor groove, in a manner analogous to that of the non-covalent binders such as berenil and netropsin. The binding site for anthramycin is three bases long, with a preference for 5'-Pu-G-Pu type sequences. Several dimers of these molecules have been designed which have exceptional cytotoxicity. The best-studied of these is DSB-120 (Bose et al., 1992; Mantaj et al., 2017a, 2017b), with an IC$_{50}$ of 0.0005 µM in some cell lines, making it some 1000 times more potent than the equivalent monomer. These cell-kill effects are probably due to inhibition of transcription, although it is not clear to what extent DNA sequence-selective effects are involved. DSB-120 (Fig. 5.63) has a site

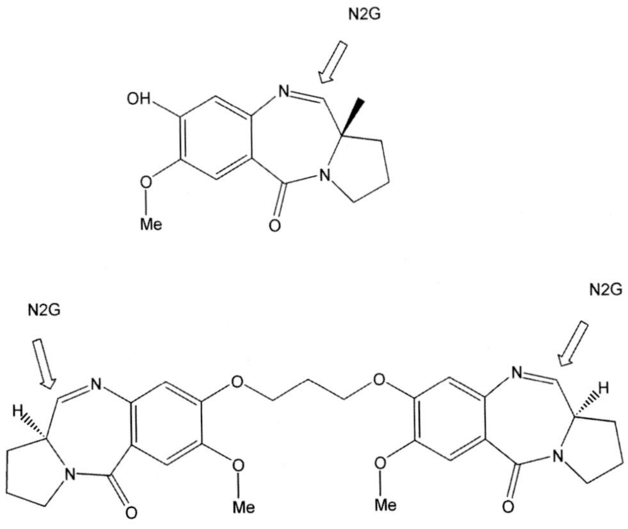

Figure 5.63 The drug anthramycin (top), showing its point of attachment to the exocyclic amino group of guanine. The anthramycin dimer DSB-120 is shown below.

size requirement of six to seven base pairs, with a sequence preference for sites of the type 5'–Pu–GATC–Py. Its ability to form inter-strand covalent cross-links to two sites in the minor groove, together with its almost ideal isohelicity, result in a tight fit to canonical B-form DNA. NMR and molecular modelling studies have shown that DSB-120 produces almost no structural perturbation to DNA upon cross-linking (Jenkins et al., 1994). This feature of the PDB dimer complexes is undoubtedly significant for their resistance to DNA repair and may contribute to the high in vivo antitumour activity shown by several members of this series, one of which proceeded to clinical trial evaluation (Alley et al., 2004). The exceptional potency of the DSB-120 class of compounds is currently being exploited in the development of novel antibody–drug conjugates (Hartley, 2020; Mantaj et al., 2017).

Mitomycin C requires metabolic activation to react with DNA (Bargonetti et al., 2010; Tomasz & Palom, 1997). Opening of the aziridine ring (Fig. 5.64) results in

Figure 5.64 The structure of mitomycin, together with a schematic diagram showing a DNA-mitomycin adduct cross-linked to two guanine bases on the opposite strands of a DNA duplex.

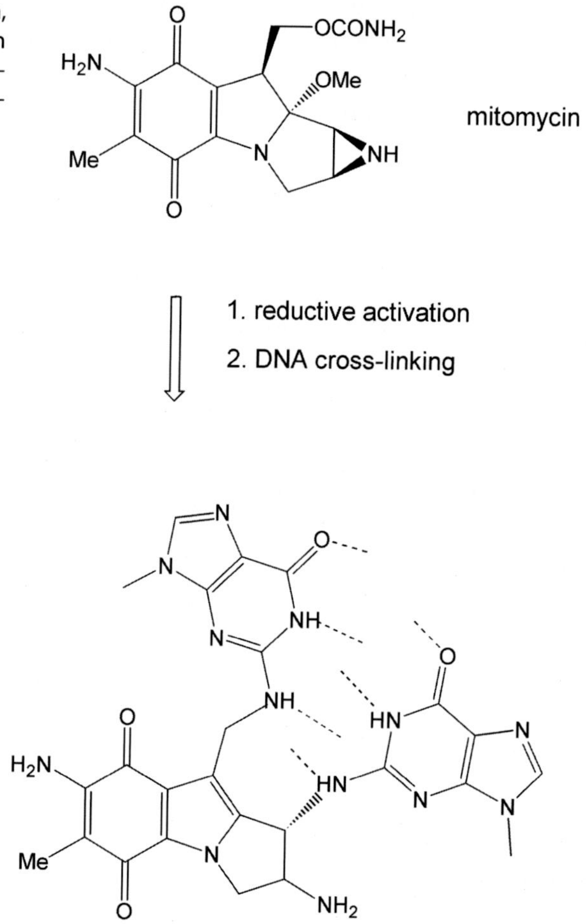

covalent attachment to one guanine in the sequence 5'-CpG. A second is through the carbamate group. The structures of both this cross-linked adduct and a mono-adduct from aziridine ring-opening have been characterised by NMR methods (Norman et al., 1990; Sastry et al., 1995), as has the very different N7 adduct, which places the bound drug in the DNA major groove (Subramanian et al., 2001), without significant perturbations from B-DNA form.

High DNA sequence selectivity is a dominant feature of several of these minor groove covalent binding molecules. This, together with their drug-like characteristics, does suggest that a fruitful future direction for the development of therapeutic sequence selectivity at the gene level may be in combining such molecules with the features of non-covalent recognition such as have been described earlier in this chapter.

References

Abe, K., Hirose, Y., Eki, H., Takeda, K., Bando, T., Endo, M., & Sugiyama, H. (2020). X-ray crystal structure of a cyclic-PIP-DNA complex in the reverse-binding orientation. *Journal of the American Chemical Society, 142*, 10544—10549.

Abu-Daya, A., Brown, P. M., & Fox, K. R. (1995). DNA sequence preferences of several AT-selective minor groove binding ligands. *Nucleic Acids Research, 17*, 3385—3392.

Adams, A., Guss, J. M., Collyer, C. A., Denny, W. A., Prakash, A. S., & Wakelin, L. P. G. (2000a). Acridinecarboxamide topoisomerase poisons: Structural and kinetic studies of the DNA complexes of 5-substituted 9-amino-(N-(2-dimethylamino)ethyl)acridine-4-carboxamides. *Molecular Pharmacology, 58*, 649—658.

Adams, A., Guss, J. M., Collyer, C. A., Denny, W. A., & Wakelin, L. P. (1999). Crystal structure of the topoisomerase II poison 9-amino-[N-(2-dimethylamino)ethyl]acridine-4-carboxamide bound to the DNA hexanucleotide d(CGTACG)₂. *Biochemistry, 38*, 9221—9233.

Adams, A., Guss, J. M., Collyer, C. A., Denny, W. A., & Wakelin, L. P. G. (2000b). A novel form of intercalation involving four DNA duplexes in an acridine-4-carboxamide complex of d(CGTACG)₂. *Nucleic Acids Research, 28*, 4244—4253.

Alcaro, S., & Musetti, C. (2013). Identification and characterization of new DNA G-quadruplex binders selected by a combination of ligand and structure-based virtual screening approaches. *Journal of Medicinal Chemistry, 56*, 843—855.

Alden, C. J., & Kim, S.-H. (1979). Solvent-accessible surfaces of nucleic acids. *Journal of Molecular Biology, 132*, 411—434.

Allen, W. J., Balius, T. E., Mukherjee, S., Brozell, S. R., Moustakas, D. T., Lang, P. T., Case, D. A., Kuntz, I. D., & Rizzo, R. C. (2015). DOCK 6: Impact of new features and current docking performance. *Journal of Computational Chemistry, 36*, 1132—1156.

Alley, M. C., Hollingshead, M. G., Pacula-Cox, C. M., Waud, W. R., Hartley, J. A., Howard, P. W., Gregson, S. J., Thurston, D. E., & Sausville, E. A. (2004). SJG-136 (NSC 694501), a novel rationally designed DNA minor groove interstrand cross-linking agent with potent and broad-spectrum antitumor activity: Part 2: Efficacy evaluations. *Cancer Research, 64*, 6700—6706.

Arai, S., Chatake, T., Ohhara, T., Kurihara, K., Tanaka, I., Suzuki, N., Fujimoto, Z., Mizuno, H., & Niimura, N. (2005). Complicated water orientations in the minor groove of the B-DNA decamer d(CCATTAATGG)₂ observed by neutron diffraction measurements. *Nucleic Acids Research, 33*, 3017—3024.

Asamitsu, S., Bando, T., & Sugiyama, H. (2019). Ligand design to acquire specificity to intended G-quadruplex structures. *Chemistry, 25*, 417—430.

Bailly, C., Chessare, G., Carrasco, C., Joubert, A., Mann, J., Wilson, W. D., & Neidle, S. (2003). Sequence-specific minor groove binding by *bis*-benzimidazoles: Water molecules in ligand recognition. *Nucleic Acids Research, 31*, 1514—1524.

Bargonetti, J., Champeil, E., & Tomasz, M. (2010). Differential toxicity of DNA adducts of mitomycin C. *Journal of Nucleic Acids, 2010,* 698960.

Bennett, M., Krah, A., Wien, F., Garman, E., McKenna, R., Sanderson, M., & Neidle, S. (2000). A DNA-porphyrin minor-groove complex at atomic resolution: The structural consequences of porphyrin ruffling. *Proceedings of the National Academy of Sciences of the United States of America, 97,* 9476—9481.

Billeter, M., Güntert, P., Luginbühl, P., & Wüthrich, K. (1996). Hydration and DNA recognition by homeodomains. *Cell, 85,* 1057—1065.

Bose, D. S., Thompson, A. S., Ching, J., Hartley, J. A., Berardini, M. D., Jenkins, T. C., Neidle, S., Hurley, L. H., & Thurston, D. E. (1992). Rational design of a highly efficient irreversible DNA inter-strand cross-linking agent based on the pyrrolobenzodiazepine ring system. *Journal of the American Chemical Society, 114,* 4939—4941.

Bostock-Smith, C. E., Laughton, C. A., & Searle, M. S. (1999). Solution structure and dynamics of the A-T tract DNA decamer duplex d(GGTAATTACC)$_2$: Implications for recognition by minor groove binding drugs. *Biochemical Journal, 342,* 125—132.

Boykin, D. W., Kumar, A., Spychala, J., Zhou, M., Lombardy, R. J., Wilson, W. D., Dykstra, C. C., Jones, S. K., Hall, J. E., Tidwell, R. R., Laughton, C., Nunn, C. M., & Neidle, S. (1995). Dicationic diarylfurans as anti-*Pneumocystis carinii* agents. *Journal of Medicinal Chemistry, 38,* 912—916.

Bremer, R. E., Wurtz, N. R., Szewczyk, J. W., & Dervan, P. B. (2001). Inhibition of major groove DNA binding bZIP proteins by positive patch polyamides. *Bioorganic & Medicinal Chemistry, 9,* 2093—2103.

Brogden, A. L., Hopcroft, N. H., Searcey, M., & Cardin, C. J. (2007). Ligand bridging of the DNA Holliday junction: Molecular recohnition of a stacked X four-way junction by a small molecule. *Angewandte Chemie International Edition, 46,* 3850—3854.

Brown, D. G., Sanderson, M. R., Garman, E., & Neidle, S. (1992). Crystal structure of a berenil-d(CGCAAATTTGCG) complex. An example of drug-DNA recognition based on sequence-dependent structural features. *Journal of Molecular Biology, 226,* 481—490.

Burnett, R., Melnder, C., Puckett, J. W., Son, L. S., Wells, R. D., Dervan, P. B., & Gottesfeld, J. M. (2006). DNA sequence-specific polyamides alleviate transcription inhibition associated with long GAA•TTC repeats in Friedreich's ataxia. *Proceedings of the National Academy of Sciences of the United States of America, 103,* 11497—11502.

Calabrese, D. R., Chen, X., Leon, E. C., Gaikwad, S. M., Phyo, Z., Hewitt, W. M., Alden, S., Hilimire, T. A., He, F., Michalowski, A. M., Simmons, J. K., Saunders, L. B., Zhang, S., Connors, D., Walters, K. J., Mock, B. A., & Schneekloth, J. S., Jr. (2018). Chemical and structural studies provide a mechanistic basis for recognition of the MYC G-quadruplex. *Nature Communications, 9,* 4229.

Campbell, N. H., Parkinson, G. N., Reszka, A. P., & Neidle, S. (2008). Structural basis of DNA quadruplex recognition by an acridine drug. *Journal of the American Chemical Society, 130,* 6722—6724.

Campbell, N. H., Patel, M., Tofa, A. B., Ghosh, R., Parkinson, G. N., & Neidle, S. (2009). Selectivity in ligand recognition of G-quadruplex loops. *Biochemistry, 48,* 1675—1680.

Canals, A., Purciolas, M., Aymami, J., & Coll, M. (2005). The anticancer agent ellipticine unwinds DNA by intercalative binding in an orientation parallel to base pairs. *Acta Crystallographica, D61,* 1009—1012.

Cassidy, S. A., Strekowski, L., Wilson, W. D., & Fox, K. R. (1994). Effect of a triplex-binding ligand on parallel and antiparallel DNA triple helixes using short unmodified and acridine-linked oligonucleotides. *Biochemistry, 33,* 15338—15347.

Cech, T. R. (2000). Life at the end of the chromosome: Telomeres and telomerase. *Angewandte Chemie International Edition, 39,* 34—43.

Chaires, J. B., Leng, F., Przewloka, T., Fokt, I., Ling, Y.-H., Perez-Soler, R., & Priebe, W. (1997). Structure-based design of a new bisintercalating anthracycline antibiotic. *Journal of Medicinal Chemistry, 40,* 261—266.

Chen, X., Ramakrishnan, B., Rao, S. T., & Sundaralingam, M. (1994). Binding of two distamycin A molecules in the minor groove of an alternating B-DNA duplex. *Nature Structural Biology, 1,* 169—175.

Chenoweth, D. M., & Dervan, P. B. (2009). Allosteric modulation of DNA by small molecules. *Proceedings of the National Academy of Sciences of the United States of America, 106,* 13175—13179.

Chien, C.-M., Wu, P.-C., Satange, R., Chang, C.-C., Lai, Z.-L., Hagler, L. D., Zimmerman, S. C., & Hou, M. H. (2020). Structural basis for targeting T:T mismatch with triaminotriazine-acridine conjugate induces a U-shaped head-to-head four-way junction in CTG repeat DNA. *Journal of the American Chemical Society, 142,* 11165—11172.

Chiu, T. K., Kaczor-Grzeskowiak, M., & Dickerson, R. E. (1999). Absence of minor groove monovalent cations in the crosslinked dodecamer C-G-C-G-A-A-T-T-C-G-C-G. *Journal of Molecular Biology, 292*, 589—608.

Chung, W. J., Heddi, B., Hamon, F., Teulade-Fichou, M.-P., & Phan, A. T. (2014). Solution structure of a G-quadruplex bound to the bisquinolinium compound Phen-DC(3). *Angewandte Chemie International Edition in English, 53*, 999—1002.

Chung, W. J., Heddi, B., Tera, M., Iida, K., Nagasawa, K., & Phan, A. T. (2013). Solution structure of an intramolecular (3 + 1) human telomeric G-quadruplex bound to a telomestatin derivative. *Journal of the American Chemical Society, 135*, 13495—13501.

Cieplak, P., Rao, S. N., Grootenhuis, P. D., & Kollman, P. A. (1990). Free energy calculation on base specificity of drug—DNA interactions: Application to daunomycin and acridine intercalation into DNA. *Biopolymers, 29*, 717—727.

Clark, G. R., Gray, E. J., Neidle, S., Ji, Y.-H., & Leupin, W. (1996). Isohelicity and phasing in drug-DNA sequence recognition: Crystal structure of a tris(benzimidazole)-oligonucleotide complex. *Biochemistry, 35*, 13745—13752.

Coll, M., Frederick, C. A., Wang, A. H.-J., & Rich, A. (1987). A bifurcated hydrogen-bonded conformation in the d(A.T) base pairs of the DNA dodecamer d(CGCAAATTTGCG) and its complex with distamycin. *Proceedings of the National Academy of Sciences of the United States of America, 84*, 8385—8389.

Collie, G. W., Promontorio, R., Hampel, S. M., Micco, M., Neidle, S., & Parkinson, G. N. (2012). Structural basis for telomeric G-quadruplex targeting by naphthalene diimide ligands. *Journal of the American Chemical Society, 134*, 2723—2731.

Coste, F., Malinge, J.-M., Serre, L., Shepard, W., Roth, M., Leng, M., & Zelwer, C. (1999). Crystal structure of a double-stranded DNA containing a cisplatin interstrand cross-link at 1.63 Å resolution: Hydration at the platinated site. *Nucleic Acids Research, 27*, 1837—1846.

Cuesta-Seijo, J. A., & Sheldrick, G. M. (2005). Structures of complexes between echinomycin and duplex DNA. *Acta Crystallographica, D61*, 442—448.

Cuesta-Seijo, J. A., Weiss, M. S., & Sheldrick, G. M. (2006). Serendipitous SAD phasing of an echinomycin-(ACGTACGT)$_2$ bisintercalation complex. *Acta Crystallographica, D62*, 417—424.

Dai, J., Carver, M., Hurley, L. H., & Yang, D. (2011). Solution structure of a 2:1 quindoline-c-MYC G-quadruplex: insights into G-quadruplex-interactive small molecule drug design. *Journal of the American Chemical Society, 133*, 17673—17680.

D'Aria, F., D'Amore, V. M., Di Leva, F. S., Amato, J., Caterino, M., Russomanno, P., Salerno, S., Barresi, E., De Leo, M., Marini, A. M., Taliani, S., Da Settimo, F., Salgado, G. F., Pompili, L., Zizza, P., Shirasawa, S., Novellino, E., Biroccio, A., Marinelli, L., & Giancola, C. (2020). Targeting the KRAS oncogene: Synthesis, physicochemical and biological evaluation of novel G-quadruplex DNA binders. *European Journal of Pharmaceutical Sciences, 149*, 105337.

de Clairac, R. P. L., Geierstanger, B. H., Mrksich, M., Dervan, P. B., & Wemmer, D. E. (1997). NMR characterization of hairpin polyamide complexes with the minor groove of DNA. *Journal of the American Chemical Society, 119*, 7909—7916.

Dervan, P. B. (1986). Design of sequence-specific DNA-binding molecules. *Science, 232*, 464—471.

Dervan, P. B. (2001). Molecular recognition of DNA by small molecules. *Bioorganic & Medicinal Chemistry, 9*, 2215—2235.

Dervan, P. B., & Bürli, R. W. (1999). Sequence-specific DNA recognition by polyamides. *Current Opinion in Structural Biology, 3*, 688—693.

Dickinson, L. A., Trauger, J. W., Eldon, E. B., Ghazal, P., Dervan, P. B., & Gottesfeld, J. M. (1999). Antirepression of RNA polymerase II transcription by pyrrole-imidazole polyamides. *Biochemistry, 38*, 10801—10807.

Dolenc, J., Oostenbrink, C., Koller, J., & van Gunsteren, W. F. (2005). Molecular dynamics simulations and free energy calculations of netropsin and distamycin binding to an AAAAA DNA binding site. *Nucleic Acids Research, 33*, 725—733.

Drew, H. R., & Dickerson, R. E. (1981). Structure of a B-DNA dodecamer: III. Geometry of hydration. *Journal of Molecular Biology, 151*, 535—556.

Duan, Y., Wilkosz, P., Crowley, M., & Rosenberg, J. M. (1997). Molecular dynamics simulation study of DNA dodecamer d(CGCGAATTCGCG) in solution: Conformation and hydration. *Journal of Molecular Biology, 272*, 553—572.

Eckdahl, T. T., Brown, A. D., Hart, S. N., Malloy, K. J., Shott, M., You, G., Hoopes, L. L., & Heyer, L. J. (2008). Microarray analysis of the in vivo sequence preferences of a minor groove binding drug. *BMC Genomics, 9*, 32.

Egli, M., Tereshko, V., Teplova, M., Minasov, G., Joachimiak, A., Sanishvili, R., Weeks, C. M., Miller, R., Maier, M. A., Ari, H., Cook, P. D., & Manoharan, M. (1998). X-ray crystallographic analysis of the hydration of A- and B-form DNA at atomic resolution. *Biopolymers, 48*, 234–252.

Egli, M., Williams, L. D., Frederick, C. A., & Rich, A. (1991). DNA-nogalamycin interactions. *Biochemistry, 30*, 1364–1372.

Eichman, B. F., Mooers, B. H. M., Alberti, M., Hearst, J. E., & Ho, P. S. (2001). The crystal structures of psoralen cross-linked DNAs: Drug-dependent formation of Holliday junctions. *Journal of Molecular Biology, 308*, 15–26.

Eis, P. S., Smith, J. A., Rydzewski, J. M., Chase, D. A., Boger, D. L., & Chazin, W. J. (1997). High resolution solution structure of a DNA duplex alkylated by the antitumor agent duocarmycin SA. *Journal of Molecular Biology, 272*, 237–252.

Escude, C., Nguyen, C. H., Kukreti, S., Janin, Y., Sun, J. S., Bisagni, E., Garestier, T., & Hélène, C. (1998). Rational design of a triple helix-specific intercalating ligand. *Proceedings of the National Academy of Sciences of the United States of America, 95*, 3591–3596.

Evans, D. A., & Neidle, S. (2006). Virtual screening of DNA minor groove binders. *Journal of Medicinal Chemistry, 49*, 4232–4238.

Farkas, M. E., Li, B. C., Dose, C., & Dervan, P. B. (2009). DNA sequence selectivity of hairpin polyamide turn units. *Bioorganic & Medicinal Chemistry Letters, 19*, 3919–3923.

Fedoroff, O. Y., Salazar, M., Han, H., Chemeris, V. V., Kerwin, S. M., & Hurley, L. H. (1998). NMR-based model of a telomerase-inhibiting compound bound to G-quadruplex DNA. *Biochemistry, 37*, 12367–12374.

Feig, M., & Pettitt, B. M. (1999). Modeling high-resolution hydration patterns in correlation with DNA sequence and conformation. *Journal of Molecular Biology, 286*, 1075–1095.

Fox, K. R., & Waring, M. J. (1986). Nucleotide sequence binding preferences of nogalamycin investigated by DNase I footprinting. *Biochemistry, 25*, 4349–4356.

Finn, P. B., Bhimsaria, D., Ali, A., Egushi, A., Ansari, A. Z., & Dervan, P. B. (2020). Single position substitution of hairpin pyrrole-imidazole polyamides imparts distinct DNA-binding profiles across the human genome. *PLOS One, 15*, e0243905.

Fox, K. R., Polucci, P., Jenkins, T. C., & Neidle, S. (1995). A molecular anchor for stabilizing triple-helical DNA. *Proceedings of the National Academy of Sciences of the United States of America, 92*, 7887–7891.

Franklin, R. E., & Gosling, R. G. (1953). The structure of sodium thymonucleate fibres. I. The influence of water content. *Acta Crystallographica, 6*, 673–677.

Fuller, W., Forsyth, T., & Mahendrasingam, A. (2004). Water-DNA interactions as studied by X-ray and neutron fibre diffraction. *Philosophical Transactions of the Royal Society of London - B, 359*, 1237–1247.

Galindo-Murillo, R., & Cheatham, T. E. (2021). Ethidium bromide interactions with DNA: An exploration of a classic DNA-ligand complex with unbiased molecular dynamics simulations. *Nucleic Acids Research.* https://doi.org/10.1093/nar/gkab143

Gao, X., Mirau, P., & Patel, D. J. (1992). Structure refinement of the chromomycin dimmer-DNA oligomer complex in solution. *Journal of Molecular Biology, 223*, 259–279.

Gao, X., & Patel, D. J. (1988). NMR studies of echinomycin bisintercalation complexes with d(A1-C2-G3-T4) and d(T1-C2-G3-A4) duplexes in aqueous solution: Sequence-dependent formation of Hoogsteen A1.T4 and Watson-Crick T1.A4 base pairs flanking the bisintercalation site. *Biochemistry, 27*, 1744–1751.

Gavathiotis, E., Heald, R. A., Stevens, M. F., & Searle, M. S. (2003). Drug recognition and stabilisation of the parallel-stranded DNA quadruplex d(TTAGGGT)4 containing the human telomeric repeat. *Journal of Molecular Biology, 334*, 25–36.

Gavathiotis, E., Sharman, G. J., & Searle, M. S. (2000). Sequence-dependent variation in DNA minor groove width dictates orientational preference of Hoechst 33258 in A-tract recognition: Solution NMR structure of the 2:1 complex with d(CTTTTGCAAAAG)2. *Nucleic Acids Research, 28*, 728–735.

Ge, W., Schneider, B., & Olson, W. K. (2005). Knowledge-based elastic potentials for docking drugs or proteins with nucleic acids. *Biophysical Journal, 88*, 1166–1190.

Gearhart, M. D., Dickinson, L., Ehley, J., nelander, C., Dervan, P. B., Wright, P. E., & Gottesfeld, J. M. (2005). Inhibition of DNA binding by human estrogen-related receptor 2 and estrogen receptor alpha with minor groove binding polyamides. *Biochemistry, 44*, 4196–4203.

Gelasco, A., & Lippard, S. J. (1998). NMR Solution structure of a DNA dodecamer duplex containing a *cis*-diammineplatinum(II) d(GpG) intrastrand cross-link, the major adduct of the anticancer drug cisplatin. *Biochemistry, 37*, 9230–9239.

Gilbert, D. E., & Feigon, J. (1992). Proton NMR study of the [d(ACGTATACGT)]$_2$-2echinomycin complex: Conformational changes between echinomycin binding sites. *Nucleic Acids Research, 20*, 2411–2420.

Glass, L. S., Nguyen, B., Goodwin, K. D., Dardonville, C., Wilson, W. D., Long, E. C., & Georgiadis, M. M. (2009). Crystal structure of a trypanocidal 4,4'-bis(imidazolinylamino)diphenylamine bound to DNA. *Biochemistry, 48*, 5943–5952.

Gochin, M. (2000). A high-resolution structure of a DNA-chromomycin-Co(II) complex determined from pseudocontact shifts in nuclear magnetic resonance. *Structure, 8*, 441–452.

Goodsell, D., & Dickerson, R. E. (1986). Isohelical analysis of DNA groove-binding drugs. *Journal of Medicinal Chemistry, 29*, 727–733.

Goodwin, K. D., Lewis, M. A., Tanious, F. A., Tidwell, R. R., Wilson, W. D., Georgiadis, M. M., & Long, E. C. (2006). A high-throughput, high-resolution strategy for the study of site-selective DNA binding agents: Analysis of a "highly twisted" benzimidazole-diamidine. *Journal of the American Chemical Society, 126*, 7846–7854.

Goodwin, K. D., Long, E. C., & Georgiadis, M. M. (2005). A host-guest approach for determining drug-DNA interactions: An example using netropsin. *Nucleic Acids Research, 33*, 4106–4116.

Gottesfeld, J. M., Melander, C., Suto, R. K., Raviol, H., Luger, K., & Dervan, P. B. (2001). Sequence-specific recognition of DNA in the nucleosome by pyrrole-imidazole polyamides. *Journal of Molecular Biology, 309*, 615–629.

Gottesfeld, J. M., Neely, L., Trauger, J. W., Baird, E. E., & Dervan, P. B. (1997). Regulation of gene expression by small molecules. *Nature, 387*, 202–205.

Guo, Q., Lu, M., Marky, L. A., & Kallenbach, N. R. (1992). Interaction of the dye ethidium bromide with DNA containing guanine repeats. *Biochemistry, 31*, 2451–2455.

Haider, S. M., Parkinson, G. N., & Neidle, S. (2003). Structure of a G-quadruplex-ligand complex. *Journal of Molecular Biology, 326*, 117–125.

Halle, B., & Denisov, V. P. (1998). Water and monovalent ions in the minor groove of B-DNA oligonucleotides as seen by NMR. *Biopolymers, 48*, 210–233.

Han, H., & Hurley, L. H. (2000). G-quadruplex DNA: A potential target for anti-cancer drug design. *Trends in Pharmacological Sciences, 21*, 136–142.

Harp, J. M., Coates, L., Sullivan, B., & Egli, M. (2021). Water structure around a left-handed Z-DNA fragment analyzed by cryo neutron crystallography. *Nucleic Acids Research.* https://doi.org/10.1093/nar/gkab264

Hartley, J. A. (2020). Antibody-drug conjugates (ADCs) delivering pyrrolobenzodiazepine (PBD) dimers for cancer therapy. *Expert Opinion on Biological Therapy, 16*, 1–13.

Hartley, J. A., Bingham, J. P., & Souhami, R. L. (1992). DNA sequence selectivity of guanine-N7 alkylation by nitrogen mustards is preserved in intact cells. *Nucleic Acids Research, 20*, 3175–3178.

Henikoff, S., & Vermaak, D. (2000). Bugs on drugs go GAGAA. *Cell, 103*, 695–698.

Hidaka, T., & Sugiyama, H. (2020). Chemical approaches to the development of artificial transcription factors based on pyrrole-imidazole polyamides. *The Chemical Record.* https://doi.org/10.1002/tcr.202000158

Hopcroft, N. H., Brogden, A. L., Searcey, M., & Cardin, C. J. (2006). X-ray crystallographic study of DNA duplex cross-linking: Simultaneous binding to two d(CGTACG)$_2$ molecules by a bis(9-aminoacridine-4-carboxamide) derivative. *Nucleic Acids Research, 34*, 6663–6672.

Hou, M.-H., Robinson, H, Gao, Y.-G., & Wang, A. H.-J. (2002). Crystal structure of actinomycin D bound to the CTG triplet repeat sequences linked to neurological diseases. *Nucleic Acids Research, 30*, 4910–4917.

Hou, M.-H., Robinson, H., Gao, Y.-G., & Wang, A. H.-J. (2004). Crystal structure of the [Mg^{2+}-(chromomycin A$_3$)$_2$]-d(TTGGCCAA)$_2$ complex reveals GGCC binding specificity of the drug dimmer chelated by a metal ion. *Nucleic Acids Research, 32*, 2214–2222.

Hu, G. G., Shui, X., Leng, F., Priebe, W., Chaires, J. B., & Williams, L. D. (1997). Structure of a DNA-bisdaunomycin complex. *Biochemistry, 36,* 5940—5946.

Huang, H., Zhu, L., Reid, B. R., Drobny, G. P., & Hopkins, P. B. (1995). Solution structure of a cisplatin-induced DNA interstrand cross-link. *Science, 270,* 1842—1845.

Hud, N. V., Sklenár, V., & Feigon, J. (1999). Localization of ammonium ions in the minor groove of DNA duplexes in solution and the origin of DNA A-tract bending. *Journal of Molecular Biology, 286,* 651—660.

Hurley, L. H., Lee, C.-S., McGovren, J. P., Mitchell, M. A., Warpehoski, M. A., Kelly, R. C., & Aristoff, P. A. (1988). Molecular basis for sequence-specific DNA alkylation by CC-1065. *Biochemistry, 27,* 3886—3892.

Iida, K., & Nagasawa, K. (2013). Macrocyclic polyoxazoles as G-quadruplex ligands. *The Chemical Record, 13,* 539—548.

Islam, M. K., Jackson, P. J., Rahman, K. M., & Thurston, D. E. (2016). Recent advances in targeting the telomeric G-quadruplex DNA sequence with small molecules as a strategy for anticancer therapies. *Future Medicinal Chemistry, 8,* 1259—1290.

Janssen, S., Cuvier, O., Müller, M., & Laemmli, U. K. (2000). Specific gain- and loss-of-function phenotypes induced by satellite-specific DNA-binding drugs fed to *Drosophila melanogaster*. *Molecular Cell, 6,* 1013—1024.

Janssen, S., Durussel, T., & Laemmli, U. K. (2000). Chromatin opening of DNA satellites by targeted sequence-specific drugs. *Molecular Cell, 6,* 999—1011.

Jenkins, T. C., Hurley, L. H., Neidle, S., & Thurston, D. E. (1994). Structure of a covalent DNA minor groove adduct with a pyrrolobenzodiazepine dimer: Evidence for sequence-specific interstrand cross-linking. *Journal of Medicinal Chemistry, 37,* 4529—4537.

Ji, Y.-H., Bur, D., Häsler, W., Schmitt, V. R., Dorn, A., Bailly, C., Waring, M. J., Hochstrasser, R., & Leupin, W. (2001). Tris-benzimidazole derivatives: Design, synthesis and DNA sequence recognition. *Bioorganic & Medicinal Chemistry, 9,* 2905—2919.

Johansson, E., Parkinson, G., & Neidle, S. (2000). A new crystal form for the dodecamer C-G-C-G-A-A-T-T-C-G-C-G: Symmetry effects on sequence-dependent DNA structure. *Journal of Molecular Biology, 300,* 551—561.

Kamitori, S., & Takusagawa, F. (1992). Crystal structure of the 2:1 complex between d(GAAGCTTC) and the anticancer drug actinomycin D. *Journal of Molecular Biology, 225,* 445—456.

Kang, S., Shafer, R. H., & Kuntz, I. D. (2003). Calculation of ligand-nucleic acid binding free energies with the generalized-born model in DOCK. *Biopolymers, 73,* 192—204.

Kelland, L. (2007). The resurgence of platinum-based cancer chemotherapy. *Nature Reviews Cancer, 7,* 573—584.

Kennard, O., Cruse, W. B. T., Nachman, J., Prange, T., Shakked, Z., & Rabinovich, D. (1986). Ordered water structure in an A-DNA octamer at 1.7 Å resolution. *Journal of Biomolecular Structure & Dynamics, 3,* 623—647.

Keppler, M. D., Read, M. A., Perry, P. J., Trent, J. O., Jenkins, T. C., Reszka, A. P., Neidle, S., & Fox, K. R. (1999). Stabilization of DNA triple helices by a series of mono- and disubstituted amidoanthraquinones. *European Journal of Biochemistry, 263,* 817—825.

Kielkopf, C. L., Baird, E. E., Dervan, P. B., & Rees, D. C. (1998a). Structural basis for G.C recognition in the DNA minor groove. *Nature Structural Biology, 5,* 104—109.

Kielkopf, C. L., Bremer, R. E., White, S., Szewczyk, J. W., Turner, J. M., Baird, E. E., Dervan, P. B., & Rees, D. C. (2000a). Structural effects of DNA sequence on T.A recognition by hydroxypyrrole/pyrrole pairs in the minor groove. *Journal of Molecular Biology, 295,* 557—567.

Kielkopf, C. L., Erkkila, K. E., Hudson, B. P., Barton, J. K., & Rees, D. C. (2000b). Structure of a photo-active rhodium complex intercalated into DNA. *Nature Structural Biology, 7,* 117—121.

Kielkopf, C. L., White, S., Szewczyk, J. W., Turner, J. M., Baird, E. E., Dervan, P. B., & Rees, D. C. (1998b). A structural basis for recognition of A.T and T.A base pairs in the minor groove of B-DNA. *Science, 282,* 111—115.

Kim, M.-Y., Vankayalapati, H., Shin-ya, K., Wierzba, K., & Hurley, L. H. (2002). Telomestatin, a potent telomerase inhibitor that interacts quite specifically with the human telomeric intramolecular G-quadruplex. *Journal of the American Chemical Society, 124,* 2098—2099.

Koeppel, F., Riou, J.-F., Laoui, A., Mailliet, P., Arimondo, P. B., Labit, D., Petitgenet, O., Hélène, C., & Mergny, J.-L. (2001). Ethidium derivatives bind to G-quartets, inhibit telomerase and act as fluorescent probes for quadruplexes. *Nucleic Acids Research, 29*, 1087–1096.

Kopka, M. L., Goodsell, D. S., Baikalov, I., Grzeskowiak, K., Cascio, D., & Dickerson, R. E. (1994). Crystal structure of a covalent DNA-drug adduct: Anthramycin bound to C-C-A-A-C-G-T-T-G-G and a molecular explanation of specificity. *Biochemistry, 33*, 13593–13610.

Kopka, M. L., Goodsell, D. S., Han, G. W., Chiu, T. K., Lown, J. W., & Dickerson, R. E. (1997). Defining GC-specificity in the minor groove: Side-by-side binding of the di-imidazole lexitropsin to C-A-T-G-G-C-C-A-T-G. *Structure, 5*, 1033–1046.

Kopka, M. L., Yoon, C., Goodsell, D., Pjura, P., & Dickerson, R. E. (1985). Binding of an antitumor drug to DNA, Netropsin and C-G-C-G-A-A-T-T-BrC-G-C-G. *Journal of Molecular Biology, 183*, 553–563.

Kotar, A., Wang, B., Shivalingam, A., Gonzalez-Garcia, J., Vilar, R., & Plavec, J. (2016). NMR structure of a triangulenium-based long-lived fluorescence probe bound to a G-quadruplex. *Angewandte Chemie International Edition in English, 55*, 12508–12511.

Kubinec, M. G., & Wemmer, D. E. (1992). NMR evidence for DNA bound water in solution. *Journal of the American Chemical Society, 114*, 8739–8740.

Kurmis, A. A., & Dervan, P. B. (2019). Sequence specific suppression of androgen receptor-DNA binding in vivo by a Py-Im polyamide. *Nucleic Acids Research, 47*, 3828–3835.

Kurmis, A. A., Yang, F., Welch, T. R., Nickols, N. G., & Dervan, P. B. (2017). A pyrrole-imidazole polyamide is active against enzalutamide-resistant prostate cancer. *Cancer Research, 77*, 2207–2212.

Laage, D., Elsaesser, T., & Hynes, J. T. (2017). Water dynamics in the hydration shells of biomolecules. *Chemical Reviews, 117*, 10694–10725.

Lane, A. N., Jenkins, T. C., Brown, T., & Neidle, S. (1991). Interaction of berenil with the EcoRI dodecamer d(CGCGAATTCGCG)$_2$ in solution studied by NMR. *Biochemistry, 30*, 1372–1385.

Laughton, C., & Luisi, B. (1999). The mechanics of minor groove width variation in DNA, and its implications for the accommodation of ligands. *Journal of Molecular Biology, 288*, 953–963.

Laughton, C. A., Tanious, F. A., Nunn, C. M., Boykin, D. W., Wilson, W. D., & Neidle, S. (1996). A crystallographic and spectroscopic study of the complex between d(CGCGAATTCGCG)$_2$ and 2,5-bis(4-guanylphenyl)furan, an analogue of berenil. Structural origins of enhanced DNA-binding affinity. *Biochemistry, 35*, 5655–5661.

Le, D. D., Di Antonio, M., Chan, L. K. M., & Balasubramanian, S. (2015). G-quadruplex ligands exhibit differential G-tetrad selectivity. *Chemical Communications (Cambridge), 51*, 8048–8050.

Lercher, L., McDonough, M. A., El-Sagheer, A. H., Thalhammer, A., Kriaucionis, S., Brown, T., & Schofield, C. J. (2014). Structural insights into how 5-hydroxymethylation influences transcription factor binding. *Chemical Communications, 50*, 1794–1796.

Lerman, L. S. (1961). Structural considerations in the interaction of DNA and acridines. *Journal of Molecular Biology, 3*, 18–30.

Li, K., Yatsunyk, L., & Neidle, S. (2021). Water spines and networks in G-quadruplex structures. *Nucleic Acids Research, 49*, 519–528.

Liepinsh, E., Otting, G., & Wüthrich, K. (1992). NMR observation of individual molecules of hydration water bound to DNA duplexes: Direct evidence for a spine of hydration water present in aqueous solution. *Nucleic Acids Research, 20*, 6549–6553.

Lin, L. Y., McCarthy, S, Powell, B. M., Manurung, Y, Xiang, I. M., Dean, W. L., ... Yatsunyk, L. A. (2020). Biophysical and X-ray structural studies of the (GGGTT)$_3$GGG G-quadruplex in complex with N-methyl mesoporphyrin IX. *PLOS One, 15*, e0241513.

Lipscomb, L. A., Zhou, F. X., Presnell, S. R., Woo, R. J., Peek, M. E., Plaskon, R. R., & Williams, L. D. (1996). Structure of a DNA-porphyrin complex. *Biochemistry, 35*, 2818–2823.

Liu, X., Chen, H., & Patel, D. J. (1991). Solution structure of actinomycin-DNA complexes: Drug intercalation at isolated G-C sites. *Journal of Biomolecular NMR, 1*, 323–347.

Liu, J., & Subirana, J. (1999). Structure of d(CGCGAATTCGCG) in the presence of Ca$^{(2+)}$ ions. *Journal of Biological Chemistry, 274*, 24749–24752.

Lo, Y.-S., Tseng, W.-H., Chuang, C.-Y., & Hou, M.-H. (2013). The structural basis of actinomycin D-binding induces nucleotide flipping out, a sharp bend and a left-handed twist in CGG triplet repeats. *Nucleic Acids Research, 41*, 4284–4294.

Lown, J. W. (1994). DNA recognition by lexitropsins, minor groove binding agents. *Journal of Molecular Recognition, 7*, 79–88.

Luo, J., Wei, W., Waldispühl, J., & Moitessier, N. (2019). Challenges and current status of computational methods for docking small molecules to nucleic acids. *European Journal of Medicinal Chemistry, 168,* 414–425.

MacMillan, K. S., & Boger, D. L. (2009). Fundamental relationships between structure, reactivity, and biological activity for the duocarmycins and CC-1065. *Journal of Medicinal Chemistry, 52,* 5771–5780.

Maeshima, K., Janssen, S., & Laemmli, U. K. (2001). Specific targeting of insect and vertebrate telomeres with pyrrole and imidazole polyamides. *The EMBO Journal, 20,* 3218–3228.

Mann, J., Baron, A., Opoku-Boahen, Y., Johansson, E., Parkinson, G., Kelland, L. R., & Neidle, S. (2001). A new class of symmetric bisbenzimidazole-based DNA minor groove-binding agents showing antitumor activity. *Journal of Medicinal Chemistry, 44,* 138–144.

Mann, J., Taylor, P. W., Dorgan, C. R., Johnson, P. D., Wilson, F. X., Vickers, R., Dale, A. G., & Neidle, S. (2015). The discovery of a novel antibiotic for the treatment of *Clostridium difficile* infections: A story of an effective academic-industrial partnership. *MedChemComm, 6,* 1420–1426.

Mantaj, J., Jackson, P. J., Rahman, K. M., & Thurston, D. E. (2017). From anthramycin to pyrrolobenzodiazepine (PBD)-containing antibody-drug conjugates (ADCs). *Angewandte Chemie International Edition in English, 56,* 462–488.

Mapp, A. K., Ansari, A. Z., Ptashne, M., & Dervan, P. B. (2000). Activation of gene expression by small molecule transcription factors. *Proceedings of the National Academy of Sciences of the United States of America, 97,* 3930–3935.

Marques, M. A., Doss, R. M., Foister, S., & Dervan, P. B. (2004). Expanding the repertoire of heterocycle ring pairs for programmable minor groove DNA recognition. *Journal of the American Chemical Society, 126,* 10339–10349.

Marzilli, L. G., Banville and, D. L., & Zon, G. (1986). Pronounced proton and phosphorus-31 NMR spectral changes on meso-tetrakis(N-methylpyridinium-4-yl)porphyrin binding to poly[d(G-C)].poly[d(G-C)] and to three tetradecaoligodeoxyribonucleotides: Evidence for symmetric, selective binding to 5'CG3' sequences. *Journal of the American Chemical Society, 108,* 4188–4192.

McConnell, K. J., & Beveridge, D. L. (2000). DNA Structure: What's in charge? *Journal of Molecular Biology, 304,* 803–820.

McDermott, M. L., Vanselous, H., Corcelli, S. A., & Petersen, P. B. (2017). DNA's chiral spine of hydration. *ACS Central Science, 3,* 708–714.

McEachern, M. J., Krauskopf, A., & Blackburn, E. H. (2000). Telomeres and their control. *Annual Review of Genetics, 34,* 331–358.

McLean, M. J., Seela, F., & Waring, M. J. (1989). Echinomycin-induced hypersensitivity to osmium tetroxide of DNA fragments incapable of forming Hoogsteen base pairs. *Proceedings of the National Academy of Sciences of the United States of America, 86,* 9687–9691.

Mergny, J.-L., Lacroix, L., Teulado-Fichou, M.-P., Hounsou, C., Guittat, L., Hoarau, M., Arimondo, P. B., Vigneron, J.-P., Lehn, J.-M., Riou, J.-F., Garestier, T., & Hélène, C. (2001). Telomerase inhibitors based on quadruplex ligands selected by a fluorescence assay. *Proceedings of the National Academy of Sciences of the United States of America, 98,* 3062–3067.

Mergny, J.-L., & Hélène, C. (1998). G-quadruplex DNA: A target for drug design. *Nature Medicine, 4,* 1366–1367.

Mergny, J.-L., Duval-Valentin, G., Nguyen, C. H., Perrouault, L., Faucon, B., Rougée, Montenay-Garestier, T., Bisagni, E., & Hélène, C. (1992). Triple helix-specific ligands. *Science, 256,* 1681–1684.

Miao, Y., Lee, M. P. H., Parkinson, G. N., Batista-Parra, A., Ismail, M., Neidle, S., Boykin, D. W., & Wilson, W. D. (2005). Out-of-shape DNA minor groove binders: Induced fit interactions of heterocyclic dictations with the DNA minor groove. *Biochemistry, 44,* 14701–14708.

Micco, M., Collie, G. W., Dale, A. G., Ohnmacht, S. A., Pazitna, I., Gunaratnam, M., Reszka, A. P., & Neidle, S. (2013). Structure-based design and evaluation of naphthalene diimide G-quadruplex ligands as telomere targeting agents in pancreatic cancer cells. *Journal of Medicinal Chemistry, 56,* 2959–2974.

Minasov, G., Tereshko, V., & Egli, M. (1999). Atomic-resolution crystal structures of B-DNA reveal specific influences of divalent metal ions on conformation and packing. *Journal of Molecular Biology, 291,* 83–99.

Monchaud, D., & Teulade-Fichou, M.-P. (2008). A hitchhiker's guide to G-quadruplex ligands. *Organic and Biomolecular Chemistry, 6,* 627–636.

Mrksich, M., & Dervan, P. B. (1993). Antiparallel side-by-side heterodimer for sequence-specific recognition in the minor groove of DNA by a distamycin/1-methylimidazole-2-carboxamide-netropsin pair. *Journal of the American Chemical Society, 115*, 2572–2576.

Monsen, R. C., & Trent, J. O. (2018). G-Quadruplex virtual drug screening: A review. *Biochimie, 152*, 134–148.

More, G. S., Thomas, A. B., Chitlange, S. S., Nanda, R. K., & Gajbhiye, R. L. (2019). Nitrogen mustards as alkylating agents: A review on chemistry, mechanism of action and current USFDA status of drugs. *Anti-Cancer Agents in Medicinal Chemistry, 19*, 1080–1102.

Neidle, S. (2016). Quadruplex nucleic acids as novel therapeutic targets. *Journal of Medicinal Chemistry, 59*, 5987–6011.

Neidle, S. (2020). Challenges in developing small-molecule quadruplex therapeutics. *Annual Reports in Medicinal Chemistry, 54*, 517–546.

Neidle, S., Achari, A., Taylor, G. L., Berman, H. M., Carrell, H. L., Glusker, J. P., & Stallings, W. C. (1977). Structure of a dinucleoside phosphate-drug complex as model for nucleic acid-drug interaction. *Nature, 269*, 304–307.

Norman, D., Live, D., Sastry, M., Lipman, R., Hingerty, B. E., Tomasz, M., Broyde, S., & Patel, D. J. (1990). NMR and computational characterization of mitomycin cross-linked to adjacent deoxyguanosines in the minor groove of the d(T-A-C-G-T-A).d(T-A-C-G-T-A) duplex. *Biochemistry, 29*, 2861–2875.

Nguyen, B., Hamelberg, D., Bailly, C., Colson, P., Stanek, J., Brun, R., Neidle, S., & Wilson, W. D. (2004). Characterization of a novel DNA minor-groove complex. *Biophysical Journal, 86*, 1028–1041.

Nguyen, B., Lee, M. P. H., Hamelberg, D., Joubert, A., Bailly, C., Brun, R., Neidle, S., & Wilson, W. D. (2002). Strong binding in the DNA minor groove by an aromatic diamidine with a shape that does not match the curvature of the groove. *Journal of the American Chemical Society, 124*, 13680–13681.

Nielsen, M. C., & Ulven, T. (2010). Macrocyclic G-quadruplex ligands. *Current Medicinal Chemistry, 17*, 3438–3448.

Nunn, C. M., Garman, E., & Neidle, S. (1997). Crystal structure of the DNA decamer d(CGCAATTGCG) complexed with the minor groove binding drug netropsin. *Biochemistry, 36*, 4792–4799.

Ohndorf, U.-M., Rould, M. A., He, Q., Pabo, C. O., & Lippard, S. J. (1999). Basis for recognition of cisplatin-modified DNA by high-mobility-group proteins. *Nature, 399*, 708–712.

Ohnmacht, S. A., Varavipour, E., Nanjunda, R., Pazitna, I., Di Vita, G., Gunaratnam, M., … Neidle, S. (2014). Discovery of new G-quadruplex binding chemotypes. *Chemical Communications, 50*, 960–963.

Oleksi, A., Blanco, A. G., Boer, R., Usón, I., Aymani, J., Rodger, A., … Coll, M. (2006). Molecular recognition of a three-way DNA junction by a metallosupramolecular helicate. *Angewandte Chemie International Edition in English, 45*, 1227–1231.

Organesian, L., & Bryan, T. M. (2007). Physiological relevance of telomeric G-quadruplex formation: A potential drug target. *BioEssays, 29*, 155–165.

Otero-Navas, I., & Seminario, J. M. (2012). Molecular electrostatic potentials of DNA base-base pairing and mispairing. *Journal of Molecular Modeling, 18*, 91–101.

Pal, S. K., Zhao, L., Xia, T., & Zewail, A. H. (2003). Site- and sequence-selective ultrafast hydration of DNA. *Proceedings of the National Academy of Sciences of the United States of America, 100*, 13746–13751.

Parkinson, G. N., Ghosh, R., & Neidle, S. (2007). Structural basis for binding of porphyrin to human telomeres. *Biochemistry, 46*, 2390–2397.

Patel, D. J., & Shapiro, L. (1986). Sequence-dependent recognition of DNA duplexes. Netropsin complexation to the AATT site of the d(G-G-A-A-T-T-C-C) duplex in aqueous solution. *Journal of Biological Chemistry, 261*, 1230–1240.

Peek, M. E., Lipscomb, L. A., Bertrand, J. A., Gao, Q., Roques, B. P., Garbay-Jaureguiberry, C., & Williams, L. D. (1994). DNA distortion in bis-intercalated complexes. *Biochemistry, 33*, 3794–3800.

Pelton, J. G., & Wemmer, D. E. (1990). Binding modes of distamycin A with d(CGCAAATTTGCG)$_2$ determined by two-dimensional NMR. *Journal of the American Chemical Society, 112*, 1393–1399.

Phan, A. T., Kuryavyi, V., Gaw, H. Y., & Patel, D. J. (2005). Small-molecule interaction with a five-guanine-tract G-quadruplex structure from the human MYC promoter. *Nature Chemical Biology, 1*, 167–173.

Phan, A. T., Leroy, J.-L., & Guéron, M. (1999). Determination of the residence time of water molecules hydrating B′-DNA and B-DNA, by one-dimensional zero-enhancement nuclear Overhauser effect spectroscopy. *Journal of Molecular Biology, 286*, 505–519.

Pierre, V. C., Kaiser, J. T., & Barton, J. K. (2007). Insights into finding a mismatch through the structure of a mispaired DNA bound by a rhodium intercalator. *Proceedings of the National Academy of Sciences of the United States of America, 104*, 429–434.

Pullman, A., & Pullman, B. (1981). Molecular electrostatic potential of the nucleic acids. *Quarterly Reviews of Biophysics, 14*, 289–380.

Quigley, G. J., Ughetto, G., van der Marel, G. A., van Boom, J. H., Wang, A. H.-J., & Rich, A. (1986). Non-Watson-Crick G.C and A.T base pairs in a DNA-antibiotic complex. *Science, 232*, 1255–1258.

Rangan, A., Fedoroff, O. Y., & Hurley, L. H. (2001). Induction of duplex to G-quadruplex transition in the c-myc promoter region by a small molecule. *Journal of Biological Chemistry, 276*, 4640–4646.

Read, M. A., Wood, A. A., Harrison, J. R., Gowan, S. M., Kelland, L. R., Dosanjh, H. S., & Neidle, S. (1999). Molecular modeling studies on G-quadruplex complexes of telomerase inhibitors: Structure-activity relationships. *Journal of Medicinal Chemistry, 42*, 4538–4546.

Read, M., Harrison, R. J., Romagnoli, B., Tanious, F. A., Gowan, S. H., Reszka, A. P., Wilson, W. D., Kelland, L. R., & Neidle, S. (2001). Structure-based design of selective and potent G quadruplex-mediated telomerase inhibitors. *Proceedings of the National Academy of Sciences of the United States of America, 98*, 4844–4849.

Read, M. A., & Neidle, S. (2000). Structural characterization of a guanine-quadruplex ligand complex. *Biochemistry, 39*, 13422–13432.

Riou, J.-F., Guittat, L., Mailliet, P., Laoui, A., Renou, E., Petitgenet, O., Mégnin-Chanet, F., Hélène, C., & Mergny, J.-L. (2002). Cell senescence and telomere shortening induced by a new series of specific G-quadruplex DNA ligands. *Proceedings of the National Academy of Sciences of the United States of America, 99*, 2672–2677.

Rohs, R., Blockh, I., Sklenar, H., & Shakked, Z. (2005). Molecular flexibility in *ab initio* drug docking to DNA: Binding-site and binding-mode transitions in all-atom Monte Carlo simulations. *Nucleic Acids Research, 33*, 7048–7057.

Rowell, K. N., Thomas, D. S., Ball, G. E., & Wakelin, L. P. G. (2020). Molecular dynamic simulations of diacridine binding to DNA: Indications that C6 diacridine can bisintercalate spanning two base pairs. *Biopolymers.* https://doi.org/10.1002/bip.23409

Rueda, M., Cubero, E., Laughton, C. A., & Orozco, M. (2004). Exploring the counterion atmosphere around DNA: What can be learned from molecular dynamics simulations? *Biophysical Journal, 87*, 800–811.

Saenger, W., Hunter, W. N., & Kennard, O. (1986). DNA conformation is determined by economics in the hydration of phosphate groups. *Nature, 324*, 385–388.

Salvati, E., Leonetti, C., Rizzo, A., Scarsella, M., Mottolese, M., Galati, R., Sperduti, I., Stevens, M. F. G., D'Incalci, M., Biasco, M., Chiorino, G., Bauwens, S., Horard, B., Gilson, E., Stoppacciaro, A., Zupi, G., & Biroccio, A. (2007). Telomere damage induced by the G-quadruplex ligand RHPS4 has an antitumor effect. *Journal of Clinical Investigation, 117*, 3236–3247.

Sastry, M., Fiala, R., Lipman, R., Tomasz, M., & Patel, D. J. (1995). Solution structure of the monoalkylated mitomycin C-DNA complex. *Journal of Molecular Biology, 247*, 338–359.

Satange, R., Chuang, C.-Y., Neidle, S., & Hou, M.-H. (2019). Polymorphic G:G mismatches act as hotspots for inducing right-handed Z DNA by DNA intercalation. *Nucleic Acids Research, 47*, 8899–8912.

Savva, L., & Georgiades, S. N. (2021). Recent developments in small-molecule ligands of medicinal relevance for harnessing the anticancer potential of G-quadruplexes. *Molecules, 26*, 841.

Schneider, B., & Berman, H. M. (1995). Hydration of the DNA bases is local. *Biophysical Journal, 69*, 2661–2669.

Schneider, B., Cohen, D., & Berman, H. M. (1992). Hydration of DNA bases: Analysis of crystallographic data. *Biopolymers, 32*, 725–750.

Schneider, B., Patel, K., & Berman, H. M. (1998). Hydration of the phosphate group in double-helical DNA. *Biophysical Journal, 75*, 2422–2434.

Sedghi Masoud, S., & Nagasawa, K. (2018). i-motif-binding ligands and their effects on the structure and biological functions of i-motif. *Chemical & Pharmaceutical Bulletin, 66*, 1091–1103.

Seeman, N. C. (2005). Structural DNA nanotechnology: An overview. *Methods in Molecular Biology, 303*, 143–166.

Seeman, N. C., Rosenberg, J. M., & Rich, A. (1976). Sequence-specific recognition of double helical nucleic acids by proteins. *Proceedings of the National Academy of Sciences of the United States of America, 73*, 804—808.

Shieh, H. S., Berman, H. M., Dabrow, M., & Neidle, S. (1980). The structure of drug-deoxydinucleoside phosphate complex; generalized conformational behavior of intercalation complexes with RNA and DNA fragments. *Nucleic Acids Research, 8*, 85—97.

Shui, X., McFail-Isom, L., Hu, G. G., & Williams, L. D. (1998a). The B-DNA dodecamer at high resolution reveals a spine of water on sodium. *Biochemistry, 37*, 8341—8355.

Shui, X., Sines, C. C., McFail-Isom, L., Van Derveer, D., & Williams, L. D. (1998b). Structure of the potassium form of CGCGAATTCGCG: DNA deformation by electrostatic collapse around inorganic cations. *Biochemistry, 37*, 16877—16887.

Silver, G. C., Sun, J.-S., Nguyen, C. H., Boutorine, A. S., Bisagni, E., & Hélène, C. (1997). Stable triple-helical DNA complexes formed by benzopyridoindole- and benzopyridoquinoxaline- oligonucleotide conjugates. *Journal of the American Chemical Society, 119*, 263—268.

Silverman, A. P., Bu, W., Cohen, S. M., & Lippard, S. J. (2002). 2.4-Å crystal structure of the asymmetric platinum complex [Pt(ammine)(cyclohexylamine)]$^{2+}$ bound to a dodecamer DNA duplex. *Journal of Biological Chemistry, 277*, 49743—49749.

Singh, R. K., Kumar, S., Prasad, D. N., & Bhardwaj, T. R. (2018). Therapeutic journey of nitrogen mustard as alkylating anticancer agents: Historic to future perspectives. *European Journal of Medicinal Chemistry, 151*, 401—433.

Smith, C. K., Davies, G. J., Dodson, E. J., & Moore, M. H. (1995). DNA-nogalamycin interactions: The crystal structure of d(TGATCA) complexed with nogalamycin. *Biochemistry, 34*, 415—425.

Soeiro, M. N., Werbovetz, K., Boykin, D. W., Wilson, W. D., Wang, M. Z., & Hemphill, A. (2013). Novel amidines and analogues as promising agents against intracellular parasites: A systematic review. *Parasitology, 140*, 929—951.

Spingler, B., Whittington, D. A., & Lippard, S. J. (2001). 2.4 Å crystal structure of an oxaliplatin 1,2-d(GpG) intrastrand cross-link in a DNA dodecamer duplex. *Inorganic Chemistry, 40*, 5596—5602.

Spink, N., Brown, D. G., Skelly, J. V., & Neidle, S. (1994). Sequence-dependent effects in drug-DNA interaction: The crystal structure of Hoechst 33258 bound to the d(CGCAAATTTGCG)$_2$ duplex. *Nucleic Acids Research, 22*, 1607—1612.

Stagno, J. R., Bhandari, Y. R., Conrad, C. E., Liu, Y., & Wang, Y. X. (2017). Real-time crystallographic studies of the adenine riboswitch using an X-ray free-electron laser. *FEBS Journal, 284*, 3374—3380.

Sterling, T., & Irwin, J. J. (2015). ZINC 15—ligand discovery for everyone. *Journal of Chemical Information and Modeling, 55*, 2324—2337.

Subramaniam, G., Paz, M. M., Kumar, G. S., Das, A., Palom, Y., Clement, C. C., Patel, D. J., & Tomasz, M. (2001). Solution structure of a guanine-N7-linked complex of the mitomycin C metabolite 2,7-diaminomitosene and DNA. Basis of sequence selectivity. *Biochemistry, 40*, 10473—10484.

Suckling, C. (2012). From multiply active natural product to candidate drug? Antibacterial (and other) minor groove binders for DNA. *Future Medicinal Chemistry, 4*, 971—989.

Sun, D., Thompson, B., Cathers, B. E., Salazar, M., Kerwin, S. M., Trent, J. O., Jenkins, T. C., Neidle, S., & Hurley, L. H. (1997). Inhibition of human telomerase by a G-quadruplex-interactive compound. *Journal of Medicinal Chemistry, 40*, 2113—2116.

Takahara, P. M., Frederick, C. A., & Lippard, S. J. (1996). Crystal structure of the anticancer drug cisplatin bound to duplex DNA. *Journal of the American Chemical Society, 118*, 12309—12321.

Tanious, F. A., Laine, W., Peixoto, P., Bailly, C., Goodwin, K. D., Lewis, M. A., Long, E. C., Georgiadis, M. M., Tidwell, R. R., & Wilson, W. D. (2007). Unusually strong binding to the DNA minor groove by a highly twisted benzimidazole diphenylether: Induced fit and bound water. *Biochemistry, 46*, 6944—6956.

Temperini, C., Cirilli, M., Aschi, M., & Ughetto, G. (2005). Role of the amino sugar in the DNA binding of disaccharide anthracyclines: Crystal structure of the complex MAR70/d(CGATCG). *Bioorganic & Medicinal Chemistry, 13*, 1673—1679.

Tereshko, V., Minasov, G., & Egli, M. (1999). The Dickerson-Drew B-DNA dodecamer revisited at atomic resolution. *Journal of the American Chemical Society, 121*, 470—471.

Teuben, J.-M., Bauer, C., Wang, A. H.-J., & Reedijk, J. (1999). Solution structure of a DNA duplex containing a *cis*-diammineplatinum(II) 1,3-d(GTG) intrastrand cross-link, a major adduct in cells treated with the anticancer drug carboplatin. *Biochemistry, 38*, 12305—12312.

Thompson, A. S., & Hurley, L. H. (1995). Solution conformation of a bizelesin A-tract duplex adduct: DNA-DNA cross-linking of an A-tract straightens out bent DNA. *Journal of Molecular Biology, 252*, 86—101.

Thorpe, J. H., Hobbs, J. R., Todd, A. K., Denny, W. A., Charlton, P., & Cardin, C. J. (2000). Topoisomerase poison in the presence of Co^{2+} ions. *Biochemistry, 39*, 15055—15061.

Tietze, L. F., Krewer, B., von Hof, J. M., Frauendorf, H., & Schuberth, I. (2009). Determination of the biological activity and structure activity relationships of drugs based on the highly cytotoxic duocarmycins and CC-1065. *Toxins (Basel), 1*, 134—150.

Todd, A. K., Adams, A., Thorpe, J. H., Denny, W. A., Wakelin, L. P. G., & Cardin, C. J. (1999). Major groove binding and 'DNA-induced' fit in the intercalation of a derivative of the mixed topoisomerase I/II poison N-(2-(dimethylamino)ethyl)acridine-4-carboxamide (DACA) into DNA: X-Ray structure complexed to d(CG(5-BrU)ACG)$_2$ at 1.3-Å resolution. *Journal of Medicinal Chemistry, 42*, 536—540.

Todd, R. C., & Lippard, S. J. (2010a). Structure of duplex DNA containing the cisplatin 1,2-{Pt(NH3)2} 2+-d(GpG) cross-link at 1.77 A resolution. *Journal of Inorganic Biochemistry, 104*, 902—908.

Todd, R. C., & Lippard, S. J. (2010b). Consequences of cisplatin binding on nucleosome structure and dynamics. *Chemical Biology, 17*, 1334—1343.

Tomasz, M., & Palom, Y. (1997). The mitomycin bioreductive antitumor agents: Cross-linking and alkylation of DNA as the molecular basis of their activity. *Pharmacology & Therapeutics, 76*, 73—87.

Van Hecke, K., Nam, P. O. C., Nguyen, M. T., & Van Meervelt, L. (2005). Netropsin interactions in the minor groove of d(GGCCAATTGG) studied by a combination of resolution enhancement and ab initio calculations. *FEBS Journal, 272*, 3531—3541.

Vega, M. C., García Sáez, I., Aymami, J., Eritja, R., Van der Marel, G. A., Van Boom, J. H., Rich, A., & Coll, M. (1994). Three-dimensional crystal structure of the A-tract DNA dodecamer d(CGCAAATTTGCG) complexed with the minor-groove-binding drug Hoechst 33258. *European Journal of Biochemistry, 222*, 721—726.

Viger, A., & Dervan, P. B. (2006). Exploring the limits of benzimidazole DNA-binding oligomers for the hypoxia inducible factor (HIF) site. *Bioorganic & Medicinal Chemistry, 14*, 8539—8549.

Wang, A. H.-J., Ughetto, G., Quigley, G. J., & Rich, A. (1987). Interactions between an anthracycline antibiotic and DNA: Molecular structure of daunomycin complexed to d(CpGpTpApCpG) at 1.2-Å resolution. *Biochemistry, 26*, 1152—1163.

Wang, A. H.-J., Ughetto, G., Quigley, G. J., Hakoshima, T., van der Marel, G. A., van Boom, J. H., & Rich, A. (1984). The molecular structure of a DNA-triostin A complex. *Science, 225*, 1115—1121.

Wang, L., Bailly, C., Kumar, A., Ding, D., Bajic, M., Boykin, D. W., & Wilson, W. D. (2000). Specific molecular recognition of mixed nucleic acid sequences: An aromatic dication that binds in the DNA minor groove as a dimer. *Proceedings of the National Academy of Sciences of the United States of America, 97*, 12—16.

Wang, H., & Laughton, C. A. (2007). Molecular modelling methods for prediction of sequence-selectivity in DNA recognition. *Methods, 42*, 196—203.

Wang, F., Wang, C., Liu, Y., Lan, W., Han, H., Wang, R., Huang, S., & Cao, C. (2020). Colchicine selective interaction with oncogene RET G-quadruplex revealed by NMR. *Chemical Communications, 56*, 2099—2102.

Waring, M. J. (1970). Variation of the supercoils in closed circular DNA by binding of antibiotics and drugs: Evidence for molecular models involving intercalation. *Journal of Molecular Biology, 54*, 247—279.

Warren, C. L., Kratochvil, N. C. S., Hauschild, K. E., Foister, S., Brezinski, M. L., Dervan, P. B., Phillips, G. N., Jr., & Ansari, A. Z. (2005). Defining the sequence-recognition profile of DNA-binding molecules. *Proceedings of the National Academy of Sciences of the United States of America, 103*, 867—872.

Wei, D., Wilson, W. D., & Neidle, S. (2013). Small-molecule binding to the DNA minor groove is mediated by a conserved water cluster. *Journal of the American Chemical Society, 135*, 1369—1377.

Wheelhouse, R. T., Sun, D., Han, H., Han, F. X., & Hurley, L. H. (1998). Cationic porphyrins as telomerase inhibitors: The interaction of tetra-(N-methyl-4-pyridyl)porphine with quadruplex DNA. *Journal of the American Chemical Society, 120*, 3261—3262.

Williams, H. E. L., & Searle, M. S. (1999). Structure, dynamics and hydration of the nogalamycin-d(ATGCAT)₂ complex determined by NMR and molecular dynamics simulations in solution. *Journal of Molecular Biology, 290*, 699–716.

Wilson, W. D., Nguyen, B., Tanious, F. A., Mathis, A., Hall, J. E., Stephens, C. A., & Boykin, D. W. (2005). Dications that target the DNA minor groove: Compound design and preparation, DNA interactions, cellular distribution and biological activity. *Current Medicinal Chemistry - Anti-Cancer Agents, 5*, 389–408.

Woods, K. C., Martin, S. S., Chu, V. C., & Baldwin, E. P. (2001). Quasi-equivalence in site-specific recombinase structure and function: Crystal structure and activity of trimeric Cre recombinase bound to a three-way Lox DNA junction. *Journal of Molecular Biology, 313*, 49–69.

Wu, Y., Bhattacharyya, D., King, C. L., Baskerville-Abraham, I., Huh, S. H., Boysen, G., Swenberg, J. A., Temple, B., Campbell, S. L., & Chasney, S. G. (2007). Solution structures of a DNA dodecamer duplex with and without a cisplatin 1,2-d(GG) intrastrand cross-link: Comparison with the same DNA duplex containing an oxaliplatin 1,2-d(GG) intrastrand cross-link. *Biochemistry, 46*, 6477–6487.

Wu, B., Davey, G. E., Nazarov, A. A., Dyson, P. J., & Davey, C. A. (2011). Specific DNA structural attributes modulate platinum anticancer drug site selection and cross-link generation. *Nucleic Acids Research, 39*, 8200–8212.

Wu, B., Droge, P., & Davey, C. A. (2008). Site selectivity of platinum anticancer therapeutics. *Nature Chemical Biology, 4*, 110–112.

Wu, P. C., Tzeng, S. L., Chang, C. K., Kao, Y. F., Waring, M. J., & Hou, M. H. (2018). Cooperative recognition of T:T mismatch by echinomycin causes structural distortions in DNA duplex. *Nucleic Acids Research, 46*, 7396–7404.

Yang, X.-L., Robinson, H., Gao, Y.-G., & Wang, A. H.-J. (2000). Binding of a macrocyclic bisacridine and ametantrone to CGTACG involves similar unusual intercalation platforms. *Biochemistry, 39*, 10950–10957.

Zahler, A. M., Williamson, J. R., Cech, T. R., & Prescott, D. M.((1991). Inhibition of telomerase by G-quartet DNA structures. *Nature, 350*, 718–720.

Zhang, Q. A., Dwyer, T. J., Tsui, V., Case, D. A., Cho, J., Dervan, P. B., & Wemmer, D. E. (2003). NMR structure of a cyclic polyamide-DNA complex. *Journal of the American Chemical Society, 126*, 7958–7966.

Zimmer, C. H., & Wähnert, U. (1986). Non-intercalating DNA-binding ligands: Specificity of the interaction and their use as tools in biophysical, biochemical and biological investigations of the genetic material. *Progress in Biophysics and Molecular Biology, 47*, 31–112.

Further reading

DNA-Water interactions

Berman, H. M. (1991). Hydration of DNA. *Current Opinion in Structural Biology, 1*, 423–427.

Berman, H. M. (1994). Hydration of DNA: Take 2. *Current Opinion in Structural Biology, 4*, 345–350.

Biedermannová, L., & Schneider, B. (2016). Hydration of proteins and nucleic acids: Advances in experiment and theory. A review. *Biochimica et Biophysica Acta, 1860*, 1821–1835.

Nguyen, B., Neidle, S., & Wilson, W. D. (2009). A role for water molecules in DNA–ligand minor groove recognition. *Accounts of Chemical Research, 42*, 11–21.

Westhof, E. (1988). Water: An integral part of nucleic acid structure. *Annual Review of Biophysics, 17*, 125–144.

Westhof, E., & Beveridge, D. L. (1990). In F. Franks (Ed.), *Water science reviews 5*. Cambridge UK: Cambridge University Press.

DNA-drug interactions

Bhaduri, S., Ranjan, N., & Arya, D. P. (2018). An overview of recent advances in duplex DNA recognition by small molecules. *Beilstein Journal of Organic Chemistry, 14*, 1051–1086.

Bolhuis, A., & Aldrich-Wright, J. R. (2014). DNA as a target for antimicrobials. *Bioorganic Chemistry, 55*, 51–59.

Denison, C., & Kodadek, T. (1998). Small-molecule-based strategies for controlling gene expression. *Chemical Biology, 5*, 129–145.

Dervan, P. B., & Edelson, B. S. (2003). Recognition of the DNA minor groove by pyrrole-imidazole polyamides. *Current Opinion in Structural Biology, 13*, 284–289.

Gale, E. F., Cundliffe, E. F., Reynolds, P. E., Richmond, M. H., & Waring, M. J. (1981). *The molecular basis of antibiotic action*. London: John Wiley.

Gao, X., & Patel, D. J. (1989). Antitumour drug-DNA interactions: NMR studies of echinomycin and chromomycin complexes. *Quarterly Reviews of Biophysics, 22*, 93–138.

Hurley, L. H. (2002). DNA and its associated processes as targets for cancer therapy. *Nature Reviews Cancer, 2*, 188–200.

Kawamoto, Y., Bando, T., & Sugiyama, H. (2018). Sequence-specific DNA binding Pyrrole-imidazole polyamides and their applications. *Bioorganic & Medicinal Chemistry, 26*, 1393–1411.

Mantaj, J., Jackson, P. J., Rahman, K. M., & Thurston, D. E. (2017). From anthramycin to pyrrolobenzodiazepine (PBD)-containing antibody-drug conjugates (ADCs). *Angewandte Chemie International Edition in English, 56*, 462–488.

Martins-Teixeira, M. B., & Carvalho, I. (2020). Antitumour anthracyclines: Progress and perspectives. *ChemMedChem, 15*, 933–948.

Neidle, S. (2001). DNA minor-groove recognition by small molecules. *Natural Product Reports, 18*, 291–309.

Neidle, S., & Nunn, C. M. (1998). Crystal structures of nucleic acids and their drug complexes. *Natural Product Reports, 15*, 1–15.

Neidle, S., & Waring, M. J. (Eds.). (1993). *Molecular basis of drug-DNA antitumour action* (Vol. 1, p. 2). London: Macmillan Press.

Pett, L., Hartley, J. A., & Kiakos, K. (2015). Therapeutic agents based on DNA sequence specific binding. *Current Topics in Medicinal Chemistry, 15*, 1293–1322.

Pratt, W. B., Ruddon, R. W., Ensminger, W. D., & Maybaum, J. (1994). *The anticancer drugs* (2nd ed.). New York: Oxford University Press.

Rajski, S. R., & Williams, R. M. (1998). DNA cross-linking agents as antitumor drugs. *Chemical Reviews, 98*, 2723–2796.

Reddy, B. S. P., Sondhi, S. M., & Lown, J. W. (1999). Synthetic DNA minor groove-binding drugs. *Pharmacology & Therapeutics, 84*, 1–111.

Ren, J., & Chaires, J. B. (1999). Sequence and structural selectivity of nucleic acid binding ligands. *Biochemistry, 38*, 16067–16075.

Rescifina, A., Zagni, C., Varrica, M. G., Pistarà, V., & Corsaro, A. (2014). Recent advances in small organic molecules as DNA intercalating agents: Synthesis, activity, and modeling. *European Journal of Medicinal Chemistry, 74*, 95–115.

Sheng, J., Gan, J., & Huang, Z. (2013). Structure-based DNA-targeting strategies with small molecule ligands for drug discovery. *Medicinal Research Reviews, 33*, 1119–1173.

Thurston, D. E., & Pysz, I. (2021). *Chemistry and pharmacology of anticancer drugs*. CRC Press.

Wemmer, D. E., & Dervan, P. B. (1997). Targeting the minor groove of DNA. *Current Opinion in Structural Biology, 7*, 355–361.

Yang, X.-L., & Wang, A. H.-J. (1999). Structural studies of atom-specific anticancer drugs acting on DNA. *Pharmacology & Therapeutics, 83*, 181–215.

Quadruplex-drug complexes

Balasubramanian, S., Hurley, L. H., & Neidle, S. (2011). Targeting G-quadruplexes in gene promoters: A novel anticancer strategy? *Nature Reviews Drug Discovery, 10*, 261–275.

Kosiol, N., Juranek, S., Brossart, P., Heine, A., & Paeschke, K. (2021). G-quadruplexes: A promising target for cancer therapy. *Molecular Cancer, 20*, 40.

Ruggiero, E., & Richter, S. N. (2020). Viral G-quadruplexes: New frontiers in virus pathogenesis and antiviral therapy. *Annual Reports in Medicinal Chemistry, 54*, 101–131.

CHAPTER 6

RNA structures and their diversity

6.1 Introduction

The preceding chapters have illustrated the structural variability displayed by DNA and its interactions with small molecules. Virtually all of this is within the boundaries imposed by the over-riding structural requirements of double-, triple- or four-stranded helices. Most RNA molecules are not subject to these limitations, and consequently their structural variety is in striking contrast to the relative predictability of many DNA tertiary structures (though the increasing range of topologies shown by quadruplex and catalytic DNAs suggests that DNA can be structurally highly versatile under appropriate circumstances when released from the constraints of the double helix). It is perhaps unsurprising (at least in retrospect!) that the ability of many RNAs to fold into compact tertiary structures, analogous to the folding of polypeptides into proteins and enzymes, is paralleled in some instances, by the possession of catalytic activity.

The need for an RNA intermediary between the code for a gene contained in DNA, and the ultimate gene product, a protein, was realised soon after the discovery of the structure of DNA itself. This category of (single-stranded) RNA molecule was initially conceived of as a carrier of the code to the ribosome, although the picture of RNA as an intermediary is somewhat complicated by the existence of double-stranded RNA in RNA viruses. The need for a second type of RNA molecule, transfer RNA (tRNA), was also recognised early in the history of molecular biology. Transfer RNAs perform the specialised function of decoding and transferring the information contained in an individual codon in a messenger RNA molecule into the correct amino acid. An organism has 20 sets of tRNA molecules, each set specific for a particular amino acid (in principle only one tRNA is required for each amino acid, but in practice there are more than this, with some redundancy being needed for several amino acids). A specific tRNA attaches a correct amino acid and transfers it to a specialised region of the ribosome, where it is incorporated into a growing protein chain. A third yet more specialised function for RNA has been found, that of RNA enzymatic activity, for catalytic cleavage of RNA (and DNA) itself. Such cleavage can be intramolecular, such as self-splicing introns in RNA editing. RNA cleavage by RNA molecules, termed ribozymes, has excited much interest with the suggestion that they could be potential key players in a prebiotic pre-DNA world. Ribozymes have also been studied for their potential therapeutic use in gene therapy, for cleavage and ultimate destruction of specific DNA sequences. A further function arises from the discovery of some complex RNA sequences that can fold and act as molecular switches (riboswitches) on binding certain small molecules. Finally, and

Principles of Nucleic Acid Structure
ISBN 978-0-12-819677-9, https://doi.org/10.1016/B978-0-12-819677-9.00002-0

most surprisingly, has been the discovery that short 20—25 nucleotide long double-stranded RNAs (small interfering RNAs, or siRNAs) can inhibit eukaryotic gene expression. They do this via the RNA interference (RNAi) pathway.

RNA crystallographic and NMR structural studies have advanced rapidly over the past few years. There have been several technical advances that have made this possible. On the one hand, solid-phase RNA chemical synthesis and purification (of sequences up to 15—20 bases in length) is now straightforward for the preparation of milligram quantities (Ahmed & Ficner, 2014; Anderson et al., 1996; Kundrot, 1997; Roy & Caruthers, 2013). Even short RNA sequences are very prone to secondary structure formation and nuclease digestion, and especial care in handling and purification is needed. Longer sequences are often prepared enzymatically, using efficient in vitro transcription of the appropriate DNA sequence with T7 RNA polymerase (Dégut et al., 2016), although there is a requirement for a purine sequence at the 5′ end so that optimal efficiency of transcription can take place. Also, random sequence nucleotides are sometimes added at the 3′ end by T7 polymerase. These problems can be overcome by methods that can post-transcriptionally excise the desired sequence (Price et al., 1995). Commercial kits are widely available for RNA transcription, which can readily produce milligram quantities of sequences up to several hundred bases in length.

Crystallisation of RNAs has traditionally been difficult, in part on account of the inherent flexibility of many RNA molecules. The introduction of sparse matrix methods coupled with robotic technology has enabled success to be achieved relatively rapidly with many RNAs (Doudna et al., 1993; Cate & Doudna, 2000; Dock-Bregeon et al., 1999; Ke & Doudna, 2004; Golden, 2007). These methods enable a very large range of crystallisation conditions to be rapidly sampled, using only small quantities of RNA. However, success by screening alone is never guaranteed, and a variety of RNA-specific methods have been successfully developed (see, for example, Zhang & Ferré-D'Amaré, 2014), which, for example, rationally engineer specific intermolecular contacts to optimise the formation of crystalline arrays of molecules (Ferre-D'Amare et al., 1998a,b). Crystal structures of the majority of small, purely helical RNAs are often routinely solved by molecular replacement methods, normally using canonical duplex RNA models as starting points. Synthetic RNAs, as with DNAs, can incorporate heavy atoms such as bromine for MAD phasing, on specific residues. These are needed when molecular replacement does not work, or with more complex structures (see, for example, Wedekind & McKay, 2000). Large, complex RNAs always require heavy-atom derivatives, and systematic approaches have been developed for rapidly homing in on those heavy-atom reagents that are most likely to produce useable derivatives, such as osmium hexamine. A useful strategy in some instances is to use a small synthetic RNA sequence containing a heavy-atom derivative, which can become an integral part of the total RNA molecule when annealed with the larger RNA component prepared by transcription. The incorporation of selenium in 2′-methylseleno derivatives, synthesised either chemically or enzymatically, is of particular use for MAD phasing (Höbartner et al., 2005). The methods developed for DNA phasing and described in Chapter 1 are also applicable to RNAs. Finally, it needs emphasising, that in common with other nucleic acid structures, errors in RNA crystallographic and NMR structures do occasionally occur. So, the reader is strongly encouraged to scrutinise the Protein Data Bank (PDB) validation reports for structures of interest, which hopefully will indicate if there is a problem with a structure (Westhof, 2016).

6.2 Fundamentals of RNA structure

6.2.1 Helical RNA conformations

RNA differs from DNA in two key chemical aspects (Fig. 6.1):

- the sugar groups in all RNAs have an OH group attached at the $2'$ position, making them ribose rather than deoxyribose sugars
- thymine bases are replaced by uracils, which have the same hydrogen bonding, but do not have a methyl group at the 5-position

The effect of the extra hydroxyl group is profound. RNA sugars are much more rigid than DNA ones, and only adopt the C3$'$-*endo* pucker type (Fig. 6.2) since otherwise the $2'$-OH group would clash with C8 (for a purine) or C6 (for a pyrimidine) of the attached base. Simulation studies, together with analysis of crystal structures, have shown that this pucker is stabilised in several positive ways (Auffinger & Westhof, 1997; Darré et al., 2016). The $2'$-OH group can be oriented in several discrete positions, resulting from steric interactions with atoms O3$'$ or O4$'$, or with accessible hydrogen-bond acceptors on the attached base. In double helical A-RNA, the important interaction is between the

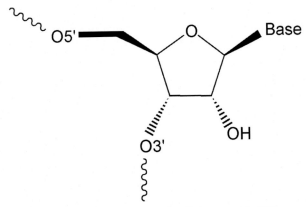

Figure 6.1 Configuration of a ribose sugar in RNA.

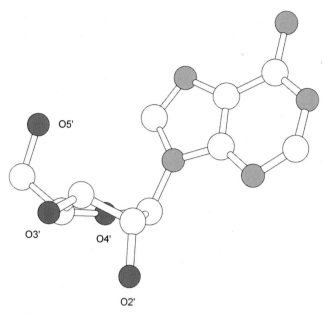

Figure 6.2 A ribonucleoside in an A-RNA conformation.

H2′ atom and O4′ of the next residue, which is not accessible for hydrogen bonding to water molecules.

Single-stranded RNAs do not adopt rigid conformations in the absence of restraining tertiary interactions. However, appropriate complementary or self-complementary sequences readily form anti-parallel double-stranded RNA structures, analogous to duplex DNA. Uracil participates in U•A base pairs that are fully isomorphous with A•T ones in duplex DNA. Duplex RNA is conformationally rather rigid, and its behaviour contrasts remarkably with the polymorphism of duplex DNA in that only one major polymorph of the RNA double helix has been observed. This has many features in common with A-DNA, and accordingly is known as A-RNA (Arnott et al., 1973). The conformational features of canonical RNA helices have been obtained from fibre-diffraction studies on duplex RNA polynucleotides from both viral and synthetic origins. A-RNA is an 11-fold helix, with a narrow and deep major groove and a wide, shallow minor groove, and base pairs inclined to and displaced from the helix axis (Fig. 6.3a and b). It is notable that the base pairs themselves in canonical A-RNA are only slightly twisted. The major A-RNA form is apparent in natural polyribonucleotide fibres, or in those formed from poly(A)• poly(T) in conditions of low ionic strength (Arnott et al., 1973). At higher ionic strength, a slightly different 12-fold helix (A′-RNA) is formed, which retains the same overall helical features, although its wider major groove resembles that of A-DNA. These are detailed in Table 6.1. Both helices have the C3′-*endo* sugar pucker characteristic of

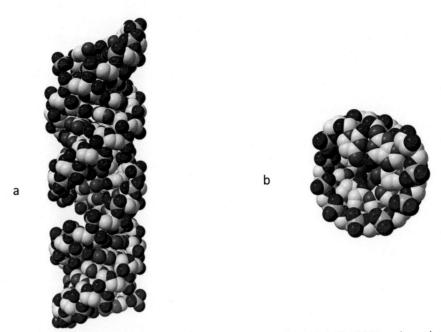

Figure 6.3 (a) Space-filling representation of a canonical A-RNA double helix. (b) View down the helix axis of the canonical A-RNA double helix. Note that the base pairs sit astride the helix axis.

Table 6.1 Structural features of RNA double helices, from fibre-diffraction studies (Arnott, 1999). (a) Conformational angles (°). Values for poly(A)• poly(T) in the A form are given. (b) Groove dimensions in Å. (c) Selected helical and base/base pair morphological parameters.

		α	β	χ	δ	ε	ξ	χ
A-RNA	Adenine	−69	179	55	82	−154	−71	−161
	Thymine	−64	178	51	83	−152	−173	−161
A′-RNA		−70	177	61	77	−153	−70	−163

	Major groove width	Major groove depth	Minor groove width	Minor groove depth
A-RNA	4.7	12.9	10.8	3.3
A′-RNA	8.9	14.4	10.5	3.4

	Base displacement (Å)	Rise (Å)	Helical twist (°)	Roll (°)	Propeller twist (°)	Inclination (°)
A-RNA	0.44	2.8	32.7	−0.8	−2.1	−15.5
A′-RNA	0.51	3.0	30.0	0.0	2.3	−10.6

ribopolynucleotides. Another difference from duplex DNA is that RNA helices, though capable of a small degree of bending (up to ca 15°), do not normally undergo the large-scale bending seen, for example, in A-tract DNA.

Numerous single-crystal and NMR studies on sequences forming perfectly base-paired duplex RNA have been reported, which have shown that the RNA geometry observed in fibres is conserved across a wide range of sequences. This conclusion assumes that there are no base mismatches, which can under some circumstances, greatly alter helical structure (see below). Due to their greater conformational rigidity RNA helices do not show the pronounced sequence-dependent microstructure of duplex DNA. This has received confirmation from several database surveys, including one of 3000 nucleotides from ribosomal crystal structures (Schneider et al., 2004), which found that most are A-type (i.e. helical). The 18 other conformational types are representative of the various non-helical motifs that are characteristic of complex RNA molecules, and which form much of the subject-matter of this chapter. A more recent survey of almost 120,000 DNA and RNA dinucleotides from the PDB (Černý et al., 2020) included riboswitches, RNA tetraloops and other complex RNA structures. Of the total of 97 dinucleotide conformational classes defined in this analysis, 28 occur predominantly in RNA non-duplex structures. One surprising conclusion from this work is that DNA molecules with more uniform, mostly double helical architecture show more complex sequence dependencies than do complex folded RNAs. An earlier analysis of RNA conformations (Richardson et al., 2008) came to broadly similar conclusions, highlighting, for example, that 75% of conformations in RNA molecules are an A-type double helical form, at least at the time of that survey. There are some differences between the conformational classes (NtCs) of Černý et al. and the consensus RNA conformers of Richardson et al., although 38 NtCs are consistent with 46 of the conformers. One factor is the greatly increased number of high-quality crystal structures that became available in the intervening 12-year period, but there are some real differences in definition, with NtCs including more detailed definitions of base/sugar orientations.

The first reported single-crystal RNA analyses were of the self-complementary ribo-dinucleosides r(AU) and r(GC) (Rosenberg et al., 1976; Seeman et al., 1976). These determinations, at atomic resolution, are of historic importance in showing Watson—Crick base pairing within the context of short helical segments, and in demonstrating that even dinucleoside duplexes have backbone conformations characteristic of helical A-RNA. They also provided the impetus for structural studies of intercalative drug binding to dinucleoside duplexes as models for the polynucleotide situation (see Chapter 5).

Crystallographic analyses of sequences such as the octanucleotide r(CCCCGGGG) have shown helices of length sufficient for a full set of helical parameters to be extracted. This sequence crystallises in two distinct crystal lattices, enabling the effects of crystal packing factors on structure to be assessed (Portmann et al., 1995). In each instance, rhombohedral and hexagonal, the RNA helices are very similar, and their features closely resemble those in fibre-diffraction canonical A-RNA, with, for example, minor groove widths of 9.8 and 9.7 Å, respectively. Visualisation of a complete turn of RNA double helix requires a longer sequence. The structure of r[U(UA)$_6$A] was the first to show this (Dock-Bregeon et al., 1989). Again, the helix is essentially classical A-RNA (Fig. 6.4), with an average helical twist of 33° and an axial rise of 2.8 Å. Interestingly the structure shows a consistent pattern on both strands, of hydrogen bonds between ribose ring sugar atom O4′ and the O2′ of the preceding nucleotide. The structure of a mixed sequence, r(UAAGGAGGUGAU) (Schindelin et al., 1995), is a more rigorous test of the conservation of RNA helical structure. This RNA dodecamer sequence closely resembles a sequence required for the initiation of translation in prokaryotics. The overall features of the two independent helices in this crystal structure are of A-RNA type rather than A′-RNA, with, for example, helical twists of 33.3 and 33.5°, respectively, corresponding to a helical repeat of 10.7−10.8 base pairs/turn.

Interpretable diffraction patterns from fibres of the polyribonucleotide poly(rA) were examined by Rich et al. (1961) and a model was proposed that was consistent with the diffraction data. This has a parallel-stranded double helix with propeller-twisted A•A

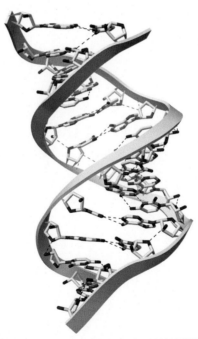

Figure 6.4 The crystal structure of the duplex of r(UUAUAUAUAUAUAA), shown in schematic form (Dock-Bregeon et al., 1989), PDB id 1RNA. Note the characteristic A-RNA inclination of base pairs with respect to the vertical helix axis.

base pairs, slightly tilted from the horizontal axis and with a helical twist of 45°. More recently, a high-resolution single-crystal analysis has been reported for r(A)$_{11}$ (Safaee et al., 2013), which unusually for an oligonucleotide crystal structure was solved by direct methods. The symmetric double helix is parallel-stranded, with symmetric A•A base pairs (Fig. 6.5a—c). There is further hydrogen bonding between N6 and a phosphate group on the opposite strand. This interaction is strengthened by the involvement of a bridging ammonium ion, between the same phosphate oxygen group and N1 of the adenine. The assignment of ammonium ions was based on their proximity to the anionic phosphate groups, although these cannot be distinguished from water molecules because of their very similar electron densities, even at the 1 Å resolution of the data. It is a tribute to the earlier fibre diffraction structure, obtained with remarkably little data, that it closely corresponds with the single-crystal structure — both are parallel duplexes and the A•A base pairing together with the involvement of a phosphate group was also included in the fibre model.

Longer RNA duplex sequences are sometimes more difficult to crystallise in ordered conformations. This is illustrated by the analysis of two duplexes, an 18- and a 19-mer,

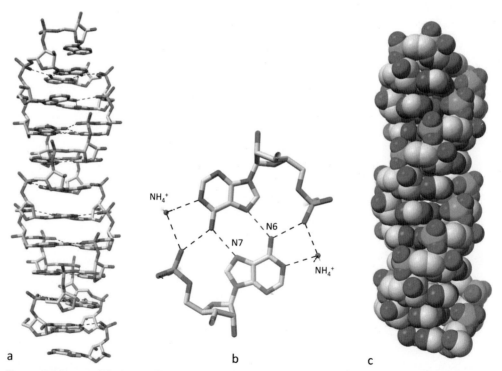

a   b   c

Figure 6.5 Views of the poly r(A) (r(A)$_{11}$) crystal structure (Safaee et al., 2013), PDB id 4JRD. (a) Cartoon view of the structure. (b) View of 1 A•A base pair, showing the two N6—N7 hydrogen bonds and the ammonium ions, each bridging between N1 and a phosphate group from the other strand. (c) A van der Waals representation, showing the smooth external surface and the absence of any substantial groove structure.

which diffract to high resolution, 1.20 and 1.55 Å, respectively (Klosterman et al., 1999). Even though this apparently indicates a high degree of internal order, the 18-mer is disordered such that each base pair is an average of the overall duplex pairs. This is closely analogous to the fibre diffraction situation and frequently occurs in both long DNA and RNA duplexes when the crystal packing motif involves continuous helices. The 19-mer is disordered in a distinct way, with two orientations being apparent. Both sequences are in classic A-RNA conformations, except for the sugar puckers, which have significantly larger pucker amplitudes than in simple ribonucleosides. Analogous static disorder has been observed in the high-resolution (1.58 Å) crystal structure (Mooers et al., 2005) of the 18-mer duplex with the sequence r[(CUG)$_6$], which is implicated in the genetic disease myotonic dystrophy. The structure contains two lattices related by a translation of one helical turn, with each double helix packing head-to-tail in a pseudo-continuous manner. The helix is in an undistorted A-form structure, even though six non-hydrogen-bonded U•U base pairs were reported, which alter the pattern of electrostatic potential in the minor groove, to one that may be important for binding to specific proteins implicated in this disease. Structure PDB id 3GM7 (Kiliszek et al., 2009) was obtained after concluding that the structure was not disordered as reported in the earlier analysis of this sequence (PDB id 1ZEV). Instead, the crystals exhibited twinning, which was deconvoluted and an unambiguous new structure was obtained. This has all the C•G base pairs being of the classic Watson–Crick type whereas the U•U base pairs are wobble pairs, each with a single hydrogen bond.

It is common for isolated non-canonical base pairs, such as A•C and G•G to be fully accommodated within an RNA double helix, without the necessity of helix disruption by, for example, bases swinging out to become extra-helical (Olson et al., 2019). Thus, all the structures listed in Table 6.2, apart from the first entry, have non-canonical base pairs, and some have several contiguous ones. The crystal structure of a 19-mer (Table 6.2), PDB id 6IA2 (Nowacka et al., 2019), is typical, with 14 canonical base pairs, 4 A•C and 1 G•G base pair, which is disordered in the crystal structure. The wobble base pairs induce a loosening up of the canonical A-RNA helical parameters, so that, for example, the helical twists vary between 22° and 38° with an average of 31°. Overall though the structure retains its A-RNA character. An extreme example of an RNA maintaining its duplex character despite non-canonical base pairs is the 18-nucleotide structure PDB id 5NXT (Garg & Heinemann, 2018), which has a stretch of six continuous wobble base pairs, 2 G•U and 4 C•A$^+$ pairs, flanked on either side by six standard Watson–Crick base pairs. Again, the overall structure is very typical of canonical A-RNA, with 11 base pairs per turn. The central wobble region with large local variations in helical twist still has an average of 12 base pairs per turn, demonstrating that the wobble central region of the helix is well accommodated within the overall A-RNA form. The RNA–DNA hybrid structures also have A-type helices, as found in the crystal structure of the RNA/DNA dodecamer corresponding to 75% of the HIV-1 polypurine tract, r(UAAAAGAAAAGG) and d(CCTTTTCTTTTA), PDB id 3SSF (Drozdzal et al., 2011).

Table 6.2 Selected RNA double helical crystal structures.

Sequence	Resolution, Å	PDB id
r(CCCCGGGG)	1.46	259D
r(UAAGGAGGUGAU)	2.00	1SRD
r[U(UA)₆A]	2.25	1RNA
r[(G)₁₁]	1.20	1QCU
r[(CUG)₆]	1.58	1ZEV
r(AGAGAACCC GGAGUUCCCU)	2.27	6IA2
r(GCCGCCGC)	1.54	4E59
r[(CUG)₆]	1.58	3GM7
r(UGUUCUCUAC GAAGAACA)	1.38	5NXT
r(GGGUGGUGC GGG) + r(CCUGC ACUGCCC)	2.60	4JRT
r(A)₁₁	1.00	4JRD

The availability of high-resolution crystal structures has enabled RNA hydration and the role of metal ions in oligonucleotides to be defined in detail (Auffinger et al., 2011; Sundaralingam & Pan, 2002). Some general rules have emerged for the water arrangements beyond the obvious finding of an inherently greater degree of hydration compared to DNA oligonucleotides, on account of the $2'$-OH group being an active and willing hydrogen bond participant. There is no analogue of the minor-groove spine of hydration seen in B-DNA. Again, this is unsurprisingly since in A-RNA the minor groove is too wide for such an arrangement. A detailed analysis (Egli et al., 1996) of the crystal structure of r(CCCCGGGG) at a resolution of 1.8 Å has shown that the base edges and phosphate groups tend to be more extensively hydrated than in DNA. This appears to be in large part due to the $2'$-OH groups, with the 16 in this duplex having 33 hydrogen bonds to water molecules. These $2'$-OH groups act as foci for the establishment of water networks in both major and minor grooves. There is, for example, a set of seven identical well-defined transverse hydrogen-bonded water linkages across the minor groove (Fig. 6.6). The deep major groove has extensive networks of linked water molecules, often with water pentagons linking phosphate groups to base edges. Similar hydration patterns have been observed in the structure of an RNA duplex with alternating CG base pairs (Adamiak et al., 2010).

Molecular dynamics simulations on equivalent DNA and RNA oligomers have confirmed the slightly greater hydration of the latter (Auffinger & Westhof, 2000), with ca 21.4 water molecules per RNA C•G base pair compared to 19.8 for a B-DNA one. Counter-ions such as sodium or potassium tend to accumulate in the major groove of RNA duplexes (Auffinger & Westhof, 2000; Cheatham & Kollman, 1997), by

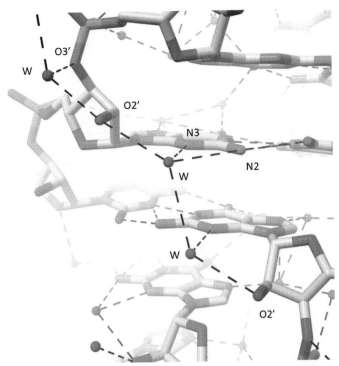

Figure 6.6 View of the transverse hydrogen-bonded water linkages across the minor groove in the structure of r(CCCCGGGG) (Egli et al., 1996), PDB id 259D, showing the inter-strand bridging between O2′ atoms achieved by two water molecules, each of which also hydrogen bonds to base edges of successive purines.

contrast with their minor-groove localisation in B-DNA. Simulations on A-RNA duplexes (Kührová et al., 2014) have also demonstrated that the choice of water model is important since different models have effects not only on helical parameters, as well as on base and base pair morphology, but also on hydration patterns in the minor groove around the 2′-OH groups. Consistent results have been obtained with the TIP3P, TIP4P/2005, and SPC/E models. Good-quality RNA force fields are now available, giving one some confidence in the veracity of long time-scale RNA molecular dynamics simulations (see, for example, Robertson et al., 2019). However, the equivocal experience with UUCG tetraloop simulations (Mrázíková et al., 2020) does suggest that caution is still needed in the interpretation of RNA simulations.

6.2.2 Mismatched and bulged RNA structures

RNAs can readily form stable base pair and triplet mismatches and extrahelical regions within the context of "normal" RNA double helices. Such features, notably the extrahelical loops (Fig. 6.7), have been the subject of intense structural study, since they are

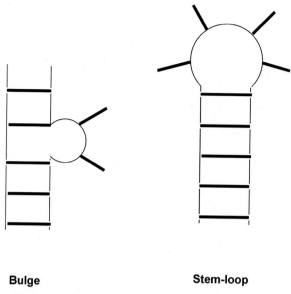

Bulge   **Stem-loop**

Figure 6.7 Two types of extra-helical features in RNA.

present in all biologically important complex RNAs (tRNAs, mRNAs, ribozymes, ribos-switches, ribosomal RNA), and together with various types of non-classical base stacking, are responsible for maintaining their tertiary structures.

The major types of RNA non-canonical mispairings are given in Table 6.3. The numbers for each type are from a survey published in 2009 (Olson et al., 2009), but more recent estimates are unlikely to significantly change the relative ratios of their occurrences.

It is paradoxically common for crystal structures (Table 6.4) of short sequences containing potential loop regions such as the UUCG "tetraloop" (which forms an especially stable extra-helical loop structure in solution), to often not to show such features. This may be a consequence of the high ionic strength of many crystallisation conditions, together with the preference of some sequences to pack in the crystal as helical arrays. So, instead of loops, these crystal structures tend to have runs of non-Watson—Crick mismatched base pairs where the loop would have formed. For example, an A-RNA double helix, albeit with G•U and U•U base pairs, is formed in the crystal (Cruse et al., 1994) by the sequence r(GCUUCGGC)d(^{Br}U). There is evidence in some structures of small deviations from the exact canonical A-RNA duplexes formed by fully Watson—Crick base pairs, since this helix has <10 base pairs per turn, although this is not always the case. The dodecamer sequence r(GGACUUCGGUCC) forms a base-paired duplex (Holbrook et al., 1991), with U•C and G•U base pairs. The A-RNA helix here has a significant

Table 6.3 The major RNA mispairings.

Base pair type	Name	Hydrogen bonds	Occurrence
G•C	Watson—Crick	N2—H … O2 O6 … H—N4 N1—H … N3	9500
A•U	Watson—Crick	N1 …. H—N3 N6—H ….O4	3069
G•U	Wobble	N1—H … O2 O6 … H—N3	1049
G•A	Sheared	N2—H … N7 N3 … H—N6	509
A•U	Hoogsteen	N6—H … O2 N7 …. H—N3	354
U•U	Wobble	O2 …. H—N3 N3—H … O4	185
G•A	Watson—Crick	N1—H … N1 O6 …. H—N6	141

Note that not all occurrences have the complete hydrogen bonding specified in column three.
Data taken from Olson, W. K., Esguerra, M., Xin, Y., & Lu, X. J. (2009). New information content in RNA base pairing deduced from quantitative analysis of high-resolution structures. *Methods*, 47, 177—186.

Table 6.4 Crystal and NMR structures of some small RNAs with non-helical features.

Sequence	Feature	Resolution in Å	PDB code no
r(GGACUUCGGUCC)	U•C, G•U base pairs	2.00	255D
r(GGCCGAAAGGCC)	G•A, A•A base pairs	2.3	283D
29-nucleotide SRP	Purine bulge	2.00	1D4R
r(GCGCGGCACCGUCC GCGGAACAAACGG)	Pseudoknot	1.60	437D
HIV-1 RRE site	G•G, A•G base pairs; A, U extrahelical	1.60	1CSL
Telomerase RNA domain	Pseudoknot	NMR	1YMO
Viral ribosomal frameshift RNA	Pseudoknot	1.25	1L2X
HIV-1 RNA dimerisation	Kissing complex	3.10	1K9W
Telomerase RNA	Pseudoknot	NMR	2K96 2K95
	Single A bulge	NMR	2JXS
Stem loop	Stem loop	NMR	2K5Z
Stem loop IV of *Tetrahymena* telomerase RNA	Stem loop	NMR	2FEY
Adenovirus virus-associated RNA	Apical loop, wobble-enriched, coaxially stacked loop, tetra-stems, pseudoknot	2.74	6OL3

increase in the width of its major groove, possibly on account of the water molecules that are strongly associated with the mismatch base pairs.

Much greater perturbations are apparent (Baeyens et al., 1996) in the structure of the dodecamer r(GGCCGAAAGGCC), PDB id 283D, where the four non–Watson–Crick base pairs in the centre of the sequence form an internal loop with sheared G•A and A•A base pairs. The resulting structure is highly distorted from A-RNA ideality (Fig. 6.8), with a compression of the major groove, an enlargement of the minor groove width to 13.5 Å, and a pronounced curvature of the resulting helix. Again, this sequence forms a tetraloop structure in solution (Heus & Pardi, 1991). The G•U base pair is a prominent and very important element of large RNA structures since it is especially stable (Varani & McClain, 2000), on account of its two hydrogen bonds (Fig. 6.9). It is known as the "wobble" pair since a G in the first position of a codon can accept either a C or a U in the third anticodon (wobble) position.

The RNA genome of the HIV-1 retrovirus contains many non-helical features, some of which have been studied by structural methods. Its dimerisation initiation site has features that act as signals for RNA packaging. Again, the expected secondary structure

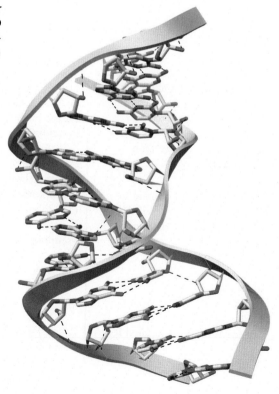

Figure 6.8 The structure of the dodecamer duplex r(GGCCGAAAGGCC), PDB id 283D (Baeyens et al., 1996), showing the distortions from regularity in the backbone, even though all bases are fully base paired.

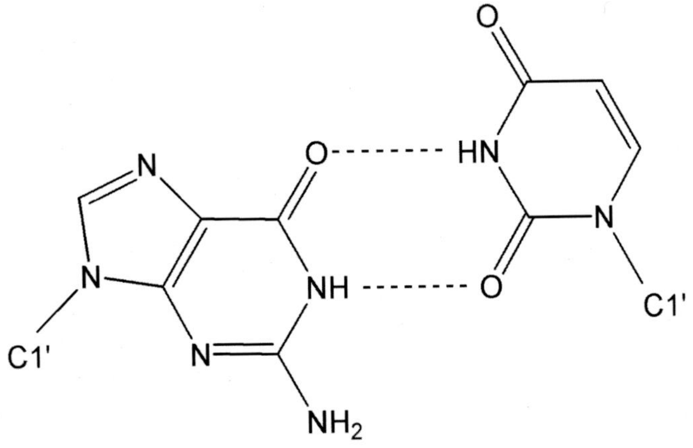

Figure 6.9 The G•U wobble base pair.

(containing two loops in a "kissing-loop" arrangement) has not always been observed. Instead, the duplex contains two A•G base pairs (Fig. 6.10), each adjacent to an extra-helical, bulged-out adenosine (Ennifar et al., 1999). The structure of a true kissing complex for this RNA has been determined (Ennifar et al., 2001), which shows loop–loop interactions (Fig. 6.11) where the two loops are in intimate contact. This involves looped-out bases from one strand forming a run of extra-helical A•U and G•C base pairs with the looped-out bases from the adjacent strand (Fig. 6.12). The tendency of short RNA sequences to maximise helical features is also apparent in the crystal structure of a 29-nucleotide fragment from the signal recognition particle (Wild et al., 1999), which forms 28-mer heteroduplexes rather than a hairpin structure. Even so, this duplex has features of wider interest, since it has a number of non-Watson–Crick base pairs such as a 5'-GGAG/3'-GGAG purine bulge. Their overall effect is to produce backbone distortions so that the helix has non-A-RNA features such as a widening of the major groove, by ~9 Å, and local under-twisting of base pairs adjacent to A•C and G•U base pairs.

NMR approaches to determining RNA structure have been fruitful in defining the conformation of double-helical RNA. However, sometimes, due to their inherent

A
5'- CUUGCUG**G**GUGCACA**C**AGCAAG

GAACGACA**C**ACGUG**G**GUCGUUC
A

Figure 6.10 The sequence of the dimerisation initiation site of the RNA genome from the HIV-1 retrovirus, studied crystallographically (Ennifar et al., 1999), PDB id 1K9W.

Figure 6.11 View of the structure of the coaxially stacked kissing complex of the HIV-1 RNA dimerisation initiation site (Ennifar et al., 2001).

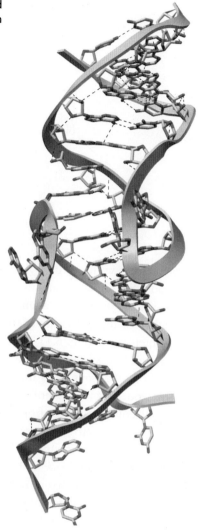

flexibility, the features of non-helical regions may be less well defined. This can lead to a lack of sufficient number of constraints so that these features of an NMR model can show several distinct conformations — such flexible features may well be functionally important. The correspondence (and complementarity) between crystallographic and NMR structures is well illustrated in studies of the high-affinity RNA-binding site of the HIV-1 Rev protein response element, whose conformation changes significantly with bound protein. The

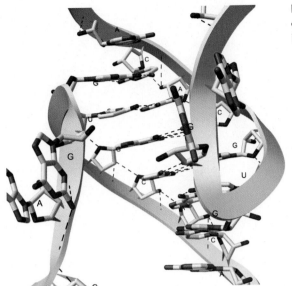

Figure 6.12 Close-up view of the kissing complex, showing the loop—loop interactions.

NMR and X-ray structures on the RNA itself agree in describing the RRE as a highly distorted helix (Ippolito & Steitz, 2000; Peterson & Feigon, 1996). The distortions are a consequence of (i) two extra-helical bases, an adenine and a uracil from one strand, which are responsible for a bulge and a kink, respectively, in the helix, and (ii) two essential purine•purine base pairs, which flank the extra-helical uracil (Fig. 6.13). Both structures show that the helix is bent by ~30°, with a narrowed major groove, and is highly distorted from canonical A-RNA. The principal difference between them is in the loop region around the extra-helical uracil. This is a consequence of the adjacent G(*syn*)•G(*anti*) base pair having an asymmetric configuration in the X-ray structure, whereas in the NMR structure it has been assigned a symmetric one, so that the backbones are forced to adopt distinct paths in this region. Whether these differences are real, or reflect the differing experimental conditions, remains to be established.

We have noted above that crystallographic studies of small RNAs containing putative loop regions very frequently result in helices that instead maximise base pairing within the duplex. The details of the geometry of the structurally important and very stable UUCG tetraloop were especially elusive for some while. Initial NMR studies proved to be only partially correct. One NMR analysis (Allain & Varani, 1995) is in close agreement with a subsequent crystallographic analysis of a tetraloop incorporated in a large RNA—protein crystal structure (Ennifar et al., 2000). Molecular dynamics simulations

Figure 6.13 The crystal structure of the RNA binding site of the HIV-1 Rev protein response element (Ippolito & Steitz, 2000), PDB id 1CSL. The G•G base pair and the extra-helical U and A are labelled.

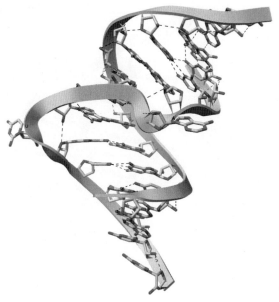

(Miller & Kollman, 1997) of the tetraloop have converged to the correct structure from an incorrect starting point, an interesting result in view of the relatively primitive nature of the RNA force fields available at that time. The essential features of the UUCG tetraloop (Fig. 6.14a) are (i) an unusual G(*syn*)•U(*anti*) base pair (Fig. 6.14b), which is seen directly in the X-ray structure, (ii) the 2′-OH group of this uridine nucleoside hydrogen bonds to the O6 atom of the guanine, (iii) the cytosine stacks on the adjacent uracil base.

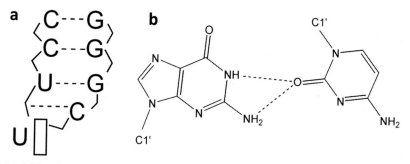

Figure 6.14 (a) A schematic view of the UUCG tetraloop, with the U base looped out from pairing. (b) The G(*syn*)•U(*anti*) base pair found in the UUCG tetraloop (Allain & Varani, 1995; Ennifar et al., 2000; Miller & Kollman, 1997).

Tetraloops, and other elusive motifs, can be captured when they interact with their biological partners, which can be either proteins or other nucleic acids. Thus, the solution structure of the AGAA tetraloop, a member of the AGNN family, has been determined bound to the duplex RNA-binding domain of the Rnt1 protein (Wu et al., 2004). Surprisingly, there are no specific hydrogen bonds between the protein and the conserved A or G bases. Instead, recognition is governed by shape complementarity, with an α-helix fitting into the minor groove of the tetraloop. The difficulties found in recent extended simulations of the UUCG tetraloop (Mráziková et al., 2020), with significant deviations from the native state, do demonstrate that accurate RNA simulations are still not routine nor always reliable.

A more complex motif in RNA is the pseudoknot, which is a prominent feature of many larger RNA structures (see Section 6.4 on ribozymes). The detailed structures of several pseudoknots have been determined by NMR methods, for example, the telomerase RNA pseudoknot domain (Theimer et al., 2005). This structure has a long triple helix close to the highly conserved pseudoknot junction (Fig. 6.15a), involving several Hoogsteen base triplets (Fig. 6.15b). Crystallographic analyses have been mostly reported for pseudoknot–containing oligoribonucleotides embedded in larger RNA structures. A

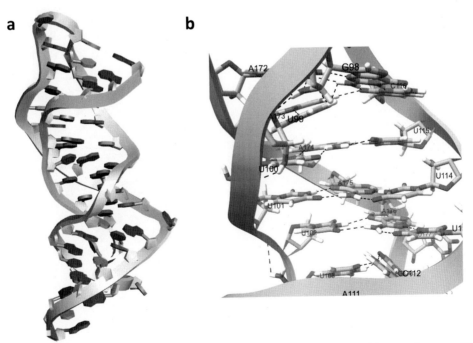

Figure 6.15 (a) One of the ensemble of NMR structures determined for the wild-type telomerase RNA pseudoknot domain (Theimer et al., 2005), PDB id 2K95. (b) Detailed view of the telomerase RNA pseudoknot domain, showing the run of Hoogsteen base triplets.

notable study has been the high-resolution analyses of the pseudoknots from potato leaf roll virus (Pallan et al., 2005) and beet western yellow virus (Egli et al., 2002), at 1.35 Å and 1.25 Å, respectively. This has enabled a high level of detail in the structures to be visualised, especially the roles played by water molecules. For instance, there are several instances of water-base "stacking" (Sarkhel et al., 2003), which involve water molecules being close to the face of an individual base such that water–π orbital interactions are likely to occur. The crystal structure of an adenovirus-associated RNA (Hood et al., 2019) shows a V-shaped molecule (Fig. 6.16), with a 22-mer long GC-rich imperfect A-RNA double-helical apical stem and a shorter helical stem, together forming a V-shape and inclined at ca. 60°. They meet in a complex arrangement involving a pseudo-knot anchored by loop–loop interactions and wobble base pairs.

The relative precision of crystallography and NMR is always hard to assess unless identical structures are compared. This has been done for two small RNAs, one of which is fully helical, and the other has a small loop in it (Rife et al., 1999). Gratifyingly, the

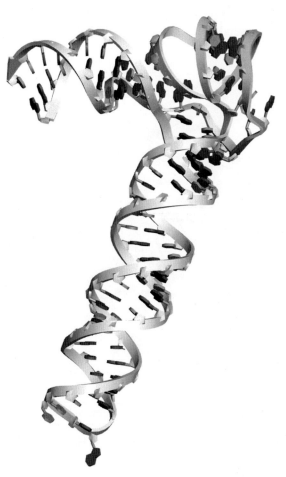

Figure 6.16 Crystal structure of an adenovirus virus-associated RNA (Hood et al., 2019), PDB id 6OL3.

differences between each pair of structures are small. The study also compared various parameter sets commonly used for nucleic acid X-ray and NMR refinements and concluded that deficiencies in them (especially in the treatment of electrostatics terms) are probably responsible for many of the (small) differences observed between solution and crystal structures. The overall message here is that it should be a goal of solution structure determination to reproduce as closely as possible the local and global structures found by high-resolution crystallography.

These relatively small RNAs have classic Watson—Crick base pairs in their helical stems. Non-helical parts show a variety of base—base interactions, which become more important with increased size and complexity of RNAs, as will be discussed in subsequent sections. In principle the total number of non-Watson—Crick base-pair possibilities is very large, as shown in Fig. 6.17, illustrating some of the possible A•G arrangements. This number is considerably increased when sugar group substituents are involved. The large number of RNA pairings that have been observed can be systematically classified according to the relative orientations of the attached nucleotide strand, and the base edges involved (Leontis & Westhof, 2001; Leontis et al., 2002). Even more base triplet possibilities can be envisaged, and very many have been observed, such as the stable G•G•A arrangement (Fig. 6.18). Of the 108 predicted base triple families, as of 2012, 68 had been experimentally observed (Abu Almakarem et al., 2012).

The crystal structure of an RNA triplex helix with 11 consecutive U•A•U base triples has been determined, within a 79-nucleotide sequence (Ruszkowska et al., 2020) at 2.5 Å resolution, representing almost a complete turn of an RNA triple helix. There are two independent molecules in the crystallographic unit, with closely similar features. The author's expectation was that there would be a limit of six consecutive RNA base triples, and therefore some structural distortions were to be expected. Instead, the triplex part of the structure (Fig. 6.19) shows remarkably little distortion from regular canonical A'-RNA (Arnott et al., 1976), with a helical twist of 30°, all sugar moieties having a C3'-*endo* conformation and a wide, deep major groove.

6.3 Transfer RNA structures

We have already seen in Chapter 4 that DNA sequences are able to participate in non-Watson—Crick base—base interactions. RNA sequences are inherently even more capable of doing so, and indeed they are the norm in large natural RNA structures, once Watson—Crick base pairing is maximised. There is a significant literature on RNA secondary structure prediction, and several computer algorithms are freely available with which one can examine RNA sequences. They are based, in large part, on the maximising of base pairing. Successful predictions also consider the existence of a number of invariant features in a sequence across species. Historically, the determination of the primary sequence of a tRNA molecule in 1964 rapidly led to suggestions for its secondary

Figure 6.17 Some of the possible G•A base pair arrangements that have been found in RNA.

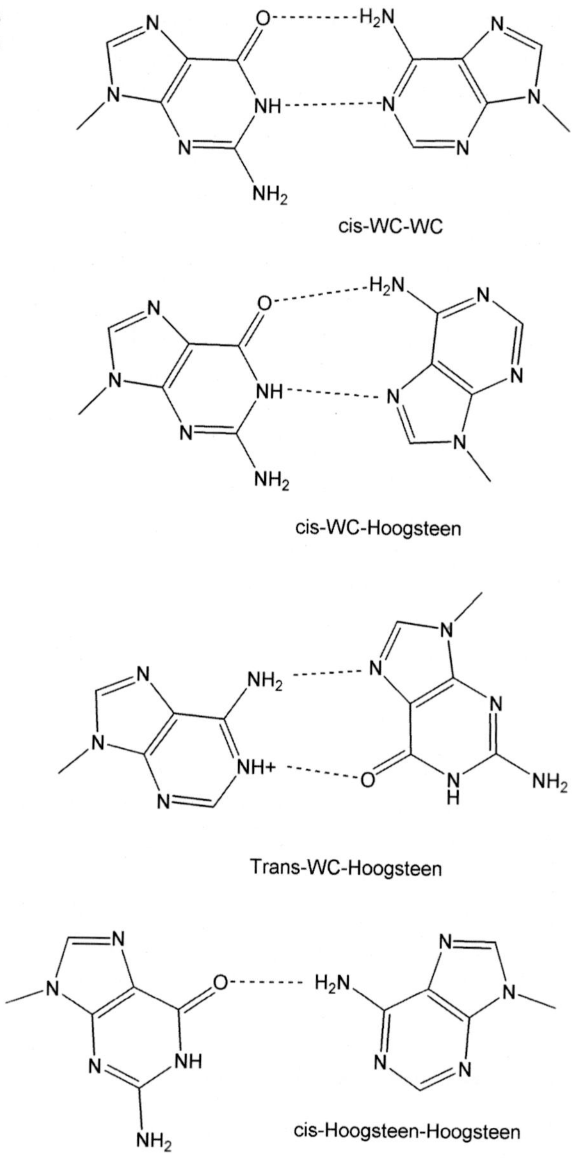

cis-WC-WC

cis-WC-Hoogsteen

Trans-WC-Hoogsteen

cis-Hoogsteen-Hoogsteen

folding into a cloverleaf structure (Levitt, 1969). This fold is a consequence of base–base hydrogen bonding between non-adjacent regions of RNA double helix and loop. It is based on the identification of the highly conserved features of acceptor, anticodon, D and T arms, and three/four loops, one of which is very variable between tRNAs (the "variable" loop). There have been numerous subsequent attempts at predicting the three-dimensional structure of a tRNA. None were completely accurate, which is

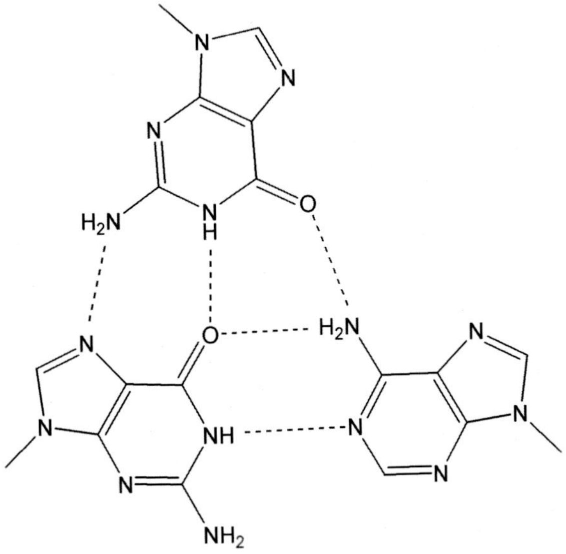

Figure 6.18 A stable G•G•A triplet arrangement.

unsurprising given the very large number of potential base–base interactions. However, the general notion that these base–base interactions are responsible for RNA folding has proved to be correct, as were the assignments of Watson–Crick base pairs. All tRNAs contain, in addition to the standard A, U, G and C nucleosides, several modified nucleosides such as N2-dimethylguanosine, N1-methyladenosine, pseudouridine (Ψ; an isomer of uridine with the glycosidic bond from the ribose sugar going to the C5 atom of the base rather than to N1) and dihydrouridine D. tRNA species have between 70 and 95 nucleosides, depending in large part on the size of the variable loop.

   The crystal structure of yeast phenylalanine tRNA was determined in the early 1970s, simultaneously by two groups (Kim et al., 1974; Robertus et al., 1974). These together with a few other tRNAs (Table 6.5) remained the sole complex RNA molecular structures available for 20 years, until the first ribozyme structure. The two independent tRNA structures, one monoclinic (Robertus et al., 1974) and the other orthorhombic (Kim et al., 1974), are closely similar (Klug et al., 1974). Both structures show that the molecule is folded into an overall L-shape, with the two arms at right angles to each other. The original predicted cloverleaf shape is still apparent, but with additional interactions between distant parts of the structure (Fig. 6.20). The arms consist of short A-RNA helices, together with these extensive base–base interactions that hold the two arms together (Fig. 6.21). The longer double helical anticodon arm has the short helix of the D stem stacked upon it. The other arm is formed by the helix of the acceptor stem, on which is stacked the four base-pair T arm helix. This key feature of helix–helix

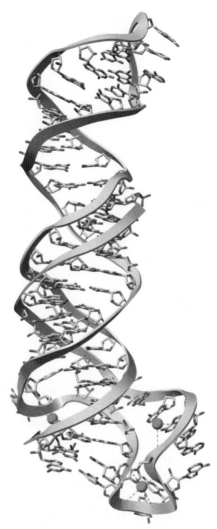

Figure 6.19 Structure of an RNA triple helix (Ruszkowska et al., 2020), PDB id 6SVS, embedded within a longer sequence. The 11-triplet triplex strand region is in the middle of the structure, and the regularity of the three backbones in the triplex region is readily apparent.

stacking has turned out to be of central importance for other complex RNAs. The nine additional base–base interactions that maintain the structural fold all tend to be in the elbow region, where the two arms are joined together. Almost all of these are of non-Watson–Crick type, and several are triplet interactions (Fig. 6.22). These are analogous to the triplets found in DNA triple helices (Chapter 4). Other subsequent crystal structures of tRNAs (Table 6.5) have shown that the overall L shape is invariant, as are many of the tertiary interactions.

The yeast tRNA^{phe} crystal structures have been re-refined several times after their initial determinations, each time using increasingly robust refinement methods, so that

Table 6.5 Selected native tRNA crystal structures.

tRNA	Resolution in Å	PDB id
Yeast Phe, monoclinic	2.00	1EVV
	1.93	1EHZ
Yeast Phe, orthorhombic	2.7	6TNA
Yeast Asp	3.0	2TRA
Yeast initiator	3.0	1YFG
Escherichia coli Phe	3.10	6Y3G
	3.00	3L0U
E. coli Asp	1.95	6UGG
E. coli initiator	3.1	3CW5

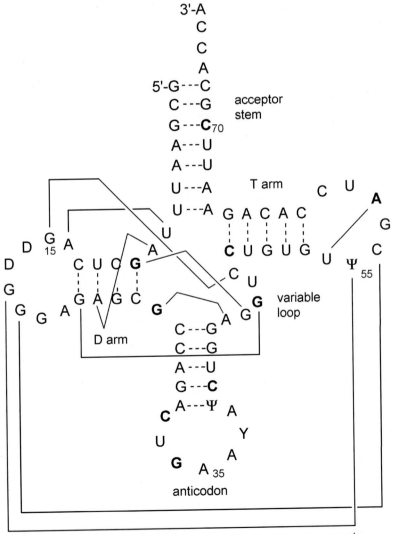

Figure 6.20 Schematic of the hydrogen-bonding interactions in the yeast tRNA^phe crystal structure (Kim et al., 1974; Robertus et al., 1974). The interactions are shown between distant nucleotides in this, the clover-leaf representation.

Figure 6.21 The structure of yeast tRNA^{phe}, showing the overall L-shape, from the 1.93 Å crystal structure (Shi & Moore, 2000), PDB id 1EHZ.

in effect these structures have been used as a test bed for the general development of nucleic acid crystal structure refinement methodology (Westhof & Sundaralingham, 1986; Westhof et al., 1988). More recently the crystal structure of the monoclinic form has been re-determined (Jovine et al., 2000; Shi & Moore, 2000) using synchrotron sources to collect data, to 1.93 and 2.0 Å resolution, respectively, from flash-frozen crystals. It is reassuring that the overall backbone geometry and fold remains as found in the earlier studies. There are corrections to the sugar pucker and conformation for a very few nucleotides, and backbone torsion angles conform much more to the canonical A-RNA values — this reflects both the improved quality of the diffraction data and the robustness of the refinement methods. Both these more recent analyses find new magnesium-binding sites, and a greater number of well-defined water molecules hydrogen bonding to the tRNA compared to the original analyses.

The notion that tRNAs are highly rigid molecules has been to some extent questioned by the crystallographic analysis of *Escherichia coli* tRNA^{Asp}, with two independent molecules in the crystallographic asymmetric unit (Chan et al., 2020). These molecules are non-identical (albeit within the overall framework of the L-shape), with a root

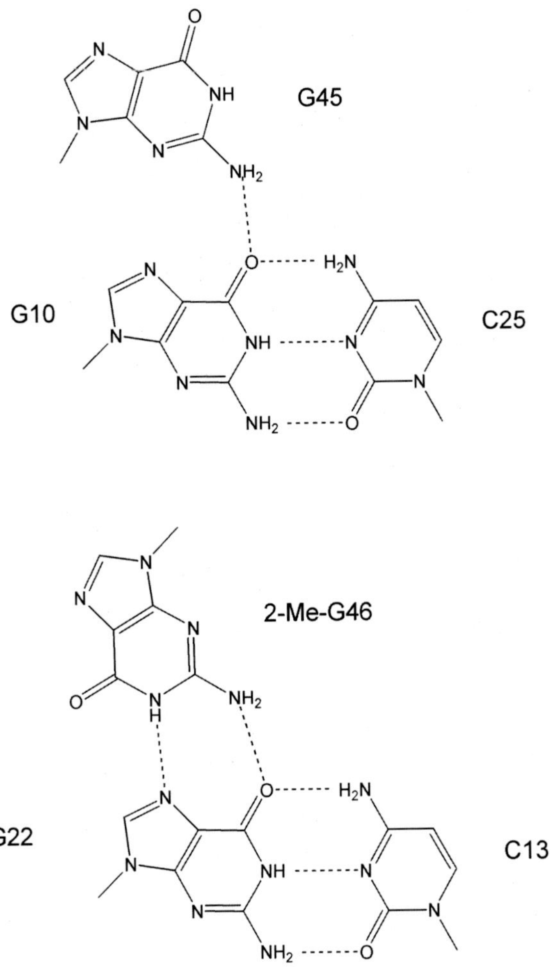

Figure 6.22 Base triplet interactions in the yeast tRNA^phe crystal structure.

mean square deviation (RMSD) of 2.52 Å for all atoms, reflected in structural differences between the nucleotides at the 3′ end, the anticodon loops, the D loops and the angle of the "elbow" region. These differences have been interpreted as manifestations of tRNA structural flexibility and their dynamic behaviour, even in the crystalline state.

6.4 Ribozymes

The ability of certain RNA molecules to catalytically cleave either themselves or other RNAs is shown by several distinct categories of RNA molecule, including:

- The RNA of self-splicing group I introns — That from *Tetrahymena* was the first ribozyme to be discovered (Cech et al., 1981; Guerrier-Takada et al., 1983). These

contain four conserved sequence elements and form a characteristic secondary structure. The initial step of the cleavage involves nucleophilic attack by a conserved
guanosine.

- The RNA of self-splicing group II introns — These also have conserved sequence elements, but a very different secondary structure and a distinct mechanism of cleavage
that involves the nucleophilic attack of a conserved adenosine, contained within the
intron sequence.
- The RNA subunit of the enzyme RNase P.
- Self-cleaving RNAs from viral and plant satellite RNAs. These are smaller ribozymes
than the group I or II intron ones, and include the hammerhead ribozyme (Fig. 6.23).
- Ribosomal RNA in the ribosome, and in particular the peptidyl transferase
machinery.
- Nucleolytic ribozymes, which tend to be small, albeit with a variety of folds, often
including a pseudoknot (Lilley, 2019), and having site-specific backbone cleavage
activity using acid—base catalysis. Members of this family include the hepatitis delta
virus (Ferre-D'Amare et al., 1998a,b), the hammerhead, the pistol (Wilson et al.,
2019) and the twister (Wilson et al., 2016) ribozymes.

6.4.1 The hammerhead ribozyme

The crystal structures of hammerhead ribozymes were the first to be determined (Table
6.6), simultaneously by two groups (Pley et al., 1994; Scott et al., 1995,1996). Both
structures are similar in their essential detail. One (Pley et al., 1994) has a DNA strand

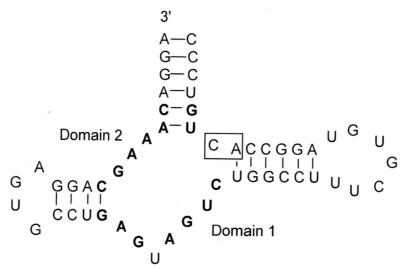

Figure 6.23 Sequence of the hammerhead ribozyme, indicating the hydrogen-bonded helical stems.
The CA cleavage site is shown boxed.

Table 6.6 Selected ribozyme crystal structures.

Ribozyme type	Resolution in Å	PDB id
Hammerhead	3.00	299D
Full length hammerhead	1.55	3ZP8
Hammerhead + vanadate ion	2.99	5EAO
P4–P6 domain	2.50	1GID
Tetrahymena ribozyme	5.00	1GRZ
	3.80	1X8W
Hepatitis delta virus	2.30	1DRZ
	1.92	3NKB
Hairpin	2.40	1M5K
RNA part of ribonuclease P	3.85	2A2E
	3.30	2A64
Group I intron	3.37	1ZZN
Ligase ribozyme	2.60	2OIU
Twister	2.30	4OJI
	2.64	5DUN
Pistol	2.97	5KTJ
	2.65	6UFJ
	3.10	6R47
Hatchet	2.06	6JQ5
Varkud Satellite (VS)	3.29	5V3I

containing the putative cleavage site. Since ribozymes do not cleave DNA, this is effectively an inhibitor complex. The other (Scott et al., 1995) is all-RNA, with a 2′-O-methyl group replacing the 2′-hydroxyl group of the active-site cytosine. The structures show three A-RNA stems connected to a central two-domain region containing the conserved residues (Fig. 6.24a). The overall arrangement resembles a wishbone. The two-residue sequence CA (the site of autocatalytic cleavage) is at the apex of two stems (Fig. 6.23), where there is a sharp turn in backbone conformation, that is focused on the uridine in domain one, formed by the invariant tetranucleotide sequence CUGA. This U-turn conformation is close to those in the anticodon and pseudouridine loops of tRNA. The second domain, at the junction of two stems, contains conserved sheared tandem A•G base pairs. This performs the key structural role of maintaining the correct geometric relationships between cleaved and catalytic strands.

The overall mechanism of hammerhead ribozyme catalysis involves nucleophilic attack by a 2′-hydroxyl group on the phosphorus atom of the adjacent backbone. However, elucidation of the details of the mechanism has not been straightforward, although it was early on recognised that metal ions often play an important role, in view of the requirement for magnesium ions. The location of magnesium ions close to the A•G base pairs suggests that they provide a directly stabilising influence on the precise conformation needed for catalysis. Several of the models for catalysis that have been proposed

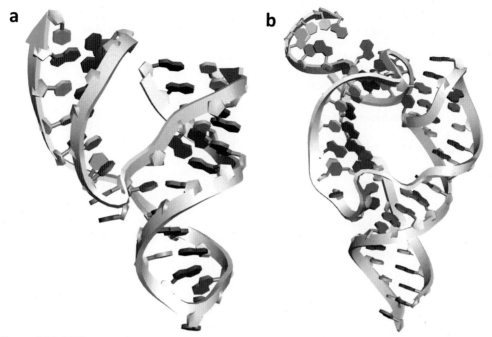

Figure 6.24 (a) The crystal structure of the hammerhead ribozyme (Scott et al., 1995), PDB id 299D. (b) The crystal structure of a vanadate complex of the hammerhead ribozyme.

(Lilley, 1999) suggested that the ions are more directly involved. Magnesium ion(s) have been located at various positions within the active site, suggestive of a role in aiding the movement of electrons, perhaps by stabilisation of leaving groups. Several crystal structures, together with evidence from other biophysical methods, have provided data relevant to the mechanistic questions (Feig et al., 1998; Murray et al., 1998, 2000).

Terbium ions inhibit catalytic cleavage, and the structure of a terbium complex (Feig et al., 1998) shows this ion to be located close to residues G5 and A6, and not within the active site. The structure of an intermediate has been captured (Murray et al., 2000), which suggests that there is a large-scale conformational change during the reaction, which brings the cleavage site and G5/A6 closer together than found in the ground-state structures (Pley et al., 1994; Scott et al., 1995); subsequent studies have shown these to be trapped in inactive conformations. Several other hammerhead ribozyme structures have been more recently reported, including a high-resolution structure of the full-length hammerhead sequence, PDB id 3ZP8 (Anderson et al., 2013). An important advance has been the determination of the crystal structure of a complex with the vanadate ion (Fig. 6.24b), acting as a transition state analogue, bound to the hammerhead ribozyme (Mir & Golden, 2016), PDB id 5EAO. There are small but significant structural differences around the active site compared with the earlier structures such as PDB id 299D. The vanadate ion acts as a surrogate for the scissile phosphate group and base G12 has moved compared to earlier structures so that it plays a pivotal role in the catalysis, hydrogen bonding to the 2′ group of the nucleophile in the cleavage reaction.

6.4.2 Complex ribozymes

The crystal structure of a Group I ribozyme domain has been determined, of the P4–P6 domain of the *Tetrahymena* intron (Cate et al., 1996) (Fig. 6.25). The structures of the

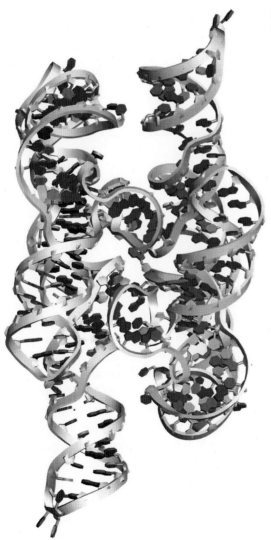

Figure 6.25 The crystal structure of the P4–P6 domain of the *Tetrahymena* intron (Cate et al., 1996), PDB id 1GID.

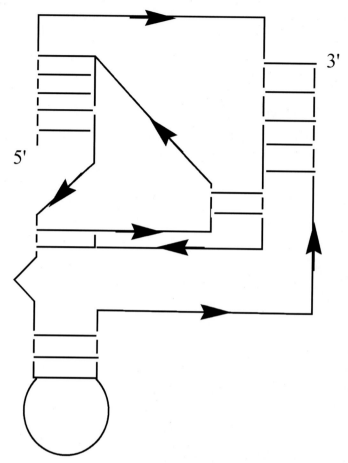

Figure 6.26 Topology of the hepatitis delta virus ribozyme (Ferré-D'Amaré et al., 1998).

complete intron from several different organisms have subsequently been determined (Adams et al., 2004; Golden et al., 2005; Guo et al., 2004) and show an architecture formed by ten helices held together by the intervening loops and junctions, with the catalytic site close to a pseudoknot and two junction regions (Vicens & Cech, 2005).

The structure of the *Tetrahymena* intron (Fig. 6.26) contains 160 nucleotides, and is arranged as two roughly colinear helical regions. Within these are several well-established RNA motifs, notably an adenosine-rich bulge, a three-way junction and a GAAA tetraloop. The close packing of the helices in this and some other large RNA structures is due to the highly favourable interactions between unpaired, flipped-out adenosines attached to one helix and the minor groove of the adjacent one (Doherty et al., 2001). This provides an explanation for the high abundance of conserved unpaired adenosine residues in several large RNA molecules.

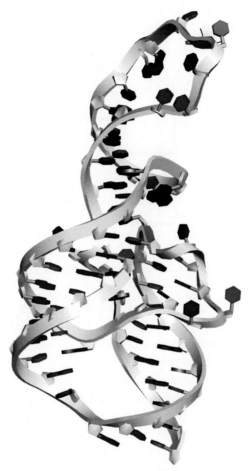

Figure 6.27 Crystal structure of the hepatitis delta virus ribozyme (Ferré-D'Amaré et al., 1998), PDB id 1DRZ.

The crystal structures of the hepatitis delta virus ribozyme and a hairpin ribozyme both show features relevant to their RNA-cleavage activities. The hepatitis delta virus one (Ferre-D'Amare et al., 1998a,b) has a very complex arrangement of five helical moieties of varying lengths forming a double pseudoknot arrangement (Fig. 6.27). (The pseudoknot is another recurrent motif in RNA structure that involves the hydrogen-bonded interaction of a single-stranded region with a loop (Fig. 6.28).) This arrangement is fundamentally distinct from that in many other ribozymes, as is the active site, which is buried within the five-helical core (Fig. 6.29). Another complex catalytic site is seen in the structure of the hairpin ribozyme (Rupert & Ferré-D'Amaré, 2001); both have features resembling enzymes that also use activated phosphates, such as ribonuclease S (Strobel and Ruder, 2001). The hairpin ribozyme has a four-way junction forming the central

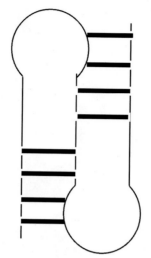

Figure 6.28 Topology of a pseudoknot.

feature of the structure. The structures of several of the individual stems forming the junction have been previously individually studied; it is a salutary lesson in the complexities of RNA structure that these have undergone large conformational changes when a part of the complete hairpin ribozyme.

Can ribozyme structures (and those of other complex RNAs) be predicted, using current knowledge of RNA motifs, base–base hydrogen bonding and their interaction modes? The RNA-Puzzles competition was established in 2011 to address this challenge (www. rnapuzzles.org). To date, four rounds of competition, currently involving four groups world-wide, have been completed. The most recent (Miao et al., 2020) round involved the prediction of four ribozyme and two riboswitch aptamer structures, each of which has a newly determined crystal structure (unpublished and unavailable to the modeller participants). It was concluded that overall, the blind predictions of 3D structures agree with crystal structures, albeit with RMSDs >3.5 Å, but <7 Å. Ligand-binding predictions are more variable. The field requires many more experimental structures as starting points for machine learning methods to be effective, which are still worse than RNA experts can achieve.

6.5 Riboswitches

The surprising finding that sequences in the 5' non–translated region of certain bacterial mRNAs can act to regulate gene expression in the absence of proteins is of major functional (and potentially therapeutic) significance (Schwalbe et al., 2007). They do this by undergoing major conformation changes under the action of appropriate endogenous small-molecule metabolites (Winkler et al., 2004). A subsequent finding has been that

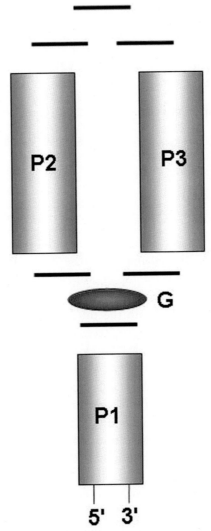

Figure 6.29 Schematic view of a simple riboswitch, with the three helices P1, P2 and P2 shown. Bound purine is also shown schematically as a shaded disc.

riboswitches may also function in eukaryotic cells (Cheah et al., 2007), where they have been shown to regulate RNA splicing and are involved in polyadenylation signalling (Spöring et al., 2020). Riboswitches can act by especially binding to purine bases, as well as some enzyme cofactors and particular amino acids.

Most riboswitches have two key domains, an aptamer component for ligand binding, and an "expression platform", which transmits the ligand-induced conformational change to result in changes in gene expression. This is shown, for example, in the crystal

Table 6.7 Selected riboswitch crystal structures.

Riboswitch type and ligand	PDB code no	Resolution in Å
S-adenosylmethionine	2GIS	2.90
Guanine riboswitch + guanine	1Y27	2.40
Guanine riboswitch + hypoxanthine	4FE5	1.32
Thiamine pyrophosphate	2CKY	2.90
riboswitch + thiamine diphosphate	2HOJ	2.50
	3D2G	2.25
Bacillus subtilis cobalamin	6VMY	3.25
Glutamine II	6QN3	2.30
Gloeobacter violaceus guanidine	5NOM	1.93
II + guanidine		
Adenine	5SWE	3.00

structure of the guanine-responsive riboswitch from the *xpt-pbuX* operon of the *Bacillus subtilis* organism, bound to the guanine-like base hypoxanthine (Batey et al., 2004) (Table 6.7). This has the loops at the top end of two parallel helices P1 and P2 linked together by several base—base hydrogen bonds; at the other end the binding pocket is formed at the three-way junction with the third helix P3 (Figs. 6.29 and 6.30). The ligand is almost totally enclosed (Fig. 6.30a and b), and as shown in the guanine-bound riboswitch (Serganov et al., 2004), is extensively hydrogen-bonded to the surrounding RNA residues, which are highly conserved. This type of riboswitch turns off gene expression. However, some riboswitches can activate gene expression, possibly by only allowing the transcription terminator stem form in the absence of, for example, adenine (Serganov et al., 2004). The guanine and adenine riboswitches share a common tertiary structure even though the extent of overall sequence identity is 59%.

A more complex category of riboswitch binds the coenzyme thiamine pyrophosphate (vitamin B1), the so-called thi-box. Several independent crystal structures have been determined for the prokaryotic riboswitch (Edwards & Ferré-D'Amaré, 2006; Serganov et al., 2006), as well as for the eukaryotic analogue (Thore et al., 2006,2008). All are closely similar (apart from small sequence changes in non-conserved regions) and show that the Y-shaped molecule comprises two irregular RNA helices/loops linked by a three-way junction to a third helix (Fig. 6.31a). The ligand-binding region is in a bulged pocket lined by conserved residues and is held in place by an intricate network of hydrogen bonds to these bases (Fig. 6.31b), as well as interactions between some of the magnesium ions in the structure and the pyrophosphate tail.

6.6 The ribosome, a ribozyme machine

The ribosome is responsible for protein synthesis in all prokaryotic and eukaryotic cells. It consists of two subunits, each a complex of proteins and ribosomal RNA. The complete

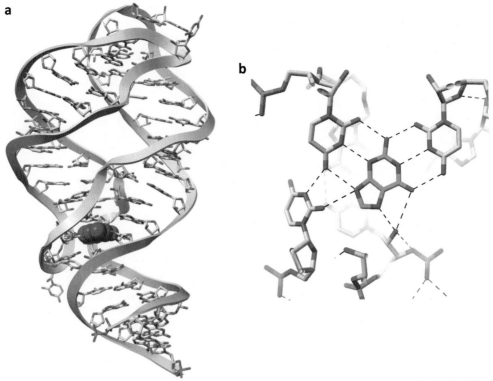

Figure 6.30 (a) The structure of the guanine riboswitch with hypoxanthine (Batey et al., 2004), PDB id 4FE5. (b) Detailed view of the bound guanine base, in the guanine riboswitch, with hydrogen bonds shown as dashed lines.

70S prokaryotic ribosome has a total molecular weight of ca 2.5 million daltons. In prokaryotes the 30S subunit contains about 20 proteins and a single RNA molecule of around 1500 nucleotides in length. The larger 50S subunit contains half as many more proteins and a large RNA of about 3000 nucleotides, together with the small (120 nucleotides) 5S RNA. The primary function of the small subunit is to control tRNA interactions with messenger RNAs. The large subunit controls peptide transfer and undertakes the catalytic function of peptide bond formation. In eukaryotes the complete 80S particle contains 40S (with an 18S RNA and 33 proteins) and 60S subunits, with two small RNAs, 5S and a 5.8S, and a 28S RNA together with 46 proteins. The discussions in this chapter are focused on the basic RNA structural features of the ribosome: there is further discussion on structure and function in Chapter 8.

Structural studies on bacterial ribosomes have been underway for over 60 years, with the goal of achieving atomic resolution to understand the mechanics of ribosome function — this appeared at first to be an impossibly difficult task. Many individual ribosomal proteins were initially studied, but their three-dimensional arrangement within the

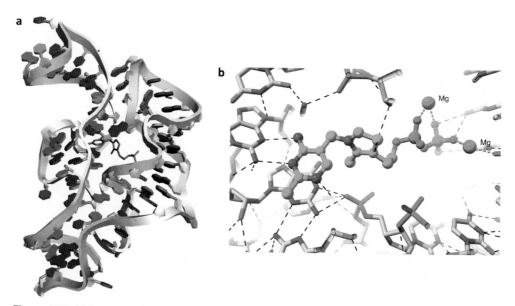

Figure 6.31 (a) Structure of the eukaryotic thiamine pyrophosphate riboswitch, with the thiamine pyrophosphate shown in ball and stick form (Thore et al., 2008), PDB id 3D2G. (b) An enlarged view of the bound thiamine pyrophosphate, with hydrogen bonds to bases and magnesium ions shown as dashed lines.

ribosome, and their functional roles, were not apparent. Electron microscopy enabled the outlines of ribosome structure to be defined, which have been of considerable help to the subsequent single-crystal and more recent cryo-EM analyses. This is now the method of choice for ribosome structural studies. Crystals of diffraction quality were first obtained well over 30 years ago, but real progress had to await the various necessary technologies become available for tackling such a massive and complex crystallographic task. It was necessary to collect huge data sets from the very large unit cells involved, with crystals that diffracted only weakly and decayed rapidly in the X-ray beam. This was achieved using synchrotron radiation together with low-temperature methods to stabilise the crystals from excessive decay.

Determining the phases of these large structures has also been a formidable challenge in the early crystallographic efforts. Rather than using single heavy atoms, the use of clusters of heavy metals was successful, especially for the initial structures, at considerably less than atomic resolution. The progress of ribosome structure determination has been rapid (Table 6.8). The first structure, at 9 Å resolution, was reported in 1998 (Ban et al., 1998), culminating in electron density maps of the small (Tocilj et al., 1999) and large (Ban et al., 2000) subunits at 3.0 and 2.4 Å resolution, respectively, at which point individual side chains and bases become apparent and can be refined. More recently, several complete ribosomal structures have been determined, of the *E. coli* ribosome

Table 6.8 Early progress in ribosome structure determination.

Subunit	Organism	Resolution (Å)	Reference
50S	*Haloarcula marismortui*	9	Ban et al. (1998)
50S	*Thermus thermophilus*	5.5	Yusupov et al. (2001)
30S	*T. thermophilus*	5.5	Clemons et al. (1999)
70S	*T. thermophilus*	7.8	Cate et al. (1999)
30S	*T. thermophilus*	4.5	Tocilj et al. (1999)
30S	*T. thermophilus*	3.0	Wimberley et al. (2000)
30S	*T. thermophilus*	3.3	Ban et al. (2000)
50S	*H. marismortui*	2.4	Ban et al. (2000)
70S	*T. thermophilus*	5.5	Carter et al. (2000)
70S	*T. thermophilus*	3.7	Korostelev et al. (2006)
70S	*T. thermophilus*	2.8	Selmer et al. (2006)
70S	*Escherichia coli*	3.5	Schuwirth et al. (2005)

(Schuwirth et al., 2005), and that from *Thermus thermophilus* complete with tRNA (Korostelev et al., 2006) and with tRNA and mRNA (Selmer et al., 2006). These structures were some of the largest determined by X-ray crystallography, and dissection of the electron density maps has only been possible with the aid of the large body of biochemical and other data accumulated on the ribosome and its subunits. It is remarkable that the secondary structures for the RNAs in these complexes are close to those predicted by analyses of RNA across species, using the methods of phylogenetic analysis, which are based on the reasonable assumption that RNA structure is largely conserved throughout evolution. More recently the structure of the complete bacterial ribosome from *E. coli* has been determined at high resolution, 1.98 Å (Fig. 6.32a—c), enabling high-quality electron density maps to be obtained that even show details of solvation (Watson et al., 2020). The structure of the 30S subunit has now been better defined, and post-transcriptionally modified nucleotides such as methylated adenosine and guanosine have been unequivocally identified. Also, key contacts between the subunits have been better resolved so that interactions between the 30S subunit and tRNAs/mRNA are now unambiguous (Fig. 6.33).

A very large number of ribosome coordinate data sets are now available from the PDB. Although only backbone C_α carbon atoms were deposited for the protein components in the earlier lower-resolution structures, protein side-chain atoms have been identified in subsequent analyses, as well as complete coordinate sets for the RNA components. The sheer size of the PDB files involved has resulted in more recent depositions being split over more than one entry. The size and complexity of ribosome structures is such that there is some advantage when viewing these very large assemblies, in restricting initial viewing to backbone atoms, cartoon or space filling representations. The necessarily brief survey below of the RNA aspects of the structures cannot do justice to their complexity, and the reader is strongly encouraged to read the primary literature on them and to spend some time viewing them in detail.

Figure 6.32 Space-filling images of the 1.98 Å structure (Watson et al., 2020) of the complete *E. coli* ribosome, PDB id 7K00. (a) The complete ribosome with RNA components mostly in blue and the associated ribosomal proteins in various colours. (b) The RNA alone. A small part of the 5S RNA can be seen on the right-hand side, coloured brown. (c) Another orientation, with the 5S RNA (coloured brown) now fully visible and the A-site tRNA^{val} shown coloured yellow, is in the centre of this orientation of the structure.

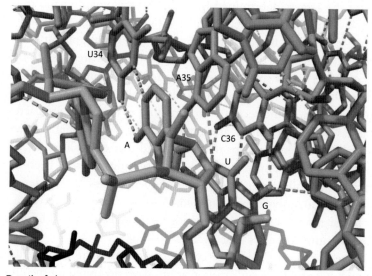

Figure 6.33 Detail of the interactions in the high-resolution *E. coli* 1.98 Å structure (Watson et al., 2020), showing the anticodon stem of the tRNA^{val} molecule (coloured light green) hydrogen bonding to codon bases in an mRNA sequence (coloured mauve), adopting an extended conformation. Three base pairs are labelled (two A•U and a C•G). Water molecules have been omitted from this view.

The initial crystal structure of the complete 70S ribosome (Yusupov et al., 2001), although at relatively low resolution, revealed much about the interactions between the subunits that are an essential element of the protein synthesis cycle (Table 6.9). It is notable that even though the resolution precluded detailed study of the RNA base–base

Table 6.9 Selected ribosome and subunit crystal structures. More extensive tables are provided in Chapter 8.

Subunit	Organism	Resolution in Å	PDB id
30S	*Thermus thermophilus*	3.0	1FJG
30S with mRNA + tRNA	*T. thermophilus*	3.1	1IBL
		3.3	1IBM
70S+ tRNA + mRNA	*T. thermophilus*	5.5	4V42
		5.0	1JGQ,1JIY
70S + tRNA	*T. thermophilus*	3.71	4V41
70S + tRNA + mRNA	*T. thermophilus*	2.8	4V51
70S	*Escherichia coli*	3.46	4V4Q
70S	*E. coli*	1.98	7K00
16S + tRNA +12S	Human mitochondrial	3.50	3J9M
21S + 15S + tRNA	Yeast mitochondrial	3.25	5MRC
80S	*Homo sapiens*	3.60	4UG0
80S	Yeast	4.00	4V7R
70S	*Staphylococcus aureus*	3.80	5LI0

interactions involved, the bound tRNA molecules were all seen to be extensively contacted by ribosomal RNA in addition to the necessary established functional interactions such as codon—anticodon recognition. Subsequent analyses, at 2.8 Å (Selmer et al., 2006) and at 3.7 Å (Korostelev et al., 2006), together with that of the complete *E. coli* ribosome at 3.5 Å (Schuwirth et al., 2005), have enabled much atomic-level details to be unambiguously defined, with even more visible in the 1.98 Å structure (Watson et al., 2020).

6.6.1 The structure of the 30S subunit

The overall shape of the 30S particle is dominated by the structure of the folded RNA, and the proteins serve in large part to fill up the gaps and hold the whole assembly together. The secondary structure of the RNA shows a very large number (>50) of helical regions. The numerous loops are mostly small and do not disrupt the runs of helix in which they are embedded. There are extensive interactions between helices, mostly involving co-axial stacking via the minor grooves. In one type of helix—helix interaction two minor grooves contact each other, with consequent distortions from A-type geometry. These distortions, which tend to involve runs of adenines, are facilitated by both extra-helical bulges and non-canonical base pairs, as have been observed in simple RNA structures. Less commonly, perpendicular packing of one helix against another (also via the minor groove) is mediated by an unpaired purine base. This mode is of especial importance since it involves the functionally significant helices in the 30S subunit. The well-established motifs of RNA tertiary structure such as non-Watson—Crick base pairs, base triplets and tetraloops, all contribute to the overall structure. The proteins in the 30S subunit play important roles in helping to stabilise the RNA folding, by essentially filling in the numerous spaces in the RNA folds.

The following are the three sites where tRNA molecules bind and function, and where the essential proofreading checks for fidelity of code-reading and translation occur:

- the P (peptidylation) site, when a tRNA anticodon base pairs with the appropriate codon in mRNA, and where the peptide chain is covalently linked to a tRNA molecule
- the A (acceptor) site, when peptide bonds are eventually formed (the actual peptidyl transferase steps occur in the 50S subunit)
- the E (exit) site, for tRNAs to be released from the subunit as part of the protein synthesis cycle

tRNA itself was not present in the first 30S crystal structure, but the RNA from a symmetry-related 30S subunit effectively served to mimic it as the anticodon stem-loop. Its interactions with the 30S RNA are also mediated via (i) minor groove surfaces, helped by some contacts with ribosomal proteins, and (ii) via backbone contacts. Interestingly, the exit site of the tRNA is almost exclusively protein-associated, whereas the other functional sites are composed of RNA and not protein. It is thus the ribosomal RNA that mediates the functions of the 30S subunit, and not the ribosomal proteins.

The determination (Ogle et al., 2001) of the crystal structure at 3.3 Å of the 30S subunit complete with the anticodon stem-loop of a tRNA and a short RNA acting as an mRNA was the first that enabled the details of codon—anticodon recognition and checking to be visualised. A key role in this structure is played by two adjacent adenines from the 16S RNA, which are in contact with the minor groove of the codon—anticodon helical stem, to verify the correctness of the first two Watson—Crick base pairs of the triplet, and selection of the correct codon. The third codon position (the "wobble" position) is not so involved in interactions with the 16S RNA, and a tRNA anticodon is thus able to cope with hydrogen bonding to a range of third-position codon bases. The codon—anticodon stringency is disrupted by the antibiotic paromomycin, which has also been co-crystallised with the 30S subunit. The high-resolution *E. coli* structure (Watson et al., 2020) shows not only the anticodon—mRNA contacts (Figs. 6.32 and 6.33), but also reveals details of base—base hydrogen bond contacts between the tRNA anticodon stem-loop, the mRNA and the 30S RNA. A paromomycin molecule was also found in this structure to be bound in the mRNA decoding site, enabling its extended conformation and the surrounding water molecules to be fully visualised (Fig. 6.34). Interestingly this conformation agrees most closely with that reported in a model oligonucleotide complex (Vicens & Westhof, 2001).

Several other antibiotics that are known to inhibit protein synthesis have been examined bound to the 30S subunit, following co-crystallisation experiments (Brodersen et al., 2000; Carter et al., 2000). For example, streptomycin has been found to bind tightly to

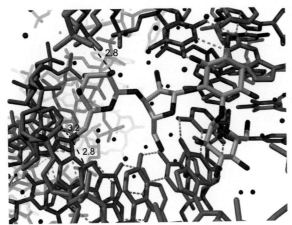

Figure 6.34 Paromomycin (shown in stick representation), bound in the mRNA decoding site, in the high-resolution *E. coli* 1.98 Å structure (Watson et al., 2020), and held in place by hydrogen bond/electrostatic interactions with phosphate groups, three of which are shown.

four distinct sites on the 16S RNA, mostly via phosphate groups. This stabilises the A site in an open conformation, thereby enabling non-cognate tRNAs to bind and reducing proofreading capability. These and other studies are relevant to the structure-based rational design of new antibiotics and are further discussed in the next section. A second crystal structure (Schlunzen et al., 2000), also at 3.3 Å resolution, has shown that the decoding region of the structure is highly conserved. Furthermore, the A and P site tRNAs and the codon part of the bound mRNA were found not to interact with any of the proteins, again emphasising the dominant role of RNA in ribosome structure and function.

6.6.2 The structure of the 50S subunit

In the 2.4 Å crystal structure (Ban et al., 2000) 2711 out of the total of 2923 ribosomal RNA nucleotides have been observed, together with 27 ribosomal proteins, and 122 nucleotides of 5S RNA, with the subunit being about 250 Å in each dimension. Again, the function of the proteins appears mainly to act as a cement, helping to maintain the integrity of the folded RNA. Although the RNA itself can be arranged into six domains based on its secondary structure, overall, the 50S subunit is remarkably globular, reflecting its greater conformational rigidity compared to the 30S subunit, to consist with its functional need to be more flexible.

The principal function of the 50S subunit is peptide bond formation. The active site where this occurs is within domain 5. Its precise location was identified by the analysis of crystals containing the antibiotic puromycin, an inhibitor of peptide synthesis. Remarkably, there is no protein structure within 18 Å of this molecule, bound in its active site, and accordingly it has been proposed that peptidyl transferase catalytic activity is entirely carried out by RNA, analogous to the function of RNAs as ribozymes. The reaction is an acid—base hydrogen transfer, with an adenine being suggested as the key residue, accepting a hydrogen atom from an aminoacylated tRNA so that it can bond to the carbonyl group of an esterified peptidyl-tRNA carrying the peptide chain to which the new amino acid is to be attached. All nucleotides involved in the active site are highly conserved in nature, strongly suggesting that RNA catalysis is a fundamental mechanism acting throughout evolution. The high-resolution *E. coli* complete ribosome structure (Watson et al., 2020) shows that the core of the 50S subunit is the most rigid part of the structure, with very many well-defined solvent molecules and counter-ions being well resolved in the electron density. These play important roles in maintaining the integrity of many RNA secondary structural features.

6.6.3 Complete ribosome structures

Structural studies on complete ribosomes are important for full understanding of mechanism since they do not function as individual sub-units, but as complete 70S particles. The crystals of the *E. coli* ribosome (Schuwirth et al., 2005) have provided two structures for this intact ribosome, since there are two independent ribosome assemblies per crystallographic asymmetric unit. There are some significant differences between the two structures, notably that the 30S sub-unit is rotated about 6° around the so-called neck region (which connects the 30S and 50S sub-units), in one structure relative to the other, and is also different from the conformation seen in the 2.8 Å crystal structure of the intact 70S *T. thermophilus* ribosome (Selmer et al., 2006), in which tRNA molecules are clearly seen together with part of the mRNA (some of it is disordered and is not visible in the electron density). Many of these discrepancies have been resolved in the more recent 1.98 Å (Watson et al., 2020) structure, as discussed above.

The determination of the structures of the complete human (Amunts et al., 2015) and yeast (Desai et al., 2017) mitochondrial ribosomes are milestones in our understanding of ribosome structure and function, even though these have not yet achieved the level of resolution seen in prokaryotic ribosomes (Bieri et al., 2018). Also, it needs to be borne in mind that mitochondrial ribosomes are quite distinct from bacterial or cytoplasmic eukaryotic ribosomes in that they have a much more specialised function, of synthesising proteins involved in cellular energy production. Consequently, their ribosomal RNA components tend to be much small than in other eukaryotic cells. The 3.5 Å structure of the human mitochondrial ribosome confirms earlier studies, that it contains three RNA species, the 12S and 16S subunits and tRNA (Fig. 6.35a and b), together with 80 proteins. The ribosomal RNA is about half the size of typical bacterial ribosomal RNAs and unsurprisingly protein—protein and protein—RNA contacts are more extensive than in the bacterial ribosomes. There are also differences in the dimensions of the mRNA entry channel, although in both bacterial and eukaryotic ribosomes the diameter

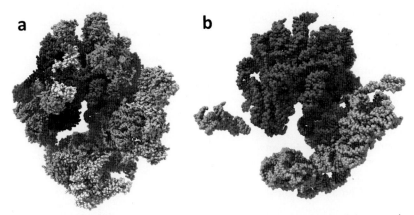

Figure 6.35 Space-filling images of the 3.5 Å structure (Amunts et al., 2015) of the complete human mitochondrial ribosome, PDB id 3J9M. (a) The complete ribosome with RNA components mostly in blue and the associated ribosomal proteins in various colours. (b) The RNA components, with the 12S RNA coloured mauve, the 16S RNA coloured blue and the tRNA coloured brown.

is not consistent with an mRNA duplex, but with the mRNA being single-stranded, as has been observed in the former structures. Complete structures for the human 80S ribosome (Khatter et al., 2015) and the yeast 80S ribosome (Ben-Shem et al., 2010) are also available.

6.7 RNA-drug complexes

A wide variety of RNA molecules and assemblies are targets for therapeutic intervention by small molecules, including RNA viral genomes, mammalian genomic RNAs and ribosomal RNAs (see Table 6.10 for a selection of these structures). An important target that has received much attention in the past is the HIV Trans-activator of transcription (Tat) protein and Trans Activation Responsive region (TAR) RNA. This has an arginine-rich group Tat recognising the base sequence and the conformation of TAR RNA. NMR studies have been extensively used for this system, and have, for example, identified the binding site of the drug acetylpromazine on TAR RNA (Du et al., 2002).

Many antibacterial antibiotics act directly at the ribosomal level, binding to RNA structural features and thereby blocking correct translation (Böttger, 2006; Poehlsgaard & Douthwaite, 2005). As indicated in the previous section, there have been many crystallographic and cryo-EM structure analyses of ribosome structures with bound drugs. For example, the crystal structures of complexes of the large ribosomal subunits from the organism *Haloarcula marismortui* (Tu et al., 2005) with bound erythromycin, azithromycin, clindamycin, virginiamycin S and telithromycin have provided some insights into the resistance mechanisms involving these antibiotics. These all contain the mutation

Table 6.10 Selected RNA-drug crystal structures.

Drug and RNA type	Resolution in Å	PDB id
Neomycin + A–site	2.40	2ET4
Paromomycin + A–site	2.50	1J7T
Streptomycin +40-mer aptamer	2.90	1NTB
Designer antibiotic 1 + 30S subunit	3.80	2F4V
Designer antibiotic 1 + A–site	2.80	2F4S
Designer antibiotic 2 + A–site	3.00	2F4T
Designer antibiotic 3 + A–site	2.60	2F4U
Streptomycin + *Thermus thermophilus* 30S ribosomal subunit	3.35	4DR3
Virginiamycin + *Escherichia coli* 50S ribosomal subunit	2.60	6PCQ
Virginiamycin analogue 47 + *E. coli* 50S ribosomal subunit	2.50	6PC6

G2099A, which confers increased affinity on erythromycin, possibly as a result of the loss of solvent at the N2 position on going from G to A. Crystallographic analyses (Hansen et al., 2002) of the four macrolide antibiotics (carbomycin A, spiramycin, and tylosin and azithromycin) with the same subunit have shown that these drugs bind in the polypeptide tunnel adjacent to the peptidyl transferase centre, and therefore that their action involves blocking a growing peptide chain from exiting. Other ribosomal antibiotics studied in situ include the *Deinococcus radiodurans* ribosome with the polyketide antibiotic lankacidin (Auerbach et al., 2010), the *E. coli* ribosome with chloramphenicol (Dunkle et al., 2010) and paromomycin at high resolution with the complete *E. coli* ribosome (Watson et al., 2020), as described in the previous section.

Since the crystal structure of the ribosomal sub-unit for a particular micro-organism has not always been available or is at only moderate resolution (ca 3 Å), a more rapid and (in principle) higher resolution approach to antibiotic drug design has been to analyse co-crystals between the antibiotic and the RNA that constitutes its minimal binding site. The power of this approach has been shown by the analyses of complexes involving the aminoglycoside antibiotics gentamicin C1A, kanamycin A, ribostamycin, lividomycin A, neomycin B (François et al., 2004) and several paromomycin derivatives (Hanessian et al., 2007). This latter study itself follows on from earlier analyses (Vicens & Westhof, 2001) of a complex of the native paromomycin antibiotic and two synthetic derivatives (François et al., 2004) with the decoding A-site RNA of *E. coli*. Paromomycin itself is in an L-shape when bound in this site, with numerous hydrogen bond and electrostatic contacts (Fig. 6.34) that are shared with most of the A-site bound aminoglycoside antibiotics. It was reasoned (François et al., 2004) that appropriate chemical modification of paromomycin would lead to a distinct bound conformation and set of contacts, with enhanced potency and activity against resistant strains, as was subsequently found experimentally and structurally. This work culminated in the design, synthesis and evaluation of a hydrophobic derivative, a (phenethylamino)ethyl ether of N1-(2S)-2-hydroxy-4-aminobutyric amide of 3′,4′-dideoxyparomomycin (Hanessian et al., 2010), which has an enhanced anti-bacterial activity profile, including against resistant bacterial strains. Structure-based methods have also been successfully employed in the design of three synthetic derivatives of the aminoglycoside antibiotic neamine, which similarly binds to the A-site (Murray et al., 2006). Confirmation of the bound conformation was obtained by crystal structure analyses of the derivatives bound to a model of the A-site (Fig. 6.36) and one to a complete 30S ribosome sub-unit, which gratifyingly confirmed the observations in the model A-site complexes, even though it is of lower resolution.

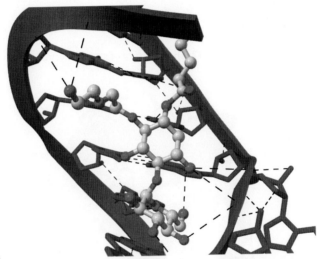

Figure 6.36 Crystal structure of a synthetic derivative of the antibiotic neamine, bound to a model sequence for the A-site (Murray et al., 2006), PDB id 2F4T. The antibiotic is shown in ball-and-stick mode and the numerous hydrogen bonds between it and the RNA are shown as dotted lines.

Streptomycin (Fig. 6.37) was the first antibiotic with clinically useful activity against tuberculosis, although it is no longer the first-line treatment due to its side effects. It acts by binding to ribosomal RNA, and a crystal structure of a complex with the 30S subunit has been reported (Carter et al., 2000), which involves the active recognition of all three rings of the drug. Its co-crystal structure with a 40-mer RNA aptamer (Tereshko et al., 2003) shows a very different arrangement, with the aptamer having an L-shaped RNA fold. Only one part of the drug molecule (the six-membered carbohydrate streptose ring and its substituents) is fully buried in the binding pocket that is formed from the two asymmetric internal loops of the RNA. The antibiotic molecule is held in place by an intricate array of hydrogen bonds and a deeply buried water molecule, in the

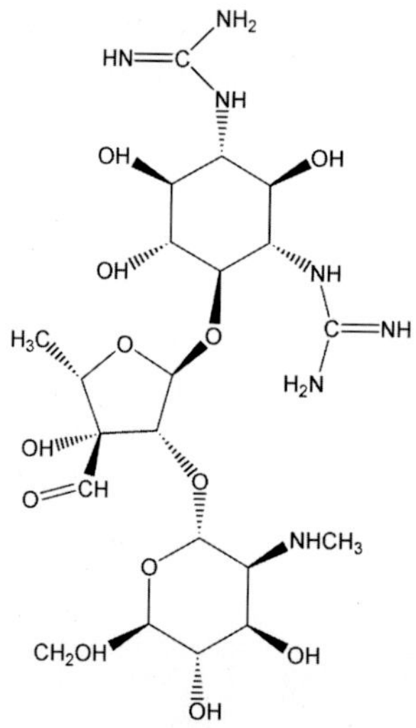

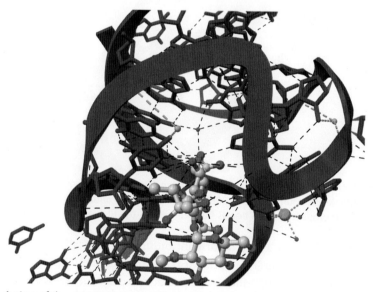

Figure 6.37 The structure of streptomycin.

Figure 6.38 A view of the streptomycin complex with a 40-mer RNA aptamer (Tereshko et al., 2003), PDB id 1NTB, showing the hydrogen bonding between the partially bound streptomycin molecule and RNA sites.

centre of the view (Fig. 6.38). A subsequent determination of the co-crystal structure with the 30S subunit from *T. thermophilus* and with several anticodon stem–loop analogues (Demirci et al., 2013), at 3.35 Å resolution, shows the streptomycin molecule, by contrast, fully bound within the 30S RNA (Fig. 6.39) where it has induced several structural distortions including a conformational change in the decoding site itself.

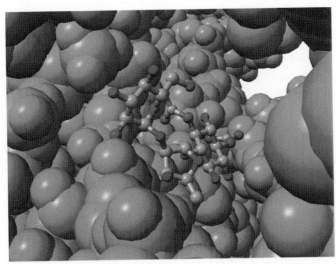

Figure 6.39 A view of the streptomycin molecule (shown in ball-and-stick form) complexed with the 30S ribosomal subunit, shown in space-filling representation (Demirci et al., 2013), PDB id 4DR3.

A structural approach has proved fruitful in developing novel analogues of the Streptogramin class of macrocyclic antibiotics, using virginiamycin as a starting point (Li et al., 2020). These compounds work by inhibiting bacterial ribosome function, and resistance to these clinically useful antibiotics arises from acetyltransferase action on them to give derivatives that sterically inhibit ribosome binding. The drug discovery programme was guided by a 2.6 Å cryo-EM structure of virginiamycin bound to the 50S *E. coli* ribosome subunit, which showed that the C3 substituent, whilst in hydrophobic contact with a uracil base, is free for potential derivatisation. A focused library of analogues was synthesised and evaluated in a panel of pathogenic cell lines. The compounds with superior activity and resistance profiles were also studied structurally with a total of nine new 50S complex structures being determined.

Genomic mRNA and viral RNA sequences (and their folded structure) are of increasing interest as drug targets. This is especially so for several neurodegenerative diseases where, for example, in Fragile X Syndrome there is a well-documented repeated r(CGG) transcript which is a validated drug target (see, for example, Falese et al., 2021). Currently the major viral RNA genome of much interest and activity is that of the SARS-CoV-2 virus, for which there are models and conjectures but as yet no specific RNA-binding compounds for use as potential drugs have been developed beyond the experimental stage (Rangan et al., 2021). Two definite quadruplex-forming sequences have been identified in the RNA genome (Zhao et al., 2021) and a quadruplex-binding pyridostatin derivative has in vitro and in vivo antiviral activity, which arises from translation inhibition of a specific viral protein, the nucleocapsid phosphoprotein.

6.8 RNA motifs

We have seen in this chapter that the variability and complexity of RNA structures is due to their ability to fully exploit the hydrogen-bonding and base-stacking potential of both the bases themselves and the ribose sugars. In hydrogen-bonding terms Watson—Crick base pairing remains the bedrock of helical stems, essential components of all RNAs. In addition, RNAs exploit the full gamut of potential base pairing presented by purines and pyrimidines, in ways that very much extend the mismatches seen with DNA, as discussed in Chapter 4. These pairings, which play key roles in stabilising loop—loop interactions, have been systematised and classified into discrete families in a rational way (Lee & Gutell, 2004; Leontis et al., 2002; Leontis & Westhof, 2001; Miao & Westhof, 2017) that emphasises the three hydrogen-bonding edges of a base — Watson—Crick, Hoogsteen and sugar/base. There is even greater diversity of possible base triplets and quartets, many of which have been observed in ribosome structures, which are the richest mines of information on RNA tertiary folds, as they have progressed to increasingly higher resolution. These interactions in turn are the basis of the various established RNA motifs, notably the adenosine platform, various tetraloops, the pseudoknot, the kink-turn and the C-loop (Lescoute et al., 2005). The sheer density of nucleotides in the ribosome results in a very large number of loop insertions into helices, primarily via their minor grooves, and novel structural motifs are involved. One, the A-minor motif, appears to be especially prevalent (Nissen et al., 2001). This involves adenines interacting in the minor groove, of which four distinct ways of doing so have been identified. The A-minor motif is an important aspect of the manner by which ribosomal RNAs preserve their functionally significant architecture — the adenines involved are often highly conserved, and so the ribosomal RNA framework is likely to be largely invariant throughout the living world. The motif also appears in several ribozyme structures (see also Section 6.4.2).

We have also seen that other fold motifs such as three-way junctions play key roles in the architecture of ribozymes and riboswitches. The various structural types of three-way junction can be categorised into distinct topological families that depend on the size of the junction, and on the relative orientations of the three stems forming the arms of the junction (Lescoute & Westhof, 2006).

RNA motifs can also be categorised, for example, based on conformational type (Schneider et al., 2004). An extension of this has been to not use single nucleotides, but libraries of pairs of structurally close nucleotides (Sykes & Levitt, 2005). From this, just 30 doublets have been found to represent most RNA structural (i.e. conformational and base pairing) features. More recently a unified dinucleotide structural alphabet has been proposed for describing RNA (and DNA) structure (Černý et al., 2020), with 28 classes mostly representing RNA structures. The complexity of RNA structures inevitably means that the motifs for recognition of RNA structures in general by proteins are much more diverse than for DNA—protein recognition, which is dominated by the regularity of the DNA double helix. RNA—protein recognition is discussed in Chapter 8.

6.9 RNA databases

The PDB is the initial depository for all RNA structures from crystallography, NMR and cryo-EM. The associated Nucleic Acid Database also lists all RNA structures. In addition there are several specialised databases for RNA and RNA—protein complexes:

http://genesilico.pl/rnabricks2 provides much useful information on all RNA and RNA—protein structures, including lists of motifs and base pairs for individual structures. It has some search facilities and as of April 2021 includes over 4300 structures.

http://rna.bgsu.edu/rna3dhub/motifs the RNA 3D Motif Atlas is a comprehensive and representative collection of internal and hairpin loop RNA 3D motifs extracted from the Representative Sets of RNA 3D structures.

https://www.bgsu.edu/research/rna/web-applications/webfr3d.html is the online version of FR3D, a suite of programs designed to search RNA 3D structures for user-specified queries.

References

Abu Almakarem, A. S., Petrov, A. I., Stombaugh, J., Zirbel, C. L., & Leontis, N. B. (2012). Comprehensive survey and geometric classification of base triples in RNA structures. *Nucleic Acids Research, 40,* 1407—1423.

Adamiak, D. A., Milecki, J., Adamiak, R. W., & Rypniewski, W. (2010). The hydration and unusual hydrogen bonding in the crystal structure of an RNA duplex containing alternating CG base pairs. *New Journal of Chemistry, 34,* 903—909.

Adams, P. L., Stahley, M. R., Kosek, A. B., Wang, J., & Strobel, S. A. (2004). Crystal structure of a self-splicing group I intron with both exons. *Nature, 430,* 45—50.

Ahmed, Y. L., & Ficner, R. (2014). RNA synthesis and purification for structural studies. *RNA Biology, 11,* 427—432.

Allain, F. H.-T., & Varani, G. (1995). Structure of the P1 helix from group I self-splicing introns. *Journal of Molecular Biology, 250,* 333—353.

Amunts, A., Brown, A., Toots, J., Scheres, S. H., & Ramakrishnan, V. (2015). The structure of the human mitochondrial ribosome. *Science, 348,* 95—98.

Anderson, A. C., Scaringe, S. A., Earp, B. E., & Frederick, C. A. (1996). HPLC purification of RNA for crystallography and NMR. *RNA, 2,* 110—117.

Anderson, M., Schultz, E., Martick, M., & Scott, W. G. (2013). Active-site monovalent cations revealed in a 1.55 A resolution hammerhead ribozyme structure. *Journal of Molecular Biology, 425,* 3790—3798.

Arnott, S., Bond, P. J., Selsing, E., & Smith, P. J. (1976). Models of triple-stranded polynucleotides with optimised stereochemistry. *Nucleic Acids Research, 3,* 2459—2470.

Arnott, S., Hukins, D. W. L., Dover, S. D., Fuller, W., & Hodgson, A. R. (1973). Structures of synthetic polynucleotides in the A-RNA and A'-RNA conformations: X-ray diffraction analyses of the molecular conformations of polyadenylic acid—polyuridylic acid and polyinosinic acid—polycytidylic acid. *Journal of Molecular Biology, 81,* 107—122.

Arnott, S. (1999). In S. Neidle (Ed.), *Oxford handbook of nucleic acid structure* (p. 1). Oxford: Oxford University Press.

Auerbach, T., Mermershtain, I., Davidovich, C., Bashan, A., Belousoff, M., Wekselman, I., Zimmerman, E., Xiong, L., Klepacki, D., Arakawa, K., Kinashi, H., Mankin, A. S., & Yonath, A. (2010). The structure of ribosome-lankacidin complex reveals ribosomal sites for synergistic antibiotics. *Proceedings of the National Academy of Sciences of the United States of America, 107*, 1983–1988.

Auffinger, P., Grover, N., & Westhof, E. (2011). Metal ion binding to RNA. *Metal Ions Life Science, 9*, 1–35.

Auffinger, P., & Westhof, E. (1997). Rules governing the orientation of the 2'-hydroxyl group in RNA. *Journal of Molecular Biology, 274*, 54–63.

Auffinger, P., & Westhof, E. (2000). Water and ion binding around RNA and DNA (C,G) oligomers. *Journal of Molecular Biology, 300*, 1113–1131.

Baeyens, K. J., de Bondt, H. L., Pardi, A., & Holbrook, S. R. (1996). A curved RNA helix incorporating an internal loop with G.A and A.A non-Watson-Crick base pairing. *Proceedings of the National Academy of Sciences of the United States of America, 93*, 12851–12855.

Ban, N., Freeborn, B., Nissen, P., Penzyek, P., Grassucci, R. A., Sweet, R. M., Frank, J., Moore, P. B., & Steitz, T. A. (1998). A 9 Å resolution X-ray crystallographic map of the large ribosomal subunit. *Cell, 93*, 1105–1115.

Ban, N., Nissen, P., Hansen, J., Moore, P. B., & Steitz, T. A. (2000). The complete atomic structure of the large ribosomal subunit at 2.4 Å resolution. *Science, 289*, 905–920.

Batey, R. T., Gilbert, S. D., & Montange, R. K. (2004). Structure of a natural guanine-responsive riboswitch complexed with the metabolite hypoxanthine. *Nature, 432*, 411–415.

Ben-Shem, A., Jenner, L., Yusupova, G., & Yusupov, M. (2010). Crystal structure of the eukaryotic ribosome. *Science, 330*, 1203–1209.

Bieri, P., Greber, B. J., & Ban, N. (2018). High-resolution structures of mitochondrial ribosomes and their functional implications. *Current Opinion in Structural Biology, 49*, 44–53.

Böttger, E. C. (2006). The ribosome as a drug target. *Trends in Biotechnology, 24*, 145–147.

Brodersen, D. E., Clemons, W. M., Carter, A. P., Morgan-Warren, R. J., Wimberley, B. T., & Ramakrishnan, V. (2000). The structural basis for the action of the antibiotics tetracycline, pactamycin, and hygromycin B on the 30S ribosomal subunit. *Cell, 103*, 1143–1154.

Carter, A. P., Clemons, W. M., Brodersen, D. E., Morgan-Warren, R. J., Wimberly, B. T., & Ramakrishnan, V. (2000). Functional insights from the structure of the 30S ribosomal subunit and its interactions with antibiotics. *Nature, 407*, 340–348.

Cate, J. H., & Doudna, J. A. (2000). Solving large RNA structures by X-ray crystallography. *Methods in Enzymology, 317*, 169–180.

Cate, J. H., Gooding, A. R., Podell, E., Zhou, K., Golden, B. L., Kundrot, C. E., Cech, T. R., & Doudna, J. A. (1996). Crystal structure of a group I ribozyme domain: Principles of RNA packing. *Science, 273*, 1678–1685.

Cate, J. H., Yusupov, M. M., Yusupova, G. Z., Earnest, T. N., & Noller, H. F. (1999). X-ray crystal structures of 70S ribosome functional complexes. *Science, 285*, 2095–2104.

Cech, T. R., Zaug, A. J., & Grabowski, P. J. (1981). In vitro splicing of the ribosomal RNA precursor of Tetrahymena: Involvement of a guanosine nucleotide in the excision of the intervening sequence. *Cell, 27*, 487–496.

Černý, J., Božíková, P., Svoboda, J., & Schneider, B. (2020). A unified dinucleotide alphabet describing both RNA and DNA structures. *Nucleic Acids Research, 48*, 6367–6381.

Chan, C. W., Badong, D., Rajan, R., & Mondragón, A. (2020). Crystal structures of an unmodified bacterial tRNA reveal intrinsic structural flexibility and plasticity as general properties of unbound tRNAs. *RNA, 26*, 278–289.

Cheah, M. T., Wachter, A., Sudarsan, N., & Breaker, R. R. (2007). Control of alternative RNA splicing and gene expression by eukaryotic riboswitches. *Nature, 447*, 497–500.

Cheatham, T. E., & Kollman, P. A. (1997). Molecular dynamics simulations highlight the structural differences among DNA:DNA, RNA:RNA, and DNA:RNA hybrid duplexes. *Journal of the American Chemical Society, 119*, 4805–4825.

Clemons, W. M., May, J. L. C., Wimberley, B. T., McCutcheon, J. P., Capel, M. S., & Ramakrishnan, V. (1999). Structure of a bacterial 30S ribosomal subunit at 5.5 Å resolution. *Nature, 400*, 833–840.

Cruse, W. B. T., Saludjian, P., Biala, E., Strazewski, P., Prangé, T., & Kennard, O. (1994). Structure of a mispaired RNA double helix at 1.6-Å resolution and implications for the prediction of RNA secondary structure. *Proceedings of the National Academy of Sciences of the United States of America, 91*, 4160–4164.

Darré, L., Ivani, I., Dans, P. D., Gómez, H., Hospital, A., & Orozco, M. (2016). Small details matter: The 2'-hydroxyl as a conformational switch in RNA. *Journal of the American Chemical Society, 138,* 16355—16363.

Dégut, C., Monod, A., Brachet, A., Crépin, T., & Tisné, C. (2016). In vitro/in vivo production of tRNA for X-ray studies. *Methods in Molecular Biology, 1320,* 37—57.

Demirci, H., Murphy, F., Murphy, E., Gregory, S. T., Dahlberg, A. E., & Jogl, G. (2013). A structural basis for streptomycin-induced misreading of the genetic code. *Nature Communications, 4,* 1355.

Desai, N., Brown, A., Amunts, A., & Ramakrishnan, V. (2017). The structure of the yeast mitochondrial ribosome. *Science, 355,* 528—531.

Dock-Bregeon, A. C., Chevrier, B., Podjarny, A., Johnson, J., de Bear, J. S., Gough, G. R., Gilham, P. T., & Moras, D. (1989). Crystallographic structure of an RNA helix: [U(UA)6A]₂. *Journal of Molecular Biology, 209,* 459—474.

Dock-Breigon, A. C., Moras, D., & Giegé, R. (1999). In A. Ducruix, & R. Giegé (Eds.), *Crystallisation of nucleic acids and proteins.* Oxford: Oxford University Press.

Doherty, E. A., Batey, R. T., Masquida, B., & Doudna, J. A. (2001). A universal mode of helix packing in RNA. *Nature Structural Biology, 8,* 339—343.

Doudna, J. A., Grosshans, C., Gooding, A., & Kundrot, C. E. (1993). Crystallization of ribozymes and small RNA motifs by a sparse matrix approach. *Proceedings of the National Academy of Sciences of the United States of America, 90,* 7829—7833.

Drozdzal, P., Michalska, K., Kierzek, R., Lomozik, L., & Jaskolski, M. (2011). Structure of an RNA/DNA dodecamer corresponding to the HIV-1 polypurine tract at 1.6 Angstrom resolution. *Acta Crystallographica, D68,* 169—175.

Du, Z., Lind, K. E., & James, T. L. (2002). Structure of TAR RNA complexed with a Tat-TAR interaction nanomolar inhibitor that was identified by computational screening. *Chemical Biology, 9,* 707—712.

Dunkle, J. A., Xiong, L., Mankin, A. S., & Cate, J. H. (2010). Structures of the *Escherichia coli* ribosome with antibiotics bound near the peptidyl transferase center explain spectra of drug action. *Proceedings of the National Academy of Sciences of the United States of America, 107,* 17152—17157.

Edwards, T. E., & Ferré-D Amaré. (2006). Crystal structures of the *thi*-box riboswitch bound to thiamine pyrophosphate analogs reveal adaptive RNA-small molecule recognition. *Structure, 14,* 1459—1468.

Egli, M., Minasov, G., Su, L., & Rich, A. (2002). Metal ions and flexibility in a viral RNA pseudoknot at atomic resolution. *Proceedings of the National Academy of Sciences of the United States of America, 99,* 4302—4307.

Egli, M., Portmann, S., & Usman, N. (1996). RNA hydration: A detailed look. *Biochemistry, 35,* 8489—8494.

Ennifar, E., Walter, P., Ehresmann, B., Ehresmann, C., & Dumas, P. (2001). Crystal structures of coaxially stacked kissing complexes of the HIV-1 RNA dimerization initiation site. *Nature Structural Biology, 8,* 1064—1068.

Ennifar, E., Yusupov, M., Walter, P., Marquet, R., Ehresmann, B., Ehresmann, C., & Dumas, P. (1999). The crystal structure of the dimerization initiation site of genomic HIV-1 RNA reveals an extended duplex with two adenine bulges. *Structure, 7,* 1439—1449.

Ennifar, E., Nikulin, A., Tishchenko, S., Serganov, A., Nevskaya, N., Garber, M., Ehresmann, B., Ehresmann, C., Nikonov, S., & Dumas, P. (2000). The crystal structure of UUCG tetraloop. *Journal of Molecular Biology, 304,* 35—42.

Falese, J. P., Donlic, A., & Hargrove, A. E. (2021). Targeting RNA with small molecules: From fundamental principles towards the clinic. *Chemical Society Reviews, 50,* 2224—2243.

Feig, A. L., Scott, W. G., & Uhlenbeck, O. C. (1998). Inhibition of the hammerhead ribozyme cleavage reaction by site-specific binding of Tb. *Science, 279,* 81—84.

Ferre-D'Amare, A. R., Zhou, K., & Doudna, J. A. (1998b). A general module for RNA crystallization. *Journal of Molecular Biology, 279,* 621—631.

Ferré-D'Amaré, A. R., Zhou, K., & Doudna, J. A. (1998a). Crystal structure of a hepatitis delta virus ribozyme. *Nature, 395,* 567—574.

François, B., Szychowski, J., Adhikari, S. S., Pachamuthu, K., Swayze, E. E., Griffey, R. H., Migawa, M. T., Westhof, E., & Hanessian, S. (2004). Antibacterial aminoglycosides with a modified mode of binding to the ribosomal-RNA decoding site. *Angewandte Chemie International Edition England, 43,* 6735—6738.

Garg, A., & Heinemann, U. (2018). A novel form of RNA double helix based on GU and C•A⁺ wobble base pairing. *RNA, 24,* 209—218.

Golden, B. L. (2007). Preparation and crystallization of RNA. *Methods in Molecular Biology, 363,* 239–257.

Golden, B. L., Kim, H., & Chase, E. (2005). Crystal structure of a phage Twort group 1 ribozyme-product complex. *Nature Structural & Molecular Biology, 12,* 82–89.

Guerrier-Takada, C., Gardiner, K., Marsh, T., Pace, N., & Altman, S. (1983). The RNA moiety of ribonuclease P is the catalytic subunit of the enzyme. *Cell, 35,* 849–857.

Guo, F., Gooding, A. R., & Cech, T. R. (2004). Structure of the *Tetrahymena* ribozyme: Base triple sandwich and metal ion at the active site. *Molecular Cell, 16,* 351–362.

Hanessian, S., Pachamuthu, K., Szychowski, J., Giguère, A., Swayze, E. E., Migawa, M. T., François, B., Kondo, J., & Westhof, E. (2010). Structure-based design, synthesis and A-site rRNA co-crystal complexes of novel amphiphilic aminoglycoside antibiotics with new binding modes: A synergistic hydrophobic effect against resistant bacteria. *Bioorganic & Medicinal Chemistry Letters, 20,* 7097–7101.

Hanessian, S., Szychowski, J., Adhikari, S. S., Vasquez, G., Kandasamy, P., Swayze, E. E., Migawa, M. T., Ranken, R., François, B., Wirmer-Bartoschek, J., Kondo, J., & Westhof, E. (2007). Structure-based design, synthesis, and A-site rRNA cocrystal complexes of functionally novel aminoglycoside antibiotics: C2'' ether analogues of paromomycin. *Journal of Medicinal Chemistry, 50,* 2352–2369.

Hansen, J. L., Ippolito, J. A., Ban, N., Nissen, P., Moore, P. B., & Steitz, T. A. (2002). The structures of four macrolide antibiotics bound to the large ribosomal subunit. *Molecular Cell, 10,* 117–128.

Heus, H. A., & Pardi, A. (1991). Structural features that give rise to the unusual stability of RNA hairpins containing GNRA loops. *Science, 253,* 191–194.

Höbartner, C., Rieder, R., Kreutz, C., Puffer, B., Lang, L., Polonskaia, A., Serganov, A., & Micura, R. (2005). Syntheses of RNAs with up to 100 nucleotides containing site-specific 2'-methylseleno labels for use in X-ray crystallography. *Journal of the American Chemical Society, 127,* 12035–12045.

Holbrook, S. R., Cheong, C., Tinoco, I., & Kim, S.-H. (1991). Crystal structure of an RNA double helix incorporating a track of non-Watson-Crick base pairs. *Nature, 353,* 579–581.

Hood, I. V., Gordon, J. M., Bou-Nader, C., Henderson, F. E., Bahmanjah, S., & Zhang, J. (2019). Crystal structure of an adenovirus virus-associated RNA. *Nature Communications, 10,* 2871.

Ippolito, J. A., & Steitz, T. A. (2000). The structure of the HIV-1 RRE high affinity rev binding site at 1.6 Å resolution. *Journal of Molecular Biology, 295,* 711–717.

Jovine, L., Djordjevic, S., & Rhodes, D. (2000). The crystal structure of yeast phenylalanine tRNA at 2.0 Å resolution: Cleavage by Mg^{2+} in 15-year old crystals. *Journal of Molecular Biology, 301,* 401–414.

Ke, A., & Doudna, J. A. (2004). Crystallization of RNA and RNA-protein complexes. *Methods, 34,* 408–414.

Khatter, H., Myasnikov, A. G., Natchiar, S. K., & Klaholz, B. P. (2015). Structure of the human 80S ribosome. *Nature, 520,* 640–645.

Kiliszek, A., Kierzek, R., Krzyzosiak, W. J., & Rypniewski, W. (2009). Structural insights into CUG repeats containing the 'stretched U-U wobble': Implications for myotonic dystrophy. *Nucleic Acids Research, 37,* 4149–4156.

Kim, S.-H., Suddath, F. L., Quigley, G. J., McPherson, A., Sussman, J. L., Wang, A. H.-J., Seeman, N. C., & Rich, A. (1974). Three-dimensional tertiary structure of yeast phenylalanine transfer RNA. *Science, 185,* 435–440.

Klosterman, P. S., Shah, S. A., & Steitz, T. A. (1999). Crystal structures of two plasmid copy control related rna duplexes: An 18 base pair duplex at 1.20 Å resolution and a 19 base pair duplex at 1.55 Å resolution. *Biochemistry, 38,* 14784–14792.

Klug, A., Robertus, J. D., Ladner, J. E., Brown, R. S., & Finch, J. T. (1974). Conservation of the molecular structure of yeast phenylalanine transfer RNA in two crystal forms. *Proceedings of the National Academy of Sciences of the United States of America, 71,* 3711–3715.

Korostelev, A., Trakhanov, S., Laurberg, M., & Noller, H. F. (2006). Crystal structure of a 70S ribosome-tRNA complex reveals functional interactions and rearrangements. *Cell, 126,* 1065–1077.

Kührová, P., Otyepka, M., Šponer, J., & Banáš, P. (2014). Are waters around RNA more than just a solvent? - an insight from molecular dynamics simulations. *Journal of Chemical Theory and Computation, 10,* 401–411.

Kundrot, C. (1997). Preparation and crystallization of RNA: A sparse matrix approach. *Methods in Enzymology, 276,* 143–156.

Lee, J. C., & Gutell, R. R. (2004). Diversity of base-pair conformations and their occurrence in rRNA structure and RNA structural motifs. *Journal of Molecular Biology, 344*, 1225–1249.

Leontis, N. B., Stombaugh, J., & Westhof, E. (2002). The non-Watson-Crick base pairs and their associated isostericity matrices. *Nucleic Acids Research, 30*, 3497–3531.

Leontis, N. B., & Westhof, E. (2001). Geometric nomenclature and classification of RNA base pairs. *RNA, 7*, 499–512.

Lescoute, A., Leontis, N. B., Massire, C., & Westhof, E. (2005). Recurrent structural RNA motifs, isostericity matrices and sequence alignments. *Nucleic Acids Research, 33*, 2395–2409.

Lescoute, A., & Westhof, E. (2006). Topology of three-way junctions in folded RNAs. *RNA, 12*, 83–93.

Levitt, M. (1969). Detailed molecular model for transfer ribonucleic acid. *Nature, 224*, 759–763.

Li, Q., Pellegrino, J., Lee, D. J., Tran, A. A., Chaires, H. A., Wang, R., Park, J. E., Ji, K., Chow, D., Zhang, N., Brilot, A. F., Biel, J. T., van Zundert, G., Borrelli, K., Shinabarger, D., Wolfe, C., Murray, B., Jacobson, M. P., Muhle, E., … Seiple, I. B. (2020). Synthetic group A streptogramin antibiotics that overcome Vat resistance. *Nature, 586*, 145–150.

Lilley, D. M. J. (1999). Structure, folding and catalysis of the small nucleolytic ribozymes. *Current Opinion in Structural Biology, 9*, 330–338.

Lilley, D. M. J. (2019). Classification of the nucleolytic ribozymes based upon catalytic mechanism. *F1000Res, 8*. F1000 Faculty Rev-1462.

Miao, Z., Adamiak, R. W., Antczak, M., Boniecki, M. J., Bujnicki, J., Chen, S. J., Cheng, C. Y., Cheng, Y., Chou, F. C., Das, R., Dokholyan, N. V., Ding, F., Geniesse, C., Jiang, Y., Joshi, A., Krokhotin, A., Magnus, M., Mailhot, O., Major, F., … Westhof, E. (2020). RNA-puzzles round IV: 3D structure predictions of four ribozymes and two aptamers. *RNA, 26*, 982–995.

Miao, Z., & Westhof, E. (2017). RNA structure: Advances and assessment of 3D structure prediction. *Annual Review of Biophysics, 46*, 483–503.

Miller, J. L., & Kollman, P. A. (1997). Theoretical studies of an exceptionally stable RNA tetraloop: Observation of convergence from an incorrect NMR structure to the correct one using unrestrained molecular dynamics. *Journal of Molecular Biology, 270*, 436–450.

Mir, A., & Golden, B. L. (2016). Two active site divalent ions in the crystal structure of the hammerhead ribozyme bound to a transition state analogue. *Biochemistry, 55*, 633–636.

Mooers, B. H. M., Logue, J. S., & Berglund, J. A. (2005). The structural basis of myotonic dystrophy from the crystal structure of CUG repeats. *Proceedings of the National Academy of Sciences of the United States of America, 102*, 16626–16631.

Mrázikova, K., Mlýnský, V., Kührová, P., Pokorná, P., Kruse, H., Krepl, M., Otyepka, M., Banáš, P., & Šponer, J. (2020). UUCG RNA tetraloop as a formidable force-field challenge for MD simulations. *Journal of Chemical Theory and Computation, 16*, 7601–7617.

Murray, J. B., Meroueh, S. O., Russell, R. J., Lentzen, G., Haddad, J., & Mobashery, S. (2006). Interactions of designer antibiotics and the bacterial ribosomal aminoacyl-tRNA site. *Chemical Biology, 13*, 129–138.

Murray, J. B., Terwey, D. P., Maloney, L., Karpeisky, A., Usman, N., Beigelman, L., & Scott, W. G. (1998). The structural basis of hammerhead ribozyme self-cleavage. *Cell, 92*, 665–673.

Murray, J. B., Szöke, H., Szöke, A., & Scott, W. G. (2000). Capture and visualization of a catalytic RNA enzyme-product complex using crystal lattice trapping and X-ray holographic reconstruction. *Molecular Cell, 5*, 279–287.

Nissen, P., Ippolito, J. A., Ban, N., Moore, P. B., & Steitz, T. A. (2001). RNA tertiary interactions in the large ribosomal subunit: The A-minor motif. *Proceedings of the National Academy of Sciences of the United States of America, 98*, 4899–4903.

Nowacka, M., Fernandes, H., Kiliszek, A., Bernat, A., Lach, G., & Bujnicki, J. M. (2019). Specific interaction of zinc finger protein Com with RNA and the crystal structure of a self-complementary RNA duplex recognized by Com. *PloS One, 14*, e0214481.

Ogle, J. M., Broderson, D. E., Clemons, W. M., Tarry, M. J., Carter, A. P., & Ramakrishnan, V. (2001). Recognition of cognate transfer RNA by the 30S ribosomal subunit. *Science, 292*, 897–902.

Olson, W. K., Esguerra, M., Xin, Y., & Lu, X. J. (2009). New information content in RNA base pairing deduced from quantitative analysis of high-resolution structures. *Methods, 47*, 177–186.

Olson, W. K., Li, S., Kaukonen, T., Colasanti, A. V., Xin, Y., & Lu, X. J. (2019). Effects of noncanonical base pairing on RNA folding: Structural context and spatial arrangements of GA pairs. *Biochemistry, 58,* 2474–2487.

Pallan, P. S., Marshall, W. S., Harp, J., Jewett, F. C., III, Wawrzak, Z., Brown, B. A., II, Rich, A., & Egli, M. (2005). Crystal structure of a luteoviral RNA pseudoknot and model for a minimal ribosomal frameshifting motif. *Biochemistry, 44,* 11315–11322.

Peterson, R. D., & Feigon, J. (1996). Structural change in Rev responsive element RNA of HIV-1 on binding rev peptide. *Journal of Molecular Biology, 264,* 863–877.

Pley, H. W., Flaherty, K. M., & McKay, D. B. (1994). Three-dimensional structure of a hammerhead ribozyme. *Nature, 372,* 68–74.

Poehlsgaard, J., & Douthwaite, S. (2005). The bacterial ribosome as a target for antibiotics. *Nature Reviews Microbiology, 3,* 870–881.

Portmann, S., Usman, N., & Egli, M. (1995). The crystal structure of r(CCCCGGGG) in two distinct lattices. *Biochemistry, 34,* 7569–7575.

Price, S. R., Ito, N., Oubridge, C., Avis, J. M., & Nagai, K. (1995). Crystallization of RNA-protein complexes. I. Methods for the large-scale preparation of RNA suitable for crystallographic studies. *Journal of Molecular Biology, 249,* 398–408.

Rangan, R., Watkins, A. M., Chacon, J., Kretsch, R., Kladwang, W., Zheludev, I. N., Townley, J., Rynge, M., Thain, G., & Das, R. (2021). De novo 3D models of SARS-CoV-2 RNA elements from consensus experimental secondary structures. *Nucleic Acids Research, 49,* 3092–3108.

Rich, A., Davies, D. R., Crick, F. H., & Watson, J. D. (1961). The molecular structure of polyadenylic acid. *Journal of Molecular Biology, 3,* 71–86.

Richardson, J. S., Schneider, B., Murray, L. W., Kapral, G. J., Immormino, R. M., Headd, J. J., Richardson, D. C., Ham, D., Hershkovits, E., Williams, L. D., Keating, K. S., Pyle, A. M., Micallef, D., Westbrook, J., & Berman, H. M. (2008). RNA backbone: Consensus all-angle conformers and modular string nomenclature (an RNA ontology consortium contribution). *RNA, 14,* 465–481.

Rife, J. P., Stallings, S. C., Correll, C. C., Dallas, A., Steitz, T. A., & Moore, P. B. (1999). Comparison of the crystal and solution structures of two RNA oligonucleotides. *Biophysical Journal, 76,* 65–75.

Robertson, M. J., Qian, Y., Robinson, M. C., Tirado-Rives, J., & Jorgensen, W. L. (2019). Development and testing of the OPLS-AA/M force field for RNA. *Journal of Chemical Theory and Computation, 15,* 2734–2742.

Robertus, J. D., Ladner, J. E., Finch, J. T., Rhodes, D., Brown, R. S., Clark, B. F. C., & Klug, A. (1974). Structure of yeast phenylalanine tRNA at 3 Å resolution. *Nature, 250,* 546–551.

Rosenberg, J. M., Seeman, N. C., Day, R. O., & Rich, A. (1976). RNA double-helical fragments at atomic resolution. II. The crystal structure of sodium guanylyl-3′,5′-cytidine nonahydrate. *Journal of Molecular Biology, 104,* 145–167.

Roy, S., & Caruthers, M. (2013). Synthesis of DNA/RNA and their analogs via phosphoramidite and H-phosphonate chemistries. *Molecules, 18,* 14268–14284.

Rupert, P. B., & Ferré-D'Amaré, A. R. (2001). Crystal structure of a hairpin ribozyme-inhibitor complex with implications for catalysis. *Nature, 410,* 780–786.

Ruszkowska, A., Ruszkowski, M., Hulewicz, J. P., Dauter, Z., & Brown, J. A. (2020). Molecular structure of a U•A-U-rich RNA triple helix with 11 consecutive base triples. *Nucleic Acids Research, 48,* 3304–3314.

Safaee, N., Noronha, A. M., Rodionov, D., Kozlov, G., Wilds, C. J., Sheldrick, G. M., & Gehring, K. (2013). Structure of the parallel duplex of poly(A) RNA: Evaluation of a 50 year-old prediction. *Angewandte Chemie International Edition England, 52,* 10370–10373.

Sarkhel, S., Rich, A., & Egli, M. (2003). Water-nucleobase "stacking": H-π and lone pair-π interactions in the atomic resolution crystal structure of an RNA pseudoknot. *Journal of the American Chemical Society, 125,* 8998–8999.

Schindelin, H., Zhang, M., Bald, R., Fürste, J.-P., Erdmann, V. A., & Heinemann, U. (1995). Crystal structure of an RNA dodecamer containing the *Escherichia coli* Shine-Dalgarno sequence. *Journal of Molecular Biology, 249,* 595–603.

Schlunzen, F., Tocilj, A., Zarivach, R., Harms, J., Gluhmann, M., Janell, D., Bashan, A., Bartels, H., Agmon, I., Franceschi, F., & Yonath, A. (2000). Structure of functionally activated small ribosomal subunit at 3.3 Å resolution. *Cell, 102,* 615–623.

Schneider, B., Morávek, Z., & Berman, H. M. (2004). RNA conformational classes. *Nucleic Acids Research, 32,* 1666–1677.

Schuwirth, B. S., Borovinskaya, M. A., Hau, C. W., Zhang, W., Vila-Sanjurjo, A., Holton, J. M., & Doudna Cate, J. H. (2005). Structures of the bacterial ribosome at 3.5 Å resolution. *Science, 310,* 827–833.

Schwalbe, H., Buck, J., Fürtig, B., Noeske, J., & Wöhnert, J. (2007). Structures of RNA switches: Insight into molecular recognition and tertiary structure. *Angewandte Chemie International Edition, 46,* 2–10.

Scott, W. G., Finch, J. T., & Klug, A. (1995). The crystal structure of an all-RNA hammerhead ribozyme: A proposed mechanism for RNA catalytic cleavage. *Cell, 81,* 991–1002.

Scott, W. G., Murray, J. B., Arnold, J. R., Stoddard, B. L., & Klug, A. (1996). Capturing the structure of a catalytic RNA intermediate: the hammerhead ribozyme. *Science, 274,* 2065–2069.

Seeman, N. C., Rosenberg, J. M., Suddath, F. L., Kim, J. J. P., & Rich, A. (1976). RNA double-helical fragments at atomic resolution. I. The crystal and molecular structure of sodium adenylyl-3′,5′-uridine hexahydrate. *Journal of Molecular Biology, 104,* 109–144.

Selmer, M., Dunham, C. M., Murphy IV, F. V., Weixlbaumer, A., Petry, S., Kelley, A. C., Weir, J. R., & Ramakrishnan, V. (2006). Structure of the 70S ribosome complexed with mRNA and tRNA. *Science, 313,* 1935–1941.

Serganov, A., Polonskaia, A., Phan, A. T., Breaker, R. R., & Patel, D. J. (2006). Structural basis for gene regulation by a thiamine pyrophosphate-sensing riboswitch. *Nature, 441,* 1067–1171.

Serganov, A., Yuan, Y. R., Pitovskaya, O., Polonskaia, A., Malinina, L., Phan, A. T., Hobartner, C., Micura, R., Breaker, R. R., & Patel, D. J. (2004). Structural basis for discriminative regulation of gene expression by adenine- and guanine-sensing mRNAs. *Chemical Biology, 11,* 1729–1741.

Shi, H., & Moore, P. B. (2000). The crystal structure of yeast phenylalanine tRNA at 1.93 Å resolution: A classic structure revisited. *RNA, 6,* 1091–1105.

Spöring, M., Boneberg, R., & Hartig, J. S. (2020). Aptamer-mediated control of polyadenylation for gene expression regulation in mammalian cells. *ACS Synthetic Biology, 9,* 3008–3018.

Sundaralingam, M., & Pan, B. (2002). Hydrogen and hydration of DNA and RNA oligonucleotides. *Biophysical Chemistry, 95,* 273–282.

Sykes, M. T., & Levitt, M. (2005). Describing RNA structure by libraries of clustered nucleotide doublets. *Journal of Molecular Biology, 351,* 26–38.

Tereshko, V., Skripkin, E., & Patel, D. J. (2003). Encapsulating streptomycin within a small 40-mer RNA. *Chemical Biology, 10,* 175–187.

Theimer, C. A., Blois, C. A., & Feigon, J. (2005). Structure of the human telomerase RNA pseudoknot reveals conserved tertiary interactions essential for function. *Molecules and Cells, 17,* 671–682.

Thore, S., Frick, C., & Ban, N. (2008). Structural basis of thiamine pyrophosphate analogues binding to the eukaryotic riboswitch. *Journal of the American Chemical Society, 130,* 8116–8117.

Thore, S., Leibundgut, M., & Ban, N. (2006). Structure of the eukaryotic thiamine pyrophosphate riboswitch with its regulatory ligand. *Science, 312,* 1208–1211.

Tocilj, A., Schlunzen, F., Janell, D., Gluhmann, M., Hansen, H. A. S., Harmo, J., Bashan, A., Bartles, H., Agmon, I., Franceschi, F., & Yonath, A. (1999). The small ribosomal subunit from Thermus thermophilus at 4.5 Å resolution: Pattern fittings and the identification of a functional site. *Proceedings of the National Academy of Sciences of the United States of America, 96,* 14252–14257.

Tu, D., Blaha, G., Moore, P. B., & Steitz, T. A. (2005). Structures of MLSBK antibiotics bound to mutated large ribosomal subunits provide a structural explanation for resistance. *Cell, 121,* 257–270.

Varani, G., & McClain, W. H. (2000). The G x U wobble base pair. A fundamental building block of RNA structure crucial to RNA function in diverse biological systems. *EMBO Reports, 1,* 18–23.

Vicens, Q., & Cech, T. R. (2005). Atomic level architecture of group I introns revealed. *Trends in Biochemical Sciences, 31,* 41–51.

Vicens, Q., & Westhof, E. (2001). Crystal structure of paromomycin docked into the eubacterial ribosomal decoding A site. *Structure, 9,* 647–658.

Watson, Z. L., Ward, F. R., Meheust, R., Ad, O., Schepartz, A., Banfield, J. F., & Cate, J. H. D. (2020). Structure of the bacterial ribosome at 2 Å resolution. *Elife, 9,* e60482.

Wedekind, J. E., & McKay, D. B. (2000). Purification, crystallization, and X-ray diffraction analysis of small ribozymes. *Methods in Enzymology, 317,* 149–168.

Westhof, E., & Sundaralingam, M. (1986). Restrained refinement of the monoclinic form of yeast phenyl-alanine transfer RNA. Temperature factors and dynamics, coordinated waters, and base-pair propeller twist angles. *Biochemistry, 25,* 4868–4878.

Westhof, E. (2016). Perspectives and pitfalls in nucleic acids crystallography. *Methods in Molecular Biology, 1320,* 3–8.

Westhof, E., Dumas, P., & Moras, D. (1988). Restrained refinement of two crystalline forms of yeast aspartic acid and phenylalanine transfer RNA crystals. *Acta Crystallographica, A44,* 112–123.

Wild, K., Weichenrieder, O., Leonard, G. A., & Cusack, S. (1999). The 2 Å structure of helix 6 of the human signal recognition particle RNA. *Structure, 7,* 1345–1352.

Wilson, T. J., Liu, Y., Li, N. S., Dai, Q., Piccirilli, J. A., & Lilley, D. M. J. (2019). Comparison of the structures and mechanisms of the pistol and hammerhead ribozymes. *Journal of the American Chemical Society, 141,* 7865–7875.

Wilson, T. J., Liu, Y., Domnick, C., Kath-Schorr, S., & Lilley, D. M. (2016). The novel chemical mechanism of the twister ribozyme. *Journal of the American Chemical Society, 138,* 6151–6162.

Wimberly, B. T., Brodersen, D. E., Clemons, W. M., Morgan-Warren, R. J., Carter, A. P., Vonrein, C., Hartsch, T., & Ramakrishnan, V. (2000). Structure of the 30S ribosomal subunit. *Nature, 407,* 327–339.

Winkler, W. C., Nahvi, A., Roth, A., Collins, J. A., & Breaker, R. R. (2004). Control of gene expression by a natural metabolite-responsive ribozyme. *Nature, 428,* 261–286.

Wu, H., Henras, A., Chanfreau, G., & Feigon, J. (2004). Structural basis for recognition of the AGNN tetraloop RNA fold by the double-stranded RNA-binding domain of Rnt1p RNase III. *Proceedings of the National Academy of Sciences of the United States of America, 101,* 8307–8312.

Yusupov, M. M., Yusupova, G. Z., Baucom, A., Lieberman, K., Earnest, T. N., Cate, J. H. D., & Noller, H. F. (2001). Crystal structure of the ribosome at 5.5 Å resolution. *Science, 292,* 883–896.

Zhang, J., & Ferré-D'Amaré, A. R. (2014). New molecular engineering approaches for crystallographic studies of large RNAs. *Current Opinion in Structural Biology, 26,* 9–15.

Zhao, C., Qin, G., Niu, J., Wang, Z., Wang, C., Ren, J., & Qu, X. (2021). Targeting RNA G-quadruplex in SARS-CoV-2: A promising therapeutic target for COVID-19? *Angewandte Chemie International Edition England, 60,* 432–438.

Further reading

RNA structures

Barnwal, R. P., Yang, F., & Varani, G. (2017). Applications of NMR to structure determination of RNAs large and small. *Archives of Biochemistry and Biophysics, 628,* 42–56.

Batey, R. T., Rambo, R. P., & Doudna, J. A. (1999). Tertiary motifs in RNA structure and folding. *Angewandte Chemie International Edition, 38,* 2326–2343.

Fürtig, B., Richter, C., Wühnert, J., & Schwalbe, H. (2003). NMR spectroscopy of RNA. *ChemBioChem, 4,* 10–36.

Kappel, K., Zhang, K., Su, Z., Watkins, A. M., Kladwang, W., Li, S., Pintilie, G., Topkar, V. V., Rangan, R., Zheludev, I. N., Yesselman, J. D., Chiu, W., & Das, R. (2020). Accelerated cryo-EM-guided determination of three-dimensional RNA-only structures. *Nature Methods, 17,* 699–707.

Latham, M. P., Brown, D. J., McCallum, S. A., & Pardi, A. (2005). NMR methods for studying the structure and dynamics of RNA. *ChemBioChem, 6,* 1492–1505.

Leontis, N. B., Altman, R. B., Berman, H. M., Brenner, S. E., Brown, J. W., Engelke, D. R., Harvey, S., C., Holbrook, S. R., Jossinet, F., Lewis, S. E., Major, F., Mathews, D. H., Richardson, J. S., Williamson, J. R., & Westhof, E. (2006). The RNA Ontology Consortium: An open invitation to the RNA community. *RNA, 12,* 533–541.

Leontis, N. B., & Westhof, E. (2003). Analysis of RNA motifs. *Current Opinion in Structural Biology, 13,* 300–308.

Shen, L. X., Cai, Z., & Tinoco, I. (1995). RNA structure at high resolution. *Federation of American Societies for Experimental Biology Journal, 9,* 1023–1033.

Šponer, J., Bussi, G., Krepl, M., Banáš, P., Bottaro, S., Cunha, R. A., Gil-Ley, A., Pinamonti, G., Poblete, S., Jurečka, P., Walter, N. G., & Otyepka, M. (2018). RNA structural dynamics as captured by molecular simulations: A comprehensive overview. *Chemical Reviews, 118*, 4177–4338.

Strobel, E. J., Yu, A. M., & Lucks, J. B. (2018). High-throughput determination of RNA structures. *Nature Reviews Genetics, 19*, 615–634.

Westhof, E. (1999). In S. Neidle (Ed.), *Oxford handbook of nucleic acid structure* (p. 533). Oxford: Oxford University Press.

Westhof, E., & Leontis, N. B. (2021). An RNA-centric historical narrative around the Protein Data Bank. *Journal of Biological Chemistry*, 100555.

tRNA

Arnez, J. G., & Moras, D. (1999). In S. Neidle (Ed.), *Oxford handbook of nucleic acid structure* (p. 603). Oxford: Oxford University Press.

Clark, B. F. (2006). The crystal structure of tRNA. *Journal of Bioscience, 31*, 453–457.

Fernández-Millán, P., Schelcher, C., Chihade, J., Masquida, B., Giegé, P., & Sauter, C. (2016). Transfer RNA: From pioneering crystallographic studies to contemporary tRNA biology. *Archives of Biochemistry and Biophysics, 602*, 95–105.

Kim, S.-H. (1981). In S. Neidle (Ed.), *Topics in nucleic acid structure* (p. 83). London: Macmillan Press.

Ribozymes

Butcher, S. E. (2001). Structure and function of the small ribozymes. *Current Opinion in Structural Biology, 11*, 315–320.

Doherty, E. A., & Doudna, J. A. (2001). Ribozyme structures and mechanisms. *Annual Review of Biophysics, 30*, 457–475.

Ferré-D'Amaré, A. R., & Doudna, J. A. (1999). RNA folds: Insights from recent crystal structures. *Annual Review of Biophysics, 28*, 57–73.

Lilley, D. M. J. (2005). Structure, folding and mechanisms of ribozymes. *Current Opinion in Structural Biology, 15*, 313–323.

Scott, W. G. (2007). Ribozymes. *Current Opinion in Structural Biology, 17*, 280–286.

Wilson, T. J., & Lilley, D. M. J. (2021). *The potential versatility of RNA catalysis* (p. e1651). Wiley Interdiscip. Rev. RNA.

Ribosome structure

Cech, T. R. (2000). Structural biology. The ribosome is a ribozyme. *Science, 289*, 878–879.

Jobe, A., Liu, Z., Gutierrez-Vargas, C., & Frank, J. (2019). New insights into ribosome structure and function. *Cold Spring Harbor Perspectives in Biology, 11*, a032615.

Maguire, B. A., & Zimmermann, R. A. (2001). The ribosome in focus. *Cell, 104*, 813–816.

Moore, P. B. (2001). The ribosome at atomic resolution. *Biochemistry, 40*, 3243–3250.

Moore, P. B. (2021). The PDB and the ribosome. *Journal of Biological Chemistry, 296*, 100561.

Moore, P. B., & Steitz, T. A. (2003). The structural basis of large ribosomal subunit function. *Annual Review of Biochemistry, 72*, 813–850.

Noller, H. F. (2005). RNA structure: Reading the ribosome. *Science, 309*, 1508–1514.

Pennisi, E. (1999). The race to the ribosome structure. *Science, 285*, 2048–2051.

Riboswitches

Garst, A. D., Edwards, A. L., & Batey, R. T. (2011). Riboswitches: Structures and mechanisms. *Cold Spring Harbor Perspectives in Biology, 3*, a003533.

Mccown, P. J., Corbino, K. A., Stav, S., Sherlock, M. E., & Breaker, R. R. (2017). Riboswitch diversity and distribution. *RNA, 23*, 995–1011.

Panchal, V., & Brenk, R. (2021). Riboswitches as drug targets for antibiotics. *Antibiotics (Basel), 10,* 45.

Schwalbe, H., Buck, J., Fürtig, B., Noeske, J., & Wöhnert, J. (2007). Structures of RNA switches: Insight into molecular recognition and tertiary structure. *Angewandte Chemie, International Edition in English, 46,* 2−10.

Tucker, B. J., & Breaker, R. R. (2005). Riboswitches as versatile gene control elements. *Current Opinion in Structural Biology, 15,* 342−348.

RNA-drug complexes

Costales, M. G., Childs-Disney, J. L., Haniff, H. S., & Disney, M. D. (2020). How we think about targeting RNA with small molecules. *Journal of Medicinal Chemistry, 63,* 8880−8900.

Gallego, J., & Varani, G. (2001). Targeting RNA with small-molecule drugs: Therapeutic promise and chemical challenges. *Accounts of Chemical Research, 34,* 836−843.

Hermann, T. (2005). Drugs targeting the ribosome. *Current Opinion in Structural Biology, 15,* 355−366.

Ursu, A., Childs-Disney, J. L., Andrews, R. J., O'Leary, C. A., Meyer, S. M., Angelbello, A. J., Moss, W. N., & Disney, M. D. (2020). Design of small molecules targeting RNA structure from sequence. *Chemical Society Reviews, 49,* 7252−7270.

Vicens, Q., & Westhof, E. (2003). RNA as a drug target: The case of aminoglycosides. *ChemBioChem, 4,* 1018−1023.

Wirmer, J., & Westhof, E. (2006). Molecular contacts between antibiotics and the 30S ribosomal particle. *Methods in Enzymology, 415,* 180−202.

Zhang, L., He, J., Bai, L., Ruan, S., Yang, T., & Luo, Y. (2021). Ribosome-targeting antibacterial agents: Advances, challenges, and opportunities. *Medicinal Research Reviews.* https://doi.org/10.1002/med.21780

CHAPTER 7

Principles of protein–DNA recognition

7.1 Introduction

A detailed knowledge of protein–DNA binding and the interactions involved has very important implications for our understanding of the processes of DNA packaging, DNA replication and transcription, DNA regulation and DNA repair. The sequence information within DNA exists to be replicated, transcribed and translated into gene products, depending on the nature of the cell type, its progression through the cell cycle and its response to external stimuli and factors. The processes of packaging DNA in the cell, regulating the expression of individual genes and then reading/translating the encoded sequences involve the interplay of large numbers of non-specific and specific proteins. Similarly, the responses to DNA damage involve the concerted interplay of numerous proteins, to first recognise the damage, then to excise it and finally to repair the damaged DNA. It is thus unsurprising that the study of DNA–protein interactions has progressively developed, initially as a major area and now into a number of specialized areas, each with a large body of structural knowledge underpinning the biology and biophysics. This chapter concentrates on the underlining principles of recognition as found by structural methods (principally X-ray crystallography, nuclear magnetic resonance [NMR] and cryo-EM), how DNA structure itself copes with a wide range of protein and functional requirements and how some of this information can be used for practical purposes, primarily in precision medicine. The reader is directed to the extensive primary literature for detailed descriptions of the individual structures.

Structural information is now available on a large (and still increasing) number of distinct DNA–protein complexes, from a wide range of eukaryotic and prokaryotic sources. As of June 2021, the Protein Data Bank (PDB) and Nucleic Acid Database have data on 5602 protein–DNA structures; in 1997, there were just 241 structures. Studies of the proteins themselves usually provide limited insight into the processes of recognition and are not discussed in detail in this book. The still relatively small number of known protein–DNA structures pales into insignificance compared to the total encoded by genomes. The determination of the human genome in 2001 provided the first insight into the number of genes encoding proteins with particular functions for this one genome. The JASPAR Transcription Factor Database www.jaspar.genereg.net (Fornes et al., 2020) now lists 1201 transcription factor entries out of a total of 20,412 protein encoding genes present in the human genome. There are also genes encoding various forms of RNA including microRNAs which have an important role in gene control. Major increases in our understanding of DNA–protein recognition

Principles of Nucleic Acid Structure
ISBN 978-0-12-819677-9, https://doi.org/10.1016/B978-0-12-819677-9.00001-9

and processing are not obtained solely from complexes with single proteins but increasingly from studies of functionally relevant multi-protein complexes, as for example, from the structure of the nucleosome and polymerase complexes.

DNA-binding proteins can be conveniently categorised in both functional and structural terms into discrete families. The principal groupings are:

- Regulatory proteins. These mostly bind to highly specific sequences of duplex DNA, to control the transcription of a particular gene or bind to signal sequences such as 5'-TATA in order to more generally initiate transcription.
- DNA cleavage proteins (nucleases). Some such as DNAase I have relatively little sequence specificity (although they may have some DNA structural selectivity). Others, such as the restriction enzymes, are highly specific for given sequences.
- Repair proteins that respond to various types of damage to DNA by recognising the lesion itself, then excising the damaged DNA and/or joining together breaks in damaged DNA.
- Proteins that resolve topological problems in DNA by unravelling or unwinding DNA prior to replication. One such family, the DNA topoisomerases, of importance as cancer and anti-microbial therapeutic targets, have been especially well studied.
- Structural proteins that maintain the integrity of folded or packaged DNA, for example, histones in chromatin.
- Processing proteins, typified by DNA and RNA polymerases, that use DNA as a template for further nucleic acid synthesis. Sequence specificity is not required for DNA recognition, rather a need to recognize a particular type of DNA duplex. These enzymes have been extensively discussed in some excellent reviews and will not further be discussed in this chapter (see, for example, Jain et al., 2018; Wolberger, 2021).

This diversity of functions shown by DNA-binding proteins is in striking contrast to the relatively few ways in which DNA structure is 'read', even though there is also a very wide variety in the folding of the proteins themselves. Direct reading of a DNA sequence generally occurs via the hydrogen-bonding edges of the bases (Seeman et al., 1976), as described in Chapter 5 for small molecules; therefore, features are required of a protein that enable the bases to be accessed through either the major or minor grooves, as we have seen in Chapter 5 with small-molecule compounds. Fig. 7.1 shows that the B-DNA major groove is richer of the two grooves of duplex DNA, both in information content per se and in its ability to facilitate discrimination between different DNA sequences, which is essential if the appropriate genes are to be transcribed. Thus, the major groove is generally the site of direct information readout. Nonetheless, the minor groove is an important target for some regulatory and structural proteins, especially those that can deform DNA so that the minor groove becomes greatly expanded. Indirect readout can occur via the sugar—phosphate backbone or solely the phosphate groups. Especially when this is sequence specific, it is the cumulative consequence of numerous individually non-specific non-bonded interactions as compared to the clearly defined

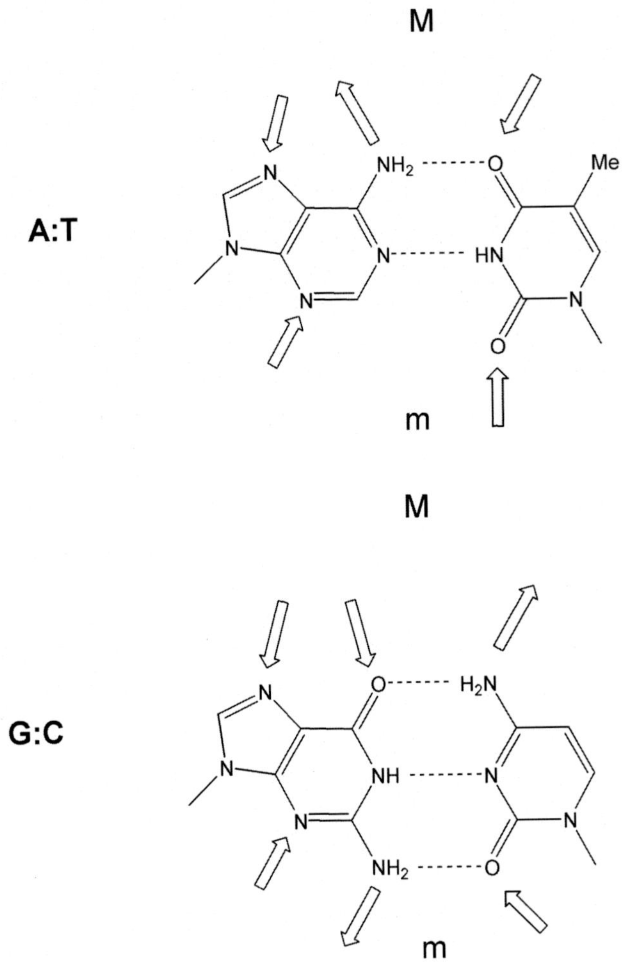

M

Figure 7.1 The pattern of hydrogen bond donors and acceptors in Watson—Crick base pairs. The direction of the *arrows* indicates the direction of hydrogen-bonding donation.

hydrogen-bonding interactions involved in direct recognition (Lin & Guo, 2019; Stracy et al., 2021). The anionic nature of phosphate groups makes them frequent targets for basic side chains, with the resulting electrostatic interactions being significant contributors to overall protein—DNA binding.

We normally assume, quite rightly, that the base pairing in B-DNA comprises solely Watson—Crick type, and therefore that proteins need to recognize Watson—Crick edges. However, this is not invariably the case as Hoogsteen base pairing has been observed in human and yeast DNA polymerase ι (iota)—DNA binding. Templating purines A and G adopt a *syn* conformation against incoming T and C, in the *anti* conformation, whereas T and C cannot present a Hoogsteen edge (Aishima et al., 2002; Nair et al., 2004, 2005, 2006; Jain et al., 2017). This is how this polymerase can read through lesions which disrupt Watson—Crick base pairs and not Hoogsteen pairing as has been shown in the structure of the complexes with the modified A templates 1, N^6 — ethenodeoxyadenosine (Nair et al., 2006) and N1-methyl-deoxyadenosine (Jain et al., 2017). The interaction of proteins with mismatched DNA sequences has been recently studied by NMR (Afek et al., 2020).

The limited number of ways by which proteins recognise DNA (Harrison, 1991; Luscombe et al., 2000, 2001) is unsurprising given that most protein—DNA complexes retain DNA in an overall B-type conformation and the deviations from ideality are not large. What is clear from analysis of many complexes in the PDB database is that in protein—DNA complexes where a protein specifically binds to its recognition site, then the interactions are with both strands and where there a range of sequences have to be accommodated in the binding site, then only one strand is specifically recognised (Tan & Takada, 2018).

The principal recognition motifs are as follows:
- α-helix-turn-α-helix (the HTH motif)
- zinc finger
- leucine zipper
- α-helix-loop-α-helix
- winged helix-turn-helix
- HMG-box
- Wor domain
- OB-fold domain
- immunoglobulin fold
- B3 domain
- TAL effectors
- RNA-guided proteins (see section on CRISPR/cas9)

7.2 Direct protein—DNA contacts

These interactions involve hydrogen bonds between amino acid side chains and the edges of the base pairs, involving their pattern of donor and acceptor molecules. The same principles of hydrogen bonding apply for small molecules binding to DNA (see Chapter 5) although the common contact areas involved in protein—DNA interfaces are often much larger (Nadassy et al., 1999). The protein backbone itself does sometimes contact with bases or phosphates, but these contacts are generally not determinants of specificity, rather they serve to enhance binding affinity. The majority of interactions involve O6 and/or N7 atoms of guanine bases forming hydrogen bonds with the charged ends of long flexible side chains from the basic residues of arginine and lysine, the amide residues glutamine and asparagine or the hydroxyl group or serine (Table 7.1 and Figs. 7.2 and 7.3) The recognition of one hydrogen-bonding site on a base, compared to two simultaneous

Table 7.1 Distribution of amino acid—base interactions in protein—DNA.

	Guanine	Cytosine	Adenine	Thymine
Arginine	98	8	19	24
Lysine	30	6	4	9
Serine	12	2	1	3
Asparagine	7	10	18	7
Glutamine	6	2	16	2
Glutamate	1	10	1	0

Adapted from Luscombe, N. M., Laskowski, R. A., & Thornton, J. M. (2001). Amino acid–base interactions: A three-dimensional analysis of protein-DNA interactions at an atomic level. *Nucleic Acids Research*, 29, 2860–2874, using a data set of 129 structures. Only those amino acids that participate in a significant number of DNA base interactions are listed.

sites on that base, actually involves only a small change in position of the amino acid side chain. In general, the latter will be energetically preferred and formed if possible. Adenine bases are recognised via their N6/or N7 atoms, although this occurs less frequently than guanine recognition. Active pyrimidine recognition is also slightly less common that guanine recognition.

There is no general 1:1 amino acid:DNA base correspondence, and recognition can sometimes occur in a wide variety of ways in addition to the simple mono- and bidentate ones. Table 7.1 also shows that hydrogen bonding to both partners in Watson—Crick base pairs is relatively uncommon, not least since binding to one usually results in sufficient discrimination for sequence selectivity to take place. Some arrangements for side chains bridging between two bases in a base pair are shown in Fig. 7.4a and b. Hydrogen bonding can also occur to two consecutive bases on the same strand (Fig. 7.5). This arrangement introduces sequence selectivity at the dinucleotide level. For example, the TG sequence (Fig. 7.5) is recognised by the lysine side chain through both carbonyl groups of the thymidine and guanine bases. Altering this to TA or TC would not put two carbonyl groups in the same correct position for interaction with lysine in the correct position for interaction with lysine. Similar arguments can be made for other dinucleotide steps involved in bi-dentate recognition with side chains.

Sometimes, a water molecule (or molecules) participates in a hydrogen-bonded bridge or itself bridges between a base and a side chain. Such water involvement was first found, for example, in the *Escherichia coli* trp repressor/operator complex (Otwinowski et al., 1988), with only two (relatively unimportant) direct (guanine ... arginine) contacts, yet three that have water-mediated contacts. The crystal structure of the excisionase—DNA complex from bacteriophage lambda (Sam et al., 2004) shows an unusual winged helix motif (α-helix + β-hairpin wing) in the excisionase protein interacting in the B-DNA major groove via an extensive network of water-mediated contacts. A further example of the important role that water molecules can play has been seen in the crystal structure of the bacteriophage l repressor—operator complex (Beamer & Pabo,

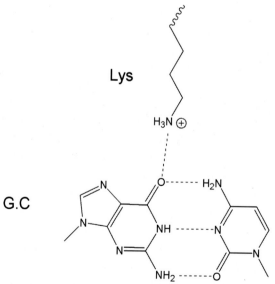

Figure 7.2 Possible patterns of hydrogen-bonding recognition of G•C base pairs by arginine and lysine side chains.

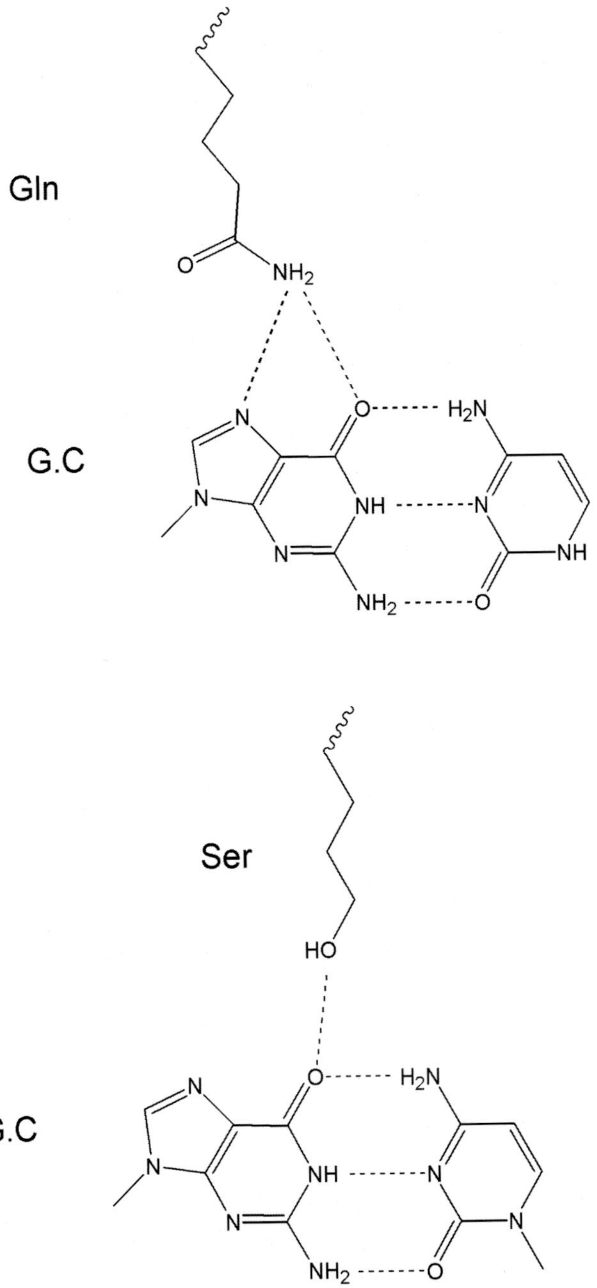

Figure 7.3 Possible patterns of hydrogen-bonding recognition of G•C base pairs by glutamine and serine side chains.

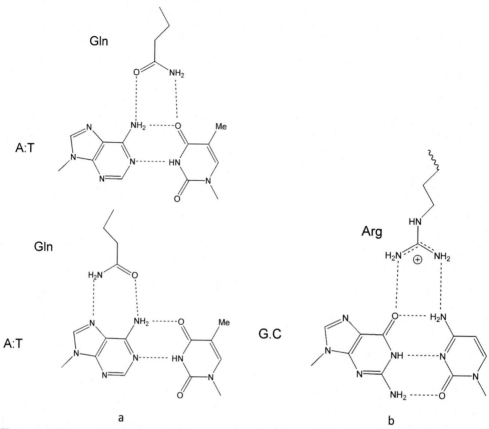

Figure 7.4 (a) Two possible patterns of hydrogen-bonding recognition of A•T base pairs by glutamine side chains. (b) The pattern of arginine recognition of both bases in a G•C base pair.

1992) where the hydroxyl group of serine-45 interacts directly with O6 and N7 of a guanine. The backbone carbonyl oxygen atom of this serine interacts via a water molecule with the N4 atom of the complementary cytosine base and through the same water molecule to O4 of the adjacent thymine.

It has not been possible to establish a pattern of preference between a particular base and an amino acid residue. This lack of a general hydrogen-bonding recognition pattern (although there are some involving very closely related proteins) reflects several factors (Mandel-Gutfreund et al., 1995; Matthews, 1988; Pabo & Nekludova, 2000):

- the effect of differences in DNA structure in various complexes
- the different types of protein motif involved in DNA recognition
- other recognition factors that can play significant roles in defining the recognition of particular sequences, notably the ability of thymine methyl groups to form close van der Waals attractive interactions with hydrophobic side chains
- in general, any one base can be recognised by several amino acids

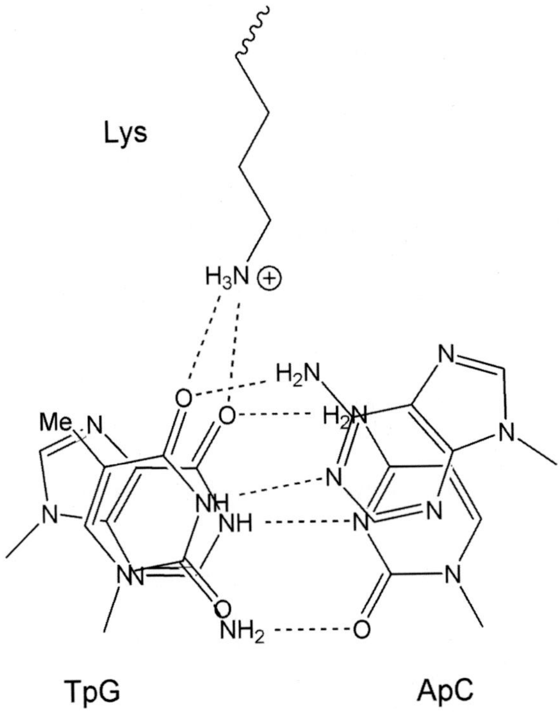

Figure 7.5 A possible recognition mode of the sequence TpG by a lysine side chain.

An example of the role of thymine methyl groups is in the *engrailed* homeodomain structure, where there is a close contact between isoleucine-47 and one such methyl group from a thymine in the TAAT recognition site (Kissinger et al., 1990). At least as important roles are played by numerous non-specific (at least in hydrogen-bonding terms) protein backbone and other interactions in stabilising a protein—DNA complex. Thus, the flexibility of amino acid side chains ensures that optimal interactions are made with all elements of a DNA structure. Taken as a whole, these enable the recognition helix to be oriented in the major groove in a manner that is distinctive for a particular protein. Such generalised hydrophobic and electrostatic interactions may also play a more active recognition role, by means of indirect readout, sensing the DNA structure and flexibility specified by a particular sequence. In view of all these factors, it is then un-surprising that there is wide variation in the number and importance of direct readout interactions that have been observed in different HTH complexes. The major role that can be occasionally played by indirect readout is strikingly demonstrated in the struc-ture of the *trp* repressor/operator complex (Otwinowski et al., 1988). This complex has many contacts between side chains and the phosphate groups of the operator DNA, yet very surprisingly has no functionally significant direct readout base amino acid interac-tions at all. In spite of this, the cumulative effect of the indirect readout contacts made by the *trp* repressor is to effectively read the particular structural details of its DNA operator sequence (Haran et al., 1992).

Readout, both direct and indirect, ensures that key residues, on both protein and DNA, are effectively recognised. The functional importance of such residues and hence the protein—DNA contacts involved can be evaluated by mutagenesis experiments, as well as by surveying their occurrence within a particular family of proteins and their consensus operator sequences. For example, residues tryptophan-48, phenylalanine-49, asparagine-51 and arginine-53 occur in every member of the large eukaryotic homeodomain family. The crystal structure of the engrailed homeodomain—DNA complex (Kissinger et al., 1990) shows that the first two of these residues play critical roles in preserving the hydrophobic core of the recognition helix. The other two directly interact with base and phosphate groups, respectively.

7.3 Major groove interactions — the α-helix as the recognition element

A protein α-helix can fit snugly only into the major groove rather than the minor groove of a B-like DNA duplex and therefore act directly as the recognition element. This element was first observed in the bacterial *cro* repressor protein (Anderson et al., 1981) as part of the HTH motif. This HTH domain pattern has subsequently been found in the crystal and solution NMR structures of many prokaryotic and eukaryotic transcription regulatory and related proteins, and by comparative sequence analysis in many others (Table 7.2). The HTH motif consists of ~20 amino acid residues with residues 1—7 forming the first α-helix and residues 12—20 the second α-helix. These are linked by a short turn, so that the two helices are inclined at 120° to each other. The second helix is the recognition one, which tends to make most of the specific contacts to DNA and lies in the major groove of a B-form duplex. The characteristic hydrophobic residues at positions 4, 8, 10, 16 and 18 of the recognition helix help to form the overall hydrophobic core of the protein. The N-terminal region of homeodomains typically binds in the DNA minor groove. The resulting side chain interactions with minor groove base edges are important contributors to overall homeodomain binding. The basic side chain of, for example, arginine closely mimics the behaviour of basic side chains in small molecules complexed to the DNA minor groove.

The HTH motif does not exist by itself but is held together in a variety of ways. Thus, the eukaryotic *engrailed* homeodomain (Kissinger et al., 1990; Wolberger et al., 1991) has a three-helix bundle (Fig. 7.6), with the HTH motif itself occurring after the first helix, whereas the phage 1 (Beamer & Pabo, 1992) and 434 repressor (Aggarwal et al., 1988) structures are five-helix bundles with the HTH motif comprising helices 2 and 3 (Fig. 7.7). An important difference between the bacterial and eukaryotic HTH proteins is that the former generally bind to their target sites as dimers, whereas the latter bind as monomers. This difference is reflected in the structures of the proteins themselves, with some aspects of the non-HTH

Table 7.2 Selected classic HTH crystal structures.

Protein	PDB id	Resolution in Å
434 repressor	2OR1	2.50
lambda operator	4CRO	3.90
Lac repressor	1EFA	2.60
Cro repressor	3ORC	3.00
Engrailed homeodomain	3HDD	2.20
Antennapedia homeodomain	9ANT	2.40
MATa1/MATα2 homeodomain	1AKH	2.50
CAP activator	1CGP	3.00
Aristaless and *Clawless* homeodomains	3A01	2.70
Homo sapiens homeodomain	3CMY	1.95
HoxB1–Pbx1 heterodimer	1B72	2.35
Hox	2R5Y	2.60
γδ-resolvase	1GDT	3.00
Rep	5KBJ	3.09

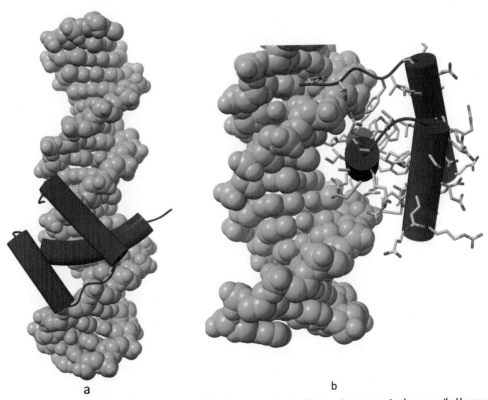

a   b

Figure 7.6 (a) The helix-turn-helix recognition of the bases in the DNA major groove in the *engrailed* homeodomain crystal structure (Fraenkel et al., 1991), PDB id 3HDD. (b) A second view, now looking down the recognition helix of the homeodomain, and showing the side chains extending into the major groove.

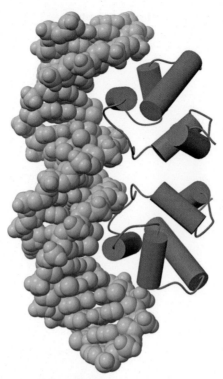

Figure 7.7 A view of the structure of the DNA-434 repressor complex (Aggarwal et al., 1998), PDB id 2OR1. The recognition helices are shown oriented into the DNA major groove.

domains being responsible for dimerisation. The features of these structures, as seen in the crystalline state, are largely preserved in solution, although side chain orientations are unsurprisingly not always identical (Fraenkel & Pabo, 1998; Neri et al., 1992).

Many homeodomains act at a DNA locus together with a second homeodomain, which can be binding in tandem at a neighbouring site. This has been studied in detail for the MATa1 and MATα2 proteins, which regulate transcription in yeast. The crystal structure of the DNA complex with MATα2 alone (Wolberger et al., 1991) shows a typical homeodomain, with a straight B-DNA helix. By contrast, the ternary complex (Li et al., 1995) has the two protein units contacting each other through the C-terminus tail of the MATα2 domain. This occurs since the DNA is bent by 60°. A subsequent analogous crystal structure (Li et al., 1998), this time using an A-tract sequence DNA, shows a very similar degree of bending, with the A-tract bent in the minor groove direction. Remarkably, the minor groove spine of hydration is preserved in both structures. Similar bending has been observed with a ternary complex of MATa2, the transcription factor MCM1 and a 26-base pair DNA sequence (Tan & Richmond, 1998). Other types of ternary complex can involve the two domains operating in close tandem to overlapping DNA sites, such as in the HoxB1–Pbx1 (Piper et al., 1999) and Ubx–Exd (Passner et al., 1999) heterodimer structures. Each protein in these complexes induces a small (10–11° in the former case) bend in the bound DNA, but since the binding sites are close and almost opposite each other, the net effect is an almost straight B-like helix. A general principle emerges from these studies that when two protein domains bind on the same side of a DNA sequence, then bending ensues; when they are opposite, then there is no net bending.

The majority of HTH proteins have numerous direct interactions between the recognition helix and the major groove of the operator sequence DNA, which maintains a B-type conformation. However, our knowledge of the detailed nature and extent of these interactions is critically dependent on reliable crystal and NMR structures. This caveat is illustrated by the re-determination at 2.2 Å resolution (Fraenkel et al., 1998) of the original 2.8 Å engrailed homeodomain crystal structure (Kissinger et al., 1990). The more recent of these two analyses confirms all major features of the complex and provides unequivocal information on the role of the important residue Gln50, which previously had been shown to interact only indirectly with a major groove base edge. This role is confirmed in the higher resolution structure, which shows several water-mediated contacts between this residue and three (T4, T5 and G7) out of the seven bases in the d(TAATTAC) recognition site. The analysis also shows that there are important contacts between homeodomain side chains and the phosphate backbone, some of which are mediated by water molecules. The case for the key role of water molecules in homeodomain recognition (Gehring et al., 1994) has been reinforced by molecular dynamics simulations (Billeter et al., 1996) and chemical modification studies (Labeots & Weiss, 1997). The simulations show that water molecules can have an appreciable residence time at the protein–DNA interface, sufficient for them to be mediating between DNA and amino acid side chains.

Most proteins which have HTH motifs bend the recognition site towards the protein (as in, for example, the *E. coli* transcription activator protein [CAP] protein). One example where the HTH motifs are located on long α-helices and bind to the distal side of the DNA is in the binding of the site-specific recombination protein γδ-resolvase, where the DNA is bent away from the catalytic domain (Fig. 7.8; Yang & Steitz, 1995).

The widespread occurrence of the HTH motif has led to studies to develop methods that can reliably locate the motif within genomic sequences using sequence data, either alone or together with a range of structural indicators such as the known structural information on HTH proteins, and electrostatic potential. The methods can achieve some success (see, for example, Pellegrini-Calace and Thornton, 2005), when both sequence and structure are taken together.

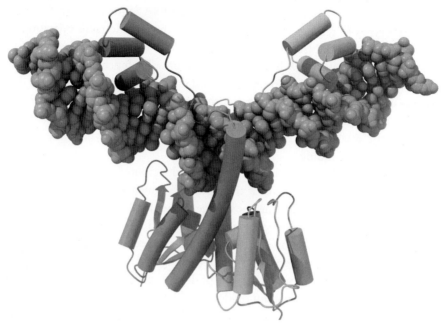

Figure 7.8 Binding of γδ-resolvase to the 34 base pair DNA cleavage site (Yang & Steitz, 1995), PDB id 1GDT.

7.4 Zinc finger recognition modes

The transcription factor TFIIIA from the clawed frog *Xenopus laevis* which controls the expression of 5S RNA in this organism was one of the first DNA-binding protein systems to be studied by Aaron Klug and colleagues (Klug & Rhodes, 1987; Klug, 2010). The protein consists of 32 repeating protein motifs, each containing zinc, as found by biochemical and X-ray structure analysis. Crystallisation of a single finger in the absence of DNA proved to be problematic and the first structure of the motif was elucidated in solution by Peter Wright's and Rachel Klevit's groups (Laity et al., 2001; Parraga et al., 1988) using NMR and was found to contain a helix loop and a β-sheet. This is a member of the family of zinc finger proteins which contain units of regularly spaced cysteine and histidine residues each coordinated to a zinc ion (Wolfe, Nekludova, & Pabo, 2000). Each finger consists of an anti-parallel β-sheet and an α-helix, held together by the zinc ion coordination (Fig. 7.9) (Klug & Rhodes, 1987). Zinc finger proteins (Table 7.3) have at least two such units ('fingers'), with each recognising a three base-pair site in the major groove of a B-DNA helix, as observed in the crystal structure of the Zif268 transcription factor (Pavletich & Pabo, 1991) (Fig. 7.10). Each finger makes direct contacts to the guanine-rich strand, with patterns of side chain interactions from the α-helix involving arginine .... guanine and histidine ... guanine recognition that are very similar to those seen in HTH proteins.

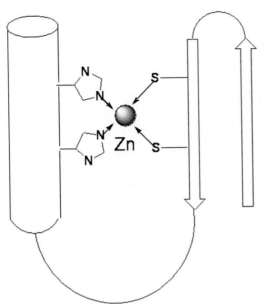

Figure 7.9 The zinc finger motif, showing the arrangement for a single finger, with the coordination to the zinc atom from histidine residues attached to the α-helix on the left, and the cysteine residues from the β-sheet on the right.

Table 7.3 Selected classic zinc finger protein–DNA crystal structures.

Protein	PBD id	Resolution in Å
Zif268	1ZAA	2.10
GAL4	1D66	2.70
Estrogen receptor	1HCQ	2.40
Retinoic acid receptor	2NLL	1.90
TATA box-zinc finger	1G2D	2.20
Designed zinc finger	1MEY	2.20

A quite distinct zinc domain is involved in the structure of the yeast transcription activator GAL4 (Marmorstein et al., 1992), which binds to DNA as a dimer. Each individual GAL4 molecule has two domains, a zinc-binding region and an α-helix, linked by an extended length of peptide (Fig. 7.11). The zinc-containing domain rests in the DNA major groove, where there is a pattern of direct interactions between basic lysine side chains and guanine/cytosine bases that exactly specifies the highly conserved sequence CCG at this point along the DNA. Nuclear receptors, such as those for steroid hormones and retinoic acid, also use zinc-containing motifs for recognition of their DNA response

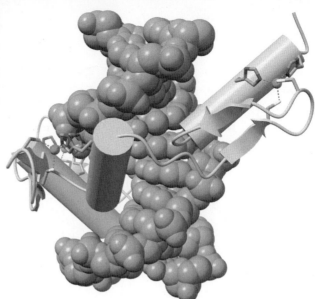

Figure 7.10 The structure of the Zif268 transcription factor–DNA complex (Pavletich & Pabo, 1991), PDB id 1ZAA, with three zinc fingers. The zinc atom in each zinc finger is shown as a red sphere, coordinated to cysteine and histidine residues.

Figure 7.11 The structure of the GAL4 transcription factor–DNA complex (Marmorstein et al., 1992), PDB id 1D66.

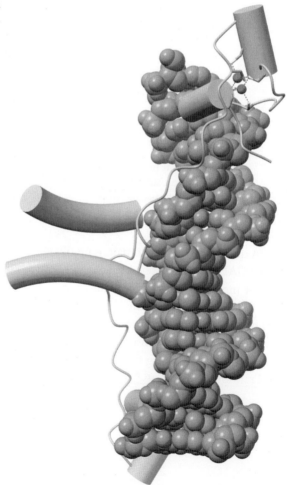

elements, and several NMR and crystal structures of such complexes have been reported. Almost all the response elements contain the half-site consensus sequences d(AGGTCA) or d(AGAACA), with the receptors binding as homo- or heterodimers. The zinc motif in them consists of a pair of α-helices linked by zinc coordination through C-terminal loops (Figs. 7.12 and 7.13). One helix is bound in the DNA major groove at each half-site (Fig. 7.14), whilst the role of the second helix is to maintain the overall structure (Luisi et al., 1991; Schwabe et al., 1993; Zhao et al., 2000). The side chains from the recognition helix participate in normal direct readout interactions; the DNA itself retains B-DNA form, though typically local distortions to roll and propeller twist have been observed.

Zinc fingers have been used in the design of synthetic DNA recognition molecules with sequence specificity that can be altered at will (Choo & Isalan, 2000; Papworth et al., 2006): this is discussed in more detail in Section 7.9 below. This approach complements that of recognition of the minor groove by dimeric polyamides (see Chapter 5) and

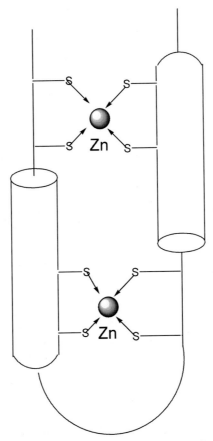

Figure 7.12 The arrangement for two zinc fingers as found in the nuclear receptor family of transcription factors. The fingers are formed between helices and loops, and the zinc atoms are co-ordinated by cysteine residues.

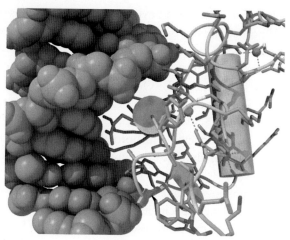

Figure 7.13 Close-up view of the structure of the oestrogen receptor–DNA complex (Schwabe et al., 1993), PDB id 1HCQ, with the zinc atoms in one of the two zinc fingers shown, as red spheres. Side chains protruding into the major groove are also shown.

has the potential advantage of greater synthetic accessibility and with potential applicability to all possible target sites with equal facility. They have the disadvantage of being protein-like in size and therefore may encounter problems in being transported into cells.

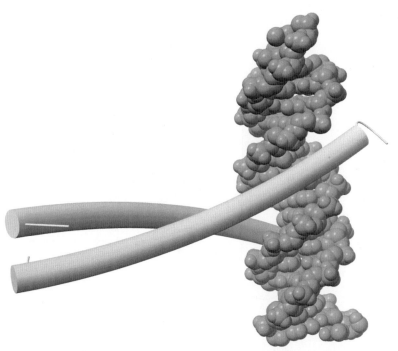

Figure 7.14 The structure of the yeast transcription factor GCN4–DNA complex (Ellenberger et al., 1992), PDB 1YSA, showing the exceptional length of the two recognition helices.

A peptide containing three zinc fingers, found by screening a library of peptides, has been shown to bind to a nine base-pair sequence in the BCR-ABL oncogene (Choo et al., 1994), with an binding constant of 6×10^{-7} M. This binding affinity is at least an order of magnitude higher than other control sequences. The original strategy did not effectively recognise all possible bases within a triplet, with the 5′ one being optimally adenine or cytosine, possibly due to sequence context effects. This problem was overcome (Isalan et al., 1998, 2001) by detailed considerations of several zinc finger crystal structures that show synergistic recognition from adjacent zinc fingers. As a result, it is possible to devise a general recognition code for zinc finger-directed DNA sequence specificity. Addition of a dimerisation motif can enable more than three zinc fingers to be assembled without loss of affinity. For example, a four-finger protein has been successfully constructed (Wolfe et al., 2000) using the GCN4 leucine zipper dimerisation motif (see below), which recognises 10 base pairs.

The use of zinc finger motifs to specifically recognise any desired DNA sequence has been applied to a wide range of problems (Papworth et al., 2006), and several web sites are now offering facilities for zinc finger design for this purpose (such as that of the Zinc Finger Consortium at http://www.zincfingers.org/). This is further discussed in Section 7.9.

7.5 Other major groove recognition motifs

Numerous eukaryotic transcription factors contain the basic 'leucine-zipper' recognition and dimerisation element, as found in the yeast transcription factor GCN4 (Ellenberger et al., 1992), which consists of a very long continuous, almost straight α-helix with regularly repeating leucine residues (Fig. 7.14). Two such helices interact together to form a parallel coiled coil. There are numerous interactions with the DNA backbone from the basic side chains, as well as specific, direct readout ones. The orientation of the α-helix within the DNA major groove is critical for these contacts to take place, with the helix being approximately parallel to the phosphodiester backbone (Table 7.4).

A distinct DNA-binding motif is used by the transcription activator E2 from papilloma virus (Hegde et al., 1992), with a dimeric anti-parallel β-barrel which delivers a pair of α-helices to the major groove. Protein–DNA contacts involve direct side chain interactions with the bases, as well as indirect backbone contacts. A distinguishing feature of the structure is the smooth bending of the DNA around the β-barrel, with a radius of curvature of 45 Å. This is less extreme than the 90° bend found (Schultz et al., 1991) in the DNA of the complex with the dimeric HTH domain *E. coli* transcription activator protein. The bending in the CAP complex, discussed further below, is due to numerous interactions with DNA phosphate groups, which serve to position the recognition helix correctly in the major groove (Fig. 7.15). A totally distinct non-helix recognition motif has been found in the structures of some bacterial repressor proteins, typified by the *met* J

Table 7.4 Crystal structures of some classic protein–DNA complexes with miscellaneous binding motifs.

Protein	Resolution in Å	PDB id code
GCN4	2.90	1YSA
CAP	3.00	1CGP
Met J	2.80	1CMA
E2	1.70	2BOP
P53	2.20	1TSR
Ku	2.50	1JEY
NF–kB	2.40	2RAM
Nucleosome	2.80	1AOI
Nucleosome with *Xenopus laevis* core histones + human DNA	1.94	1KX5
Human nucleosome	1.99	6IPU
Tetra-nucleosome	5.77	5OY7
Hexa-nucleosome	9.7	6HKT

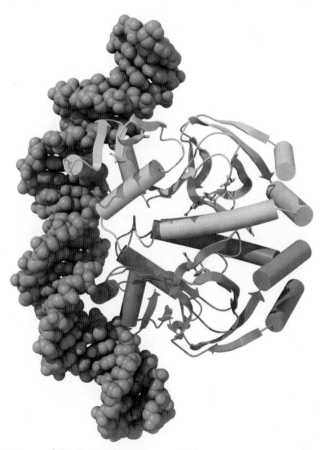

Figure 7.15 The structure of the *Escherichia coli* transcription activator protein (CAP)–DNA complex, showing one CAP unit with its recognition helix in the major groove of the bent DNA (Schultz et al., 1991), PDB id 1CGP.

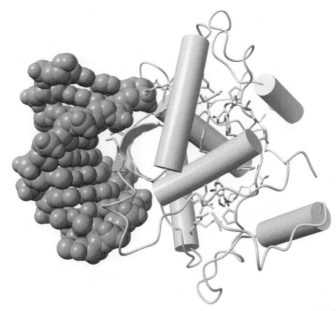

Figure 7.16 The structure of the *Met* J protein bound to its consensus operator DNA (Garvie & Phillips, 2000), PDB id 1CMA. Note the two β-sheets fitting snugly into the DNA major groove. Two molecules of the co-factor S-adenosylmethionine are also shown.

repressor/operator complex (Garvie & Phillips, 2000; Somers & Phillips, 1992), where double-stranded anti-parallel β-ribbons are the major groove recognition elements (Fig. 7.16). Side chains from the ribbons interact directly with the DNA phosphates and bases, although the side chain … base direct readout mechanism remains the same as that observed with the other repressor/operator complexes outlined above. Other analogous types of folds have been found in the structures of several DNA repair enzymes.

7.6 Other DNA binding domains

7.6.1 The winged-helix-turn-helix domain

In the winged HTH DNA-binding motif, the 'wings', or loops, are small β-sheets. Sequence-specific DNA contacts are made by the DNA recognition helix with the major groove of DNA, whilst the wings can make different DNA contacts with the minor groove or the backbone of DNA. Typically, the wings mediate protein—protein interactions through hydrophobic residues on their surface.

7.6.2 The Wor3 domain

Wor3 DNA-binding domains, which are DNA sequence specific with a strong preference for the sequence d(ATAACC), have been found in a small number of fungi and are called Wor3 after the White–Opaque Regulator 3 in *Candida albicans* which has evolved most recently in terms of evolutionary time compared with other DNA-binding motifs (Lohse et al., 2013).

7.6.3 The OB-fold domain

OB folds bind single-stranded DNA and are named OB due to their oligonucleotide/oligosaccharide binding properties. They range in length between 70 and 150 amino acids. OB-fold proteins have been identified as critical for DNA replication, DNA recombination, DNA repair, transcription, translation, cold shock response and telomere maintenance.

7.6.4 The immunoglobulin fold

The immunoglobulin domain (www.ebi.ac.uk/interpro/entry/InterPro/IPR013783) consists of a β-sheet structure with large connecting loops, which serve to recognise either DNA major grooves or antigens. Usually found in immunoglobulin proteins, they are also present in Stat proteins of the cytokine pathway. This is likely because the cytokine pathway evolved relatively recently and has made use of systems that were already functional, rather than creating its own.

7.6.5 B3 domains

In *Arabidopsis thaliana*, for example, there are three main families of transcription factors that contain the B3 domain: ARF (Auxin Response Factors), ABI3 (ABscisic acid Insensitive3) and RAV (Related to ABI3/VP1). The B3 DNA-binding domain is a highly conserved motif found in transcription factors in more than 40 species, in combination with other domains. In this domain, there are about 100–120 residues. It consists of seven β-strands and two α-helices forming a DNA-binding pseudo-barrel protein fold which binds to the major groove of DNA.

7.7 Minor groove recognition

7.7.1 Recognition of B-DNA

Sequence-specific proteins generally do exploit the major groove to a greater extent than the minor one. This is not always solely on account of its greater information potential. The ability of an α-helix to snugly fit into the major groove is also a factor. This led to a view in the past that the minor groove is of little importance for protein recognition. However, there are many examples in which interactions of side chains in the minor groove are significant contributors to overall protein–DNA stabilisation. For example, an arginine side chain of the 434 repressor (Aggarwal et al., 1988) bridges to bases and phosphate groups via water molecules. Homeodomain structures (see, for example,

Kissinger et al., 1990; Li et al., 1995; Li et al., 1998; Tan & Richmond, 1998; Wolberger et al., 1991) have extended N-terminal arms which lie in the DNA minor groove, making extensive base and backbone contacts to the A/T regions of the operator sequence. The hydrogen-bonding interactions, which the arginine side chains in these complexes make with the O2 atom of a thymine base, are closely analogous to the mode of interaction shown by minor-groove binding drugs (Morávek et al., 2002). These together with the indirect readout of the dimensions of the groove itself by van der Waals interactions involving the side chains are a significant factor in the preference shown by homeodomains for A/T-rich sites on DNA, again analogous to the preferences shown by the drugs. These N-terminal sequences are also related to the proposal (Suzuki, 1989) that the repeating sequence SPKK (serine—proline—lysine/arginine—lysine/arginine), which occurs in, for example, the N-terminus of some histone proteins, also binds in the minor groove at A/T-rich regions.

The crystal structure of the HTH-motif *Hin* recombinase enzyme bound to a 14 base-pair A/T-rich DNA sequence (Feng et al., 1994) shows that the straight B-DNA helix has a narrow minor groove in the A/T region, in which the N-terminus of the enzyme resides (Fig. 7.17). There are several amino acid—base edge-specific contacts, such as an arginine to N3 of an adenine. More unusually, the main chain amide of this arginine is in hydrogen-bonding contact with O2 of a thymine. The pattern of hydrogen bonding in this A/T-rich region is reminiscent of netropsin binding. The C-terminus of this enzyme also lies in the minor groove, entering at the other end of the DNA duplex (Table 7.5).

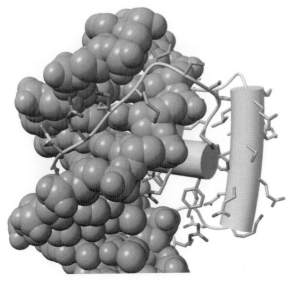

Figure 7.17 A view of the structure of the *hin* recombinase–DNA complex, showing the N-terminus residing in the minor groove surface of the DNA (Feng et al., 1994), PDB id 1HCR.

Table 7.5 Selected crystal structures of protein–DNA complexes involving minor-groove recognition.

Protein	Resolution in Å	PDB ID code
Hin recombinase	2.30	1HCR
TBP	1.86	1QN4
TBP–TFIIB complex	2.70	1VOL
HMG-D	2.20	1QRV
PCG2	2.40	4UX5
Transposase	3.50	6XGW

Loops can also bind in the minor groove, as found in the solution (NMR) structure of the *Mu* repressor protein–DNA complex (Wojciak et al., 2001), which has an HTH motif containing an additional 'wing' loop between helix 2 and 3. The wing is flexible in the absence of DNA, but in the complex is found inserted in the minor groove, where two lysine side chains make extensive contacts with, in particular, adenine and thymine base edge atoms. Another example is shown by the PCG2 protein, which is an orthologue of the MBF family of transcription factors. The DNA-binding domain of PCG2 in its DNA complex (Liu et al., 2015) comprises a winged HTH arrangement. This binds to its core d(CGCG) sequence by a loop and the edge of an α-helix being inserted into the minor groove so that two conserved glutamine residues, Q82 and Q89, can make specific hydrogen bond contacts with base-pair edges in the recognition sequence (Fig. 7.18a and b).

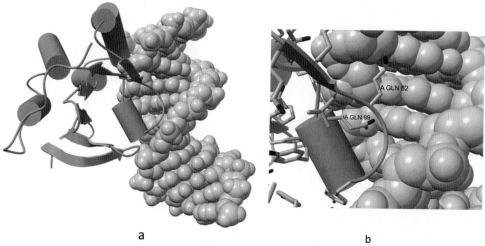

a   b

Figure 7.18 (a) The structure of the PCG2–DNA complex (Liu et al., 2015), PDB id 4UX5, showing the insertion of part of the winged helix-turn-helix motif into the DNA minor groove. (b) Close-up view showing the two glutamine residues in contact with base-pair edges.

7.7.2 The opening-up of the minor groove by TBP

The general transcription factor complex TFIID plays a key role in the initiation of transcription in eukaryotic cells. It functions by binding a component protein, TBP, to the 'TATA box' sequence upstream of the start of transcription. This has been shown by chemical protection studies to bind in the minor groove at the 5′-TATA site itself, as well as in the major groove on the 3′ side of this site. There is a strong preference for 5′-TATA as compared to other A/T-containing sequences. The crystal structure of the TBP protein (Nikolov et al., 1992), in the absence of DNA, shows a novel DNA-binding fold, with a symmetric a/b arrangement. It was initially suggested that the saddle-shaped arrangement of the a/b structure, with an extended concave surface, is ideally complementary to a B-form DNA duplex and would be effectively wrapped around it. Mutagenesis studies have identified several key bases and base pairs. These bases become unstacked from the DNA helix and result in the observed kinking. There are few direct readout base-side chain contacts, but numerous amino acid contacts with phosphate groups. Many of these, especially with lysine side chains, are water mediated.

However, the actual structures of the TBP–DNA complexes show a very different arrangement for the DNA compared to this initial model (Kim, Nikolov et al., 1993; Kim, Geiger et al., 1993). The DNA is dramatically bent, by ca 80° so as to follow the concave curvature of the β-sheet, i.e. it is at right angles to the earlier model (Fig. 7.19). This structure has been observed in TATA-containing DNA sequences complexed with TBP from a wide range of organisms, so it is clearly a general feature of TATA box recognition (Patikoglou et al., 1999). Eight base pairs are in contact with the β-sheet saddle and are in a non-B-DNA conformation. By contrast, analysis of complexes with longer DNA sequences shows that the DNA flanking the central site

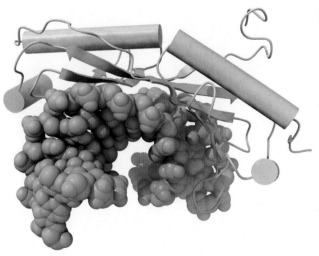

Figure 7.19 The structure of the TBP–DNA complex (Kim & Burley, 1994), PDB id 1Q4N. Note the high degree of curvature of the DNA duplex and the complementary curvature of the β-sheet in the major groove of the DNA.

is in a B-like form, on both 3′ and 5′ sides. Structure determination at high resolution (1.9 Å) of a TBP complex with the sequence d(TATAAAAG) has enabled a detailed view of the bent DNA structure to be obtained (Kim & Burley, 1994). The DNA is unwound by 105° over the seven base pairs and has a greatly enlarged and flattened minor groove (with a maximum width of over 9 Å) to accommodate the β-sheet saddle. The major groove is highly compressed. There are large positive rolls, of up to 40° at the TA step in the 5′-TATA sequence, together with large propeller twists, of up to −39° in the 5′-AAAA sequence. These large twists result in a pattern of A tract major-groove bifurcated hydrogen bonds. Sugar puckers are in the C3′-*endo* family. The overall impression of the DNA conformation in the TATA box region is of distorted A-type form, with abrupt changes to B-DNA morphology immediately outside the box. The bending is achieved by pairs of phenylalanine residues that are inserted into the ends of the TATA box and are stacked with several bases and base pairs. These bases become unstacked from the DNA helix and result in the observed kinking. There are few direct readout base-side chain contacts, but numerous amino acid contacts with phosphate groups. Many of these, especially with lysine side chains, are water mediated.

Several ternary complexes of general transcription factors with the TBP–TATA complex have been reported. Those with TFIIA (Nikolov et al., 1995) and TFIIB (Tan et al., 1996) show that the TBP–TATA box structure seen in the binary complexes is fully retained, strongly suggesting that these transcription factors bind to a preformed TATA-box complex (Fig. 7.20), as does the negative cofactor NC2 in its TBP–DNA complex (Kamada et al., 2001).

Figure 7.20 Structure of the TFIIB–TBP–DNA complex (Nikolov et al., 1996), PDB id 1VOL. The DNA structure reverts to canonical B-form outside the immediate environment of the TBP saddle structure. The TFIIB protein is coloured pink.

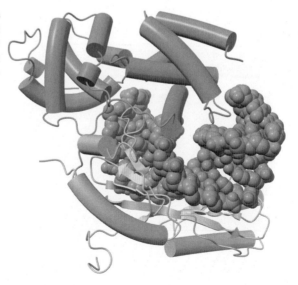

7.7.3 Other proteins that induce bending of DNA

The ability of TBP to bend DNA on binding to its recognition sequence is shared by several other minor-groove regulatory proteins notably those containing the so-called HMG (high mobility group) box sequence-neutral DNA-binding domains. Such proteins include SRY, which determines the expression of human male-specific genes. NMR studies of HMG boxes (Allain et al., 1999; Weir et al., 1993; Werner et al., 1995) show that the BOX structure consists of three α-helices arranged in an L-shape, with conserved basic arginines and lysines on the inner face of the L. The helices do not correspond to those in any (major-groove binding) HTH arrangement, and thus constitute a novel DNA-binding motif. A crystal structure of a complex between HMG-D and an A/T-rich duplex decamer (Murphy et al., 1999) shows that the bound DNA is very distorted from canonical B-DNA and smoothly bent (Fig. 7.21). The protein binds to the minor groove, which is underwound, widened and flattened so that the overall shape resembles A-DNA. This is reminiscent of the distortions induced by the TBP protein, and indeed, there are significant similarities, although here the minor groove is even wider, 12.4 Å. The crystal structures of the DNA complexes, with the bacterial proteins HU (histone-like) and IHF (integration host factor), show that they are also analogous in inducing very large bends in the DNA even though they are structurally dissimilar (Swinger & Rice, 2004, 2007).

The organisation of almost 2 metres of DNA in the eukaryotic chromosome requires that it is packed up in a very tight manner (Luger & Richmond, 1998). The basic unit of packaging is the nucleosome core particle, which consists of an octamer of pairs of the four histone proteins H2A, H2B, H3 and H4, together with between ca 150 and 240 DNA base pairs, wound around the core. The nucleosome particle itself is the repeating

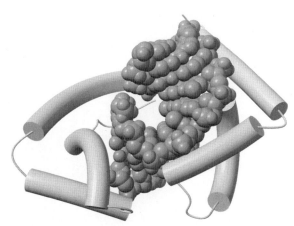

Figure 7.21 Structure of the HMG–D–DNA complex (Murphy et al., 1999), PDB id 1QRV. Note the extreme bending of the DNA.

unit for the further level of supercoiled structure in chromosomes. The crystal structure of the nucleosome core particle was initially determined at 7 Å resolution (Richmond et al., 1984), subsequently, at 2.8 Å (Luger et al., 1997), at 1.94 Å (Davey et al., 2002; Richmond & Davey, 2003) and most recently at 1.99 Å (Sharma et al., 2019) (Table 7.4), due to the availability of highly diffracting crystals. Many other nucleosome structures are now available in the PDB, with features such as histone variants and DNA mutations. The higher resolution native structures have enabled features such as DNA bending, to be described in detail. In all these structures, there are ca 145–146 base pairs, which are wound around the octameric histone core, forming a flattened disk-like superhelix (Fig. 7.22), such that the structure overall has pseudo twofold symmetry. There are

Figure 7.22 Two views of the 1.99 Å nucleosome structure (Sharma et al., 2019), PDB id 6IPU, showing (a) the histone octamer surrounded by supercoiled DNA. (b) A view at right angles highlighting the disc-like nature of the structure, which has almost two complete turns of bound supercoiled DNA. The DNA double helix, coloured light blue, is shown in van der Waals mode, and the histone proteins are shown in cartoon form.

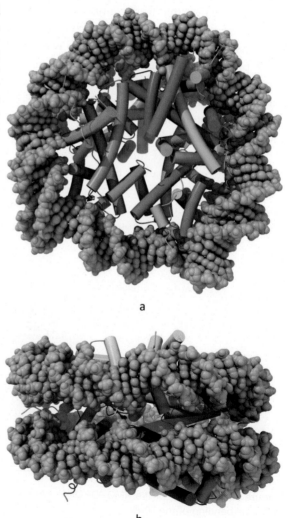

a

b

10.2 base pairs per turn and the bending at low resolution appears to be altogether regular, with the DNA retaining B-like features. The smooth bending is accomplished by relatively small but numerous deformations from ideality, in a sequence-dependent manner, analogous to changes observed in crystal structures of some short yet bent oligonucleotides. The high-resolution (1.9—2.0 Å) structures (Richmond & Davey, 2003; Sharma et al., 2019) are comparable in resolution to that reported for many native oligonucleotides, and thus has enabled an altogether more detailed comparative picture of local DNA deformations to be obtained than was possible at lower resolution. These show that nucleosomal DNA is indeed B-form overall, but bending is not uniform, due to (a) the effects of the histone proteins at various points, (b) anisotropic flexibility of the DNA and (c) local sequence-dependent structural features, some of which have not been observed on native oligonucleotide crystal (or NMR) structures. This irregularity is apparent in Fig. 7.23, where we see almost continually micro-deformed DNA. The base-step morphology parameters such as roll, helical twist and tilt vary in a sinusoidal manner along the 147-nucleotide length, with surprisingly large variations at different points (Fig. 7.24). There is a small preference for bending in the direction of the major groove, characterised by smooth DNA deformations; by contrast, the minor groove bending tends to have more extensive kinking. This sequence-dependent effect occurs at CA base steps and is associated with significant deviations from B-form DNA mean values for parameters describing base-step and base-pair morphology. The crystal structures of a human DNA sequence in a nucleosome core structure (Sharma et al., 2019; Tsunaka et al., 2005) show small changes in the DNA path as it winds around the histone core compared to nucleosomal DNA from other organisms, although the overall arrangement is closely similar in all nucleosome structures, studied from a range of eukaryotic organisms (Fig. 7.25).

The next level of organisation in eukaryotic chromosomes is the chromatin fibre. The crystal structure, initially at low (9 Å) resolution (Schalch et al., 2005) and more recently at 5.77 Å (Ekundayo et al., 2017) for the tetranucleosome, has revealed details about how

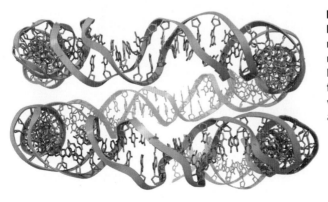

Figure 7.23 A cartoon view of the DNA alone, taken from a high-resolution crystal structure of the nucleosome (Sharma et al., 2019), PDB id 6IPU, showing the non-uniform nature of the DNA backbone with numerous small local kinks and deformation.

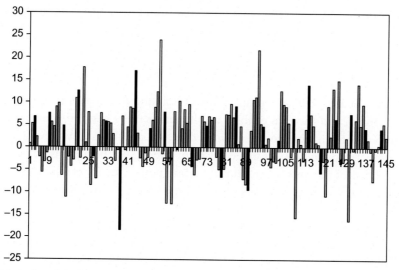

Figure 7.24 A plot of the base-step roll angles for the DNA in the 1.94 Å high-resolution nucleosome crystal structure (Davey et al., 2002), PDB id 1KX5.

Figure 7.25 A view of the crystal structure of the tetranucleosome (Ekundayo et al., 2017), PDB id 5OY7, with the DNA coloured blue.

individual nucleosomes may be organised in the fibre, although the tetranucleosome structure appears to be less compacted than the fibre and its exposed octamer faces may have relevance to inter-fibre contacts. The structure of a hexanucleosome crystal structure at 9.7 Å resolution (Boopathi et al., 2020; Garcia-Saez et al., 2018) and cryo-EM structures of short reconstituted sections of the 30 nm fibre (Song et al., 2014; Zhu & Li, 2016a,b) support the two-start nucleosomal model. Each nucleosome is connected by 20 base-pair linkers, in a supercoiled zig-zag-like arrangement. The overall structure has been used to build plausible models for the chromatin fibre with a continuous array of nucleosome units.

7.8 DNA bending and protein recognition

X-ray crystallography, NMR and other techniques have shown that DNA structures in the absence of bound drug or protein can have a significant degree of flexibility. This is reflected in sequence-dependent backbone, sugar pucker, base step and base pair features, although the classic features of canonical forms are often observed in both native and protein-bound oligonucleotide structures.

There is considerably greater diversity of DNA structure in protein complexes than in native DNA structures, largely because the binding energy for a protein—DNA interaction is sufficient to readily overcome the barriers to local DNA deformability. Hence, these can lead to even large-scale changes in DNA structure such as bending, to optimise interactions, should such changes be needed. Many proteins, notably HTH and zinc fingers, do not need to distort to produce the required protein contacts. Sometimes, though when distortions do occur, they can blur the distinctions between B and A-form DNA structural types so that it is no longer possible to unequivocally assign the resulting DNA structure to one or the other form. This is especially apparent when large changes in parameters such as groove width, helical twist and base-pair inclination to the helix axis are observed. For example, the nature of the structure of DNA in the Zif268—DNA zinc finger complex has been controversial, with both A- and B-type helices being suggested. The crystal structure of the complex (Pavletich & Pabo, 1991) shows that the DNA adopts an overall B-type form, although there are several local features that are intermediate between A and B. In such a case, the simple canonical DNA classifications are of limited use by themselves and are best accompanied by more detailed information describing particular features.

The DNA in the *met* J complex (Somers & Phillips, 1992) has a straight 10 base-pair central segment, with bends at each end due to groove width changes necessary for both effective protein dimer formation and protein—DNA interaction. In the structure of the *trp* repressor/operator complex (Haran et al., 1992; Otwinowski et al., 1988), the DNA is appreciably distorted from canonical B-form due to numerous small, cumulative local changes, especially to roll and slide. These together with changes in groove width and backbone geometry from canonical B-DNA are necessary to achieve the large number of indirect readout contacts that produce specificity. The DNA is bent and its surface closely follows the contours of the protein surface. Rather greater DNA bending is required in the papillomavirus-1 E2 DNA complex (Hegde et al., 1992), where there is smooth DNA bending over the β-barrel recognition motif, to expose the major groove to protein side chains. Groove widths are compressed on the side of the DNA that faces protein, with alterations in propeller twist and roll angles contributing to these width changes. The prokaryotic transcription activator CAP, which acts as a dimer, has a consensus DNA binding site 22 base pairs long with the essential bases towards the end of the sequence. This is far too long a DNA sequence to be contacted by the relatively small CAP dimer for the DNA to remain linear. The structure of the CAP–DNA complexes shows (Benoff et al., 2002; Schultz et al., 1991) that the DNA has solved this problem by bending by ~90° to effectively wrap around the protein dimer. This bending, or kinking, is produced almost entirely at two base-pair steps, one on each side of the twofold axis of the complex, by roll angles of ~40° and a twist angle of ~20° (Napoli et al., 2006). It is likely that this bending is an important aspect of transcriptional activation by CAP and other analogous proteins. There is a very close structural correspondence between CAP and other architectural proteins (Werner & Burley, 1997) such as TBP and the globular domain of histone H5 (Ramakrishnan et al., 1993), which is involved in higher-level nucleosome organisation. These suggest that the ability of this type of HTH protein to bend DNA sequences is of wide structural and functional significance.

In the light of the observations of DNA deformability in these and other protein complexes, the question arises as to their relevance to the numerous studies on native DNA structures. The answer is not clear cut and depends on the structures involved. It is apparent though that native DNA structures do sometimes demonstrate inherent sequence dependency, which can also be found in their protein complexes, provided the changes in structure are not profound. Examples are provided by the *trp* operator site (Shakked et al., 1994), which shares common features of structure and hydration in the native DNA structure and in this protein–DNA complex. This is also shown by the human papillomavirus type 6 E2 protein, which shows considerable bending in its DNA complex (Hooley et al., 2006), with the A-tract-containing DNA sequence similarly showing a predisposition to bending in the native structure. Larger-scale DNA conformational differences, such as are seen in the TBP and other minor-groove-binding protein complexes, can still be related to native DNA structures since these are often describable in terms of simple concerted changes from standard forms of DNA. Thus, the structure of the DNA in the TATA box may be thought of as a canonical A-form with a change in glycosidic angles (Guzikevich-Guerstein & Shakked, 1996). It is striking that this and other small variants of the A-form (Lu et al., 2000), when embedded within B-DNA, can produce bent and curved structures suitable for binding to a wide range of proteins (and small molecules such as cis-platinum — see Chapter 5 and the next section). Bending was foreseen over 40 years ago, as a natural consequence of junctions between canonical A and B forms (Selsing et al., 1979). It was less obvious until comparatively, recently, that both forms would play key roles in the functioning of DNA.

7.9 Protein–DNA small-molecule recognition

Chapter 5 has detailed the many of the binary structures between small molecules (frequently useful as drugs) and DNA. These large numbers of structures have often provided insights into biological activity, but only comparatively rarely have structures with protein partners in ternary complexes been determined. These can enable a fuller picture of biological function to be obtained. This is the case for the sequence-specific polyamides (Section 5.5.3), which are one of the few classes of DNA-binding molecules for which there is structural information directly relevant to chromatin. The crystal structures of four pyrrole–imidazole polyamide ligand–nucleosome complexes have been determined (Edayathumangalam et al., 2004; Suto et al., 2004), one at a resolution of 2.03 Å. These polyamides form anti-parallel hairpin structures, recognising DNA sequence via the minor groove, as has been observed in their binary DNA complexes. The same arrangement is seen in their nucleosome complexes (Fig. 7.26), with several polyamide molecules bound to each nucleosome on the exterior face, away from the histone octamer core. Firstly, the observation of bound ligand is itself significant, demonstrating that such molecules can indeed bind to nucleosomal DNA when it is associated with histones. Some distortions to the native nucleosome structure are apparent, but only to the DNA and not to the histone core. These changes are largely observed as increases in minor groove width, which at some points are local and at others are transmitted along the sequence, so that some DNA–octamer contacts become loosened. One significant effect of polyamide binding is to block temperature-induced histone repositioning — this is likely to have consequences for chromatin function, not least re-modelling during transcription. Polyamides have also been engineered as dimers to target the gap between the two DNA coils in a nucleosome, termed the 'supergroove' (Edayathumangalam et al., 2004). Crystallography has shown that this hairpin dimer (with two polyamide dimers linked by a short polyethylene glycol chain) does bind in the predicted manner, in accord with the finding that this ligand hinders nucleosome dissociation (Table 7.6).

Figure 7.26 A view of the crystal structure (Suto et al., 2004) of a nucleosome–polyamide complex (PDB id 1M19), with several polyamide molecules shown in orange space-filling representation. The distinctive dimeric side-by-side arrangement of the polyamide ligand at each binding site is clearly visible.

Table 7.6 Crystal structures of selected protein—DNA—drug complexes.

Protein—DNA	Small molecule/lesion	PDB ID code
Nucleosome	ImPyPyPy-γ- PyPyPyPy-β-Dp	1M19
Nucleosome	Polyamide hairpin dimer	1S32
HMG domain	Cis-platinum	1CKT
Human topoisomerase I	Topotecan	1K4T
Human topoisomerase I	Indenoisoquinoline MJ—II—38	1SC7
Human topoisomerase I	Camptothecin	1T8I
Topoisomerase IV from *Streptococcus pneumoniae*	Clinafloxacin	3FOE
Topoisomerase IV from *Streptococcus pneumoniae*	Novel 7,8-bridged fluoroquinolone	4KPE

Structural information is available on several other categories of drug—DNA—protein ternary complexes (see, for example, Section 5.6.1 for details of platinum—nucleosome complexes). The HMG domain complexed with DNA bound to the anti-cancer drug - cis-platinum (see Chapter 5) shows a highly bent DNA (Fig. 7.27), with the minor groove resembling A-DNA (Ohndorf et al., 1999).

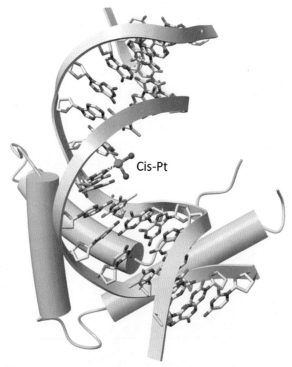

Figure 7.27 A view of the crystal structure of the HMG—DNA—cis-platinum ternary complex (Ohndorf et al., 1999), PDB id 1CKT. The platinum drug is shown in ball-and-stick representation, bound to two adjacent guanines.

The DNA topoisomerase enzymes are responsible in both prokaryotics and eukaryotics for changing the supercoiling of DNA after replication or transcription. They do this by first cleaving one or both DNA strands, then producing the desired topological change in DNA and finally resealing the breaks. Type I topoisomerases, such as topoisomerase I, catalyse single-stranded breaks, whereas type II topoisomerases, such as human topoisomerase II, and prokaryotic gyrase and topoisomerase IV, catalyse double-stranded DNA breaks. A variety of drugs, ranging from anti-bacterial agents to anti-cancer drugs, can interfere with topoisomerase function, often by preventing the relegation step from occurring. The crystal structure of human topoisomerase I in a ternary complex with the anti-cancer drug topotecan, bound to a 22-mer DNA (Staker et al., 2002), shows the drug bound in an intercalation pocket (Fig. 7.28a and b) of a straight B-form DNA helix. It thus acts as a conventional intercalating molecule in moving one part of the DNA sequence relative to the other, down by one base pair, and thus interfering with normal topoisomerase substrate function. Crystal structures of ternary complexes, involving the related drug camptothecin with topoisomerase I mutants that confer resistance to the drug (Chrencik et al., 2004), have revealed changes in drug–enzyme contacts, for example, that have been able to rationalise the resistance observations. Further crystal structure analyses of topoisomerase I ternary complexes, with a structurally diverse

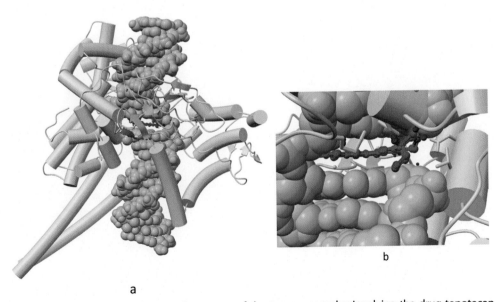

a

b

Figure 7.28 (a) A view of the crystal structure of the ternary complex involving the drug topotecan bound to a 22-mer DNA duplex and DNA topoisomerase I (Staker et al., 2002), PDB id 1K4T. The drug molecule is in the middle of the DNA helix and is coloured magenta in ball-and-stick mode. (b) A detailed view of the binding site in the ternary topoisomerase complex. The break in the DNA strand on the right-hand side is clearly visible.

range of topoisomerase inhibitors (Staker et al., 2005), have provided data for the rational design of novel types of anti-cancer agent of this general class.

The first type II topoisomerase—DNA—drug complex crystal structure was elucidated initially at 4 Å resolution and then at 3 Å for the topoisomerase IV from *Streptococcus pneumoniae*. There are now a wide range of type II topoisomerase—DNA—drug complexes both for gyrases (in gram-negative bacteria), and for topoisomerase IVs from gram-positive bacteria, organisms, as well as human topoisomerases IIs, solved to resolutions of up to 2.5 Å. These structures have aided the development of new fluoroquinolones and other type II topoisomerase targeted drugs, and improved our understanding of the action of these drugs. Fig. 7.29 shows the structure of the complex of a bacterial topoisomerase with DNA and the quinolone drug clinafloxacin, which is bound in a hemi-intercalated manner in between adjacent base pairs, at the DNA gate site of strand cleavage. For these type II topoisomerase complexes, the conformation of the bound DNA changes from B-form to a B—A—B form on binding to the protein complex. There is now a structure of the gyrase holoenzyme with both G-gate and T-gate DNA bound solved by cryo-EM showing the binding of the T-gate DNA to the β-pinwheel structure of the C-terminal domain of the A subunit of the A_2B_2 tetramer (van den Broeck et al., 2019).

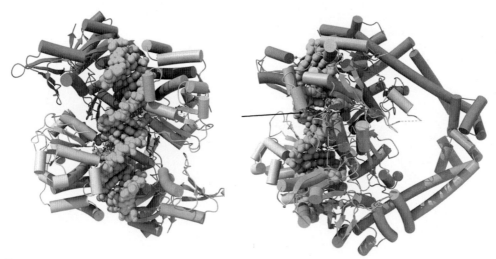

Figure 7.29 Two views of the crystal structure of the ternary complex involving the ParC breakage reunion and ParE TOPRIM domains of topoisomerase IV from *Streptococcus pneumoniae*, a 34 base pair DNA duplex and the quinolone drug clinafloxacin (Laponogov et al., 2009), PDB id 3FOE. The hemi-intercalated drug molecule is indicated by an arrow in the right-hand figure. The DNA exit gate is on the extreme right-hand side of the right-hand figure.

7.10 DNA-binding proteins and their role in genomics

One of the first examples of a site-specific tailored DNA-binding protein DNA interaction was for a four-finger protein that successfully constructed (Wolfe, Ramme, & Pabo, 2000) using the GCN4 leucine zipper dimerisation motif (see below), which recognises 10 base pairs.

7.10.1 Zinc finger proteins

Although the structures of most DNA-binding motifs and their mode of binding to DNA were elucidated in the 1980 and 1990s, primarily by X-ray crystallography and NMR as discussed above, a most interesting current aspect of this class of proteins is the development of several of them for gene control (both activation and repression) and gene editing. These developments will have profound future impact in the areas of medicine and plant agronomy.

7.10.2 The development of zinc finger targeting

The early structural work on zinc fingers has been discussed in Section 7.4 above. Klug and coworkers (Isalan et al., 1997, 1998, 2001; Choo et al., 1994; Choo & Isalan, 2000; Klug, 2010) developed the zinc finger proteins as reagents, which could specifically target given DNA sequences with high fidelity. It had been found in Klug's group that in addition to the important single amino interactions with the adjacent DNA strand which were first shown from the three finger Zif268—DNA complex solved by Pavletich and Pabo (Pavletich & Pabo, 1991) there are in addition very important cross-strand interactions between amino acids and bases (Fig. 7.30). These became apparent from the analysis of the single crystal X-ray structure zinc finger—DNA complex determined by Lu et al. (2003).

The cross-strand interactions were incorporated into the phage display libraries developed by Isalan et al. in Klug's group (Klug, 2010). Phage display permits the display of randomised sequences of zinc fingers onto the pole of the phage and by interacting them with a given DNA sequence that has a terminal magnetic bead allowing the binding zinc finger and associated phage to be 'fished' out (Fig. 7.32). In this way, a key of specific amino acids in given fingers associating with specific bases on adjacent strands and also cross-strand interactions can be constructed (Fig. 7.30). A peptide containing three zinc fingers, found by screening a library of peptides, has been shown to bind to a nine base-pair sequence in the BCR-ABL oncogene (Choo et al., 1994), with an equilibrium constant of 6×10^{-7} M. This binding affinity is at least an order of magnitude higher than to other control sequences. The original strategy did not effectively recognise all possible bases within a triplet, with the 5′ one being optimally adenine or cytosine, possibly due to sequence context effects. This problem was

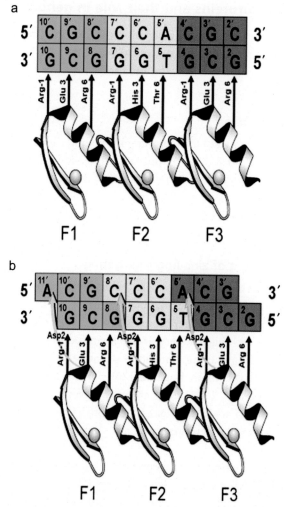

Figure 7.30 (a) Schematic diagram (Jamieson et al., 2003) of a model for modular recognition of DNA by a three zinc finger peptide, illustrating the results of the first crystal structure determination of the complex between the DBD of the transcription factor Zif268 and an optimised DNA-binding site (Choo & Klug, 1994). Each finger interacts with a 3-base pair sub-site on one strand of the DNA, using amino acid residues in helical positions x 1, 3 and 6. (b) Refined model of DNA recognition (Jamieson et al., 2003; Reynolds et al., 2003). View of the potential hydrogen bonds to the second strand of the DNA ('cross-strand interactions') emanating from position 2 on the recognition helix. This is based on the crystal structure of the tramtrack—DNA complex (Choo & Klug, 1994), mutagenesis analysis (Jamieson et al., 2003) and the phage display selection studies of Isalan and co-workers (Reynolds et al., 2003), together with the refined structure of the Zif268—DNA complex and variants by Pabo and his colleagues (Elrod-Erickson et al., 1998). The fingers ideally bind 4-base-pair overlapping sub-sites, so that adjacent fingers are functionally synergistic though structurally independent. *(Reproduced with permission from Klug, A. (2010). The discovery of zinc fingers and their development for practical applications in gene regulation and genome manipulation. Quarterly Reviews of Biophysics, 43, 1—21.)*

overcome (Isalan et al., 1998, 2001) by detailed considerations of several zinc finger crystal structures that show synergistic recognition from adjacent zinc fingers. As a result, it is possible to start devising a general recognition code for zinc finger-directed DNA sequence specificity. Addition of a dimerisation motif can enable more than three zinc fingers to be assembled without loss of affinity. A biotech company, Gendaq, was set up by Klug, Choo and Isalan to fully develop and exploit this technology in the most productive fashion. It was subsequently acquired by Sangamo, a Californian-based company who also acquired all Gendaq's zinc finger phage libraries.

NMR studies (Nakaseko et al., 1992; Neuhaus et al., 1992) of a pair of zinc fingers interacting with a DNA duplex in solution have demonstrated the flexibility of the joining linker. The important question was how to bind a DNA sequence with a tight binding constant in the nanomolar range, as well as discriminating the recognition sequence against a background of the 5×10^9 base pairs present in the human genome. It was found by constructing phage libraries that the best arrangement was a $3 \times 2F$ (three pairs of zinc fingers) construct with suitable spacing linkers and not a $2 \times 3F$ (two pairs of triple fingers) or a 6F (a run of six fingers) constructs (Fig. 7.31).

These libraries have been exploited in several ways. The use of zinc finger motifs to specifically recognise any desired DNA sequence is increasingly being applied to a wide range of problems (Papworth et al., 2006), and several web sites are now offering facilities for zinc finger design for this purpose (such as that of the Zinc Finger Consortium at http://www.zincfingers.org/).

7.10.3 Repair of DNA lesions by site-specific cutting and DNA recombination

Zinc finger proteins (ZFPs) coupled to FokI nuclease (Fig. 7.32) have been exploited to recognise DNA sequences around a mutation and enable the insertion of an unmutated DNA strand, thus returning the gene to the wild type. It has been long known that DNA recombination is highly active in regions of DNA which undergo DNA breaks. By combining the action of the zinc finger cutter and an unmutated strand, the DNA mutation can be repaired. This was first shown in frog oocytes (Bibikova et al., 2001; Carroll, 2017) and therefore boded well for future studies in humans. The Sangamo Corporation and their collaborators then showed that the editing and recombination technique could be applied to the treatment of SCID (Severe Combined Immunodeficiency) patients as well as to the treatment of AIDS patients by deleting the CCR5 receptor in the white cells of AIDS patients and then re-introducing these into the patients (Didigu et al., 2014). The white cells without the CCR5 co-receptors then evolve to be dominant in the patient by natural selection and the patients are effectively

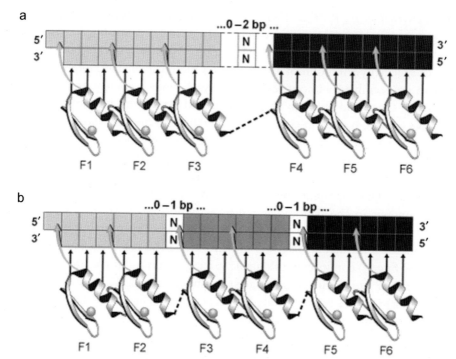

Figure 7.31 Two modes of generating six ZFPs for specific recognition of 18-base pair sequences. (a) Two three-finger peptides fused together using an extended canonical linker (2r3F scheme). (b) Three two-finger peptides linked using canonical linkers extended by an insertion of either a glycine residue or a glycine–serine–glycine sequence in the canonical linkers between fingers 2 and 3 and fingers 4 and 5, respectively. *(Reproduced with permission from Klug, A. (2010). The discovery of zinc fingers and their development for practical applications in gene regulation and genome manipulation. Quarterly Reviews of Biophysics, 43, 1–21.)*

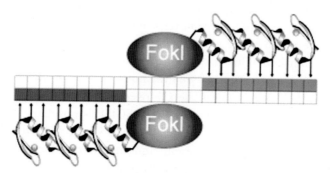

Figure 7.32 Gene correction using a pair of three ZFNs (Urnov et al., 2005) to produce a double-stranded DNA break. The zinc finger peptides are linked to the non-specific catalytic domain of the Fok1 endonuclease by a short amino acid linker. In the 3D structure, the two catalytic domains form a dimeric association. *(Reproduced with permission from Klug, A. (2010). The discovery of zinc fingers and their development for practical applications in gene regulation and genome manipulation. Quarterly Reviews of Biophysics, 43, 1–21.)*

cured of AIDS. This was the case for the Berlin patient who was an AIDS sufferer with leukaemia and who then was transfused by a bone marrow donor who lacked the CCR5 co-receptor. The AIDS patient was 'cured' of AIDS since cellular entry of HIV was blocked (Jilg & Li, 2019).

At the time of writing, there is still a debate as to whether ZFPs are as good as clustered regularly inter-spaced short palindromic repeat (CRISPR)/Cas9 (see below) for gene editing. ZFPs have the advantage that the proteins are human in origin, whereas CRISP/Cas9 are of bacterial in origin so immunogenicity may present as a problem. Also, the question of off target effects of one against the other is not fully resolved and this problem could be system specific. The second generation of ZFPs are looking very good and when they are being employed the problems of possible off target effects are being fully investigated to prevent this being a problem in a clinical setting.

7.10.4 The regulation of gene control by ZFPs

Where ZFPs come into their own is in systems where they are exploited to repress or activate genes, for example, in the silencing of Huntingtin from the Huntington gene, where ZFP constructs can be made to bind to the Huntington gene preventing the translation of the RNA and production of the protein with the polyglutamine repeat from the repeating CAG triplet in the gene (Fig. 7.33). ZFPs are being developed for many clinical treatments, for example, in the editing of genes in β-thalassemia and sickle cell disease.

7.10.5 CRISPR-associated nuclease Cas9

Bacteria have an immune system for recognising foreign DNA, either phage or bacterial entering their cells incorporating this DNA into 'clustered regularly inter-spaced short palindromic repeats' (CRISPR) (Adli, 2018). There are a wide range of CRISPR/Cas endonuclease systems in different bacterial systems, but it is through the functional and structural work primarily of the groups of Charpentier and Doudna (Charpentier and

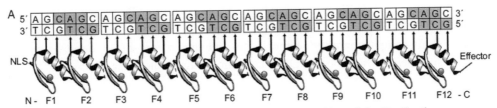

Figure 7.33 An extended zinc finger array used in recognition of the Huntingtin gene.

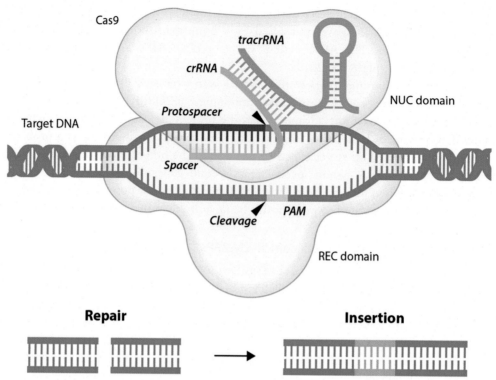

Figure 7.34 Schematic illustrating the mechanism of action of Crispr Cas9. We thank Anncharlott Ber-glar, https://www.scivislab.com/ for kindly producing this diagram.

Doundna, 2013; Collins et al., 2021; Doudna & Charpentier, 2014; Jiang & Doudna, 2017; Jinek et al., 2014; Lapinaite et al., 2020) that the CRISPR/Cas9 system from *Streptococcus pyogenes* has become a well-developed and widely used tool for genetic editing. By fully understanding the molecular biology of the system, which in its natural form requires two RNA molecules (Fig. 7.34) it was possible to fuse these two RNA molecules into one 20-base RNA with a hairpin. This designed RNA incorporates immediately upstream of a 5′-NGG-3′ or a 5′-NAG-3′ triplet with the CRISPR/Cas9 as a very powerful DNA recognition and cleavage system (Fig. 7.34). The genomic DNA is scanned until complementarity of the RNA and one of the DNA duplex strands is reached and there is then specific DNA cleavage of the duplex (Fig. 7.35).

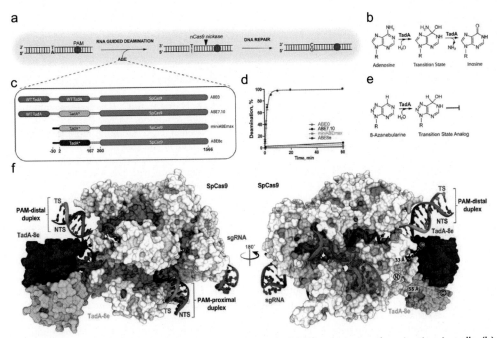

Figure 7.35 (a) Schematic representation of RNA-guided DNA adenosine deamination in cells. (b) Mechanism of *Escherichia coli* TadA-catalyzed tRNA adenosine deamination. (c) Domain architecture of four generations of ABEs encoding WT TadA, evolved TadA7.10 (TadA*, yellow), or evolved TadA8e (TadA*, red) N-terminally linked to SpCas9. (d) Single-turnover dsDNA deamination kinetics of targeting ABE RNPs measured using dsDNA containing a single adenine. The fraction of deaminated dsDNA is plotted as a function of time and fitted to a single-phase exponential equation. The extracted apparent deamination rates of ABE7.10 (black), mini-ABEmax (orange) and ABE8e (red) are $0.0010 \pm 3 \times 10^{-4}$ min^{-1}, $0.0005 \pm 1 \times 10^{-4}$ min^{-1} and 0.585 ± 0.034 min^{-1}, respectively. Data are represented as the mean $\pm$ SD from three independent experiments. The deamination data of ABE8e and ABE7.10 were originally published in (Richter et al., 2020). (e) Mechanism of inhibition of adenosine deamination by 8Az that mimics the adenosine deamination reaction intermediate. (f) The 3.2 Å resolution cryo-EM structure of the SpCas9—ABE8e complex, PDB id 6VPC. Subunits are coloured as follows: SpCas9, gray; sgRNA, purple; TS, teal; NTS, blue and TadA-8e dimer, red and pink. The direct distances between the Cas9 N-terminus and either TadA-8e C-termini are shown as *dashed orange lines. (Reproduced with permission from Lapinaite, A., Knott, G. J., Palumbo, C. M., Lin-Shiao, E., Richter, M. F., Zhao, K. T., Beal, P. A., Liu, D. R., & Doudna, J. A. (2020). DNA capture by a CRISPR-Cas9-guided adenine base editor.* Science, 369, 566—571.)

References

Adli, M. (2018). The CRISPR tool kit for genome editing and beyond. *Nature Communications, 9,* 1911.

Afek, A., Shi, H., Rangadurai, A., Sahay, H., Senitzki, A., Xhani, S., Fang, M., Salinas, R., Mielko, Z., Pufal, l M. A., Poon, G. M. K., Haran, T. E., Schumacher, M. A., Al-Hashimi, H. M., & Gordan, R. (2020). DNA mismatches reveal widespread conformational penalties in protein-DNA recognition. *Nature, 587,* 291–296.

Aggarwal, A. K., Rodgers, D. W., Drottar, M., Ptashne, M., & Harrison, S. C. (1988). Recognition of a DNA operator by the repressor of phage 434: A view at high resolution. *Science, 242,* 899–907.

Aishima, J., Gitti, R. K., Joan, J. E., Gan, H. H., Schlick, T., & Wolberger, C. (2002). A Hoogsteen base pair embedded in undistorted B-DNA. *Nucleic Acids Research, 23,* 5244–5252.

Allain, F. H.-T., Yen, Y.-M., Masse, J. E., Schultze, P., Dieckmann, T., Johnson, R. C., & Feigon, J. (1999). Solution structure of the HMG protein NHP6A and its interaction with DNA reveals the structural determinants for non-sequence-specific binding. *The EMBO Journal, 18,* 2563–2579.

Anderson, W. F., Ohlendorf, D. H., Takeda, Y., & Matthews, B. W. (1981). Structure of the cro repressor from bacteriophage lambda and its interaction with DNA. *Nature, 290,* 754–758.

Beamer, L. J., & Pabo, C. O. (1992). Refined 1.8 Å crystal structure of the lambda repressor-operator complex. *Journal of Molecular Biology, 227,* 177–196.

Benoff, B., Yang, H., Lawson, C. L., Parkinson, G., Liu, J., Blatter, E., Ebright, Y. W., Berman, H. M., & Ebright, R. H. (2002). Structural basis of transcription activation: The CAP-αCTD-DNA complex. *Science, 297,* 1562–1566.

Bibikova, M., Carroll, D., Segal, D. J., Trautman, J. K., Smith, J., Kim, Y. G., & Chandrasegaran, S. (2001). Stimulation of homologous recombination through targeted cleavage by chimeric nucleases. *Molecular and Cellular Biology, 21,* 289–297.

Billeter, M., Güntert, P., Luginbühl, P., & Wüthrich, K. (1996). Hydration and DNA recognition by homeodomains. *Cell, 85,* 1057–1065.

Boopathi, R., Dimitrov, S., Hamiche, A., Petosa, C., & Bednar, J. (2020). Cryo-electron microscopy of the chromatin fiber. *Current Opinion in Structural Biology, 64,* 97–103.

Carroll, D. (2017). Genome editing: Past, present, and future. *Yale Journal of Biology & Medicine, 90,* 653–659.

Charpentier, E., & Doudna, J. A. (2013). Biotechnology: Rewriting a genome. *Nature, 495,* 50–51.

Choo, Y., & Isalan, M. (2000). Advances in zinc finger engineering. *Current Opinion in Structural Biology, 10,* 411–416.

Choo, Y., & Klug, A. (1994). Selection of DNA binding sites for zinc fingers using rationally randomized DNA reveals coded interactions. *Proceedings of the National Academy of Sciences of the United States of America, 372,* 642–645.

Choo, Y., Sánchez-García, I., & Klug, A. (1994). In vivo repression by a site-specific DNA-binding protein designed against an oncogenic sequence. *Nature, 372,* 642–645.

Chrencik, J. E., Staker, B. L., Burgin, A. B., Pourquier, P., Pommier, Y., Stewart, L., & Redinbo, M. R. (2004). Mechanisms of camptothecin resistance by human topoisomerase I mutations. *Journal of Molecular Biology, 339,* 773–784.

Collins, F. S., Doudna, J. A., Lander, E. S., & Rotimi, C. N. (2021). Human molecular genetics and genomics - important advances and exciting possibilities. *New England Journal of Medicine, 384,* 1–4.

Davey, C. A., Sargent, D. F., Luger, K., Maeder, A. W., & Richmond, T. J. (2002). X-ray structure of the nucleosome core particle, NCP147, at 1.9 Å resolution. *Journal of Molecular Biology, 319,* 1097–1113.

Didigu, C. A., Wilen, C. B., Wang, J., Duong, J., Secreto, A. J., Danet-Desnoyers, G. A., Riley, J. L., Gregory, P. D., June, C. H., Holmes, M. C., & Doms, R. W. (2014). Simultaneous zinc-finger nuclease editing of the HIV coreceptors ccr5 and cxcr4 protects CD4+ T cells from HIV-1 infection. *Blood, 123,* 61–69.

Doudna, J. A., & Charpentier, E. (2014). Genome editing. The new frontier of genome engineering with CRISPR-Cas9. *Science, 346,* 1258096.

Edayathumangalam, R. S., Weyermann, P., Gottesfeld, J. M., Dervan, P. B., & Luger, K. (2004). Molecular recognition of the nucleosomal "supergroove". *Proceedings of the National Academy of Sciences of the United States of America, 101,* 6844–6869.

Ellenberger, T. E., Brandl, C. J., Struhl, K., & Harrison, S. C. (1992). The GCN4 basic region leucine zipper binds DNA as a dimer of uninterrupted alpha helices: Crystal structure of the protein-DNA complex. *Cell, 71*, 1223—1237.

Ekundayo, B., Richmond, T. J., & Schalch, T. (2017). Capturing structural heterogeneity in chromatin fibers. *Journal of Molecular Biology, 429*, 3031—3042.

Elrod-Erickson, M., Benson, T. E., & Pabo, C. O. (1998). High-resolution structures of variant Zif268-DNA complexes: implications for understanding zinc finger-DNA recognition. *Structure, 6*, 451—464.

Feng, J. A., Johnson, R. C., & Dickerson, R. E. (1994). Hin recombinase bound to DNA: The origin of specificity in major and minor groove interactions. *Science, 263*, 348—535.

Fornes, O., Castro-Mondragon, J. A., Khan, A., van der Lee, R., Zhang, X., Richmond, P. A., Modi, B. P., Correard, S., Gheorghe, M., Baranašić, D., Santana-Garcia, W., Tan, G., Chèneby, J., Ballester, B., Parcy, F., Sandelin, A., Lenhard, B., Wasserman, W. W., & Mathelier, A. (2020). JASPAR 2020: Update of the open-access database of transcription factor binding profiles. *Nucleic Acids Research, 48*, D87—D92.

Fraenkel, E., & Pabo, C. O. (1998). Comparison of X-ray and NMR structures for the antennapedia homeodomain-DNA complex. *Nature Structural Biology, 5*, 692—697.

Fraenkel, E., Rould, M. A., Chambers, K. A., & Pabo, C. O. (1998). Engrailed homeodomain-DNA complex at 2.2 Å resolution: A detailed view of the interface and comparison with other engrailed structures. *Journal of Molecular Biology, 284*, 351—361.

Garcia-Saez, I., Menoni, H., Boopathi, R., Shukla, M. S., Soueidan, L., Noirclerc-Savoye, M., Le Roy, A., Skoufias, D. A., Bednar, J., Hamiche, A., Angelov, D., Petosas, C., & Dimitrov, S. (2018). Structure of an H1-bound 6-nucleosome array reveals an untwisted two-start chromatin fibre conformation. *Molecular Cell, 72*, 902—915.

Garvie, C. W., & Phillips, S. E. V. (2000). Direct and indirect readout in mutant Met repressor-operator complexes. *Structure, 8*, 905—914.

Gehring, W. J., Qian, Y. Q., Billeter, M., Furukubo-Tokunaga, K., Schier, A., Resendez-Perez, D., Affolter, M., Otting, G., & Wüthrich, K. (1994). Homeodomain-DNA recognition. *Cell, 78*, 211—223.

Guzikevich-Guerstein, G., & Shakked, Z. (1996). A novel form of the DNA double helix imposed on the TATA-box by the TATA-binding protein. *Nature Structural Biology, 3*, 32—37.

Haran, T. E., Joachimiak, A., & Sigler, P. B. (1992). The DNA target of the trp repressor. *The EMBO Journal, 11*, 3021—3030.

Harrison, S. C. (1991). A structural taxonomy of DNA-binding domains. *Nature, 353*, 715—719.

Hegde, R. S., Grossman, S. R., Laimonis, L. A., & Sigler, P. B. (1992). Crystal structure at 1.7 Å of the bovine papillomavirus-1 E2 DNA-binding domain bound to its DNA target. *Nature, 359*, 505—512.

Hooley, E., Fairweather, V., Clarke, A. R., Gaston, K., & Brady, R. L. (2006). The recognition of local DNA conformation by the human papillomavirus type 6 E2 protein. *Nucleic Acids Research, 34*, 3897—3908.

Isalan, M., Choo, Y., & Klug, A. (1997). Synergy between adjacent zinc fingers in sequence-specific DNA recognition. *Proceedings of the National Academy of Sciences of the United States of America, 94*, 5617—5621.

Isalan, M., Choo, Y., & Klug, A. (1998). Comprehensive DNA recognition through concerted interactions from adjacent zinc fingers. *Biochemistry, 37*, 12026—12033.

Isalan, M., Klug, A., & Choo, Y. (2001). A rapid, generally applicable method to engineer zinc fingers illustrated by targeting the HIV-1 promoter. *Nature Biotechnology, 19*, 656—660.

Jain, R., Aggarwal, A. K., & Rechkoblit, O. (2018). Eukaryotic DNA polymerases. *Current Opinion in Structural Biology, 53*, 77—87.

Jain, R., Choudhury, J. R., Buku, A., Johnson, R. E., Prakash, L., Prakash, S., & Aggarwal, A. K. (2017). Mechanism of error-free DNA synthesis across N1-methyl-deoxyadenosine by human DNA polymerase-iota. *Scientific Reports, 7*, 43904.

Jamieson, A. C., Miller, J. C., & Pabo, C. O. (2003). Drug discovery with engineered zinc-finger proteins. *Nature Reviews Drug Discovery, 2*, 361—368.

Jiang, F., & Doudna, J. A. (2017). CRISPR-Cas9 structures and mechanisms. *Annual Review of Biophysics, 46*, 505—529.

Jilg, N., & Li, J. Z. (2019). One patient has been cured of HIV — will there ever be more? *Infectious Disease Clinics of North America, 33*, 857—868.

Jinek, M., Jiang, F., Taylor, D. W., Sternberg, S. H., Kaya, E., Ma, E., Anders, C., Hauer, M., Zhou, K., Lin, S., Kaplan, M., Iavarone, A. T., Charpentier, E., Nogales, E., & Doudna, J. A. (2014). Structures of Cas9 endonucleases reveal RNA-mediated conformational activation. *Science, 343*, 1247997.

Kamada, K., Shu, F., Chen, H., Malik, S., Stelzer, G., Roeder, R. G., Meisterernst, M., & Burley, S. K. (2001). Crystal structure of negative cofactor 2 recognizing the TBP-DNA transcription complex. *Cell, 106*, 71–81.

Kim, J. L., & Burley, S. K. (1994). 1.9 Å resolution refined structure of TBP recognizing the minor groove of TATAAAAG. *Nature Structural Biology, 1*, 638–653.

Kim, J. L., Nikolov, D. B., & Burley, S. K. (1993). Co-crystal structure of TBP recognizing the minor groove of a TATA element. *Nature, 365*, 520–527.

Kim, Y., Geiger, J. H., Hahn, S., & Sigler, P. B. (1993). Crystal structure of a yeast TBP/TATA-box complex. *Nature, 365*, 512–520.

Kissinger, C. R., Liu, B., Martin-Blanco, E., Kornberg, T. B., & Pabo, C. O. (1990). Crystal structure of an engrailed homeodomain-DNA complex at 2.8 Å resolution: A framework for understanding homeodomain-DNA interactions. *Cell, 63*, 579–590.

Klug, A., & Rhodes, D. (1987). Zinc fingers: A novel protein motif for nucleic acid recognition. *Trends in Biochemical Sciences, 12*, 464–469.

Klug, A. (2010). The discovery of zinc fingers and their development for practical applications in gene regulation and genome manipulation. *Quarterly Reviews of Biophysics, 43*, 1–21.

Labeots, L. A., & Weiss, M. A. (1997). Electrostatics and hydration at the homeodomain-DNA interface: Chemical probes of an interfacial water cavity. *Journal of Molecular Biology, 269*, 113–128.

Laity, J. H., Lee, B. M., & Wright, P. E. (2001). Zinc finger proteins: New insights into structural and functional diversity. *Current Opinion in Structural Biology, 11*, 39–46.

Lapinaite, A., Knott, G. J., Palumbo, C. M., Lin-Shiao, E., Richter, M. F., Zhao, K. T., Beal, P. A., Liu, D. R., & Doudna, J. A. (2020). DNA capture by a CRISPR-Cas9-guided adenine base editor. *Science, 369*, 566–571.

Laponogov, I., Sohi, M. K., Veselkov, D. A., Pan, X.-S., Sawhney, R., Thompson, A. W., McAuley, K. E., Fisher, L. M., & Sanderson, M. R. (2009). Structural insight into the quinolone-DNA cleavage complex of type IIA topoisomerases. *Nature Structural & Molecular Biology, 16*, 667–669.

Li, T., Jin, Y., Vershon, A. K., & Wolberger, C. (1998). Crystal structure of the MATa1/MATalpha2 homeodomain heterodimer in complex with DNA containing an A-tract. *Nucleic Acids Research, 26*, 5707–5718.

Li, T., Stark, M. R., Johnson, A. D., & Wolberger, C. (1995). Crystal structure of the MATa1/MAT alpha 2 homeodomain heterodimer bound to DNA. *Science, 270*, 262–269.

Lin, M., & Guo, J.-T. (2019). New insights into protein-DNA binding specificity from hydrogen bond based comparative study. *Nucleic Acids Research, 47*, 11103–11113.

Liu, J., Huang, J., Zhao, Y., Liu, H., Wang, D., Yang, J., Zhao, W., Taylor, I. A., & Peng, Y. L. (2015). Structural basis of DNA recognition by PCG2 reveals a novel DNA binding mode for winged helix-turn-helix domains. *Nucleic Acids Research, 43*, 1231–1240.

Lohse, M. B., Hernday, A. D., Fordyce, P. M., Noiman, L., Sorrells, T. R., Hanson-Smith, V., Nobile, C. J., DeRisi, J. L., & Johnson, A. D. (2013). Identification and characterization of a previously undescribed family of sequence-specific DNA-binding domains. *Proceedings of the National Academy of Sciences of the United States of America, 110*, 7660–7665.

Lu, D., Searles, M. A., & Klug, A. (2003). Crystal structure of a zinc-finger-RNA complex reveals two modes of molecular recognition. *Nature, 426*, 96–100.

Lu, X.-J., Shakked, Z., & Olson, W. K. (2000). A-form conformational motifs in ligand-bound DNA structures. *Journal of Molecular Biology, 300*, 819–840.

Luger, K., & Richmond, T. J. (1998). DNA binding within the nucleosome core. *Current Opinion in Structural Biology, 8*, 33–40.

Luger, K., Mäder, A. W., Richmond, R. K., Sargent, D. F., & Richmond, T. J. (1997). Crystal structure of the nucleosome core particle at 2.8 Å resolution. *Nature, 389*, 251–260.

Luisi, B., Xu, W. X., Otwinowski, Z., Freedman, L. P., Yamamoto, K. R., & Sigler, P. B. (1991). Crystallographic analysis of the interaction of the glucocorticoid receptor with DNA. *Nature, 352*, 497–505.

Luscombe, N. M., Austin, S. E., Berman, H. M., & Thornton, J. M. (2000). An overview of the structures of protein-DNA complexes. *Genome Biology, 1*, 1.

Luscombe, N. M., Laskowski, R. A., & Thornton, J. M. (2001). Amino acid-base interactions: A three-dimensional analysis of protein-DNA interactions at an atomic level. *Nucleic Acids Research, 29*, 2860–2874.

Mandel-Gutfreund, Y., Schueler, O., & Margalit, H. (1995). Comprehensive analysis of hydrogen bonds in regulatory protein DNA-complexes: In search of common principles. *Journal of Molecular Biology, 253*, 370–382.

Marmorstein, R., Carey, M., Ptashne, M., & Harrison, S. C. (1992). DNA recognition by GAL4: Structure of a protein-DNA complex. *Nature, 356*, 408–414.

Matthews, B. W. (1988). Protein-DNA interaction. No code for recognition. *Nature, 335*, 294–295.

Morávek, Z., Neidle, S., & Schneider, B. (2002). Protein and drug interactions in the minor groove of DNA. *Nucleic Acids Research, 30*, 1182–1191.

Murphy, F. V., Sweet, R. M., & Churchill, M. E. A. (1999). The structure of a chromosomal high mobility group protein-DNA complex reveals sequence-neutral mechanisms important for non-sequence-specific DNA recognition. *The EMBO Journal, 18*, 6610–6618.

Nadassy, K., Wodak, S. J., & Janin, J. (1999). Structural features of protein-nucleic acid recognition sites. *Biochemistry, 38*, 1999–2017.

Nair, D. T., Johnson, R. E., Prakash, S., & Aggarwal, A. K. (2005). Human DNA polymerase iota incorporates dCTP opposite template G via a G.C + Hoogsteen base pair. *Structure, 13*, 1569–1577.

Nair, D. T., Johnson, R. E., Prakash, S., Prakash, L., & Aggarwal, A. K. (2004). Replication by human DNA polymerase-ι occurs by Hoogsteen base-pairing. *Nature, 430*, 377–380.

Nair, D. T., Johnson, R. E., Prakash, L., Prakash, S., & Aggarwal, A. K. (2006). An incoming nucleotide imposes an anti to syn conformational change on the templating purine in the human DNA polymerase-ι active site. *Structure, 14*, 749–755.

Nakaseko, Y., Neuhaus, D., Klug, A., & Rhodes, D. (1992). Adjacent zinc-finger motifs in multiple zinc-finger peptides from SWI5 form structurally independent, flexibly linked domains. *Journal of Molecular Biology, 228*, 619–636.

Napoli, A. A., Lawson, C. L., Ebright, R. H., & Berman, H. M. (2006). Indirect readout of DNA sequence at the primary-kink site in the CAP-DNA complex: Recognition of pyrimidine-purine and purine-purine steps. *Journal of Molecular Biology, 357*, 173–183.

Neri, D., Billeter, M., & Wüthrich, K. (1992). Determination of the nuclear magnetic resonance solution structure of the DNA-binding domain (residues 1 to 69) of the 434 repressor and comparison with the X-ray crystal structure. *Journal of Molecular Biology, 223*, 743–767.

Neuhaus, D., Nakaseko, Y., Schwabe, J. W., & Klug, A. (1992). Solution structures of two zinc-finger domains from SWI5 obtained using two-dimensional 1H nuclear magnetic resonance spectroscopy. A zinc-finger structure with a third strand of β-sheet. *Journal of Molecular Biology, 228*, 637–651.

Nikolov, D. B., Chen, H., Halay, E. D., Usheva, A. A., Hisatake, K., Lee, D. K., Roeder, R. G., & Burley, S. K. (1995). Crystal structure of a TFIIB-TBP-TATA-element ternary complex. *Nature, 377*, 119–128.

Nikolov, D. B., Hu, S.-H., Lin, J., Gasch, A., Hoffmann, A., Horikoshi, M., Chua, N.-H., Roeder, R. G., & Burley, S. K. (1992). Crystal structure of TFIID TATA-box binding protein. *Nature, 360*, 40–46.

Ohndorf, U.-M., Rould, M. A., He, Q., Pabo, C. O., & Lippard, S. J. (1999). Basis for recognition of cisplatin-modified DNA by high-mobility-group proteins. *Nature, 399*, 708–712.

Otwinowski, Z., Schevitz, R. W., Zhang, R.-G., Lawson, C. L., Joachimiak, A., Marmorstein, R. Q., Luisi, B. F., & Sigler, P. B. (1988). Crystal structure of trp repressor/operator complex at atomic resolution. *Nature, 335*, 321–329.

Pabo, C. O., & Nekludova, L. (2000). Geometric analysis and comparison of protein-DNA interfaces: Why is there no simple code for recognition? *Journal of Molecular Biology, 301*, 597–624.

Párraga, G., Horvath, S. J., Eisen, A., Taylor, W. E., Hood, L., Young, E. T., & Klevit, R. E. (1988). Zinc-dependent structure of a single-finger domain of yeast ADR1. *Science, 241*, 1489–1492.

Passner, J. M., Ryoo, H. D., Shen, L., Mann, R. S., & Aggarwal, A. K. (1999). Structure of a DNA-bound ultrabithorax-extradenticle homeodomain complex. *Nature, 397*, 714–719.

Patikoglou, G. A., Kim, J. L., Sun, L., Yang, S.-H., Kodadek, T., & Burley, S. K. (1999). TATA element recognition by the TATA box-binding protein has been conserved throughout evolution. *Genes & Development, 13*, 3217–3230.

Papworth, M., Kolasinska, P., & Minczuk, M. (2006). Designer zinc-finger proteins and their applications. *Gene, 366*, 27–38.

Pavletich, N. P., & Pabo, C. O. (1991). Zinc finger-DNA recognition: Crystal structure of a Zif268-DNA complex at 2.1 Å. *Science, 252*, 809–817.

Pellegrini-Calace, M., & Thornton, J. M. (2005). Detecting DNA-binding helix-turn-helix structural motifs using sequence and structure information. *Nucleic Acids Research, 33*, 2129–2140.

Piper, D. E., Batchelor, A. H., Chang, C.-P., Cleary, M. L., & Wolberger, C. (1999). Structure of a HoxB1-Pbx1 heterodimer bound to DNA: Role of the hexapeptide and a fourth homeodomain helix in complex formation. *Cell, 96*, 587–597.

Ramakrishnan, V., Finch, J. T., Graziano, V., Lee, P. L., & Sweet, R. M. (1993). Crystal structure of globular domain of histone H5 and its implications for nucleosome binding. *Nature, 362*, 219–223.

Reynolds, L., Ullman, C., Moore, M., Isalan, M., West, M. J., Clapham, P., ... Choo, Y. (2003). Repression of the HIV-1 5' LTR promoter and inhibition of HIV-1 replication by using engineered zinc-finger transcription factors. *Proceedings of the National Academy of Sciences of the United States of America, 100*, 1615–1620.

Richmond, T. J., & Davey, C. A. (2003). The structure of DNA in the nucleosome core. *Nature, 423*, 145–149.

Richmond, T. J., Finch, J. T., Rushton, B., Rhodes, D., & Klug, A. (1984). Structure of the nucleosome core particle at 7 Å resolution. *Nature, 311*, 532–537.

Richter, M. F., Zhao, K. T., Eton, E., Lapinaite, A., Newby, G. A., Thuronyi, B. W., Wilson, C., Koblan, L. W., Zeng, J., Bauer, D. E., Doudna, J. A., & Liu, D. R. (2020). Phage-assisted evolution of an adenine base editor with enhanced Cas domain compatibility and activity. *Nature Biotechnology, 38*, 883–891.

Sagendorf, J. M., Markarian, N., Berman, H. M., & Rohs, R. (2020). DNAproDB: An expanded database and web-based tool for structural analysis of DNA-protein complexes. *Nucleic Acids Research, 48*, D277–D287.

Sam, M. D., Cascio, D., Johnson, R. C., & Clubb, R. T. (2004). Crystal structure of the excisionase-DNA complex from bacteriophage lambda. *Journal of Molecular Biology, 338*, 229–240.

Schalch, T., Duda, S., Sargent, D. F., & Richmond, T. J. (2005). X-ray structure of a tetranucleosome and its implications for the chromatin fibre. *Nature, 436*, 138–141.

Schultz, S. C., Shields, G. C., & Steitz, T. A. (1991). Crystal structure of a CAP-DNA complex: The DNA is bent by 90 degrees. *Science, 253*, 1001–1007.

Schwabe, J. W. R., Chapman, L., Finch, J. T., & Rhodes, D. (1993). The crystal structure of the estrogen receptor DNA-binding domain bound to DNA: How receptors discriminate between their response elements. *Cell, 75*, 567–578.

Seeman, N. C., Rosenberg, J. M., & Rich, A. (1976). Sequence-specific recognition of double helical nucleic acids by proteins. *Proceedings of the National Academy of Sciences of the United States of America, 73*, 804–808.

Selsing, E., Wells, R. D., Alden, C. J., & Arnott, S. (1979). Bent DNA: Visualization of a base-paired and stacked A-B conformational junction. *Journal of Biological Chemistry, 254*, 5417–5422.

Shakked, Z., Guzikevich-Guerstein, G., Frolow, F., Rabinovich, D., Joachimiak, A., & Sigler, P. B. (1994). Determinants of repressor/operator recognition from the structure of the trp operator binding site. *Nature, 368*, 469–473.

Sharma, D., De Falco, L., Padavattan, S., Rao, C., Geifman-Shochat, S., Liu, C. F., & Davey, C. A. (2019). PARP1 exhibits enhanced association and catalytic efficiency with gamma H2A.X-nucleosome. *Nature Communications, 10*, 5751.

Somers, W. S., & Phillips, S. E. V. (1992). Crystal structure of the met repressor-operator complex at 2.8 Å resolution reveals DNA recognition by β-strands. *Nature, 359*, 387–393.

Song, F., Chen, P., Sun, D., Wang, M., Dong, L., Liang, D., Xu, R.-M., Zhu, P., & Li, G. (2014). Cryo-EM study of the chromatin fiber reveals a double helix twisted by tetranucleosomal units. *Science, 344*, 376–380.

Staker, B. L., Feese, M. D., Cushman, M., Pommier, Y., Zembower, D., Stewart, L., & Burgin, A. G. (2005). Structures of three classes of anticancer agents bound to the human topoisomerase I-DNA covalent complex. *Journal of Medicinal Chemistry, 48*, 2336—2345.

Staker, B. L., Hjerrild, K., Feese, M. D., Behnke, C. A., Burgin, A. B., & Stewart, L. (2002). The mechanism of topoisomerase I poisoning by a camptothecin analog. *Proceedings of the National Academy of Sciences of the United States of America, 99*, 15387—15392.

Stracy, M., Schweizer, J., Sherrat, D. J., Kapanidis, A. N., Uphoff, S., & Lesterlin, C. (2021). Transient non-specific DNA binding dominates the target search of bacterial DNA-binding proteins. *Molecular Cell, 81*, 1499—1514.e6.

Suto, R. K., Edayathumangalam, R. S., White, C. L., Melander, C., Gottesfeld, J. M., Dervan, P. B., & Luger, K. (2004). Crystal structures of nucleosome core particles in complex with minor groove DNA-binding ligands. *Journal of Molecular Biology, 326*, 371—380.

Suzuki, M. (1989). SPKK, a new nucleic acid-binding unit of protein found in histone. *The EMBO Journal, 8*, 797—804.

Swinger, D. K., & Rice, P. A. (2004). IHF and HU: Flexible architects of bent DNA. *Current Opinion in Structural Biology, 14*, 28—35.

Swinger, D. K., & Rice, P. A. (2007). Structure-based analysis of HU-DNA binding. *Journal of Molecular Biology, 365*, 1005—1016.

Tan, C., & Takada, S. (2018). Dynamic and structural modelling of the specificity in protein-DNA interactions guided by binding assay and structural data. *Journal of Chemical Theory and Computation, 14*, 3877—3889.

Tan, S., & Richmond, T. J. (1998). Crystal structure of the yeast MATalpha2/MCM1/DNA ternary complex. *Nature, 391*, 660—666.

Tan, S., Hunziker, Y., Sargent, D. F., & Richmond, T. J. (1996). Crystal structure of a yeast TFIIA/TBP/DNA complex. *Nature, 381*, 127—151.

Tsunaka, T., Kajimure, N., Tate, S., & Morikawa, K. (2005). Alteration of the nucleosomal DNA path in the crystal structure of a human nucleosome core particle. *Nucleic Acids Research, 33*, 3424—3434.

Urnov, F. D., Miller, J. C., Lee, Y.-L., Beausejour, C. M., Rock, J. M., Augustus, S., … Holmes, M. C. (2005). Highly efficient endogenous human gene correction using designed zinc-finger nucleases. *Nature, 435*, 646—651.

Van den Broeck, A., Lotz, C., Ortiz, J., & Lamour, V. (2019). Cryo-EM structure of the complete *E. coli* DNA gyrase nucleoprotein complex. *Nature Communications, 10*, 4935.

Weir, H. M., Kraulis, P. J., Hill, C. S., Raine, A. R. C., Laue, E. D., & Thomas, J. O. (1993). Structure of the HMG box motif in the B-domain of HMG1. *The EMBO Journal, 12*, 1311—1319.

Werner, M. H., & Burley, S. K. (1997). Architectural transcription factors: Proteins that remodel DNA. *Cell, 88*, 733—736.

Werner, M. H., Huth, J. R., Gronenborn, A. M., & Clore, G. M. (1995). Molecular basis of human 46X,Y sex reversal revealed from the three-dimensional solution structure of the human SRY-DNA complex. *Cell, 81*, 705—714.

Wojciak, J. M., Iwahara, J., & Clubb, R. T. (2001). The Mu repressor-DNA complex contains an immobilized "wing" within the minor groove. *Nature Structural Biology, 8*, 84—90.

Wolberger, C. (2021). How structural biology transformed studies of transcription regulation. *Journal of Biological Chemistry, 296*, 100741.

Wolberger, C., Vershon, A. K., Liu, B., Johnson, A. D., & Pabo, C. O. (1991). Crystal structure of a MAT alpha 2 homeodomain-operator complex suggests a general model for homeodomain-DNA interactions. *Cell, 67*, 517—528.

Wolfe, S. A., Nekludova, L., & Pabo, C. O. (2000). DNA recognition by Cys2His2 zinc finger proteins. *Annual Review of Biophysics, 29*, 183—212.

Wolfe, S. A., Ramm, E. I., & Pabo, C. O. (2000). Combining structure-based design with phage display to create new Cys(2)His(2) zinc finger dimmers. *Structure, 8*, 739—750.

Yang, W., & Steitz, T. A. (1995). Crystal structure of the site-specific recombinase γδ-resolvase complexed with a 34 bp cleavage site. *Cell, 82*, 193—207.

Zhao, Q., Chasse, S. A., Devarakonda, S., Sierk, M. L., Ahvazi, B., & Rastinejad, F. (2000). Structural basis of RXR-DNA interactions. *Journal of Molecular Biology, 296*, 509—520.

Zhu, P., & Li, G. (2016a). Structural insights of nucleosome and 30-nm chromatin fibre. *Current Opinion in Structural Biology, 26*, 106–115.

Zhu, P., & Li, G. (2016b). Higher-order structure of the 30-nm chromatin fiber revealed by cryo-EM. *IUBMB Life, 68*, 873–878.

Further reading

Branden, C. I., & Tooze, J. (1999). *Introduction to protein structure* (2nd ed.). Garland Press.

Freemont, P. S., Lane, A. N., & Sanderson, M. R. (1991). Structural aspects of protein-DNA recognition. *Biochemical Journal, 278*, 1–23.

Gardini, S., Furini, S., Santucci, A., & Niccolai, N. (2017). A structural bioinformatics investigation on protein-DNA complexes delineates their modes of interaction. *Molecular BioSystems, 13*, 1010–1017.

Harrison, S. C., & Aggarwal, A. K. (1990). DNA recognition by proteins with the helix-turn-helix motif. *Annual Review of Biochemistry, 59*, 933–969.

Hitomi, K., Iwai, S., & Tainer, J. A. (2007). The intricate structural chemistry of base excision repair machinery: Implications for DNA damage recognition, removal, and repair. *DNA Repair, 6*, 410–428.

Joyce, A. P., Zhang, C., & Havranek, J. J. (2014). Structure-based modelling of protein: DNA specificity. *Briefings in Functional Genomics, 14*, 39–49.

Kulandaisamy, A., Srivastava, A., Nagarajan, R., & Gromiha, M. M. (2017). Dissecting and analyzing key residues in protein-DNA complexes. *Journal of Molecular Recognition, 31*, 1–10.

Lambert, S. A., Jolma, A., Campitelli, L. F., Das, P. K., Yin, Y., Albu, M., Chen, X., Taipale, J., Hughes, T. R., & Weirauch, M. T. (2018). The human transcription factors. *Cell, 172*, 650–665.

Liljas, A., Liljas, L., Ash, M.-R., Lindblom, G., Nissen, P., & Kjeldgaard, M. (2018). In *Textbook of structural biology* (2nd ed.). World Scientific.

Norambuena, T., & Melo, F. (2010). The protein-DNA Interface database. *BMC Bioinformatics, 11*, 262.

Pabo, C. O., & Sauer, R. T. (1992). Transcription factors: Structural families and principles of DNA recognition. *Annual Review of Biochemistry, 61*, 1053–1095.

Patikoglou, G., & Burley, S. K. (1997). Eukaryotic transcription factor-DNA complexes. *Annual Review of Biophysics, 26*, 289–325.

Qiu, J., Bernhofer, M., Heinzinger, M., Kemper, S., Norambuena, T., Melo, F., & Rost, B. (2020). ProNA2020 predicts protein-DNA, protein-RNA, and protein-protein binding proteins and residues from sequence. *Journal of Molecular Biology, 432*, 2428–2443.

Redinbo, J. R., Champoux, J. J., & Hol, G. J. W. (1999). Structural insights into the function of type IB topoisomerases. *Current Opinion in Structural Biology, 9*, 29–36.

Sarai, A., & Kono, H. (2005). Protein-DNA recognition patterns and predictions. *Annual Review of Biophysics and Biomolecular Structure, 34*, 379–398.

Schneider, B., Černý, J., Svozil, D., Čech, P., Gelly, J.-C., & de Brevern, A. G. (2014). Bioinformatic analysis of the protein/DNA interface. *Nucleic Acids Research, 42*, 3381–3394.

Steitz, T. A. (1990). Structural studies of protein-nucleic acid interaction: The sources of sequence-specific binding. *Quarterly Reviews of Biophysics, 23*, 205–280.

Travers, A. A., & Buckle, M. (Eds.). (2000). *DNA-protein interactions, a practical approach*. Oxford: Oxford University Press.

Useful web sites

https://www.dnaprodb.usc.edu/ Provides tools for comprehensive structural analysis of DNA protein complexes (Sagendorf et al., 2020).

https://www.npidb.belozersky.msu.ru/ A database of DNA-protein structures. As of 22nd April 2021, a total of 8410 structures have been included, together with several analytical tools.

https://www.elixir-europe.org/communities/3d-bioinfo The portal of the ELIXIR 3DBioInfo Community, for validating structural data and developing various tools for working with them.

CHAPTER 8

RNA binding proteins

Full justice cannot be done in this chapter to the very large area of RNA binding proteins, and only specific aspects will be highlighted. The reader is referred to the bibliography at the end of this chapter for books and articles covering this rapidly expanding field. The high-resolution structural biology of this field started with the structural elucidation of the tRNA synthetase—tRNA complexes that for glutamine (Rould et al., 1989) and this was followed by the structural elucidation of the replicative enzymes which use RNA as a template, such as the HIV reverse transcriptase (RT) (Jacobo-Molina et al., 1991; Kohlstaedt et al., 1992), Polio virus polymerase (Hansen et al., 1997) and other RNA viral polymerases. This was followed by the structure of the first structure of a small nuclear ribonucleoprotein (snRP), U1A complexed with RNA (Oubridge et al., 1994) and later by the high-resolution structures of the ribosome and more detailed structures of the spliceosome.

8.1 Aminoacyl synthetase—tRNA complexes

Aminoacyl tRNA synthetases (aaRSs) play a central role in the translation of the genetic message encoded in mRNA into protein. They are fascinating enzymes since they precisely select from the cellular pool, members of two classes of very chemically diverse molecules, amino acids and tRNAs, bringing them together and joining them in an acylation reaction. aaRSs are present in all organisms usually in 20 different varieties within the cell, one for each of the 20 amino acids. They charge the cognate tRNAs with their correct amino acids and this has to occur with very high accuracy, since if this goes awry, corrections cannot be made downstream at the ribosomal stage. Moreover, the aaRSs have to be able to identify the correct cognate tRNA from a pool of cognate and non-cognate tRNAs within the cell and charge it with the appropriate amino acid. Hence, they have an editing and filter system, so that incorrectly acylated tRNAs are hydrolysed and correctly acylated once released into the cytosol. Protein sequence and structural analysis now show that the aaRSs fall into two classes of synthetases, Class I and Class II, specific for 10 amino acids each. These categories are further sub-divided into three subclasses as shown in Table 8.1. Class I are usually monomeric enzymes and Class II tend to be multimeric (dimers or tetramers).

Co-crystallisation of tRNA synthetase—tRNA complexes had been hampered, as had tRNA crystallisation initially by the purity of the tRNAs. In the case of the tRNA structure solution, obtaining highly diffracting crystals was alleviated by the selection of a suitable organism as a source and by carefully exploiting a many column chromatography

Principles of Nucleic Acid Structure
ISBN 978-0-12-819677-9, https://doi.org/10.1016/B978-0-12-819677-9.00006-8

Table 8.1 Classification of the tRNA synthetases.

Class of aaRS	Class I	Class II
Sub-class	A. C, I, L, M, R B. E, K, Q C. W, Y	A, G, H, P, S, T D, K, N, F
Fold of ATP domain	Rossmann fold with parallel β sheets	Anti-parallel β sheet
Location of aminoacylation on the 3' CCA stem	2'-OH	3'-OH
Amino acid binding	On the surface	In a deep pocket
Conformation of tRNA acceptor end	Bent	Straight

purification. This was executed successfully in Rich and Klug's groups who successfully solved yeast tRNA^{phe} (Kim et al., 1974; Robertus et al., 1974). They decided to isolate phenylalanine—tRNA from yeast as a source rather than using *Escherichia coli (E. coli)*, since it was possible to obtain tRNA to much greater purity from yeast. *E. coli* was used by many groups, some of whom did obtain crystals, but which did not diffract to a required resolution of 3 Å and above. The purity of the tRNA was greatly enhanced by the development of tRNA cloning (Perona et al., 1988; Swanson et al., 1988) and by the use of anion exchange HPLC in the purification. This development of having extremely pure tRNA synthetase from an over-expressing clone and very pure tRNA led to the first successful co-crystallisation and structure solution of glutaminyl tRNA synthetase—tRNA^{gln} complex from data collected from highly diffracting crystals (Rould et al., 1989). Subsequently, many tRNA synthetase—tRNA complexes were solved by the Strasbourg group headed by Dino Moras, Richard Giegé and Jean-Pierre Ebel, the Grenoble group of Stephen Cusack, the consortium at RIKEN headed by S. Yokoyama, Charlie Carter's group at the University of North Carolina and by other groups. There has been great interest in using LysRS as an anti-bacterial drug target and many structures of LysRSs have been solved by the Seattle Consortium from human pathogens such as *Cryptosporidium parvum*, *Mycobacterium ulcerans* and *Chlamydia trachomatis*.

tRNA synthetases catalyse a two-step reaction. The first stage is the binding of the amino acid and activation with ATP and the second the acylation of the tRNA by the activated amino acid.

$$aa + ATP \rightarrow aa\text{-acyl-AMP} + PP_i$$

$$aa\text{-acyl-AMP} + tRNA^{aa} \rightarrow aa\text{-acyl-}tRNA^{aa} + AMP$$

giving an overall 'charging' reaction

$$aa + ATP + tRNA^{aa} \rightarrow aa\text{-acyl-}tRNA^{aa} + AMP + PPi$$

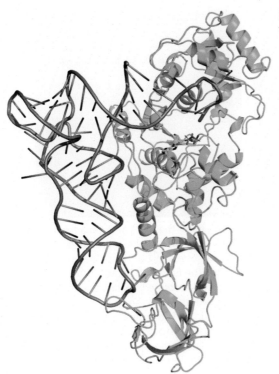

Figure 8.1 Structure of the glutaminyl synthetase—tRNA complex (PDB id 1GTR).

The structure of the glutaminyl synthetase—RNA^{gln} complex which is a member of Class I is illustrated in Fig. 8.1. The tRNA interacts with the tRNA synthetase along one side of the L and the 3′ CCA acceptor stem makes a hairpin turn to the inside of the L. At the other end, the anticodon adopts a conformation not seen in free tRNA. A number of key features are present in the complex, notably the enzyme contacting the 2-amino group of guanine via the tRNA minor group in the acceptor stem at G2 and G3. There are a range of interactions between the enzyme and the anticodon nucleotides. G73 and U1·A72 of the cognate tRNA adopt a conformation which is stabilised by protein at a lower free energy cost compared with non-cognate sequences.

Looking in more detail at the structure of the tRNA as compared to free tRNA, it is apparent that base pairs between 1 and 72 are disrupted and the 3′ CCA end of the tRNA forms a hairpin in the direction of the anticodon. Bases A76, C75 and G73 are stacked on each other and C74 is looped out and stabilised in a protein pocket (Fig. 8.2a and b). This alternate structure is stabilised by a hydrogen bond from G73 to the phosphate of A72 and by numerous interactions of the sugar phosphate backbone with protein and the 3′ end in the ATP and glutamine binding site. Other features are that the tRNA^{gln} base U35 is stacked beneath A37 and C34, with C36 unstacked and interacting with

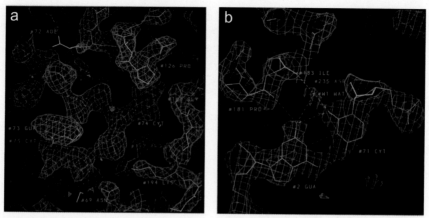

Figure 8.2 Interaction between the bases of the 3′ CCA end of the tRNA and glutaminyl tRNA synthetase amino acid residues. (a) A section through the 2.8 Å resolution solvent flattened, MIR electron density map in the region of the hairpin acceptor strand, contoured at 2 σ. The protein model fitted to the map is coloured yellow and the tRNA model is orange. The base of C74, with its exocyclic atoms clearly visible, is looped out and fits snugly into a binding pocket formed by the enzyme. The bases of G73, C75 and A76 are seen to be stacked on each other with the 2-amino group of G73 well within bonding distance to the phosphate group of A72. (b) Electron density from the solvent flattened, MIR map around base pair G2•C71, contoured at 2 σ. A well-ordered buried water molecule (Wi WAT) is bound to the protein backbone amide of residue 183 and the carboxylate side chain of Asp235 (in the background). The remaining hydrogen bond donor and acceptor from this water molecule together with the backbone carbonyl oxygen of Pro81 form a hydrogen-bonding surface that is specifically complementary to the G2•C71 base pair in the minor groove. *(Figure and legend reproduced with permission from Rould, M. A., Perona, J. J., Söll, D., & Steitz, T. A. (1989). Structure of E. coli glutaminyl-tRNA synthetase complexed with tRNA(Gln) and ATP at 2.8 A resolution. Science, 246, 1135–1142.)*

protein. The overall differences in conformation of the tRNA^{gln} compared with free tRNA^{phe} were clearly seen in the least-squares fit between these two structures with distinct differences in the area of stem-loop. Within the structure of the enzyme, ATP is bound into a Rossmann (or dinucleotide) fold consisting of β-sheets and α-helices (Table 8.2) and the glutamine in a neighbouring pocket.

An example of the Class II enzyme is shown for Aspartyl tRNA synthetase–tRNA interaction (Fig. 8.3), where the complex is dimeric (1ASY) (Ruff et al., 1991).

In general, the aaRS selects the tRNA on the basis of interactions with the region around the 3′ CCA terminus and the anticodon and stem. In contrast for some amino acids, this is not the case; for Ser and Leu, where there are six different codons, for Leu the second base is U in the anticodon, but for Ser all these positions can be different. Instead of relying on the sensing of the anticodon, a different mechanism is used by SerRS, exploiting an extended variable loop and a long hairpin region in the SerRS which is disordered in the absence of tRNA and becomes ordered and nestles between the TΨC loop and long variable region (Biou et al., 1994), PDB id 1SER (Fig. 8.4).

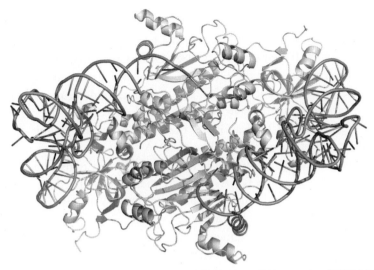

Figure 8.3 Structure of the aspartyl tRNA synthetase–tRNA complex (PDB id 1ASY).

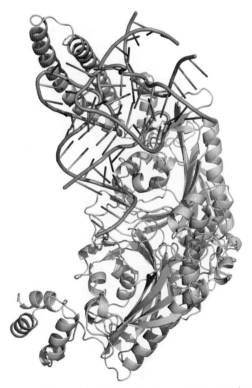

Figure 8.4 Structure of the seryl tRNA synthetase–tRNA complex (PDB id 1SER).

8.2 Structure of reverse transcriptases

Following hard on the structure solution of the first DNA polymerases and polymerase complexes with DNA (Freemont et al., 1988; Ollis et al., 1985; Steitz et al., 1987) was the structure determination of the HIV RT. There was a medical imperative that this target for nucleoside drugs, such as AZT and non-nucleoside drugs such as nevirapine, had to be solved ever since the start of the outbreak of AIDS in the early 1980s. HIV has an RNA genome and replication by the RT produces a double-stranded DNA from reverse transcription of RNA. The dsDNA is then integrated into the human genome by the HIV integrase, where it resides and is then transcribed by the human RNA polymerase to produce RNA which is then encapsulated in coat proteins to make new virus that then buds out of the human cells. HIV RT is expressed as a homo-dimer of two subunits of 66,000 Da molecular weight. One of these subunits is proteo-lytically cleaved to a 51,000 Da fragment (p51) to create a p66/p51 heterodimer. Due to the large domain rearrangement that occurs in the p51 subunit, computational models of the heterodimer calculated at the start were incorrect. Crystallisation of well-diffracting crystals was highly problematic (protein even went up twice on the Space Shuttle in an attempt to crystallise it under zero gravity conditions without convection), but after 10 years of crystallisation trials, success was finally achieved on Earth by two groups. One group locked down a mobile domain by co-crystallising with nevirapine (Kohlstaedt et al., 1992) (Fig. 8.5) and another group by complexing the enzyme with a DNA sub-strate (Jacobo-Molina et al., 1991) (Fig. 8.6) and a monoclonal antibody. The Harrison group also solved a HIV RT structure in an unliganded form (Rodgers et al.,1995). The

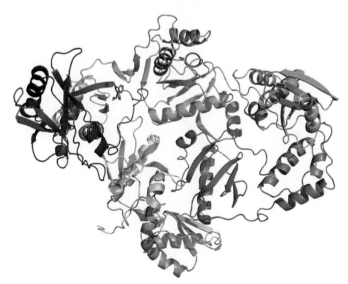

Figure 8.5 Complex of HIV reverse transcriptase (RT) with nevirapine (PDB id 3HVT).

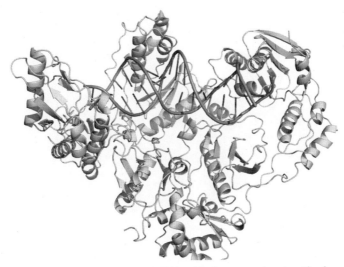

Figure 8.6 Early complex between HIV RT and DNA, with the monoclonal antibody omitted for clarity (PDB id 2HMI).

structural configuration of the *E. coli* Klenow fragment (Joyce & Steitz, 1994) may be described as a right hand with a thumb (a cluster of α-helices), fingers (α-helices) and palm region (a sheet formed of β-sheet and α-helices). This spatial arrangement of thumb, fingers and palm also holds for RT even though there is low sequence similarity between DNA Klenow fragment and RT due to divergence over the long evolutionary period separating these two enzymes; the structural framework is still maintained. RT triple drug targeting against both RT (using two drugs which may be either two nucleoside inhibitors or a nucleoside and a non-nucleoside inhibitor such as nevirapine) and the HIV protease (using protease inhibitors) forms the backbone of highly successful HAART 'Highly Active Anti-retroviral therapy' which has considerably extended the life expectancy of AIDS sufferers.

In order to more deeply understand the mechanism of binding of retroviral drugs to the nucleic acid duplex RT complex, knowledge of the structure of a drug-long duplex nucleic acid—RT complex is essential. Many attempts at obtaining this have been thwarted by the low binding constant of DNA for RT. Using elegant enzymology experiments and tethering a long DNA double helix to the RT in this complex, with TTP in the nucleotide binding site, a successful structure solution was achieved (Huang et al., 1998) by Harrison's group. More recently an RT—DNA complex with bound monophosphorylated AZT (Fig. 8.7, PDB id 3V6D) has been solved by Arnold's group. Development of new RT drugs (Kang et al., 2019) is paramount as resistance may re-emerge. Pharmaceutical companies use very high-resolution X-ray crystallography as a

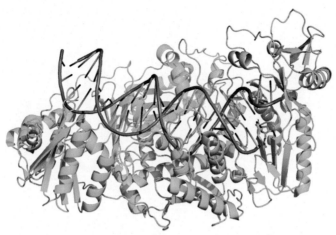

Figure 8.7 HIV RT in complex with DNA and AZT bound in the nucleotide binding site (PDB id 3V6D).

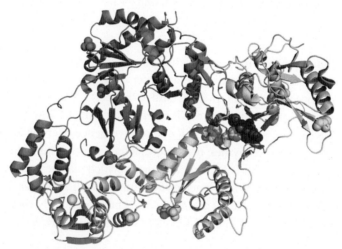

Figure 8.8 HIV RT in complex with a novel non-nucleoside anti-AIDS drug, 4-[(4-{4-[(E)-2-cyanoethenyl]-2,6-dimethylphenoxy}thieno[3,2-d]pyrimidin-2-yl)amino]-2-fluorobenzonitrile (PDB id 6UL5).

cornerstone in rational drug design programs for developing new drugs. A new non-nucleoside RT inhibitor bound to RT is illustrated in Fig. 8.8 (Kang et al., 2020).

8.3 snRPs (small nuclear ribonucleoproteins)

The mid-1970s witnessed the stunning discovery that in contrast to prokaryotic transcription where an uninterrupted transcript is translated into protein, in eukaryotes including *homo sapiens*, long mRNA transcripts are made up of exons and introns with only the exon regions being expressed, whereas the introns are excised (Berget et al., 1977, 1978; Breathnach et al., 1977; Broker et al., 1978; Chambon, 1978; Chow et al., 1977). How these introns are excised and the introns are joined was a puzzle for which the first evidence of the mechanism and the molecules involved came initially from the serendipitous study of the antibodies from patients suffering from systemic lupus erythematosus, within a laboratory deeply involved in researching the role of small RNAs (Lerner & Steitz, 1979; Lerner et al., 1980). They found that these antibodies would recognise snRPs involved in the splicing reaction and allow them to be identified. Fig. 8.9 shows the overall mechanism of how snRPs excise the introns. The first stage is the binding of the U1 snRP to the 5′ splice site of the intervening sequence, between upstream exon 1 and the intron, which has the sequence r(GU). The border between the intron and downstream exon 2 has a 3′ splice site with the sequence r(AG). At between 15 and 45 nucleotides upstream of the 3′ splice site is a sequence known as the 'branch point sequence' which contains an adenine which is all important in the subsequent reaction. This is often embedded in a sequence of the form YNCURAY found in mammals where Y is a pyrimidine base, R a purine and N is any base. U2 snRP binds to the A of the branch point sequence. Once the U1 and U2 recognise their DNA binding sites, then there they are joined by the so called tri-snRP consisting of snRP proteins U4, U5 and U6 which binds and brings together U1 and U2 and in the process loops the intron DNA to produce an active spliceosome. The 5′ join between the upstream exon (exon 1) and the intron is cleaved and the 5′ G of the GU sequence attacks the A of the branch point sequence to create a lariat (like a cowboy's lasso). At this point, U4 leaves and upstream exon1 covalently bonds to the downstream one exon2 completing the reaction. There is now a wealth of structural detail of the stages of the mechanism: in this section, full coverage cannot be given, but some excellent recent reviews have been written. The focus here will be on only two of the complexes. Structural information was first obtained by X-ray crystallography for U1A and then subsequently using the same structural technique for U1A with RNA. Due to the difficulty in crystallising these complexes, cryogenic electron microscopy (cryo-EM) is now the method of choice for solving their structures.

The first stage in pre-RNA processing involves the binding of U1 protein to the 5′ end and U2 binding to the branch point as shown in Fig. 8.9. Human snRP U1 consists of two proteins U1 70K and U1A proteins and they contain the RNP motif. U1 70K binds to stem-loops I and U1A protein to stem-loop II.

The structure of the N-terminal domain of the U1A snRP protein, which along with U1 70K protein is a constituent part of U1 snRP, gave the first structural information on a snRP to high resolution (2.8 Å) (Nagai et al., 1990). The structure is a four-stranded β-sheet with a pair of α-helices on the same side of the sheet (Fig. 8.10).

There are two consensus regions in common in U1A and U2B2″ (a constituent of U2 snRP which binds to stem-loop IV of U2 in the presence of the other component U2A′ protein) as shown in the sequence alignment data which can be mapped onto the region

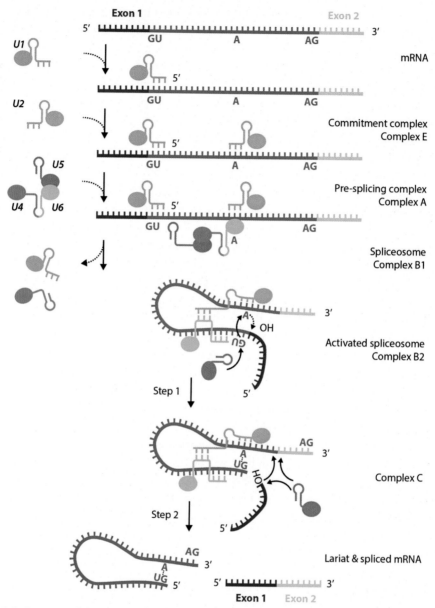

Figure 8.9 Overview of the action of spliceosome in the removal of introns, detailing the snRNAs and snRNPs involved. *(Figure designed by A.V. Berglar, PhD (Scientific Visualisation Lab, www.scivislab.com).)*

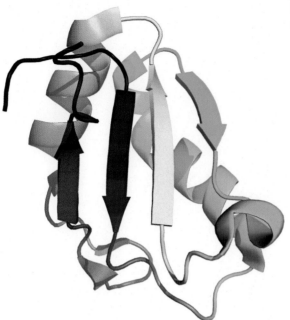

Figure 8.10 Structure of UA1 snRP, domain 1-95 (PDB id 1OIA).

of β-sheet. From U1 RNA binding studies on mutant proteins, it was possible to map onto the structure the key charged residues forming the U1A binding surface for stem-loop II of U1 RNA. It took 4 years of very hard work, screening different RNAs and different crystallisation conditions by the same group in order to crystallise the U1A protein with RNA and obtain very well diffracting crystals (1.92 Å) (Oubridge et al., 1994). U1A protein was co-crystallised, complexed to a 21-mer RNA which has same sequence as the U1 snRP for the loop region and all but one base at the base of the stem. The sequence was designed to have an overhanging U at the base (Fig. 8.11). There were three very similar molecules in the asymmetric unit with almost identical structure. Fig. 8.11 shows the overall structure of one of the monomers. The loop of the RNA extends at close to 90° from the stem with the bases splayed out and forming interactions with a large area of the four-stranded β-sheet, which was first observed in the structure of native U1A. Previous structure prediction of how the RNA may bind based on the earlier U1A structure was found not to be correct in detail: the β2—β3 protein loop sticks through the RNA loop at a position close to the stem and prevents base pairing within the loop (Fig. 8.11). Lysine and arginine residues bind the RNA loop through interaction with the negatively charged phosphate backbone.

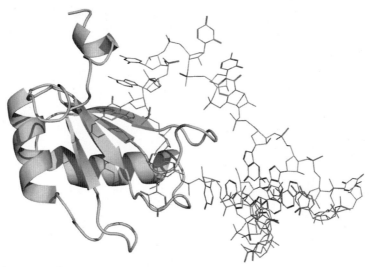

Figure 8.11 Interaction between human U1A snRP and hairpin II of U1A snRNA, bound chloride ion and glycerol also shown (PDB id 1URN).

On account of the length of time it took to go from the initial structure of the protein on its own to that of the protein–RNA complex and the difficulty in obtaining well-diffracting crystals for parts of the spliceosome, cryo-EM has become the method of choice for solving the structure of regions of the spliceosome. Initially, there was a 7 Å cryo-EM structure of the human tri-snRP consisting of the snRPs U4,U5 and U6 and the constituent snRNAs (Agafonov et al., 2016) which allowed fitting of the proteins and the RNA and some very interesting biological conclusions to be drawn. When the resolution was increased to 2.9 Å in the cryo-EM structure determination of complex of the tri-snRP with U1 and U2 bound to RNA 5′ splice site, the pre-B spliceosome complex, fine details in the lower resolution structure were found to be incorrect in regions of the RNA trace. This pre-B spliceosome is a very intricate structure of inter-linking folded RNAs and bound proteins (Charenton et al., 2019) giving a wealth of detail of the protein, RNA interactions and RNA–RNA associations. Since the resolution is above the base-pair separation in an RNA helix, separate bases and backbone can be unambiguously positioned, and at this resolution, side chains of amino acids can be placed with certainty. Fig. 8.12 shows views within the complex of the tri-snRP and a front and side view of the pre-B complex with the proteins and RNAs indicated. For further details, the reader is encouraged to read the original publication (Charenton et al., 2019), view the associated films (movies) and read the recent reviews from Kiyoshi Nagai's group (Wilkinson et al., 2020) and also that from Reinhard Lehrmann and Holger Stark's groups (Kastner et al., 2019).

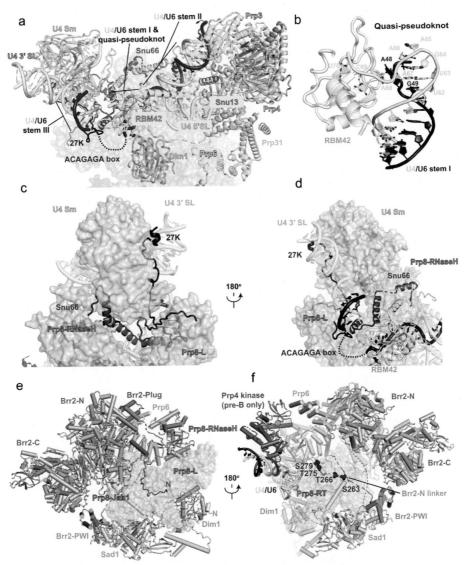

Figure 8.12 Structure of the human tri-snRP (PDB id 6QW6) viewed from different orientations to focus on detailed RNA—RNA and RNA—protein interactions. Structural features of the human tri-snRNP. (a) Organisation of U4 snRNA in the tri-snRNP before Brr2 relocation. SL, stem-loop. The mobile ACAGAGA loop is depicted as a red dashed line. (b) Magnified view of U4/U6 stem I capped by the quasi-pseudoknot and its stabilisation by RBM42 RRM. (c and d) The U4 region in two orientations. Snu66 and SNRNP-27K wrap around the U4 Sm ring, the Prp8 Endo and RNase H domains, U4/U6 stem III, and the quasi-pseudoknot, thereby solidifying the organisation of the U4 region before Brr2 relocation. (e) Complex interaction network of the Brr2 N-terminal domain with tri-snRNP components. Brr2-C, C-terminal helicase cassette of Brr2; Brr2-N, N-terminal helicase cassette of Brr2; Prp8-L, large domain of Prp8. (f) Another view of the complex interaction network of the Brr2 N-terminal domain with tri-snRNP components. The extended N-terminal domain of Prp6 is apparent. Shown are the Prp4 kinase (active site is represented as an orange circle) as it interacts in pre-B and its phosphorylation targets on Prp6 (magenta spheres) (Charenton et al., 2019).

8.4 The ribosome

In this section, the role of proteins within the ribosome will be described, complementing the section in Chapter 6 where the focus is on the role of RNA within the ribosome. The ribosome is a complex playing a central role in all prokaryotic and eukaryotic cells and translates the information contained within messenger RNA into protein. For prokaryotes, translation may be divided into three stages 1) Initiation, 2) Elongation and 3) Termination. During 1) Initiation, mRNA is bound to the 30S subunit along with the protein initiator factor 3, IF3 (Fig. 8.13). The role of IF3 is to prevent the 50S subunit prematurely associating with the 30S and mRNA, hence making sure that the assembly of the ribosomal complex is sequentially orchestrated. It also guides the initiator tRNA into the P site. A pyrimidine-rich 3' region of the 16S RNA of the 30S subunit known as the Shine−Dalgarno sequence base-pairs with a purine-rich 5' region of the leader of the translated mRNA ensuring the correct positioning of the AUG start codon on the ribosome. IF1 then binds in the A site abrogating fMet-tRNA incorrectly binding to the A site and assuring its correct binding to the P site (Fig. 8.16). fMet-tRNA forms a ternary complex with IF2 and GTP and binds to the AUG start codon. GTP is hydrolysed to GDP and the initiator factors IF1, IF2 and IF3 dissociate and the 50S subunit binds to the 30S complex to form the 70S ribosomal initiation complex. The 70S complex has fMet-tRNA correctly positioned in the P site and the A and E sites vacant (Fig. 8.13).

The 30 and 50S subunits form the 70S ribosome which has a molecular weight of around 2.6 MDa (in the following text, the abbreviations 30S, 50S and 70S have been used to stand for 30S subunit, 50S subunit and 70S subunit to avoid continuous repetition of 'subunit' and similarly for the eukaryotic 40S, 60S and 80S subunits). The 30S consists of a 16S ribosomal RNA (1500 nucleotides in length) and around 20 proteins, the 50S consists of a 23S ribosomal RNA (3000 nucleotides in length), a 5S RNA (120 nucleotides) and around 30 proteins. The biochemistry of translation was worked out in the 1960 and 1970s in many laboratories, resulting in an understanding of the steps and factors involved in the prokaryotic and eukaryotic ribosomal cycles. During this period, many cross-linking experiments were also conducted on the prokaryotic systems in order to attempt to define which regions of the ribosomal proteins are close to regions of the ribosomal RNA (Smith, 1976 and references therein).

In order to gain a deeper understanding of the mechanism of operation of the ribosome, it has been intensely studied by structural techniques for 60 years, starting with lower resolution EM studies, neutron diffraction studies and culminating with the successful high resolution X-ray crystallographic studies, and now with the advent of the CMOS, direct electron detectors (Henderson, 2015, 2018), high-resolution cryo-electron microscopy (cryo-EM) studies.

Prior to the advent of high-resolution cryo-EM, the structure solution of the ribosome has naturally been dependent on the availability of well-diffracting crystals of the prokaryotic ribosomal units 30S and 50S and intact 70S. Negative stain EM studies first by Breck Byers at Harvard using cooled chicken embryos and then by Unwin's group

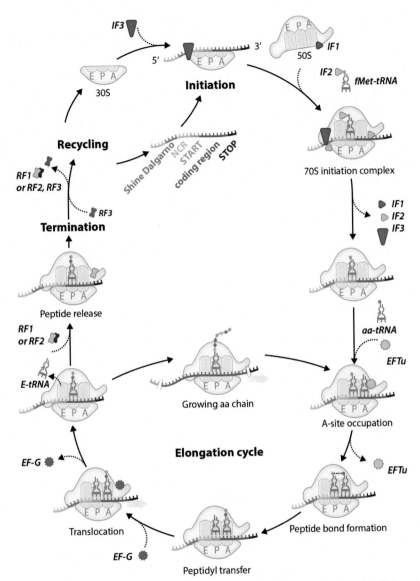

Figure 8.13 The prokaryotic ribosomal cycle. *(Figure designed by A.V. Berglar, PhD (Scientific Visualisation Lab, www.scivislab.com).)*

(Kühlbrandt & Unwin, 1982) on ribosomes from lizard oocytes had shown that from both of these systems ordered 2D ordered arrays were formed, implying that crystallisation of ribosomes in 3D crystalline arrays may be achievable. Unwin and Kühlbrandt also determined the RNA and protein distribution in these eukaryotic ribosomes, their attachment to the endoplasmic reticulum and later that there was a protein exit tunnel on the opposite side of the ribosome to the decoding region, supporting Jim Lake's EM finding using an antibody raised against a protein exiting through this tunnel.

EM structural studies by Jim Lake's group at UCLA using antibodies and negative stain had clearly defined the outline of the ribosomal subunits and the placing of a number of ribosomal proteins through the use of antibody labelling as well as determining the position of the protein exit tunnel. Attempts at ribosome crystallisation of 3D crystals date back to the late 1960s and early 1970s, when a biochemistry group purified large amounts of purified 70S *E. coli* ribosomes and supplied them to an X-ray crystallographer at King's College London. Growth of diffracting crystals at this stage was not successful (the late Prof. Henry Arnstein, personal communication). Not until 40 years later were 70S *E. coli* ribosomes successfully crystallised and their structures solved by Jamie Cate's group at Berkeley (Schuwirth et al., 2005). Whilst many researchers left this X-ray crystallographic area assuming the challenge to be too difficult, Ada Yonath from the Weizmann Institute in collaboration with Heinz-Günter Wittmann's large ribosomal research group in Berlin persevered over many, many years in the quest to obtain crystals, which diffracted to around 3 Å and above. Initially, crystals were obtained from *Bacillus stearothermophilus* 50S and 30S ribosomal subunits which diffracted to below 3 Å resolution (Glotz et al., 1987; Yonath et al., 1983), then finally the Yonath and Wittmann's groups obtained crystals of 50S which diffracted to 3 Å from ribosomes isolated from the Dead Sea bacteria *Haloarcula marismortui* (von Böhlen et al., 1991). In Marina Garber's group (which included Marat Yusupov, Gulnara Yusupova, Vladimir Barynin and Sergei Trakhanov) working in Alexander Spirin's Protein Institute in Pushchino, crystals were grown of the 30 and 70S from *Thermus thermophilus* (Trakhanov et al., 1987). Ada Yonath and Heinz-Günter Wittmann's groups were also able to crystallise these subunits from *T. thermophilus* (Volkmann et al., 1990; Yonath et al., 1988).

8.4.1 The ribosomal subunits

8.4.1.1 The 30S subunit

The crystal structure of 30S consisting of 16S RNA and 20 proteins from *T. thermophilus* at 5.5 Å resolution was solved by Venki Ramakrishnan's group (Clemons et al., 1999). They had a preparation of 30S from which they separated the S1 using column chromatography. The electron density map was very well phased, which allowed them to place many of the proteins for this subunit for which there were independently solved ribosomal protein structures and they were also able to identify, and also see clearly the long helix H44 and many segments of A-form RNA structure which could be placed in the electron density map. Structures at 3 Å (Wimberly et al., 2000) and 3.3 Å (Schluenzen et al., 2000) then followed, determined by Ramakrishnan's and the Yonath and Franceschi's (Wittmann successor) groups from *T. thermophilus*. The structure revealed the overall architecture of the 30S subunit which were designated the Head H, Beak Be, Neck N Platform P, Shoulder Sh, Spur Sp and Body Bo domains described and illustrated in Chapter 6 with tRNA bound to the 30S subunit.

8.4.1.2 The 50S subunit

Phasing the structure of the 50S from the X-ray diffraction data for the ribosome from *H. marismortui* using native and derivative crystals proved to be far from easy. Due to its very high molecular weight, heavy atom clusters were employed, in contrast to compounds with single heavy atoms used for solving proteins. The problem was that these crystals suffered from crystal twinning, dependent on their salt concentration and this had to be resolved before Patterson maps could be correctly solved and the structure elucidated. Also due to radiation sensitivity, the crystals had to be cryo-cooled to liquid nitrogen temperature for successful data collection as had the 30S subunit. It was found that for the first 50S crystallisations, two morphologically identical crystal forms existed with very similar cell dimensions, one of which was twinned. This twinning problem was resolved by the Yale group, so that meaningful phases could be calculated to produce an interpretable electron density map. The 50S structure was initially solved at 9 Å (Ban et al., 1998) by Tom Steitz's and Peter Moore's groups and at successively higher resolution (5 Å) (Ban et al., 1999) and finally (2.4 Å) (Ban et al., 2000) resolution using diffraction data collected at the Brookhaven synchrotron, with the highest resolution data being collected at the APS synchrotron at the Argonne National Laboratory.

The high-resolution 50S structure showed starkly that this ribosomal subunit consists of large coils of RNA with proteins embedded in the surface and distributed evenly over the surface. It is a very large particle, approximately 66% RNA and 34% protein by weight. The overall structure, in the 'crown' view, shows elements that had already been shown in earlier EM photographs from Jim Lake's group at UCLA with what is known as the central protuberance and on the wings the L1 and the L7/L11 regions (Fig. 8.14). The width across the crown of the 50S subunit is around 250 Å. The overall architecture of the 50S is a solid core of irregularly packed RNA like a large 3D jigsaw resulting in a very large isometric structure. The surface is surrounded by evenly distributed and separated proteins 17 of which are globular and 13 (Fig. 8.15) of which either have globular heads and disordered tails or are fully extended. The tails penetrate into the core of the RNA and help to bind it together in a mortar-like manner, hence enhancing interactions between domains of RNA.

8.4.1.3 The entire 70S ribosomal structure

Following on from the structure solution of the 30 and 50S subunits at high resolution, there were several rounds of 70S structures being published with increasing resolution following the initial 70S structure from Harry Noller's group at 7.8Å then at 5.5 Å resolution (Cate et al., 1999; Yusupov et al., 2001). The publication of the *E. coli* 70S structure at 3.5 Å resolution by Jamie Cate's group (Schuwirth et al., 2005) was followed by 70S structures from the Ramakrishnan and Yonath groups for *T. thermophilus* and an increased resolution structure to 3.7 Å structure from the Noller group. Subsequently, there was a very high-resolution structure at 2.8 Å resolution from the Ramakrishnan

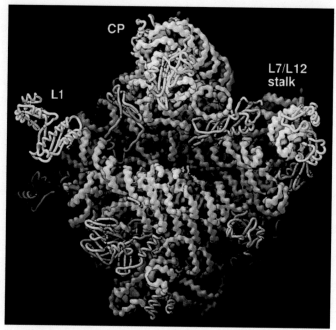

Figure 8.14 Structure of the 50S ribosomal subunit from *Haloarcula marismortui*. *(With permission taken from Ban, N., Nissen, P., Hansen, J., Moore, P. B., & Steitz, T. A. (2000). The complete atomic structure of the large ribosomal subunit at 2.4 Å resolution. Science, 289, 905—920.)*

group with two very similar molecules in the asymmetric unit. These higher resolution structures extended detailed aspects of mRNA and tRNA positioning within the complete context of the 70S subunit (Selmer et al., 2006). There were subsequent complexes with elongation factors EFTu and EF-G important in the ribosomal cycle namely, which gave a deeper insight into the action of the decoding region (Schmeing et al., 2009) and also the elongation step and will be discussed below.

8.4.2 Other protein GTPases acting in the ribosomal cycle

Elongation factor EFTu is the most prevalent protein within bacterial cells with around 100,000 copies and comprising around 5% of the total protein complement in the cell. The concentration of EFTu far outstrips the concentration of tRNAs, and hence, most tRNAs in the cell exist as aminoacyl—tRNA—EFTu complexes. EFTu is a GTPase which recognises features on the $3'$ CCA stem of the tRNA, in contrast to the tRNA synthetases discussed in Section 8.1 above, where structural features in both the stem and anticodon loop structure are mostly sensed.

EFTu also binds to a charged tRNA with an amino acid and not to free tRNA, so it is very discriminatory. Following the formation of the initiation complex with the fMet-tRNA positioned in the P site of the 70s (discussed in Section 8.4.1 above), the EFTu—aatRNA^{aa}—GTP enters the T site (adjacent to the A site) and is prevented from going

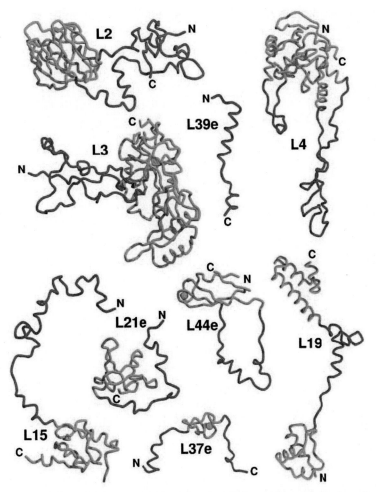

Figure 8.15 Conformations of the 50S ribosomal proteins. *(Taken with permission from Ban, N., Nissen, P., Hansen, J., Moore, P. B., & Steitz, T. A. (2000). The complete atomic structure of the large ribosomal subunit at 2.4 Å resolution. Science, 289, 905–920.)*

further by the interaction of EFTu and the factor site on the ribosome. This initial event is also mRNA independent. The amino acid on the tRNA CCA stem cannot move into the peptidyl transferase site. The anticodon loop samples the anticodon—codon interaction, and if the tRNA is found to be cognate, a stable codon/anticodon duplex is then formed. The only way this base pairing can form between codon and anticodon is to induce a bend in the tRNA between the ASL (anticodon stem-loop) and the D stem. At this stage, the tRNA in the EFTu—aa—tRNA^{aa}—GTP is in a bent conformation with the strain transferred to the GTPase site 70 Å away from the anticodon loop inducing GTPase activity. For cognate—anticodon recognition, the rate of GTP hydrolysis is accelerated by 5×10^4. EFTu then catalyses the GTP hydrolysis to GDP and P$_i$ and EFTu is released, the aa-tRNA^{aa} binds fully into the A site ready for peptide formation. For near-cognate inter-actions the EFTu—aa—tRNA^{aa}GTP readily dissociates from the ribosome. The dissocia-tion of cognate is 350-fold faster than non-cognate tRNA. The conformational changes at the decoding centre arise from changes in the 16S RNA with residues A1492, A1493 and G530 of the 16S RNA involved in RNA—RNA interactions, as outlined in Chapter 6. For the liberated EFTu bound to GDP, the GDP is exchanged for GTP by the exchange factor EFTs and can bind another aminoacylate tRNA from the cellular pool.

8.4.2.1 EF-G

GTP is hydrolysed and the peptidyl transferase centre catalyses the transfer of the amino acid fMet to form a bond with the incoming amino acid in the A site. The first step in elongation is translocation and a 180° rotation, creating a hybrid A/P site where the pep-tidyl tRNA (the bulk of the tRNA) is still in the A site. The EF-G binds and undergoes a conformational change which allows it to push the peptidyl tRNA fully into the P site, akin to a shovel head pushing snow. The deacylated tRNA moves from the P site to the E site and the mRNA is ratcheted along to present a new codon to a new incoming tRNA. EF-G is essentially a molecular mimic of EFTu—aa—tRNA^{aa}—GTP, with the form of the tRNA sculpted in protein. It has five domains and is an extended molecule. The G$'$ domain is inserted into the G domain, domain II is an anti-parallel β-barrel related to domain II in IF-2 and EFTu, domains III and V have the topology of ribosomal protein S6 and many RNA-binding proteins. tRNA is mimicked by domains II, IV and V. In contrast to EFTu, EF-G does not require an exchange factor due to its lower nucleotide affinity. After the entire mRNA is translated, the final step of termination is catalysed by three release factors: RF1, RF2 and RF3. The complete prokaryotic cycle is shown in Fig. 8.13.

8.4.3 Further ribosome studies

Following on from the very successful structural, mechanistic and drug developmental studies on the subunits of the prokaryotic ribosomes, there has been a focus to understand the structures of ribosomes from eukaryotes, such as yeast and *homo sapiens*, mitochondrial ribosomes and ribosomes from chloroplasts. The method of choice for structure determi-nation is now primarily cryo-EM.

8.4.4 Eukaryotic ribosomes

The ribosome cycle for eukaryotes is much more complex compared with prokaryotes and involves many more factors (Fig. 8.16). For example, the alignment of the start codon by the base pairing of the Shine—Dalgarno sequence with the anti-Shine—Dalgarno sequence does not exist in eukaryotic translation, in place there is a 'scanning mechanism' which scans from the 5$'$ cap to the start codon. Some viruses high-jack the cellular translation mechanism by suppressing cap-dependent translation, by uti-lising their IRESs (internal ribosomal entry sites) to start ribosomal translation. Fig. 8.16

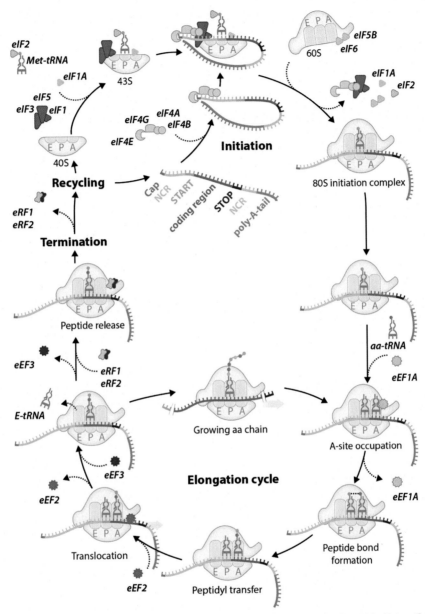

Figure 8.16 The eukaryotic ribosomal cycle. *(Figure designed by A.V. Berglar, PhD (Scientific Visualisation Lab, www.scivislab.com).)*

shows the stages in eukaryotic translation and Table 8.2 compares and contrasts the ribosomal factors present in prokaryotes and eukaryotes cycles. Eukaryotic chromosomes are 80S particles made up of 40S and 60S subunits. The 40S subunit consists of an 18S RNA and 33 proteins, whereas the 60S subunit consists of a 25S or 28S RNA, a 5S (120

Table 8.2 Comparison of factors involved in the eukaryotic and prokaryotic ribosomal cycle. Prokaryotic factors are in brackets.

Name of the protein factor	Role
EF1A (EFTu)	GTPase which is involved in the binding on the aminoacyl–tRNA to the ribosome
EF1B (EFTs)	Nucleotide exchange factor for EF1A once dissociated from the ribosome as EF1A–GDP to exchange GDP to GTP and reconstitute EF1A–GTP Similar role to EFTs. EF1B has two subunits α and γ, α has the exchange factor activity
EF2 (EF-G)	GTPase involved in the translocation of the peptidyl–tRNA from the A site to the P site
EF3	Facilitates the release of E-site deacylated tRNA in eukaryotes
EF4	Behaves like a back translocase Moves the tRNAs and mRNA in the opposite direction to EF2 (EF-G)

nucleotides), a 5.8S (154 nucleotides) and 46 proteins. The molecular weight of the 40S subunit is 1.2 MegaDa. and the 60S is 2.0 MegaDa., a total of 3.1 MegaDa., for the 80S, so that they are 40% larger than prokaryotic ribosomes. Contributing to this mass differential are 25 proteins uniquely found in eukaryotes and extensions in the RNA and extensions primarily at the N′ and C′ termini of the conserved proteins.

There were early structures of eukaryotic ribosomes determined by cryo-EM, at 11.7 Å resolution (Sengupta et al., 2004), but these studies could not fit amino acid side chains and precisely fit bases into the EM maps of the constituent proteins and RNA but could only locate enzymes such as RACK1 (Receptor for activated C kinase). The *Tetrahymena thermophila* 40S subunit (3.9 Å resolution) and the 60S (3.5 Å resolution) ribosomal subunit were the first to be crystallised and solved at high resolution by X-ray crystallography by Nenad Ban's group at the ETH. The 40S was complexed with initiation factor eIF1 (Rabl et al., 2011) and the 60S with eIF6 (Klinge et al., 2011). This was followed by high-resolution structures of the 80S ribosome and complexes from yeast by Marat Yusupov's group in Strasbourg, solved first at 4 Å (Ben-Shem et al., 2010) and then at 3 Å (Garreau de Loubresse et al., 2014), see Table 8.3. The major findings were of extension regions in a large number of ribosomal proteins that do not have prokaryotic homologues, as well as extensions primarily at the termini of the RNA and that the majority of the changes are on the surface of the ribosome. The amount of protein–protein coverage in the eukaryotic ribosome is 2.4 times greater than that found for prokaryotes with many more protein–protein contacts being formed. Many structures of yeast ribosomal complexes have been solved (Table 8.3), and recently, there has been a very high-resolution structure of the 80S human ribosome (Khatter et al., 2015) and complexes (Table 8.4), which are important studies in the mapping of human diseases

Table 8.3 Yeast ribosomal structures (first and last authors given in each reference).

Year	Details of the complex	PDB code	Technique	Resolution (Å)	Reference	Laboratory location
2004	80S with identifying of RACK1 protein	1TRJ	cryo-EM	11.7	Sengupta, J.… Frank, J. Nat. Struct. Mol. Biol. **11**, 957–962 (2004)	Wadsworth Center, NY Dept. Health, Upstate NY, USA
2004	80S–eEF2 and antibiotic sodarin	4V4B	cryo-EM	11.7	Spahn, C.M.… Frank, J. EMBO J **23**, 1008–1019 (2004)	Wadsworth Center, NY Dept. Health, Upstate NY, USA
2009	80S bound to idle Ssh1 complex	2WWA	cryo-EM	8.9	Becker, T.… Beckmann, R. Science **326**, 1369–1373 (2009)	Gene Centre, Munich, LMU, Munich, Germany
2009	80S bound to active Ssh1 complex	3WW9	cryo-EM	8.6	Becker, T.… Beckmann, R. Science **326**, 1369–1373 (2009)	Gene Centre, Munich, LMU, Munich, Germany
2010	80S ribosome	4V7R	X-ray	4	Ben-Shem, A.… Yusupov, M. Science **330**, 1203–1209 (2010)	IGBMC, Illkirch, France
2014	80S with anisomycin bound	4U3M	X-ray	3	Garreau de Loubresse, N.… Yusupov, M. Nature **513**, 517–522 (2014)	IGBMC, Illkirch, France
2014	80S with edeine bound	4U3M	X-ray	3.1	Garreau de Loubresse, N.… Yusupov, M. Nature **513**, 517–522 (2014)	IGBMC, Illkirch, France
2014	80S with geneticin bound	4U4O	X-ray	3.6	Garreau de Loubresse, N.… Yusupov, M. Nature **513**, 517–522 (2014)	IGBMC, Illkirch, France

Continued

Table 8.3 Yeast ribosomal structures (first and last authors given in each reference).—cont'd

Year	Details of the complex	PDB code	Technique	Resolution (Å)	Reference	Laboratory location
2014	80S with verrucarin bound	4U50	X-ray	3.2	Garreau de Loubresse, N.... Yusupov, M. Nature **513**, 517–522 (2014)	IGBMC, Illkirch, France
2014	80S with narciclasine bound	4U51	X-ray	3.2	Garreau de Loubresse, N.... Yusupov, M. Nature **513**, 517–522 (2014)	IGBMC, Illkirch, France
2014	80S with homoharringtonine bound	4U4Q	X-ray	3.0	Garreau de Loubresse, N.... Yusupov, M. Nature **513**, 517–522 (2014)	IGBMC, Illkirch, France
2014	80S with lactimidomycin bound	4U4R	X-ray	2.8	Garreau de Loubresse, N.... Yusupov, M. Nature **513**, 517–522 (2014)	IGBMC, Illkirch, France
2014	80S with deoxyvalenol bound	4U53	X-ray	3.3	Garreau de Loubresse, N.... Yusupov, M. Nature **513**, 517–522 (2014)	IGBMC, Illkirch, France
2014	80S with cycloheximide bound	4U3U	X-ray	2.9	Garreau de Loubresse, N.... Yusupov, M. Nature **513**, 517–522 (2014)	IGBMC, Illkirch, France
2014	80S with cryptopleurine bound	4U55	X-ray	3.2	Garreau de Loubresse, N.... Yusupov, M. Nature **513**, 517–522 (2014)	IGBMC, Illkirch, France
2014	80S with lycorine bound	4U4U	X-ray	3.0	Garreau de Loubresse, N.... Yusupov, M. Nature **513**, 517–522 (2014)	IGBMC, Illkirch, France

Year	Description	PDB ID	Method	Resolution	Reference	Institution
2014	80S with pactamicine bound	4U4Y	X-ray	3.2	Garreau de Loubresse, N.... Yusupov, M. Nature **513**, 517-522 (2014)	IGBMC, Illkirch, France
2014	80S with phyllanthoside bound	4U4Z	X-ray	3.1	Garreau de Loubresse, N.... Yusupov, M. Nature **513**, 517-522 (2014)	IGBMC, Illkirch, France
2014	80S with CCA trinucleotide bound	4U3N	X-ray	3.2	Garreau de Loubresse, N.... Yusupov, M. Nature **513**, 517-522 (2014)	IGBMC, Illkirch, France
2014	80S with T-2 toxin bound	4U6F	X-ray	3.1	Garreau de Loubresse, N.... Yusupov, M. Nature **513**, 517-522 (2014)	IGBMC, Illkirch, France
2014	80S with tRNA complexes in rotated and non-rotated conform[n] Class II rotated with 1 tRNAs	3J77	cryo-EM	6.2	Svidritskiy, E.... Korostelev, A.A. Structure **22**, 1210-1218 (2014)	RNA therapeutics Institute, UMass, Worcester, MA, USA
2014	80S with tRNA complexes in rotated and non-rotated conform[n] Class I non-rotated with 2 tRNAs	3J78	cryo-EM	6.3	Svidritskiy, E.... Korostelev, A.A. Structure **22**, 1210-1218 (2014)	RNA therapeutics Institute, UMass, Worcester, MA, USA
2016	80S ribosomal complex with non-modified eIF5a	5DC3	X-ray cryst	3.25	Melnikov, S.... Yusupov, M. JMB **428**, 3570-3576 (2016)	IGBMC, Illkirch, France
2016	80S bound to amicoumacin	5I4L	X-ray	3.1	Prokhorova, I.V.... Yusupov, M. Sci. Rep. **6**, 27720 (2016)	IGBMC, Illkirch, France
2017	80S with chlorolissoclimide bound	5TBW	X-ray	3.0	Konst, Z.A.... Vanderwal, C.D. Nat. Chem. **9**, 1140-1149 (2017)	Dept of Chemistry, UC Irvine, CA, USA

Continued

Table 8.3 Yeast ribosomal structures (first and last authors given in each reference).—cont'd

Year	Details of the complex	PDB code	Technique	Resolution (Å)	Reference	Laboratory location
2017	80S with paromomysin bound	5NDV	X-ray	3.3	Prokhorova, I.V.... Yusupova, G. PNAS **114**, E10899-E10908 (2017)	IGBMC, Illkirch, France
2017	80S with gentamisin bound	5OBM	X-ray	3.4	Prokhorova, I.V.... Yusupova, G. PNAS **114**, E10899-E10908 (2017)	IGBMC, Illkirch, France
2017	80S with geneticin (G418) bound	5NDG	X-ray	3.7	Prokhorova, I.V.... Yusupova, G. PNAS **114**, E10899-E10908 (2017)	IGBMC, Illkirch, France
2017	80S with aminoglycoside TC007 bound	5NDW	X-ray	3.7	Prokhorova, I.V.... Yusupova, G. PNAS **114**, E10899-E10908 (2017)	IGBMC, Illkirch, France
2018	80S with mRNA, tRNA, and EF2 (GMPPCP) bound	6GQV	cryo-EM	4.0	Pellegrino, S.... Hashem, Y. JMB **430**, 2677-2687 (2018)	IGBMC, Illkirch, France
2018	80S with mRNA, tRNA, and EF2 (GMPPCP/sordarin) bound	6GQ1	cryo-EM	4.0	Pellegrino, S.... Hashem, Y. JMB **430**, 2677-2687 (2018)	IGBMC, Illkirch, France
2018	80S with mRNA, tRNA, and EF2 (GDP+AlF$_4$/sordarin) bound	6GQB	cryo-EM	3.9	Pellegrino, S.... Hashem, Y. JMB **430**, 2677-2687 (2018)	IGBMC, Illkirch, France
2019	80S with C45 bound	6HHQ	X-ray	3.1	Pellegrino, S.... Yusupov, M. NAR **47**, 3223-3232, 2687 (2019)	IGBMC, Illkirch, France

Year	PDB ID	Description	Method	Resolution	Reference	Institution
2019	6Q8Y	80S-Xrn1 nuclease, mRNA translating, and degrading complex	cryo-EM	3.1	Tesina, P.... Beckmann, R. Nat. Struct. Mol. Biol. **26**, 275–280 (2019)	Gene Centre, Munich, LMU, Munich, Germany
2020	6SNT	80S ribosome stalled on SDD1 mRNA	cryo-EM	2.8	Tesina, P.... Beckmann, R. Nat. Struct. Mol. Biol. **27**, 323–332 (2020)	Gene Centre, Munich, LMU, Munich, Germany
2020	6T7T	80S ribosome stalled on poly(A)tract	cryo-EM	3.1	Tesina, P.... Beckmann, R. EMBO J. e103365 (2020)	Gene Centre, Munich, LMU, Munich, Germany
2020	6T4Q	80S ribosome stalled on CGA-CCG inhibitor codon	cryo-EM	2.6	Tesina, P.... Beckmann, R. EMBO J. e103365 (2020)	Gene Centre, Munich, LMU, Munich, Germany
2020	6T7I	80S ribosome stalled on CGA-CGA inhibitor codon	cryo-EM	3.2	Tesina, P.... Beckmann, R. EMBO J. e103365 (2020)	Gene Centre, Munich, LMU, Munich, Germany
2020	6TB3	80S ribosome bound to the Not5 subunit of the CCR4–NOT complex	cryo-EM	2.8	Buschauer, R.... Beckmann, R. Science **368**, eaay6912 (2020)	Gene Centre, Munich, LMU, Munich, Germany
2020	6TNU	80S with eIF5A and A-site and P-site tRNAs	cryo-EM	3.1	Buschauer, R.... Beckmann, R. Science **368**, eaay6912 (2020)	Gene Centre, Munich, LMU, Munich, Germany
2020	6Z6K	80S complexed with yeast reconstituted Lso2 bound	cryo-EM	3.4	Wells, J.N.... Beckmann, R. PLoS Biology, e3000780 (2020)	Gene Centre, Munich, LMU, Munich, Germany

Table 8.4 Human ribosomal structures (first and last authors given in each reference).

Year	Details of the complex	PDB code	Technique	Resolution (Å)	Reference	Laboratory location
2013	80S ribosome	4V6X	cryo-EM	5	Anger, A.M.… Beckmann, R. Nature **497**, 80–85 (2013)	Gene Centre, Munich, LMU, Munich, Germany
2015	80S with Hepatitis C virus IRES	2AGN	cryo-EM	15	Boehringer, D.… Stark, H. Structure **13**, 1625–1726 (2005)	Max Planck Institute for Biophysical Chemistry, Goettingen, Germany
2015	80S translating polysomes (Post)	5AJO	cryo-EM	3.5	Behrmann, E.… Spahn, C.M.T. Cell **161**, 845–857 (2015)	Inst. for Medicinal Physics and Biophysics, Charité Univ. Berlin, Germany
2015	80S with no factors	4UG0	cryo-EM	3.6	Khatter, H.… Klaholz, B.P. Nature **520**, 640–645 (2015)	IGMBC, Illkirch, France
2016	80S comparison with Leishmania and stabilisation of kinetoplastid specific split rRNA	5T2C	cryo-EM	3.6	Zhang, X.… Zhou, Z.H. Nat. Commun. **7**, 13223 (2016)	Centre for Cryo-EM, Zheijang Univ. Zhejiang, China
2016	80s as a target for antibiotics	5LKS	cryo-EM	3.6	Myasnikov, A.G.… Klaholz, B. Nat. Commun. **7**, 12856 (2016)	IGBMC, Illkirch, France
2017	80S with not factors	6EK0	cryo-EM	2.9	Natchiar, S.K.… Klaholz, B. Nature **551**, 472–477 (2017)	IGBMC, Illkirch, France
2017	80S with not factors	6QZP	cryo-EM	2.9	Natchiar, S.K.… Klaholz, B. Nature **551**, 472–477 (2017)	IGBMC, Illkirch, France

Year	Description	PDB ID	Method	Resolution	Reference	Institution
2018	80S with double translocated CrPV IRES P-site tRNA and eRF1	6D90	cryo-EM	3.2	Pisarev, V.P ... Fernandez, I.S. eLife .34062.001 (2018)	SUNY Downstate Medical Centre, Columbia Univ. NY, USA
2019	80S CMV stalled structure (ii)	6IP5	cryo-EM	3.9	Yokoyama, T.... Ito, T. Mol. Cell **74**, 1205–1214 (2019)	RIKEN Centre, Yokohama, Japan
2019	80S CMV stalled structure (iii)	6IP6	cryo-EM	4.5	Yokoyama, T.... Ito, T. Mol. Cell **74**, 1205–1214 (2019)	RIKEN Centre, Yokohama, Japan
2019	80S CMV stalled structure (iv)	6IP8	cryo-EM	3.9	Yokoyama, T.... Ito, T. Mol. Cell **74**, 1205–1214 (2019)	RIKEN Centre, Yokohama, Japan
2019	80S complex with Israeli Acute paralysis virus IREs	6P5N	cryo-EM	3.2	Acosta-Reyes, F.... Fernandez, I.S. EMBO J. e10226 (2019)	Dept. of Biophysics and Biochemistry, Columbia University, NY. USA
2019	80S complex with the Israeli Acute Paralysis Virus IRES (Class 2)	6P5I	cryo-EM	3.1	Acosta-Reyes, F.... Fernandez, I.S. EMBO J. e10226 (2019)	Dept. of Biophysics and Biochemistry, Columbia University, NY. USA
2019	80S complex with the Israeli Acute Paralysis Virus IRES (Class 2)	6P5J	cryo-EM	3.1	Acosta-Reyes, F.... Fernandez, I.S. EMBO J. e10226 (2019)	Dept. of Biophysics and Biochemistry, Columbia University, NY. USA
2019	80S complex with the Israeli Acute Paralysis Virus IRES (Class 3)	6P5K	cryo-EM	3.1	Acosta-Reyes, F.... Fernandez, I.S. EMBO J. e10226 (2019)	Dept. of Biophysics and Biochemistry, Columbia University, NY. USA
2019	80S nascent chain complex (PCSK9-RNC) stalled by a drug-like molecule with PP tRNA	6OLZ	cryo-EM	3.9	Li, W.... Cate, J.H.D. Nat. Struct. Mol. Biol. **26**, 501–509 (2019)	Dept of Molecular and Cellular Biology UC Berkeley, USA
2019	80S nascent chain complex (PCSK9-RNC) stalled by a drug-like molecule with AP and PE tRNAs	6OM0	cryo-EM	3.1	Li, W.... Cate, J.H.D. Nat. Struct. Mol. Biol. **26**, 501–509 (2019)	Dept of Molecular and Cellular Biology UC Berkeley, USA
2019	80S nascent chain complex (PCSK9-RNC) stalled by a drug-like small molecule with AA and PE tRNAs	6OM7	cryo-EM	3.7	Li, W.... Cate, J.H.D. Nat. Struct. Mol. Biol. **26**, 501–509 (2019)	Dept of Molecular and Cellular Biology UC Berkeley, USA

Continued

Table 8.4 Human ribosomal structures (first and last authors given in each reference).—cont'd

Year	Details of the complex	PDB code	Technique	Resolution (Å)	Reference	Laboratory location
2019	80S nascent chain complex selectively stalled by a drug-like small molecule (USO1-RNC)	6OLI	cryo-EM	3.5	Li, W.... Cate, J.H.D. Nat. Struct. Mol. Biol. **26**, 501-509 (2019)	Dept of Molecular and Cellular Biology UC Berkeley, USA
2019	80S nascent chain complex (CDH1-RNC) stalled by a drug-like molecule with AA and PE tRNAs	6OLF	cryo-EM	3.9	Li, W.... Cate, J.H.D. Nat. Struct. Mol. Biol. **26**, 501-509 (2019)	Dept of Molecular and Cellular Biology UC Berkeley, USA
2019	80S nascent chain complex (CDH1-RNC) stalled by a drug-like molecule with AP and PE tRNAs	6OLE	cryo-EM	3.1	Li, W.... Cate, J.H.D. Nat. Struct. Mol. Biol. **26**, 501-509 (2019)	Dept of Molecular and Cellular Biology UC Berkeley, USA
2019	80S nascent chain complex stalled by a drug-like small molecule (CDH1_RNC with PP tRNA)	6OLG	cryo-EM	3.4	Li, W.... Cate, J.H.D. Nat. Struct. Mol. Biol. **26**, 501-509 (2019)	Dept of Molecular and Cellular Biology UC Berkeley, USA
2019	80S and double translocated CrPV IRES P-site tRNA and eRF1	6D9J	cryo-EM	3.2	Paraseva, V.... Fernandez, I.S. eLife. 34062 (2019)	Dept. of Biophysics and Biochemistry, Columbia University, NY, USA
2019	80S XBP1u-paused ribosome nascent chain complex (post-state)	6R5Q	cryo-EM	3.0	Shanmuganathan, V.... Beckmann, R. eLife. 46267 (2019)	Gene Centre, Munich, LMU, Munich, Germany
2019	80S XBP1u-paused ribosome nascent chain complex with Sec61	6R7Q	cryo-EM	3.9	Shanmuganathan, V.... Beckmann, R. eLife. 46267 (2019)	Gene Centre, Munich, LMU, Munich, Germany
2019	80S XBP1u-paused ribosome nascent chain complex (rotated state)	6R6P	cryo-EM	3.1	Shanmuganathan, V.... Beckmann, R. eLife. 46267 (2019)	Gene Centre, Munich, LMU, Munich, Germany

Year	Description	PDB ID	Method	Resolution	Reference	Location
2019	80S with yeast Lso2 bound	6Z6J	cryo-EM	3.4	Wells, J.N.... Beckmann, R. PloS Biol. **18**, e3000780 (2020)	Gene Centre, Munich, LMU, Munich, Germany
2019	80S with yeast reconstituted Lso2 bound	6Z6K	cryo-EM		Wells, J.N.... Beckmann, R. PloS Biol. **18**, e3000780 (2020)	Gene Centre, Munich. LMU, Munich, Germany
2020	Human and Human CCD124	6Z6L	cryo-EM	3.0	Wells, J.N.... Beckmann, R. PloS Biol. **18**, e3000780 (2020)	Gene Centre, Munich, LMU, Munich, Germany
2020	80S bound to EBP1, eEF2 abd SERBP1	6Z6M	cryo-EM	3.1	Wells, J.N.... Beckmann, R. PloS Biol. **18**, e3000780 (2020)	Gene Centre, Munich, LMU, Munich, Germany.
2020	80S bound to EBP1 (focus on EBP1)	6Z6N	cryo-EM	2.9	Wells, J.N.... Beckmann, R. PloS Biol. **18**, e3000780 (2020)	Gene Centre, Munich, LMU, Munich, Germany
2020	80S inhibition by drug like molecule(PF846)	6XA1	cryo-EM	2.8	Li, W.... Cate, J. Nat. Comm. **11**, 4941 (2020)	Dept of Molecular and Cellular Biology UC Berkeley, USA
2020	80S in classic prestate	6Y0G	cryo-EM	3.2	Bhaskar, V.... Chao, J.A. Cell Rep. **31**, 107473 (2020)	Friedrich Miescher Institute for Biomedical Research, Basel, Switzerland
2020	80S in hybrid prestate	6Y5G	cryo-EM	3.5	Bhaskar, V.... Chao, J.A. Cell Rep. **31**, 107473 (2020)	Friedrich Miescher Institute for Biomedical Research, Basel, Switzerland
2020	80S in poststate	6Y5L	cryo-EM	3.0	Bhaskar, V.... Chao, J.A. Cell Rep. **31**, 107473 (2020)	Friedrich Miescher Institute for Biomedical Research, Basel, Switzerland

originating from mutations in components of the ribosome. Tables 8.2 and 8.3 list the recent structures for yeast and human ribosomal studies and the factors involved, together with the technique used in the structural elucidation, and the resolution achieved. There are now a large number of mitochondrial ribosomal structures and complexes from different organisms, plants, eukaryotes and mammals, which are not listed here due to space restrictions. An excellent overview of ribosomal structures held at the PDB is now available (Moore, 2021).

References

tRNA-synthetases

Biou, V., Yaremchuk, A., Tukalo, M., & Cusack, S. (1994). The 2.9 Å crystal structure of *T. thermophilus* seryl-tRNA synthetase complexed with tRNA(Ser). *Science, 263*, 1404–1410.

Hansen, J. L., Long, A. M., & Schultz, S. C. (1997). Structure of the RNA-dependent RNA polymerase of poliovirus. *Structure, 5*, 1109–11022.

Kim, S. H., Suddath, F. L., Quigley, G. J., McPherson, A., Sussman, J. L., Wang, A. H. J., Seeman, N. C., & Rich, A. (1974). Three-dimensional tertiary structure of yeast phenylalanine transfer RNA. *Science, 185*, 435–440.

Perona, J. J., Swanson, R., Steitz, T. A., & Söll, D. (1988). Overproduction and purification of Escherichia coli tRNA(2Gln) and its use in crystallization of the glutaminyl-tRNA synthetase-tRNA(Gln) complex. *Journal of Molecular Biology, 202*, 121–126.

Robertus, J. D., Ladner, J. E., Finch, J. T., Rhodes, D., Brown, R.,S., Clark, B. F., & Klug, A. (1974). Structure of yeast phenylalanine tRNA at 3 Å resolution. *Nature, 250*, 546–551.

Rould, M. A., Perona, J. J., Söll, D., & Steitz, T. A. (1989). Structure of E. coli glutaminyl-tRNA synthetase complexed with tRNA(Gln) and ATP at 2.8 A resolution. *Science, 246*, 1135–1142.

Ruff, M., Krishnaswamy, S., Boeglin, M., Poterszman, A., Mitschler, A., Podjarny, A., Rees, B., Thierry, J. C., & Moras, D. (1991). Class II aminoacyl transfer RNA synthetases: Crystal structure of yeast aspartyl-tRNA synthetase complexed with tRNA(Asp). *Science, 252*, 1682–1689.

Swanson, R., Hoben, P., Sumner-Smith, M., Uemura, H., Watson, L., & Söll, D. (1988). Accuracy of in vivo aminoacylation requires proper balance of tRNA and aminoacyl-tRNA synthetase. *Science, 242*, 1548–1551.

Reverse Transcriptases (RT)

Freemont, P. S., Friedman, J. M., Beese, L. S., Sanderson, M. R., & Steitz, T. A. (1988). Cocrystal structure of an editing complex of Klenow fragment with DNA. *Proceedings of the National Academy of Sciences of the United States of America, 85*, 8924–8928.

Huang, H., Chopra, R., Verdine, G. L., & Harrison, S. C. (1998). Structure of a covalently trapped catalytic complex of HIV-1 reverse transcriptase: Implications for drug resistance. *Science, 282*, 1669–1675.

Jacobo-Molina, A., Clark, A. D., Jr., Williams, R. L., Nanni, R. G., Clark, P., Ferris, A. L., Hughes, S. H., & Arnold, E. (1991). Crystals of a ternary complex of human immunodeficiency virus type 1 reverse transcriptase with a monoclonal antibody Fab fragment and double-stranded DNA diffract X-rays to 3.5-Å resolution. *Proceedings of the National Academy of Sciences of the United States of America, 88*, 10895–10899.

Joyce, C. M., & Steitz, T. A. (1994). Function and structure relationships in DNA polymerases. *Annual Review of Biochemistry, 63*, 777–822.

Kang, D., Ruiz, F. X., Feng, D., Pilch, A., Zhao, T., Wei, F., Wang, Z., Sun, Y., Fang, Z., De Clercq, E., Pannecouque, C., Arnold, E., Liu, X., & Zhan, P. (2020). Discovery and characterization of fluorine-substituted diarylpyrimidine derivatives as novel HIV-1 NNRTIs with highly improved resistance profiles and low activity for the hERG ion channel. *Journal of Medicinal Chemistry, 63*, 1298–1312.

Kang, D., Zhang, H., Wang, Z., Zhao, T., Ginex, T., Javier Luque, F., Yang, Y., Wu, G., Feng, D., Wei, F., Zhang, J., De Clercq, E., Pannecouque, C., Chen, C. H., Lee, K.-H., Arul Murugan, N., Steitz, T. A., Zhan, P., & Liu, X. (2019). Identification of dihydrofuro[3,4- d]pyrimidine derivatives as novel HIV-1 non-nucleoside reverse transcriptase inhibitors with promising antiviral activities and desirable physico-chemical properties. *Journal of Medicinal Chemistry, 62,* 1484–1501 (and references therein).

Kohlstaedt, L. A., Wang, J., Friedman, J. M., Rice, P. A., & Steitz, T. A. (1992). Crystal structure at 3.5 Å resolution of HIV-1 reverse transcriptase complexed with an inhibitor. *Science, 256,* 1783–1790.

Ollis, D. L., Brick, P., Hamlin, R., Xuong, N. G., & Steitz, T. A. (1985). Structure of large fragment of *Escherichia coli* DNA polymerase I complexed with dTMP. *Nature, 313,* 762–766.

Rodgers, D. W., Gamblin, S. J., Harris, B. A., Ray, S., Culp, J. S., Hellmig, B., Woolf, D. J., Debouck, C., & Harrison, S. C. (1995). The structure of unliganded reverse transcriptase from the human immunodeficiency virus type 1. *Proceedings of the National Academy of Sciences of the United States of America, 92,* 1222–1226.

Steitz, T. A., Beese, L., Freemont, P. S., Friedman, J. M., & Sanderson, M. R. (1987). Structural studies of Klenow fragment: An enzyme with two active sites. *Cold Spring Harbor Symposia on Quantitative Biology, 152,* 465–471.

snRPs

Agafonov, D. E., Kastner, B., Dybkov, O., Hofele, R. V., Liu, W. T., Urlaub, H., Lührmann, R., & Stark, H. (2016). Molecular architecture of the human U4/U6.U5 tri-snRNP. *Science, 351,* 1416–1420.

Berget, S. M., Berk, A. J., Harrison, T., & Sharp, P. A. (1978). Spliced segments at the 5' termini of adenovirus-2 late mRNA: A role for heterogeneous nuclear RNA in mammalian cells. *Cold Spring Harbor Symposia on Quantitative Biology, 42,* 523–529.

Berget, S. M., Moore, C., & Sharp, P. A. (1977). Spliced segments at the 5' terminus of adenovirus 2 late mRNA. *Proceedings of the National Academy of Sciences of the United States of America, 74,* 3171–3175.

Breathnach, R., Mandel, J. L., & Chambon, P. (1977). Ovalbumin gene is split in chicken DNA. *Nature, 270,* 314–319.

Broker, T. R., Chow, L. T., Dunn, A. R., Gelinas, R. E., Hassell, J. A., Klessig, D. F., Lewis, J. B., Roberts, R. J., & Zain, B. S. (1978). Adenovirus-2 messengers—an example of baroque molecular architecture. *Cold Spring Harbor Symposia on Quantitative Biology, 42,* 531–553.

Chambon, P. (1978). The molecular biology of the eukaryotic genome is coming of age. *Cold Spring Harbor Symposia on Quantitative Biology, 42,* 1209–1234.

Charenton, C., Wilkinson, M. E., & Nagai, K. (2019). Mechanism of 5' splice site transfer for human spliceosome activation. *Science, 364,* 362–367.

Chow, L. T., Gelinas, R. E., Broker, T. R., & Roberts, R. J. (1977). An amazing sequence arrangement at the 5' ends of adenovirus 2 messenger RNA. *Cell, 12,* 1–8.

Kastner, B., Will, C. L., Stark, H., & Lührmann, R. (2019). Structural insights into nuclear pre-mRNA splicing in higher eukaryotes. *Cold Spring Harbor Perspectives in Biology, 11,* a032417.

Lerner, M. R., Boyle, J. A., Mount, S. M., Wolin, S. L., & Steitz, J. A. (1980). Are snRNPs involved in splicing? *Nature, 283,* 220–224.

Lerner, M. R., & Steitz, J. A. (1979). Antibodies to small nuclear RNAs complexed with proteins are produced by patients with systemic lupus erythematosus. *Proceedings of the National Academy of Sciences of the United States of America, 76,* 5495–5499.

Nagai, K., Oubridge, C., Jessen, T. H., Li, J., & Evans, P. R. (1990). Crystal structure of the RNA-binding domain of the U1 small nuclear ribonucleoprotein A. *Nature, 6,* 515–520.

Oubridge, C., Ito, N., Evans, P. R., Teo, C. H., & Nagai, K. (1994). Crystal structure at 1.92 Å resolution of the RNA-binding domain of the U1A spliceosomal protein complexed with an RNA hairpin. *Nature, 372,* 432–438.

Wilkinson, M. E., Charenton, C., & Nagai, K. (2020). RNA splicing by the spliceosome. *Annual Review of Biochemistry, 89,* 359–388.

The Ribosome

Ban, N., Freeborn, B., Nissen, P., Penzyek, P., Grassucci, R. A., Sweet, R. M., Frank, J., Moore, P. B., & Steitz, T. A. (1998). A 9 Å resolution X-ray crystallographic map of the large ribosomal subunit. *Cell, 93,* 1105–1115.

Ban, N., Nissen, P., Hansen, J., Capel, M., Moore, P. B., & Steitz, T. A. (1999). Placement of protein and RNA structures into a 5 Å-resolution map of the 50S ribosomal subunit. *Nature, 400*, 841—847.

Ban, N., Nissen, P., Hansen, J., Moore, P. B., & Steitz, T. A. (2000). The complete atomic structure of the large ribosomal subunit at 2.4 Å resolution. *Science, 289*, 905—920.

Ben-Shem, A., Jenner, L., Yusupova, G., & Yusupov, M. (2010). Crystal structure of the eukaryotic ribosome. *Science, 330*, 1203—12099.

Cate, J. H., Yusupov, M. M., Yusupova, G. Z., Earnest, T. N., & Noller, H. F. (1999). X-ray crystal structures of 70S ribosome functional complexes. *Science, 285*, 2095—2104.

Clemons, W. M., May, J. L. C., Wimberley, B. T., McCutcheon, J. P., Capel, M. S., & Ramakrishnan, V. (1999). Structure of a bacterial 30S ribosomal subunit at 5.5 Å resolution. *Nature, 400*, 833—840.

Garreau de Loubresse, N., Prokhorova, I., Holtkamp, W., Rodnina, M. V., Yusupova, G., & Yusupov, M. (2014). Structural basis for the inhibition of the eukaryotic ribosome. *Nature, 513*, 517—522.

Glotz, C., Müssig, J., Gewitz, H. S., Makowski, I., Arad, T., Yonath, A., & Wittmann, H. G. (1987). Three-dimensional crystals of ribosomes and their subunits from eu- and archaebacteria. *Biochemistry International, 15*, 953—960.

Henderson, R. (2015). Overview and future of single particle electron cryomicroscopy. *Archives of Biochemistry and Biophysics, 581*, 19—24.

Henderson, R. (2018). From electron crystallography to single particle Cryo-EM (Nobel Lecture). *Angewandte Chemie International Edition in English, 57*, 10804—10825.

Khatter, H., Myasnikov, A. G., Natchiar, S. K., & Klaholz, B. P. (2015). Structure of the human 80S ribosome. *Nature, 520*, 640—645.

Klinge, S., Voigts-Hoffmann, F., Leibundgut, M., Arpagaus, S., & Ban, N. (2011). Crystal structure of the eukaryotic 60S ribosomal subunit in complex with initiation factor 6. *Science, 334*, 941—948.

Kühlbrandt, W., & Unwin, P. N. J. (1982). Distribution of RNA and protein in crystalline eukaryotic ribosomes. *Journal of Molecular Biology, 156*, 431—448.

Moore, P. B. (2021). The PDB and the ribosome. *Journal of Biological Chemistry, 296*(100561), 1—11.

Rabl, J., Leibundgut, M., Ataide, S. F., Haag, A., & Ban, N. (2011). Crystal structure of the eukaryotic 40S ribosomal subunit in complex with initiation factor 1. *Science, 331*, 730—736.

Schluenzen, F., Tocilj, A., Zarivach, R., Harms, J., Gluehmann, M., Janell, D., Bashan, A., Bartels, H., Agmon, I., Franceschi, F., & Yonath, A. (2000). Structure of functionally activated small ribosomal subunit at 3.3 angstroms resolution. *Cell, 102*, 615—623.

Schmeing, T. M., Voorhees, R. M., Kelley, A. C., Gao, Y. G., Murphy, F. V. 4th, Weir, J. R., & Ramakrishnan, V. (2009). The crystal structure of the ribosome bound to EF-Tu and aminoacyl-tRNA. *Science, 326*, 688—694.

Schuwirth, B. S., Borovinskaya, M. A., Hau, C. W., Zhang, W., Vila-Sanjurjo, A., Holton, J. M., & Cate, J. H. (2005). Structures of the bacterial ribosome at 3.5 Å resolution. *Science, 310*, 827—833.

Selmer, M., Dunham, C. M., Murphy IV, F. V., Weixlbaumer, A., Petry, S., Kelley, A. C., Weir, J. R., & Ramakrishnan, V. (2006). Structure of the 70S ribosome complexed with mRNA and tRNA. *Science, 313*, 1935—1941.

Sengupta, J., Nilsson, J., Gursky, R., Spahn, C. M., Nissen, P., & Frank, J. (2004). Identification of the versatile scaffold protein RACK1 on the eukaryotic ribosome by cryo-EM. *Nature Structural & Molecular Biology, 11*, 957—962.

Trakhanov, S. D., Yusupov, M. M., Agalarov, S. C., Garber, M. B., Ryazantsev, S. N., Tischenko, S. V., & Shirokov, V. A. (1987). Crystallization of 70 S ribosomes and 30 S ribosomal subunits from Thermus thermophilus. *FEBS Letters, 220*, 319—322.

Volkmann, N., Hottenträger, S., Hansen, H. A., Zayzsev-Bashan, A., Sharon, R., Berkovitch-Yellin, Z., Yonath, A., & Wittmann, H. G. (1990). Characterization and preliminary crystallographic studies on large ribosomal subunits from Thermus thermophilus. *Journal of Molecular Biology, 216*, 239—241.

von Böhlen, K., Makowski, I., Hansen, H. A., Bartels, H., Berkovitch-Yellin, Z., Zaytzev-Bashan, A., Meyer, S., Paulke, C., Franceschia, F., & Yonath, A. (1991). Characterization and preliminary attempts for derivatization of crystals of large ribosomal subunits from Haloarcula marismortui diffracting to 3 A resolution. *Journal of Molecular Biology, 222*, 11—15.

Wimberly, B. T., Brodersen, D. E., Clemons, W. M., Morgan-Warren, R. J., Carter, A. P., Vonrein, C., Hartsch, T., & Ramakrishnan, V. (2000). Structure of the 30S ribosomal subunit. *Nature, 407*, 327—339.

Yonath, A., Glotz, C., Gewitz, H. S., Bartels, K. S., von Böhlen, K., Makowski, I., & Wittmann, H. G. (1988). Characterization of crystals of small ribosomal subunits. *Journal of Molecular Biology, 203*, 831–834.

Yonath, A., Tesche, B., Lorenz, S., Müssig, J., Erdmann, V. A., & Wittmann, H. G. (1983). Several crystal forms of the Bacillus stearothermophilus 50 S ribosomal particles. *FEBS Letters, 54*, 15–20.

Yusupov, M. M., Yusupova, G. Z., Baucom, A., Lieberman, K., Earnest, T. N., Cate, J. H. D., & Noller, H. F. (2001). Crystal structure of the ribosome at 5.5 Å resolution. *Science, 292*, 883–896.

Further reading

Books

Bloomberg, V. A., Crothers, D. M., & Tinoco, I. (2000). *Nucleic acids: Structures, properties, and functions*. University Science Books.

Liljas, A., & Ehrenberg, M. (2013). In A. Liljas (Ed.). *Structural aspects of protein synthesis* (2nd ed., Vol. 2). World Scientific.

Liljas, A., Liljas, L., Ash, M.-R., Lindblom, G., Nissen, P., & Kjeldgaard, M. (2018). *Textbook of structural biology* (2nd ed.). World Scientific.

Ramakrishnan, V. (2018). *Gene Machine. The race to decipher the secrets of the Ribosome*. Oneworld Publications Ltd.

Smith, A. E. (1976). *Protein biosynthesis (outline studies in biology)*. Chapman and Hall Ltd.

Steitz, T. A., Moore, P. B., & Eatherton, P. (2021). *Structural insights into gene expression and protein synthesis* (series ed., Vol. 12) (Liljas, A World Scientific. A compilation of original papers from the Steitz laboratory, contains many seminal papers on RNA-protein interactions, with reflections from laboratory members).

Zhang, S. (Ed.). (2018). *The excitement of discovery: Selected papers of Alexander rich: A tribute to Alexander rich*. World Scientific.

Index

'Note: Page numbers followed by "f" indicate figures and "t" indicate tables.'

Printed in the United States
by Baker & Taylor Publisher Services